SCHAUM'S OUTLINE OF

THEORY AND PROBLEMS

OF

DIFFERENTIAL EQUATIONS

•

BY

FRANK AYRES, JR., Ph.D.

Professor and Head, Department of Mathematics
Dickinson College

•

SCHAUM'S OUTLINE SERIES
McGRAW-HILL BOOK COMPANY
New York, St. Louis, San Francisco, Toronto, Sydney

ISBN 07-002654-8

8 9 10 11 12 13 14 15 SH SH 7 5 4 3 2 1

Preface

This book is designed primarily to supplement standard texts in elementary differential equations. All types of ordinary and partial differential equations found in current texts, together with the various procedures for solving them, are included. Since the beginning student must be concerned largely with mastering the methods of solving a variety of different type equations, it is felt that there is need for a comprehensive problem book such as this. It should prove also of equal service to practicing engineers and scientists who feel the need for a review of the theory and problem work in this increasingly important field.

Each chapter, except for the third which is entirely expository, begins with a brief statement of definitions, principles, and theorems, followed by a set of solved and supplementary problems. These solved problems have been selected to make a careful study of each as rewarding as possible. Equal attention has been given to the chapters on applications, which include a wide variety of problems from geometry and the physical sciences.

Much more material is presented here than can be taken up in most first courses. This is done not only to meet any choice of topics which the instructor may make, but also to stimulate further interest in the subject and to provide a handy book of reference. However, this book is definitely not a formal textbook and, since there is always a tendency to "get on" with the problems, those being introduced to the subject for the first time are warned against using it as a means of avoiding a thorough study of the regular text.

The author is pleased to acknowledge his indebtedness to Mr. Louis Sandler, associate editor of the publishers, for invaluable suggestions and critical review of the entire manuscript.

FRANK AYRES, JR.

Carlisle, Pa.
September, 1952

Contents

CHAPTER 1

Origin of Differential Equations

A DIFFERENTIAL EQUATION is an equation which involves derivatives. For example,

1) $\dfrac{dy}{dx} = x + 5$

2) $\dfrac{d^2y}{dx^2} + 3\dfrac{dy}{dx} + 2y = 0$

3) $xy' + y = 3$

4) $y''' + 2(y'')^2 + y' = \cos x$

5) $(y'')^2 + (y')^3 + 3y = x^2$

6) $\dfrac{\partial z}{\partial x} = z + x\dfrac{\partial z}{\partial y}$

7) $\dfrac{\partial^2 z}{\partial x^2} + \dfrac{\partial^2 z}{\partial y^2} = x^2 + y.$

If there is a single independent variable, as in 1)-5), the derivatives are ordinary derivatives and the equation is called an *ordinary differential equation*.

If there are two or more independent variables, as in 6)-7), the derivatives are partial derivatives and the equation is called a *partial differential equation*.

The *order* of a differential equation is the order of the highest derivative which occurs. Equations 1), 3), and 6) are of the first order; 2), 5), and 7) are of the second order; and 4) is of the third order.

The *degree* of a differential equation which can be written as a polynomial in the derivatives is the degree of the highest ordered derivative which then occurs. All of the above examples are of the first degree except 5) which is of the second degree.

A discussion of partial differential equations will be given in Chapter 28. For the present, only ordinary differential equations with a single dependent variable will be considered.

ORIGIN OF DIFFERENTIAL EQUATIONS.

a) Geometric Problems. See Problems 1 and 2 below.

b) Physical Problems. See Problems 3 and 4 below.

c) Primitives. A relation between the variables which involves n essential arbitrary constants, as $y = x^4 + Cx$ or $y = Ax^2 + Bx$, is called a *primitive*. The n constants, always indicated by capital letters here, are called *essential* if they cannot be replaced by a smaller number of constants. See Problem 5.

In general, a primitive involving n essential arbitrary constants will give rise to a differential equation, of order n, free of arbitrary constants. This equation is obtained by eliminating the n constants between the $(n+1)$ equations consisting of the primitive and the n equations obtained by differentiating the primitive n times with respect to the independent variable. See Problems 6-14 below.

SOLVED PROBLEMS

1. A curve is defined by the condition that at each of its points (x,y), its slope dy/dx is equal to twice the sum of the coordinates of the point. Express the condition by means of a differential equation.

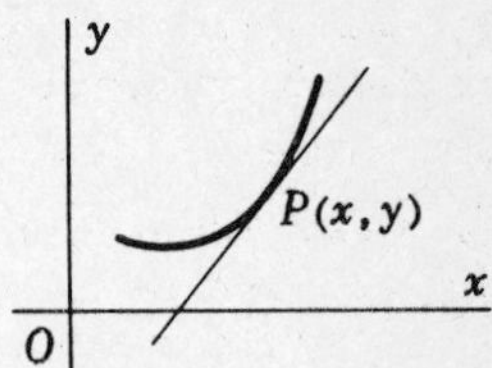

The differential equation representing the condition is $\dfrac{dy}{dx} = 2(x+y)$.

2. A curve is defined by the condition that the sum of the x- and y-intercepts of its tangents is always equal to 2. Express the condition by means of a differential equation.

The equation of the tangent at (x,y) on the curve is $Y - y = \dfrac{dy}{dx}(X - x)$ and the x- and y-intercepts are respectively $X = x - y\dfrac{dx}{dy}$ and $Y = y - x\dfrac{dy}{dx}$. The differential equation representing the condition is $X + Y = x - y\dfrac{dx}{dy} + y - x\dfrac{dy}{dx} = 2$ or $x(\dfrac{dy}{dx})^2 - (x+y-2)\dfrac{dy}{dx} + y = 0$.

3. One hundred grams of cane sugar in water are being converted into dextrose at a rate which is proportional to the amount unconverted. Find the differential equation expressing the rate of conversion after t minutes.

Let q denote the number of grams converted in t minutes. Then $(100-q)$ is the number of grams unconverted and the rate of conversion is given by $\dfrac{dq}{dt} = k(100-q)$, k being the constant of proportionality.

4. A particle of mass m moves along a straight line (the x-axis) while subject to 1) a force proportional to its displacement x from a fixed point O in its path and directed toward O and 2) a resisting force proportional to its velocity. Express the total force as a differential equation.

The first force may be represented by $-k_1x$ and the second by $-k_2\dfrac{dx}{dt}$, where k_1 and k_2 are factors of proportionality.

The total force (mass × acceleration) is given by $m\dfrac{d^2x}{dt^2} = -k_1x - k_2\dfrac{dx}{dt}$.

5. In each of the equations $a)$ $y = x^2 + A + B$, $b)$ $y = Ae^{x+B}$, $c)$ $y = A + \ln Bx$ show that only one of the two arbitrary constants is essential.

$a)$ Since $A+B$ is no more than a single arbitrary constant, only one essential arbitrary constant is involved.

$b)$ $y = Ae^{x+B} = Ae^xe^B$, and Ae^B is no more than a single arbitrary constant.

$c)$ $y = A + \ln Bx = A + \ln B + \ln x$, and $(A + \ln B)$ is no more than a single constant.

6. Obtain the differential equation associated with the primitive $y = Ax^2 + Bx + C$.

Since there are three arbitrary constants, we consider the four equations

$$y = Ax^2 + Bx + C, \qquad \frac{dy}{dx} = 2Ax + B, \qquad \frac{d^2y}{dx^2} = 2A, \qquad \frac{d^3y}{dx^3} = 0.$$

The last of these $\dfrac{d^3y}{dx^3}$, being free of arbitrary constants and of the proper order, is the required equation.

Note that the constants could not have been eliminated between the first three of the above equations. Note also that the primitive can be obtained readily from the differential equation by integration.

7. Obtain the differential equation associated with the primitive $x^2y^3 + x^3y^5 = C$.

Differentiating once with respect to x, we obtain $(2xy^3 + 3x^2y^2\frac{dy}{dx}) + (3x^2y^5 + 5x^3y^4\frac{dy}{dx}) = 0$

or, when $xy \neq 0$, $(2y + 3x\frac{dy}{dx}) + xy^2(3y + 5x\frac{dy}{dx}) = 0$ as the required equation.

When written in differential notation, these equations are

$$1)\quad (2xy^3\,dx + 3x^2y^2\,dy) + (3x^2y^5\,dx + 5x^3y^4\,dy) = 0$$

and

$$2)\quad (2y\,dx + 3x\,dy) + xy^2(3y\,dx + 5x\,dy) = 0.$$

Note that the primitive can be obtained readily from 1) by integration but not so readily from 2). Thus, to obtain the primitive when 2) is given, it is necessary to determine the factor xy^2 which was removed from 1).

8. Obtain the differential equation associated with the primitive $y = A\cos ax + B\sin ax$, A and B being arbitrary constants, and a being a fixed constant.

Here $$\frac{dy}{dx} = -Aa\sin ax + Ba\cos ax$$

and $$\frac{d^2y}{dx^2} = -Aa^2\cos ax - Ba^2\sin ax = -a^2(A\cos ax + B\sin ax) = -a^2y.$$

The required differential equation is $\frac{d^2y}{dx^2} + a^2y = 0$.

9. Obtain the differential equation associated with the primitive $y = Ae^{2x} + Be^x + C$.

Here $\frac{dy}{dx} = 2Ae^{2x} + Be^x$, $\frac{d^2y}{dx^2} = 4Ae^{2x} + Be^x$, $\frac{d^3y}{dx^3} = 8Ae^{2x} + Be^x$.

Then $\frac{d^3y}{dx^3} - \frac{d^2y}{dx^2} = 4Ae^{2x}$, $\frac{d^2y}{dx^2} - \frac{dy}{dx} = 2Ae^{2x}$, and $\frac{d^3y}{dx^3} - \frac{d^2y}{dx^2} = 2(\frac{d^2y}{dx^2} - \frac{dy}{dx})$.

The required equation is $\frac{d^3y}{dx^3} - 3\frac{d^2y}{dx^2} + 2\frac{dy}{dx} = 0$.

10. Obtain the differential equation associated with the primitive $y = C_1e^{3x} + C_2e^{2x} + C_3e^x$.

Here $\frac{dy}{dx} = 3C_1e^{3x} + 2C_2e^{2x} + C_3e^x$, $\frac{d^2y}{dx^2} = 9C_1e^{3x} + 4C_2e^{2x} + C_3e^x$,

and $\frac{d^3y}{dx^3} = 27C_1e^{3x} + 8C_2e^{2x} + C_3e^x$.

The elimination of the constants by elementary methods is somewhat tedious. If three of the equations are solved for C_1, C_2, C_3, using determinants, and these substituted in the fourth equation, the result may be put in the form (called the eliminant):

$$\begin{vmatrix} e^{3x} & e^{2x} & e^{x} & y \\ 3e^{3x} & 2e^{2x} & e^{x} & y' \\ 9e^{3x} & 4e^{2x} & e^{x} & y'' \\ 27e^{3x} & 8e^{2x} & e^{x} & y''' \end{vmatrix} = e^{6x}\begin{vmatrix} 1 & 1 & 1 & y \\ 3 & 2 & 1 & y' \\ 9 & 4 & 1 & y'' \\ 27 & 8 & 1 & y''' \end{vmatrix} = e^{6x}(-2y''' + 12y'' - 22y' + 12y) = 0.$$

The required differential equation is $\dfrac{d^3y}{dx^3} - 6\dfrac{d^2y}{dx^2} + 11\dfrac{dy}{dx} - 6y = 0.$

11. Obtain the differential equation associated with the primitive $y = Cx^2 + C^2$.

Since $\dfrac{dy}{dx} = 2Cx$, $C = \dfrac{1}{2x}\dfrac{dy}{dx}$ and $y = Cx^2 + C^2 = \dfrac{1}{2x}\dfrac{dy}{dx}x^2 + \dfrac{1}{4x^2}(\dfrac{dy}{dx})^2$.

The required differential equation is $(\dfrac{dy}{dx})^2 + 2x^3\dfrac{dy}{dx} - 4x^2y = 0.$

Note. The primitive involves one arbitrary constant of degree two and the resulting differential equation is of order one and degree two.

12. Find the differential equation of the family of circles of fixed radius r with centers on the x-axis.

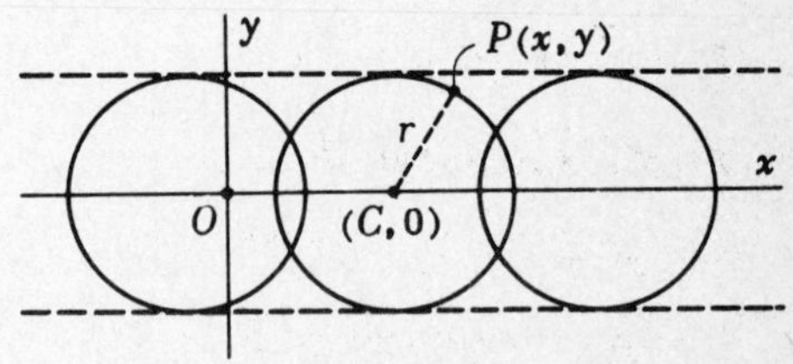

The equation of the family is $(x-C)^2 + y^2 = r^2$, C being an arbitrary constant.

Then $(x-C) + y\dfrac{dy}{dx} = 0$, $x - C = -y\dfrac{dy}{dx}$, and the differential equation is $y^2(\dfrac{dy}{dx})^2 + y^2 = r^2$.

13. Find the differential equation of the family of parabolas with foci at the origin and axes along the x-axis.

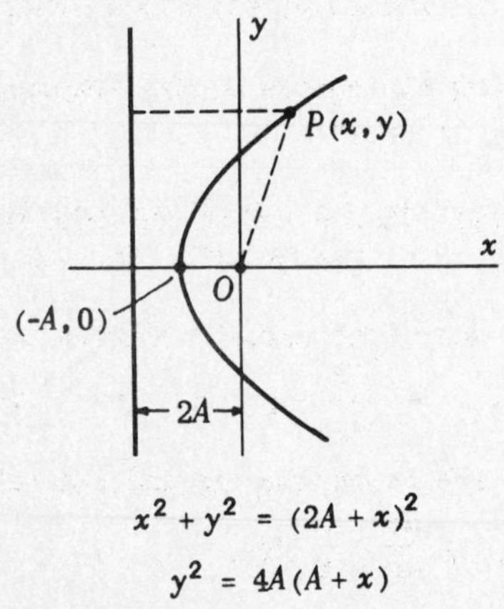

$$x^2 + y^2 = (2A+x)^2$$
$$y^2 = 4A(A+x)$$

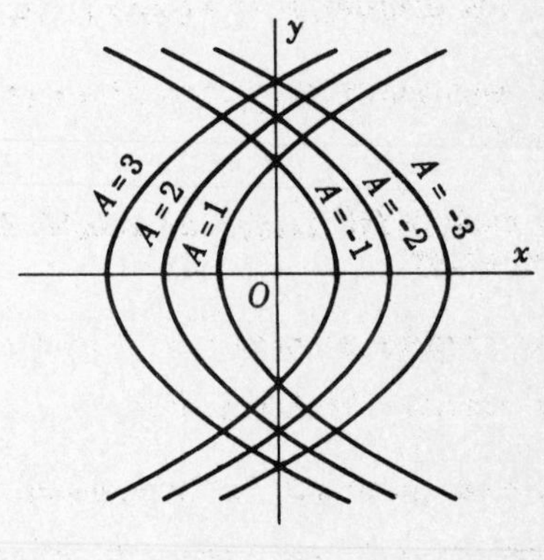

$$y^2 = 4A(A+x)$$

The equation of the family of parabolas is $y^2 = 4A(A+x)$.
Then $yy' = 2A$, $A = \frac{1}{2}yy'$, and $y^2 = 2yy'(\frac{1}{2}yy' + x)$.

The required equation is $y(\dfrac{dy}{dx})^2 + 2x\dfrac{dy}{dx} - y = 0.$

14. Form the differential equation representing all tangents to the parabola $y^2 = 2x$.

At any point (A,B) on the parabola, the equation of the tangent is $y-B = (x-A)/B$ or, since $A = \frac{1}{2}B^2$, $By = x + \frac{1}{2}B^2$. Eliminating B between this and $By' = 1$, obtained by differentiation with respect to x, we have as the required differential equation $2x(y')^2 - 2yy' + 1 = 0$.

SUPPLEMENTARY PROBLEMS

15. Classify each of the following equations as to order and degree.

a) $dy + (xy - \cos x)dx = 0$ *Ans.* Order one; degree one

b) $L\dfrac{d^2Q}{dt^2} + R\dfrac{dQ}{dt} + \dfrac{Q}{C} = 0$ *Ans.* Order two; degree one

c) $y''' + xy'' + 2y(y')^2 + xy = 0$ *Ans.* Order three; degree one

d) $\dfrac{d^2v}{dx^2}\dfrac{dv}{dx} + x\left(\dfrac{dv}{dx}\right)^2 + v = 0$ *Ans.* Order two; degree one

e) $\left(\dfrac{d^3w}{dv^3}\right)^2 - \left(\dfrac{d^2w}{dv^2}\right)^4 + vw = 0$ *Ans.* Order three; degree two

f) $e^{y'''} - xy'' + y = 0$ *Ans.* Order three; degree does not apply

g) $\sqrt{\rho' + \rho} = \sin\theta$ *Ans.* Order one; degree one

h) $y' + x = (y - xy')^{-3}$ *Ans.* Order one; degree four

i) $\dfrac{d^2\rho}{d\theta^2} = \sqrt[4]{\rho + \left(\dfrac{d\rho}{d\theta}\right)^2}$ *Ans.* Order two; degree four

16. Write the differential equation for each of the curves determined by the given conditions.

a) At each point (x,y) the slope of the tangent is equal to the square of the abscissa of the point. *Ans.* $y' = x^2$

b) At each point (x,y) the length of the subtangent is equal to the sum of the coordinates of the point. *Ans.* $y/y' = x+y$ or $(x+y)y' = y$

c) The segment joining $P(x,y)$ and the point of intersection of the normal at P with the x-axis is bisected by the y-axis. *Ans.* $y + x\dfrac{dx}{dy} = \frac{1}{2}y$ or $yy' + 2x = 0$

d) At each point (ρ,θ) the tangent of the angle between the radius vector and the tangent is equal to 1/3 the tangent of the vectorial angle. *Ans.* $\rho\dfrac{d\theta}{d\rho} = \dfrac{1}{3}\tan\theta$

e) The area bounded by the arc of a curve, the x-axis, and two ordinates, one fixed and one variable, is equal to twice the length of the arc between the ordinates.

Hint: $\displaystyle\int_a^x y\,dx = 2\int_a^x \sqrt{1+(y')^2}\,dx$. *Ans.* $y = 2\sqrt{1+(y')^2}$

17. Express each of the following physical statements in differential equation form.

a) Radium decomposes at a rate proportional to the amount Q present. *Ans.* $dQ/dt = -kQ$

b) The population P of a city increases at a rate proportional to the population and to the difference between 200,000 and the population. *Ans.* $dP/dt = kP(200{,}000 - P)$

c) For a certain substance the rate of change of vapor pressure (P) with respect to temperature (T) is proportional to the vapor pressure and inversely proportional to the square of the temperature. *Ans.* $dP/dT = kP/T^2$

d) The potential difference E across an element of inductance L is equal to the product of L and the time rate of change of the current i in the inductance. *Ans.* $E = L\dfrac{di}{dt}$

e) Mass × acceleration = net force. *Ans.* $m\dfrac{dv}{dt} = F$ or $m\dfrac{d^2s}{dt^2} = F$

18. Obtain the differential equation associated with the given primitive, A and B being arbitrary constants.

a) $y = Ax$ *Ans.* $y' = y/x$

b) $y = Ax + B$ *Ans.* $y'' = 0$

c) $y = e^{x+A} = Be^x$ *Ans.* $y' = y$

d) $y = A \sin x$ *Ans.* $y' = y \cot x$

e) $y = \sin(x + A)$ *Ans.* $(y')^2 = 1 - y^2$

f) $y = Ae^x + B$ *Ans.* $y'' = y'$

g) $x = A \sin(y + B)$ *Ans.* $y'' = x(y')^3$

h) $\ln y = Ax^2 + B$ *Ans.* $xyy'' - yy' - x(y')^2 = 0$

19. Find the differential equation of the family of circles of variable radii r with centers on the x-axis. (Compare with Problem 12.)
Hint: $(x - A)^2 + y^2 = r^2$, A and r being arbitrary constants. *Ans.* $yy'' + (y')^2 + 1 = 0$

20. Find the differential equation of the family of cardioids $\rho = a(1 - \cos\theta)$.
Ans. $(1 - \cos\theta)d\rho = \rho \sin\theta\, d\theta$

21. Find the differential equation of all straight lines at a unit distance from the origin.
Ans. $(xy' - y)^2 = 1 + (y')^2$

22. Find the differential equation of all circles in the plane.
Hint: Use $x^2 + y^2 - 2Ax - 2By + C = 0$. *Ans.* $[1 + (y')^2]y''' - 3y'(y'')^2 = 0$

CHAPTER 2

Solutions of Differential Equations

THE PROBLEM in elementary differential equations is essentially that of recovering the primitive which gave rise to the equation. In other words, the problem of solving a differential equation of order n is essentially that of finding a relation between the variables involving n independent arbitrary constants which together with the derivatives obtained from it satisfy the differential equation. For example:

	Differential Equation	Primitive	
1)	$\frac{d^3y}{dx^3} = 0$	$y = Ax^2 + Bx + C$	(Prob. 6, Chap. 1)
2)	$\frac{d^3y}{dx^3} - 6\frac{d^2y}{dx^2} + 11\frac{dy}{dx} - 6y = 0$	$y = C_1e^{3x} + C_2e^{2x} + C_3e^x$	(Prob. 10, Chap. 1)
3)	$y^2(\frac{dy}{dx})^2 + y^2 = r^2$	$(x-C)^2 + y^2 = r^2$	(Prob. 12, Chap. 1)

THE CONDITIONS under which we can be assured that a differential equation is solvable are given by *Existence Theorems*.

For example, a differential equation of the form $y' = g(x,y)$ for which

a) $g(x,y)$ is continuous and single valued over a region R of points (x,y),

b) $\frac{\partial g}{\partial y}$ exists and is continuous at all points in R,

admits an infinity of solutions $f(x,y,C) = 0$ (C, an arbitrary constant) such that through each point of R there passes one and only one curve of the family $f(x,y,C) = 0$. See Problem 5.

A PARTICULAR SOLUTION of a differential equation is one obtained from the primitive by assigning definite values to the arbitrary constants. For example, in 1) above $y = 0$ ($A = B = C = 0$), $y = 2x + 5$ ($A = 0$, $B = 2$, $C = 5$), and $y = x^2 + 2x + 3$ ($A = 1$, $B = 2$, $C = 3$) are particular solutions.

Geometrically, the primitive is the equation of a family of curves and a particular solution is the equation of some one of the curves. These curves are called *integral curves* of the differential equation.

As will be seen from Problem 6, a given form of the primitive may not include all of the particular solutions. Moreover, as will be seen from Problem 7, a differential equation may have solutions which cannot be obtained from the primitive by any manipulation of the arbitrary constant as in Problem 6. Such solutions, called *singular solutions*, will be considered in Chapter 10.

The primitive of a differential equation is usually called *the general solution* of the equation. Certain authors, because of the remarks in the paragraph above, call it *a general solution* of the equation.

A DIFFERENTIAL EQUATION $\frac{dy}{dx} = g(x,y)$ associates with each point (x_0, y_0) in the region R of the above existence theorem a direction $m = \left.\frac{dy}{dx}\right|_{(x_0,y_0)} = g(x_0, y_0)$. The direction at each such point is that of the tangent to the curve of the family $f(x,y,C) = 0$, that is, the primitive, passing through the point.

The region R with the direction at each of its points indicated is called a *direction field*. In the adjoining figure, a number of points with the direction at each is shown for the equation $dy/dx = 2x$. The integral curves of the differential equation are those curves having at each of their points the direction given by the equation. In this example, the integral curves are parabolas.

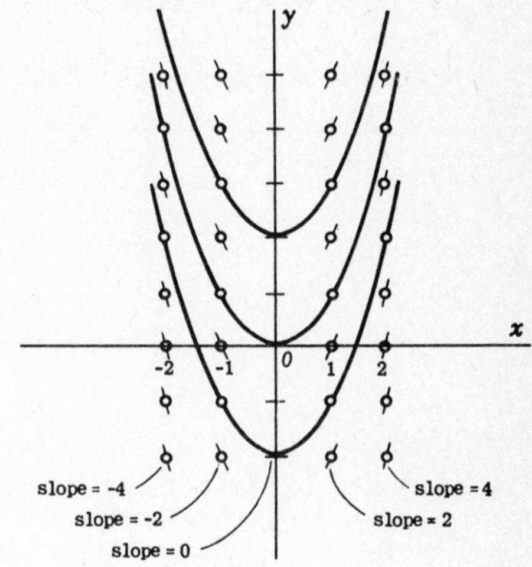

Such diagrams are helpful in that they aid in clarifying the relation between a differential equation and its primitive, but since the integral curves are generally quite complex, such a diagram does not aid materially in obtaining their equations.

SOLVED PROBLEMS

1. Show by direct substitution in the differential equation and a check of the arbitrary constants that each primitive gives rise to the corresponding differential equation.

a) $y = C_1 \sin x + C_2 x$ $\qquad (1 - x\cot x)\frac{d^2y}{dx^2} - x\frac{dy}{dx} + y = 0$

b) $y = C_1 e^x + C_2 xe^x + C_3 e^{-x} + 2x^2 e^x$ $\qquad \frac{d^3y}{dx^3} - \frac{d^2y}{dx^2} - \frac{dy}{dx} + y = 8e^x$

a) Substitute $y = C_1 \sin x + C_2 x$, $\frac{dy}{dx} = C_1 \cos x + C_2$, $\frac{d^2y}{dx^2} = -C_1 \sin x$ in the differential equation to obtain

$$(1 - x\cot x)(-C_1 \sin x) - x(C_1 \cos x + C_2) + (C_1 \sin x + C_2 x) =$$
$$-C_1 \sin x + C_1 x\cos x - C_1 x \cos x - C_2 x + C_1 \sin x + C_2 x = 0.$$

The order of the differential equation (2) and the number of arbitrary constants (2) agree.

b)
$$y = C_1 e^x + C_2 xe^x + C_3 e^{-x} + 2x^2 e^x,$$
$$y' = (C_1 + C_2)e^x + C_2 xe^x - C_3 e^{-x} + 2x^2 e^x + 4xe^x,$$
$$y'' = (C_1 + 2C_2)e^x + C_2 xe^x + C_3 e^{-x} + 2x^2 e^x + 8xe^x + 4e^x,$$
$$y''' = (C_1 + 3C_2)e^x + C_2 xe^x - C_3 e^{-x} + 2x^2 e^x + 12xe^x + 12e^x,$$

and $y''' - y'' - y' + y = 8e^x$. The order of the differential equation and the number of arbitrary constants agree.

2. Show that $y = 2x + Ce^x$ is the primitive of the differential equation $\frac{dy}{dx} - y = 2(1-x)$ and find the particular solution satisfied by $x = 0$, $y = 3$ (*i.e.*, the equation of the integral curve through $(0,3)$).

Substitute $y = 2x + Ce^x$ and $\frac{dy}{dx} = 2 + Ce^x$ in the differential equation to obtain $2 + Ce^x - (2x + Ce^x) = 2 - 2x$. When $x = 0$, $y = 3$, $3 = 2\cdot 0 + Ce^0$ and $C = 3$. The particular solution is $y = 2x + 3e^x$.

3. Show that $y = C_1 e^x + C_2 e^{2x} + x$ is the primitive of the differential equation $\frac{d^2y}{dx^2} - 3\frac{dy}{dx} + 2y = 2x - 3$ and find the equation of the integral curve through the points (0,0) and (1,0).

Substitute $y = C_1 e^x + C_2 e^{2x} + x$, $\frac{dy}{dx} = C_1 e^x + 2C_2 e^{2x} + 1$, $\frac{d^2y}{dx^2} = C_1 e^x + 4C_2 e^{2x}$ in the differential equation to obtain $C_1 e^x + 4C_2 e^{2x} - 3(C_1 e^x + 2C_2 e^{2x} + 1) + 2(C_1 e^x + C_2 e^{2x} + x) = 2x - 3$.

When $x = 0$, $y = 0$: $C_1 + C_2 = 0$. When $x = 1$, $y = 0$: $C_1 e + C_2 e^2 = -1$.

Then $C_1 = -C_2 = \frac{1}{e^2 - e}$ and the required equation is $y = x + \frac{e^x - e^{2x}}{e^2 - e}$.

4. Show that $(y - C)^2 = Cx$ is the primitive of the differential equation $4x(\frac{dy}{dx})^2 + 2x\frac{dy}{dx} - y = 0$ and find the equations of the integral curves through the point (1,2).

Here $2(y - C)\frac{dy}{dx} = C$ and $\frac{dy}{dx} = \frac{C}{2(y - C)}$.

Then $$4x\frac{C^2}{4(y-C)^2} + 2x\frac{C}{2(y-C)} - y = \frac{C^2x + Cx(y-C) - y(y-C)^2}{(y-C)^2} = \frac{y[Cx - (y-C)^2]}{(y-C)^2} = 0.$$

When $x = 1$, $y = 2$: $(2 - C)^2 = C$ and $C = 1, 4$.

The equations of the integral curves through (1,2) are $(y - 1)^2 = x$ and $(y - 4)^2 = 4x$.

5. The primitive of the differential equation $\frac{dy}{dx} = \frac{y}{x}$ is $y = Cx$. Find the equation of the integral curve through *a*) (1,2) and *b*) (0,0).

a) When $x = 1$, $y = 2$: $C = 2$ and the required equation is $y = 2x$.

b) When $x = 0$, $y = 0$: C is not determined, that is, all of the integral curves pass through the origin. Note that $g(x,y) = y/x$ is not continuous at the origin and hence the existence theorem assures one and only one curve of the family $y = Cx$ through each point of the plane except the origin.

6. Differentiating $xy = C(x - 1)(y - 1)$ and substituting for C, we obtain the differential equation

$$x\frac{dy}{dx} + y = C\{(x-1)\frac{dy}{dx} + y - 1\} = \frac{xy}{(x-1)(y-1)}\{(x-1)\frac{dy}{dx} + y - 1\}$$

or 1) $$x(x-1)\frac{dy}{dx} + y(y-1) = 0.$$

Now both $y = 0$ and $y = 1$ are solutions of 1), since, for each, $dy/dx = 0$ and 1) is satisfied. The first is obtained from the primitive by setting $C = 0$, but the second $y = 1$ cannot be obtained by assigning a finite value to C. Similarly, 1) may be obtained from the primitive in the form $Bxy = (x - 1)(y - 1)$. Now the solution $y = 1$ is obtained by setting $B = 0$ while the solution $y = 0$ cannot be obtained by assigning a finite value to B. Thus, the given form of a primitive may not include all of the particular solutions of the differential equation. (Note that $x = 1$ is also a particular solution.)

7. Differentiating $y = Cx + 2C^2$, solving for $C = \frac{dy}{dx}$, and substituting in the primitive yields the differential equation

$$1) \qquad 2(\frac{dy}{dx})^2 + x(\frac{dy}{dx}) - y = 0.$$

Since $y = -\frac{1}{8}x^2$, $\frac{dy}{dx} = -\frac{1}{4}x$ satisfies 1), $x^2 + 8y = 0$ is a solution of 1).

Now the primitive is represented by a family of straight lines and it is clear that the equation of a parabola cannot be obtained by manipulating the arbitrary constant. Such a solution is called a singular solution of the differential equation.

8. Verify and reconcile the fact that $y = C_1 \cos x + C_2 \sin x$ and $y = A\cos(x+B)$ are primitives of $\frac{d^2y}{dx^2} + y = 0$.

From $y = C_1\cos x + C_2 \sin x$, $y' = -C_1 \sin x + C_2 \cos x$ and

$$y'' = -C_1 \cos x - C_2 \sin x = -y \quad \text{or} \quad \frac{d^2y}{dx^2} + y = 0.$$

From $y = A\cos(x+B)$, $y' = -A\sin(x+B)$ and $y'' = -A\cos(x+B) = -y$.

Now $y = A\cos(x+B) = A(\cos x \cos B - \sin x \sin B)$

$$= (A\cos B)\cos x + (-A\sin B)\sin x = C_1\cos x + C_2 \sin x.$$

9. Show that $\ln x^2 + \ln\frac{y^2}{x^2} = A + x$ may be written as $y^2 = Be^x$.

$$\ln x^2 + \ln\frac{y^2}{x^2} = \ln(x^2\frac{y^2}{x^2}) = \ln y^2 = A + x. \quad \text{Then} \quad y^2 = e^{A+x} = e^A\cdot e^x = Be^x.$$

10. Show that $\text{Arc}\sin x - \text{Arc}\sin y = A$ may be written as $x\sqrt{1-y^2} - y\sqrt{1-x^2} = B$.

$\sin(\text{Arc}\sin x - \text{Arc}\sin y) = \sin A = B$.

Then $\sin(\text{Arc}\sin x)\cos(\text{Arc}\sin y) - \cos(\text{Arc}\sin x)\sin(\text{Arc}\sin y) = x\sqrt{1-y^2} - y\sqrt{1-x^2} = B$.

11. Show that $\ln(1+y) + \ln(1+x) = A$ may be written as $xy + x + y = C$.

$\ln(1+y) + \ln(1+x) = \ln(1+y)(1+x) = A$.

Then $(1+y)(1+x) = xy + x + y + 1 = e^A = B$ and $xy + x + y = B - 1 = C$.

12. Show that $\sinh y + \cosh y = Cx$ may be written as $y = \ln x + A$.

Here $\sinh y + \cosh y = \frac{1}{2}(e^y - e^{-y}) + \frac{1}{2}(e^y + e^{-y}) = e^y = Cx$.

Then $y = \ln C + \ln x = A + \ln x$.

SUPPLEMENTARY PROBLEMS

Show that each of the following expressions is a solution of the corresponding differential equation. Classify each as a particular solution or general solution (primitive).

13. $y = 2x^2$,	$xy' = 2y$.	Particular solution
14. $x^2 + y^2 = C$,	$yy' + x = 0$.	Primitive
15. $y = Cx + C^4$,	$y = xy' + (y')^4$.	Primitive
16. $(1-x)y^2 = x^3$,	$2x^3y' = y(y^2 + 3x^2)$.	Particular solution
17. $y = e^x(1+x)$,	$y'' - 2y' + y = 0$.	Particular solution
18. $y = C_1x + C_2e^x$,	$(x-1)y'' - xy' + y = 0$.	General solution
19. $y = C_1e^x + C_2e^{-x}$,	$y'' - y = 0$.	General solution
20. $y = C_1e^x + C_2e^{-x} + x - 4$,	$y'' - y = 4 - x$.	General solution
21. $y = C_1e^x + C_2e^{2x}$,	$y'' - 3y' + 2y = 0$.	General solution
22. $y = C_1e^x + C_2e^{2x} + x^2e^x$,	$y'' - 3y' + 2y = 2e^x(1-x)$.	General solution

CHAPTER 3

Equations of First Order and First Degree

A DIFFERENTIAL EQUATION of the first order and first degree may be written in the form

1) $$M(x,y)\,dx + N(x,y)\,dy = 0.$$

EXAMPLE 1. *a*) $\dfrac{dy}{dx} + \dfrac{y+x}{y-x} = 0$ may be written as $(y+x)\,dx + (y-x)\,dy = 0$ in which $M(x,y) = y+x$ and $N(x,y) = y-x$.

b) $\dfrac{dy}{dx} = 1 + x^2 y$ may be written as $(1+x^2 y)\,dx - dy = 0$ in which $M(x,y) = 1 + x^2 y$ and $N(x,y) = -1$.

If $M(x,y)\,dx + N(x,y)\,dy$ is the complete differential of a function $\mu(x,y)$, that is, if

$$M(x,y)\,dx + N(x,y)\,dy = d\mu(x,y).$$

1) is called exact and $\mu(x,y) = C$ is its primitive or general solution.

EXAMPLE 2. $3x^2y^2\,dx + 2x^3y\,dy = 0$ is an exact differential equation since $3x^2y^2\,dx + 2x^3y\,dy = d(x^3y^2)$. Its primitive is $x^3y^2 = C$.

If 1) is not exact but

$$\xi(x,y)\{M(x,y)\,dx + N(x,y)\,dy\} = d\mu(x,y),$$

$\xi(x,y)$ is called an integrating factor of 1) and $\mu(x,y) = C$ is its primitive.

EXAMPLE 3. $3y\,dx + 2x\,dy = 0$ is not an exact differential equation but when multiplied by $\xi(x,y) = x^2 y$, we have $3x^2y^2\,dx + 2x^3y\,dy = 0$ which is exact. Hence, the primitive of $3y\,dx + 2x\,dy = 0$ is $x^3y^2 = C$. See Example 2.

If 1) is not exact and no integrating factor can be found readily, it may be possible by a change of one or both of the variables to obtain an equation for which an integrating factor can be found.

EXAMPLE 4. The transformation $x = t - y$, $dx = dt - dy$, (i.e., $x + y = t$), reduces the equation $(x+y+1)\,dx + (2x+2y+3)\,dy = 0$

to $(t+1)(dt - dy) + (2t+3)\,dy = 0$

or $(t+1)\,dt + (t+2)\,dy = 0.$

By means of the integrating factor $\dfrac{1}{t+2}$ the equation takes the form

$$dy + \frac{t+1}{t+2}dt = dy + dt - \frac{1}{t+2}dt = 0.$$

Then $y + t - \ln(t+2) = C$

and, since $t = x+y$, $2y + x - \ln(x+y+2) = C$.

Note. The transformation $x+y+1 = t$ or $2x+2y+3 = 2s$ is also suggested by the form of the equation.

A DIFFERENTIAL EQUATION for which an integrating factor is found readily has the form

2) $$f_1(x)\cdot g_2(y)\,dx + f_2(x)\cdot g_1(y)\,dy = 0.$$

By means of the integrating factor $\dfrac{1}{f_2(x)\cdot g_2(y)}$, 2) is reduced to

2') $$\frac{f_1(x)}{f_2(x)}\,dx + \frac{g_1(y)}{g_2(y)}\,dy = 0$$

whose primitive is

$$\int\frac{f_1(x)}{f_2(x)}\,dx + \int\frac{g_1(y)}{g_2(y)}\,dy = C.$$

Equation 2) is typed as *Variables Separable* and in 2') the variables are separated.

EXAMPLE 5. When the differential equation

$$(3x^2y - xy)\,dx + (2x^3y^2 + x^3y^4)\,dy = 0$$

is put in the form $\quad y(3x^2 - x)\,dx + x^3(2y^2 + y^4)\,dy = 0$

it is seen to be of the type Variables Separable. The integrating factor $\dfrac{1}{yx^3}$ reduces it to $(\dfrac{3}{x} - \dfrac{1}{x^2})\,dx + (2y + y^3)\,dy = 0$ in which the variables are separated. Integrating, we obtain the primitive

$$3\ln x + \frac{1}{x} + y^2 + \frac{1}{4}y^4 = C.$$

IF EQUATION 1) admits a solution $f(x,y,C) = 0$, where C is an arbitrary constant, there exist infinitely many integrating factors $\xi(x,y)$ such that

$$\xi(x,y)\{M(x,y)\,dx + N(x,y)\,dy\} = 0$$

is exact. Also, there exist transformations of the variables which carry 1) into the type Variables Separable. However, no general rule can be stated here for finding either an integrating factor or a transformation. Thus we are limited to solving certain types of differential equations of the first order and first degree, i.e., those for which rules may be laid down for determining either an integrating factor or an effective transformation.

Equations of the type Variables Separable, together with equations which can be reduced to this type by a transformation of the variables are considered in Chapter 4.

Exact differential equations and other types reducible to exact equations by means of integrating factors are treated in Chapter 5.

The linear equation of order one

3) $$\frac{dy}{dx} + P(x)\cdot y = Q(x)$$

and equations reducible to the form 3) by means of transformations are considered in Chapter 6.

These groupings are a matter of convenience. A given equation may fall into more than one group.

EXAMPLE 6. The equation $x\,dy - y\,dx = 0$ may be placed in any one of the groups since

a) by means of the integrating factor $1/xy$ the variables are separated; thus, $dy/y - dx/x = 0$ so that $\ln y - \ln x = \ln C$ or $y/x = C$.

b) by means of the integrating factor $1/x^2$ or $1/y^2$ the equation is made exact; thus, $\dfrac{x\,dy - y\,dx}{x^2} = 0$ and $\dfrac{y}{x} = C$ or $\dfrac{x\,dy - y\,dx}{y^2} = 0$ and $-\dfrac{x}{y} = C_1$, $\dfrac{y}{x} = -\dfrac{1}{C_1} = C$.

c) when written as $\dfrac{dy}{dx} - \dfrac{1}{x}y = 0$, it is a linear equation of order one.

Attention has been called to the fact that the form of the primitive is not unique. Thus, the primitive in Example 6 might be given as

a) $\ln y - \ln x = \ln C$, *b*) $y/x = C$, *c*) $y = Cx$, *d*) $x/y = K$, etc.

It is usual to accept any one of these forms with the understanding, already noted, that thereby certain particular solutions may be lost. There is an additional difficulty!

EXAMPLE 7. It is clear that $y = 0$ is a particular solution of $dy/dx = y$ or $dy - y\,dx = 0$. When $y \neq 0$, we may write $dy/y - dx = 0$ and obtain $\ln y - x = \ln C$ with $C \neq 0$; in turn, this may be written as $y = Ce^x$, $C \neq 0$. Thus, to include all solutions, we should write $y = 0$; $y = Ce^x$, $C \neq 0$. But note that $y = Ce^x$, free of the restrictions imposed on y and C, includes *all* solutions.

This situation will arise repeatedly as we proceed but, as is customary, we shall refrain from pointing out the restrictions; that is, we shall write the primitive as $y = Ce^x$, with C completely arbitrary. In defense, we offer the following observation. Let us multiply the given equation by e^{-x} to obtain $e^{-x}dy - ye^{-x}dx = 0$ from which, by integration, we get $e^{-x}y = C$ or $y = Ce^x$. In this procedure, it has not been necessary to impose any restriction on y or C.

CHAPTER 4

Equations of First Order and First Degree

VARIABLES SEPARABLE AND REDUCTION TO VARIABLES SEPARABLE

VARIABLES SEPARABLE. The variables of the equation $M(x,y)\,dx + N(x,y)\,dy = 0$ are separable if the equation can be written in the form

1) $$f_1(x)\cdot g_2(y)\,dx + f_2(x)\cdot g_1(y)\,dy = 0.$$

The integrating factor $\dfrac{1}{f_2(x)\cdot g_2(y)}$, found by inspection, reduces 1) to the form

$$\frac{f_1(x)}{f_2(x)}\,dx + \frac{g_1(y)}{g_2(y)}\,dy = 0$$

from which the primitive may be obtained by integration.

For example, $(x-1)^2 y\,dx + x^2(y+1)\,dy = 0$ is of the form 1). The integrating factor $\dfrac{1}{x^2 y}$ reduces the equation to $\dfrac{(x-1)^2}{x^2}\,dx + \dfrac{(y+1)}{y}\,dy = 0$ in which the variables are separated. See Problems 1-5.

HOMOGENEOUS EQUATIONS. A function $f(x,y)$ is called homogeneous of degree n if

$$f(\lambda x, \lambda y) = \lambda^n f(x,y).$$

For example:

a) $f(x,y) = x^4 - x^3 y$ is homogeneous of degree 4 since
$$f(\lambda x,\lambda y) = (\lambda x)^4 - (\lambda x)^3(\lambda y) = \lambda^4(x^4 - x^3 y) = \lambda^4 f(x,y).$$

b) $f(x,y) = e^{y/x} + \tan\frac{y}{x}$ is homogeneous of degree 0 since
$$f(\lambda x,\lambda y) = e^{\lambda y/\lambda x} + \tan\frac{\lambda y}{\lambda x} = e^{y/x} + \tan\frac{y}{x} = \lambda^0 f(x,y).$$

c) $f(x,y) = x^2 + \sin x \cos y$ is not homogeneous since
$$f(\lambda x,\lambda y) = \lambda^2 x^2 + \sin(\lambda x)\cos(\lambda y) \neq \lambda^n f(x,y).$$

The differential equation $M(x,y)\,dx + N(x,y)\,dy = 0$ is called homogeneous if $M(x,y)$ and $N(x,y)$ are homogeneous and of the same degree. For example, $x\ln\frac{y}{x}\,dx + \frac{y^2}{x}\arcsin\frac{y}{x}\,dy = 0$ is homogeneous of degree 1, but neither $(x^2+y^2)\,dx - (xy^2 - y^3)\,dy = 0$ nor $(x+y^2)\,dx + (x-y)\,dy = 0$ is a homogeneous equation.

The transformation
$$y = vx, \qquad dy = v\,dx + x\,dv$$
will reduce any homogeneous equation to the form
$$P(x,v)\,dx + Q(x,v)\,dv = 0$$
in which the variables are separable. After integrating, v is replaced by y/x to recover the original variables. See Problems 6-11.

EQUATIONS IN WHICH $M(x,y)$ AND $N(x,y)$ ARE LINEAR BUT NOT HOMOGENEOUS.

a) The equation $(a_1x + b_1y + c_1)\,dx + (a_2x + b_2y + c_2)\,dy = 0$, $(a_1b_2 - a_2b_1 = 0)$, is reduced by the transformation
$$a_1x + b_1y = t, \qquad dy = \frac{dt - a_1\,dx}{b_1}$$
to the form
$$P(x,t)\,dx + Q(x,t)\,dt = 0$$
in which the variables are separable. See Problem 12.

b) The equation $(a_1x + b_1y + c_1)\,dx + (a_2x + b_2y + c_2)\,dy = 0$, $(a_1b_2 - a_2b_1 \neq 0)$, is reduced to the homogeneous form
$$(a_1x' + b_1y')\,dx' + (a_2x' + b_2y')\,dy' = 0$$
by the transformation
$$x = x' + h, \qquad y = y' + k$$
in which $x = h$, $y = k$ are the solutions of the equations
$$a_1x + b_1y + c_1 = 0 \quad \text{and} \quad a_2x + b_2y + c_2 = 0.$$
See Problems 13-14.

EQUATIONS OF THE FORM $y\cdot f(xy)\,dx + x\cdot g(xy)\,dy = 0$. The transformation
$$xy = z, \qquad y = \frac{z}{x}, \qquad dy = \frac{x\,dz - z\,dx}{x^2}$$
reduces an equation of this form to the form
$$P(x,z)\,dx + Q(x,z)\,dz = 0$$
in which the variables are separable. See Problems 15-17.

OTHER SUBSTITUTIONS. Equations, not of the types discussed above, may be reduced to a form in which the variables are separable by means of a properly chosen transformation. No general rule of procedure can be given; in each case the form of the equation suggests the transformation. See Problems 18-22.

SOLVED PROBLEMS

VARIABLES SEPARABLE.

1. Solve $x^3\,dx + (y+1)^2\,dy = 0$.

The variables are separated. Hence, integrating term by term,
$$\frac{x^4}{4} + \frac{(y+1)^3}{3} = C_1 \quad \text{or} \quad 3x^4 + 4(y+1)^3 = C.$$

2. Solve $x^2(y+1)dx + y^2(x-1)dy = 0$.

The integrating factor $\dfrac{1}{(y+1)(x-1)}$ reduces the equation to $\dfrac{x^2}{x-1}dx + \dfrac{y^2}{y+1}dy = 0$.

Then, integrating $(x+1+\dfrac{1}{x-1})dx + (y-1+\dfrac{1}{y+1})dy = 0$,

$$\tfrac{1}{2}x^2 + x + \ln(x-1) + \tfrac{1}{2}y^2 - y + \ln(y+1) = C_2,$$

$$x^2 + y^2 + 2x - 2y + 2\ln(x-1)(y+1) = C_1,$$

and $$(x+1)^2 + (y-1)^2 + 2\ln(x-1)(y+1) = C.$$

3. Solve $4x\,dy - y\,dx = x^2dy$ or $y\,dx + (x^2-4x)dy = 0$.

The integrating factor $\dfrac{1}{y(x^2-4x)}$ reduces the equation to $\dfrac{dx}{x(x-4)} + \dfrac{dy}{y} = 0$ in which the variables are separated.

The latter equation may be written as $\dfrac{\frac{1}{4}dx}{x-4} - \dfrac{\frac{1}{4}dx}{x} + \dfrac{dy}{y} = 0$ or $\dfrac{dx}{x-4} - \dfrac{dx}{x} + 4\dfrac{dy}{y} = 0$.

Integrating, $\ln(x-4) - \ln x + 4\ln y = \ln C$ or $(x-4)y^4 = Cx$.

4. Solve $\dfrac{dy}{dx} = \dfrac{4y}{x(y-3)}$ or $x(y-3)dy = 4y\,dx$.

The integrating factor $\dfrac{1}{xy}$ reduces the equation to $\dfrac{y-3}{y}dy = \dfrac{4}{x}dx$.

Integrating, $y - 3\ln y = 4\ln x + \ln C_1$ or $y = \ln(C_1x^4y^3)$.

This may be written as $C_1x^4y^3 = e^y$ or $x^4y^3 = Ce^y$.

5. Find the particular solution of $(1+x^3)dy - x^2y\,dx = 0$ satisfying the initial conditions $x = 1$, $y = 2$.

First find the primitive, using the integrating factor $\dfrac{1}{y(1+x^3)}$.

Then $\dfrac{dy}{y} - \dfrac{x^2}{1+x^3}dx = 0$, $\ln y - \dfrac{1}{3}\ln(1+x^3) = C_1$, $3\ln y = \ln(1+x^3) + \ln C$, $y^3 = C(1+x^3)$.

When $x = 1$, $y = 2$: $2^3 = C(1+1)$, $C = 4$, and the required particular solution is $y^3 = 4(1+x^3)$.

HOMOGENEOUS EQUATIONS.

6. When $M dx + N dy = 0$ is homogeneous, show that the transformation $y = vx$ will separate the variables.

When $M dx + N dy = 0$ is homogeneous of degree n, we may write

$$M\,dx + N\,dy = x^n\{M_1(\tfrac{y}{x})\,dx + N_1(\tfrac{y}{x})\,dy\} = 0 \quad \text{whence} \quad M_1(\tfrac{y}{x})\,dx + N_1(\tfrac{y}{x})dy = 0.$$

The transformation $y = vx$, $dy = v\,dx + x\,dv$ reduces this to

$$M_1(v)\,dx + N_1(v)\{v\,dx + x\,dv\} = 0 \quad \text{or} \quad \{M_1(v) + vN_1(v)\}dx + xN_1(v)\,dv = 0$$

or, finally, $\dfrac{dx}{x} + \dfrac{N_1(v)\,dv}{M_1(v) + vN_1(v)} = 0$ in which the variables are separated.

7. Solve $(x^3+y^3)dx - 3xy^2dy = 0$.

The equation is homogeneous of degree 3. We use the transformation $y = vx$, $dy = v\,dx + x\,dv$ to obtain

1) $x^3\{(1+v^3)dx - 3v^2(v\,dx + x\,dv)\} = 0$ or $(1-2v^3)dx - 3v^2x\,dv = 0$

in which the variables are separable.

Upon separating the variables, using the integrating factor $\dfrac{1}{x(1-2v^3)}$, $\dfrac{dx}{x} - \dfrac{3v^2\,dv}{1-2v^3} = 0$,

and $\ln x + \frac{1}{2}\ln(1-2v^3) = C_1$, $2\ln x + \ln(1-2v^3) = \ln C$, or $x^2(1-2v^3) = C$.

Since $v = y/x$, the primitive is $x^2(1-2y^3/x^3) = C$ or $x^3 - 2y^3 = Cx$.

Note that the equation is of degree 3 and that after the transformation x^3 is a factor of the left member of 1). This factor may be removed when making the transformation.

8. Solve $x\,dy - y\,dx - \sqrt{x^2-y^2}\,dx = 0$.

The equation is homogeneous of degree 1. Using the transformation $y = vx$, $dy = v\,dx + x\,dv$ and dividing by x, we have

$$v\,dx + x\,dv - v\,dx - \sqrt{1-v^2}\,dx = 0 \quad \text{or} \quad x\,dv - \sqrt{1-v^2}\,dx = 0.$$

When the variables are separated, using the integrating factor $\dfrac{1}{x\sqrt{1-v^2}}$, $\dfrac{dv}{\sqrt{1-v^2}} - \dfrac{dx}{x} = 0$.

Then $\arcsin v - \ln x = \ln C$ or $\arcsin v = \ln(Cx)$, and returning to the original variables, using $v = y/x$, $\arcsin\dfrac{y}{x} = \ln(Cx)$ or $Cx = e^{\arcsin y/x}$.

9. Solve $(2x\sinh\dfrac{y}{x} + 3y\cosh\dfrac{y}{x})dx - 3x\cosh\dfrac{y}{x}\,dy = 0$.

The equation is homogeneous of degree 1. Using the standard transformation and dividing by x, we have

$$2\sinh v\,dx - 3x\cosh v\,dv = 0.$$

Then, separating the variables, $2\dfrac{dx}{x} - 3\dfrac{\cosh v}{\sinh v}dv = 0$.

Integrating, $2\ln x - 3\ln\sinh v = \ln C$, $x^2 = C\sinh^3 v$, and $x^2 = C\sinh^3\dfrac{y}{x}$.

10. Solve $(2x+3y)dx + (y-x)dy = 0$.

The equation is homogeneous of degree 1. The standard transformation reduces it to

$$(2+3v)dx + (v-1)(v\,dx + x\,dv) = 0 \quad \text{or} \quad (v^2+2v+2)dx + x(v-1)dv = 0.$$

Separating the variables, $\dfrac{dx}{x} + \dfrac{v-1}{v^2+2v+2}dv = \dfrac{dx}{x} + \frac{1}{2}\dfrac{2v+2}{v^2+2v+2}dv - \dfrac{2\,dv}{(v+1)^2+1} = 0$.

Integrating, $\ln x + \frac{1}{2}\ln(v^2+2v+2) - 2\arctan(v+1) = C_1$,

$\ln x^2(v^2+2v+2) - 4\arctan(v+1) = C$, and $\ln(y^2+2xy+2x^2) - 4\arctan\dfrac{x+y}{x} = C$.

11. Solve $(1+2e^{x/y})dx + 2e^{x/y}(1-\dfrac{x}{y})dy = 0$.

The equation is homogeneous of degree 0. The appearance of x/y throughout the equation suggests the use of the transformation $x = vy$, $dx = v\,dy + y\,dv$.

Then $(1+2e^v)(v\,dy + y\,dv) + 2e^v(1-v)dy = 0$, $(v+2e^v)dy + y(1+2e^v)dv = 0$,

and
$$\frac{dy}{y} + \frac{1+2e^v}{v+2e^v}\,dv = 0.$$

Integrating and replacing v by x/y, $\ln y + \ln(v+2e^v) = \ln C$ and $x + 2ye^{x/y} = C$.

LINEAR BUT NOT HOMOGENEOUS.

12. Solve $(x+y)dx + (3x+3y-4)dy = 0$.

The expressions $(x+y)$ and $(3x+3y)$ suggest the transformation $x+y = t$.

We use $y = t-x$, $dy = dt-dx$ to obtain $t\,dx + (3t-4)(dt-dx) = 0$

or $(4-2t)dx + (3t-4)dt = 0$

in which the variables are separable.

Then $2\,dx + \dfrac{3t-4}{2-t}\,dt = 2\,dx - 3\,dt + \dfrac{2}{2-t}\,dt = 0.$

Integrating and replacing t by $x+y$, we have

$2x-3t-2\ln(2-t) = C_1$, $2x-3(x+y)-2\ln(2-x-y) = C_1$, and $x+3y+2\ln(2-x-y) = C$.

13. Solve $(2x-5y+3)dx - (2x+4y-6)dy = 0$.

First solve $2x-5y+3 = 0$, $2x+4y-6 = 0$ simultaneously to obtain $x = h = 1$, $y = k = 1$.

The transformation
$$x = x'+h = x'+1,\quad dx = dx'$$
$$y = y'+k = y'+1,\quad dy = dy'$$

reduces the given equation to $(2x'-5y')dx' - (2x'+4y')dy' = 0$

which is homogeneous of degree 1. (Note that this latter equation can be written down without carrying out the details of the transformation.)

Using the transformation $y' = vx'$, $dy' = v\,dx' + x'\,dv$,

we obtain $(2-5v)dx' - (2+4v)(v\,dx' + x'\,dv) = 0$, $(2-7v-4v^2)dx' - x'(2+4v)dv = 0$,

and finally
$$\frac{dx'}{x'} + \frac{4}{3}\frac{dv}{4v-1} + \frac{2}{3}\frac{dv}{v+2} = 0.$$

Integrating, $\ln x' + \dfrac{1}{3}\ln(4v-1) + \dfrac{2}{3}\ln(v+2) = \ln C_1$ or $x'^3(4v-1)(v+2)^2 = C$.

Replacing v by y'/x', $(4y'-x')(y'+2x')^2 = C$,

and replacing x' by $x-1$ and y' by $y-1$, we obtain the primitive $(4y-x-3)(y+2x-3)^2 = C$.

14. Solve $(x-y-1)dx + (4y+x-1)dy = 0$.

Solving $x-y-1=0$, $4y+x-1=0$ simultaneously, we obtain $x=h=1$, $y=k=0$.

The transformation
$$x = x'+h = x'+1,\quad dx = dx'$$
$$y = y'+k = y',\quad dy = dy'$$

reduces the given equation to $(x'-y')dx' + (4y'+x')dy' = 0$ which is homogeneous of degree 1. (Note that this transformation $x-1 = x'$, $y = y'$ could have been obtained by inspection, that is, by examining the terms $(x-y-1)$ and $(4y+x-1)$.)

Using the transformation $y' = vx', \quad dy' = v\,dx' + x'\,dv$

we obtain $(1-v)dx' + (4v+1)(v\,dx' + x'\,dv) = 0.$

Then $$\frac{dx'}{x'} + \frac{4v+1}{4v^2+1}\,dv = \frac{dx'}{x'} + \tfrac{1}{2}\,\frac{8v}{4v^2+1}\,dv + \frac{dv}{4v^2+1} = 0,$$

$$\ln x' + \tfrac{1}{2}\ln(4v^2+1) + \tfrac{1}{2}\arctan 2v = C_1, \qquad \ln x'^2(4v^2+1) + \arctan 2v = C,$$

$$\ln(4y'^2 + x'^2) + \arctan\frac{2y'}{x'} = C, \quad \text{and} \quad \ln[4y^2 + (x-1)^2] + \arctan\frac{2y}{x-1} = C.$$

FORM $y\,f(xy)\,dx + x\,g(xy)\,dy = 0.$

15. Solve $y(xy+1)dx + x(1+xy+x^2y^2)dy = 0.$

The transformation $xy = v, \quad y = v/x, \quad dy = \dfrac{x\,dv - v\,dx}{x^2}$

reduces the equation to $$\frac{v}{x}(v+1)dx + x(1+v+v^2)\frac{x\,dv - v\,dx}{x^2} = 0$$

which, after clearing of fractions and rearranging, becomes $v^3\,dx - x(1+v+v^2)dv = 0.$

Separating the variables, we have $$\frac{dx}{x} - \frac{dv}{v^3} - \frac{dv}{v^2} - \frac{dv}{v} = 0.$$

Then $$\ln x + \frac{1}{2v^2} + \frac{1}{v} - \ln v = C_1, \qquad 2v^2\ln\left(\frac{v}{x}\right) - 2v - 1 = Cv^2,$$

and $$2x^2y^2\ln y - 2xy - 1 = Cx^2y^2.$$

16. Solve $(y - xy^2)dx - (x + x^2y)dy = 0$ or $y(1-xy)dx - x(1+xy)dy = 0.$

The transformation $xy = v, \; y = v/x, \; dy = \dfrac{x\,dv - v\,dx}{x^2}$ reduces the equation to

$$\frac{v}{x}(1-v)dx - x(1+v)\frac{x\,dv - v\,dx}{x^2} = 0 \quad \text{or} \quad 2v\,dx - x(1+v)dv = 0.$$

Then $$2\frac{dx}{x} - \frac{1+v}{v}\,dv = 0, \quad 2\ln x - \ln v - v = \ln C, \quad \frac{x^2}{v} = Ce^v, \quad \text{and} \quad x = Cye^{xy}.$$

17. Solve $(1 - xy + x^2y^2)dx + (x^3y - x^2)dy = 0$ or $y(1 - xy + x^2y^2)dx + x(x^2y^2 - xy)dy = 0.$

The transformation $xy = v, \; y = v/x, \; dy = \dfrac{x\,dv - v\,dx}{x^2}$ reduces the equation to

$$\frac{v}{x}(1 - v + v^2)dx + x(v^2 - v)\frac{x\,dv - v\,dx}{x^2} = 0 \quad \text{or} \quad v\,dx + x(v^2 - v)dv = 0.$$

Then $$\frac{dx}{x} + (v-1)dv = 0, \quad \ln x + \tfrac{1}{2}v^2 - v = C, \quad \text{and} \quad \ln x = xy - \tfrac{1}{2}x^2y^2 + C.$$

MISCELLANEOUS SUBSTITUTIONS.

18. Solve $\dfrac{dy}{dx} = (y-4x)^2$ or $dy = (y-4x)^2\,dx$.

The suggested transformation $y - 4x = v$, $dy = 4dx + dv$ reduces the equation to

$$4\,dx + dv = v^2 dx \quad \text{or} \quad dx - \frac{dv}{v^2-4} = 0.$$

Then $x + \frac{1}{4}\ln\dfrac{v+2}{v-2} = C_1$, $\ln\dfrac{v+2}{v-2} = \ln C - 4x$, $\dfrac{v+2}{v-2} = Ce^{-4x}$, and $\dfrac{y-4x+2}{y-4x-2} = Ce^{-4x}$.

19. Solve $\tan^2(x+y)dx - dy = 0$.

The suggested transformation $x + y = v$, $dy = dv - dx$ reduces the equation to

$$\tan^2 v\,dx - (dv - dx) = 0, \quad dx - \frac{dv}{1+\tan^2 v} = 0, \quad \text{or} \quad dx - \cos^2 v\,dv = 0.$$

Integrating, $x - \frac{1}{2}v - \frac{1}{4}\sin 2v = C_1$ and $2(x-y) = C + \sin 2(x+y)$.

20. Solve $(2 + 2x^2y^{\frac{1}{2}})y\,dx + (x^2y^{\frac{1}{2}} + 2)x\,dy = 0$.

The suggested transformation $x^2y^{\frac{1}{2}} = v$, $y = \dfrac{v^2}{x^4}$, $dy = \dfrac{2v}{x^4}dv - \dfrac{4v^2}{x^5}dx$ reduces the equation to

$$(2+2v)\frac{v^2}{x^4}dx + x(v+2)\left(\frac{2v}{x^4}dv - \frac{4v^2}{x^5}dx\right) = 0 \quad \text{or} \quad v(3+v)dx - x(v+2)dv = 0.$$

Then $\dfrac{dx}{x} - \dfrac{2}{3}\dfrac{dv}{v} - \dfrac{1}{3}\dfrac{dv}{v+3} = 0$, $3\ln x - 2\ln v - \ln(v+3) = \ln C_1$, $x^3 = C_1v^2(v+3)$,

and $$1 = C_1xy(x^2y^{\frac{1}{2}} + 3) \quad \text{or} \quad xy(x^2y^{\frac{1}{2}} + 3) = C.$$

21. Solve $(2x^2 + 3y^2 - 7)x\,dx - (3x^2 + 2y^2 - 8)y\,dy = 0$.

The suggested transformation $x^2 = u$, $y^2 = v$ reduces the equation to

$$(2u + 3v - 7)du - (3u + 2v - 8)dv = 0$$

which is linear but not homogeneous.

The transformation $u = s + 2$, $v = t + 1$ yields the homogeneous equation $(2s+3t)ds - (3s+2t)dt = 0$, and the transformation $s = rt$, $ds = r\,dt + t\,dr$ yields $2(r^2-1)dt + (2r+3)t\,dr = 0$.

Separating the variables, we have $2\dfrac{dt}{t} + \dfrac{2r+3}{r^2-1}dr = 2\dfrac{dt}{t} - \dfrac{1}{2}\dfrac{dr}{r+1} + \dfrac{5}{2}\dfrac{dr}{r-1} = 0.$

Then $$4\ln t - \ln(r+1) + 5\ln(r-1) = \ln C,$$

$$\frac{t^4(r-1)^5}{r+1} = \frac{(s-t)^5}{s+t} = \frac{(u-v-1)^5}{u+v-3} = \frac{(x^2-y^2-1)^5}{x^2+y^2-3} = C, \quad \text{and} \quad (x^2-y^2-1)^5 = C(x^2+y^2-3).$$

22. Solve $x^2(x\,dx + y\,dy) + y(x\,dy - y\,dx) = 0$.

Here $x\,dx + y\,dy = \frac{1}{2}d(x^2+y^2)$ and $x\,dy - y\,dx = x^2 d(y/x)$ suggest $x^2 + y^2 = \rho^2$, $y/x = \tan\theta$, or $x = \rho\cos\theta$, $y = \rho\sin\theta$, $dx = -\rho\sin\theta\,d\theta + \cos\theta\,d\rho$, $dy = \rho\cos\theta\,d\theta + \sin\theta\,d\rho$.

The given equation takes the form $\rho^2\cos^2\theta\,(\rho\,d\rho) + \rho\sin\theta\,(\rho^2\,d\theta) = 0$
or $d\rho + \tan\theta\sec\theta\,d\theta = 0$.

Then $\rho + \sec\theta = C_1$, $\sqrt{x^2+y^2}\left(\dfrac{x+1}{x}\right) = C_1$, and $(x^2+y^2)(x+1)^2 = Cx^2$.

SUPPLEMENTARY PROBLEMS

23. Determine whether or not each of the following functions is homogeneous and, when homogeneous, state the degree.

a) $x^2 - xy$,	homo. of degree two.	e) arc sin xy,	not homo.
b) $\dfrac{xy}{x+y^2}$,	not homo.	f) $xe^{y/x} + ye^{x/y}$,	homo. of degree one.
c) $\dfrac{xy}{x^2+y^2}$;	homo. of degree zero.	g) $\ln x - \ln y$ or $\ln\dfrac{x}{y}$,	homo. of degree zero.
		h) $\sqrt{x^2+2xy+3y^2}$,	homo. of degree one.
d) $x + y\cos\dfrac{y}{x}$,	homo. of degree one.	i) $x\sin y + y\sin x$,	not homo.

Classify each of the equations below in one or more of the following categories:

(1) Variables separable
(2) Homogeneous equations
(3) Equations in which $M(x,y)$ and $N(x,y)$ are linear but not homogeneous
(4) Equations of the form $y\,f(xy)dx + x\,g(xy)dy = 0$
(5) None of the above apply.

24. $4y\,dx + x\,dy = 0$ *Ans.* (1); (2), of degree one

25. $(1+2y)dx + (4-x^2)dy = 0$ (1)

26. $y^2\,dx - x^2\,dy = 0$ (1); (2), of degree two

27. $(1+y)dx - (1+x)dy = 0$ (1); (3)

28. $(xy^2+y)dx + (x^2y-x)dy = 0$ (4)

29. $(x\sin\dfrac{y}{x} - y\cos\dfrac{y}{x})dx + x\cos\dfrac{y}{x}\,dy = 0$ (2), of degree one

30. $y^2(x^2+2)dx + (x^3+y^3)(y\,dx - x\,dy) = 0$ (5)

31. $y\sqrt{x^2+y^2}\,dx - x(x+\sqrt{x^2+y^2})dy = 0$ (2), of degree two

32. $(x+y+1)dx + (2x+2y+1)dy = 0$ (3)

33. Solve each of the above equations (Problems 24-32) which fall in categories (1)-(4).

Ans. 24. $x^4y = C$ 28. $y = Cxe^{xy}$

25. $(1+2y)^2 = C\dfrac{2-x}{2+x}$ 29. $x\sin\dfrac{y}{x} = C$

26. $y = x + Cxy$ 31. $Cx - \sqrt{x^2+y^2} = x\ln(\sqrt{x^2+y^2} - x)$

27. $(1+y) = C(1+x)$ 32. $x + 2y + \ln(x+y) = C$

Solve each of the following equations.

34. $(1+2y)dx - (4-x)dy = 0$ *Ans.* $(x-4)^2(1+2y) = C$

35. $xy\,dx + (1+x^2)dy = 0$ *Ans.* $y^2(1+x^2) = C$

36. $\cot\theta\,d\rho + \rho\,d\theta = 0$ *Ans.* $\rho = C\cos\theta$

37. $(x+2y)dx + (2x+3y)dy = 0$ *Ans.* $x^2+4xy+3y^2 = C$

38. $2x\,dy - 2y\,dx = \sqrt{x^2+4y^2}\,dx$ *Ans.* $1 + 4Cy - C^2x^2 = 0$

39. $(3y-7x+7)dx + (7y-3x+3)dy = 0$ *Ans.* $(y-x+1)^2(y+x-1)^5 = C$

40. $xy\,dy = (y+1)(1-x)dx$ *Ans.* $y+x = \ln Cx(y+1)$

41. $(y^2-x^2)dx + xy\,dy = 0$ *Ans.* $2x^2y^2 = x^4 + C$

42. $y(1+2xy)dx + x(1-xy)dy = 0$ *Ans.* $y = Cx^2e^{-1/xy}$

43. $dx + (1-x^2)\cot y\,dy = 0$ *Ans.* $\sin^2 y = C\dfrac{1-x}{1+x}$

44. $(x^3+y^3)dx + 3xy^2dy = 0$ *Ans.* $x^4 + 4xy^3 = C$

45. $(3x+2y+1)dx - (3x+2y-1)dy = 0$ *Ans.* $\ln(15x+10y-1) + \frac{5}{2}(x-y) = C$

In each of the following, find the particular solution indicated.

46. $x\,dy + 2y\,dx = 0$; when $x = 2,\ y = 1$. *Ans.* $x^2y = 4$

47. $(x^2+y^2)dx + xy\,dy = 0$; when $x = 1,\ y = -1$. *Ans.* $x^4 + 2x^2y^2 = 3$

48. $\cos y\,dx + (1+e^{-x})\sin y\,dy = 0$; when $x = 0,\ y = \pi/4$. *Ans.* $(1+e^x)\sec y = 2\sqrt{2}$

49. $(y^2+xy)dx - x^2dy = 0$; when $x = 1,\ y = 1$. *Ans.* $x = e^{1-x/y}$

50. Solve the equation of Problem 30 using the substitution $y = vx$.

Ans. $x^2y\ln x - y + x^3 - \frac{1}{2}y^3 = Cx^2y$

51. Solve $y' = -2(2x+3y)^2$ using the substitution $z = 2x+3y$.

Ans. $\dfrac{1+\sqrt{3}(2x+3y)}{1-\sqrt{3}(2x+3y)} = Ce^{4\sqrt{3}\,x}$

52. Solve $(x - 2\sin y + 3)dx + (2x - 4\sin y - 3)\cos y\,dy = 0$ using the substitution $\sin y = z$.

Ans. $8\sin y + 4x + 9\ln(4x - 8\sin y + 3) = C$

CHAPTER 5

Equations of First Order and First Degree

EXACT EQUATIONS AND REDUCTION TO EXACT EQUATIONS

THE NECESSARY AND SUFFICIENT CONDITION that

1) $$M(x,y)\,dx + N(x,y)\,dy = 0$$

be exact is

2) $$\frac{\partial M}{\partial y} = \frac{\partial N}{\partial x}.$$

At times an equation may be seen to be exact after a regrouping of its terms. The equation in the regrouped form may then be integrated term by term.

For example, $(x^2 - y)\,dx + (y^2 - x)\,dy = 0$ is exact since

$$\frac{\partial M}{\partial y} = \frac{\partial}{\partial y}(x^2 - y) = -1 = \frac{\partial}{\partial x}(y^2 - x) = \frac{\partial N}{\partial x}.$$

This may also be seen after regrouping thus: $x^2 dx + y^2 dy - (y\,dx + x\,dy) = 0$. This equation may be integrated term by term to obtain the primitive $x^3/3 + y^3/3 - xy = C$. The equation $(y^2 - x)\,dx + (x^2 - y)\,dy = 0$, however, is not exact since $\frac{\partial M}{\partial y} = 2y \neq 2x = \frac{\partial N}{\partial x}$. See also Problem 1.

IF 1) IS THE EXACT DIFFERENTIAL of the equation $\mu(x,y) = C$,

$$d\mu = \frac{\partial \mu}{\partial x}dx + \frac{\partial \mu}{\partial y}dy = M(x,y)\,dx + N(x,y)\,dy.$$

Then $\frac{\partial \mu}{\partial x}dx = M(x,y)\,dx$ and $\mu(x,y) = \int^x M(x,y)\,dx + \phi(y)$,

where $\int^x$ indicates that in the integration y is to be treated as a constant and $\phi(y)$ is the constant (with respect to x) of integration. Now

$$\frac{\partial \mu}{\partial y} = \frac{\partial}{\partial y}\{\int^x M(x,y)\,dx\} + \frac{d\phi}{dy} = N(x,y)$$

from which $\frac{d\phi}{dy} = \phi'(y)$ and, hence, $\phi(y)$ can be found. See Problems 2-3.

INTEGRATING FACTORS. If 1) is not exact, an integrating factor is sought.

a) If $\dfrac{\frac{\partial M}{\partial y} - \frac{\partial N}{\partial x}}{N} = f(x)$, a function of x alone, then $e^{\int f(x)\,dx}$ is an integrating factor of 1).

If $\dfrac{\dfrac{\partial M}{\partial y} - \dfrac{\partial N}{\partial x}}{M} = -g(y)$, a function of y alone, then $e^{\int g(y)\,dy}$ is an integrating factor of 1).

See Problems 4-6.

b) If 1) is homogeneous and $Mx + Ny \neq 0$, then $\dfrac{1}{Mx+Ny}$ is an integrating factor.

See Problems 7-9.

c) If 1) can be written in the form $y\,f(xy)\,dx + x\,g(xy)\,dy = 0$, where $f(xy) \neq g(xy)$, then $\dfrac{1}{xy\{f(xy) - g(xy)\}} = \dfrac{1}{Mx - Ny}$ is an integrating factor.

See Problems 10-12.

d) At times an integrating factor may be found by inspection, after regrouping the terms of the equation, by recognizing a certain group of terms as being a part of an exact differential. For example:

Group of Terms	Integrating Factor	Exact Differential
$x\,dy - y\,dx$	$\dfrac{1}{x^2}$	$\dfrac{x\,dy - y\,dx}{x^2} = d(\dfrac{y}{x})$
$x\,dy - y\,dx$	$\dfrac{1}{y^2}$	$-\dfrac{y\,dx - x\,dy}{y^2} = d(-\dfrac{x}{y})$
$x\,dy - y\,dx$	$\dfrac{1}{xy}$	$\dfrac{dy}{y} - \dfrac{dx}{x} = d(\ln \dfrac{y}{x})$
$x\,dy - y\,dx$	$\dfrac{1}{x^2+y^2}$	$\dfrac{x\,dy - y\,dx}{x^2+y^2} = \dfrac{\dfrac{x\,dy - y\,dx}{x^2}}{1+(\dfrac{y}{x})^2} = d(\arctan \dfrac{y}{x})$
$x\,dy + y\,dx$	$\dfrac{1}{(xy)^n}$	$\dfrac{x\,dy + y\,dx}{(xy)^n} = d\{\dfrac{-1}{(n-1)(xy)^{n-1}}\}$, if $n \neq 1$ $\dfrac{x\,dy + y\,dx}{xy} = d\{\ln(xy)\}$, if $n = 1$
$x\,dx + y\,dy$	$\dfrac{1}{(x^2+y^2)^n}$	$\dfrac{x\,dx + y\,dy}{(x^2+y^2)^n} = d\{\dfrac{-1}{2(n-1)(x^2+y^2)^{n-1}}\}$, if $n \neq 1$ $\dfrac{x\,dx + y\,dy}{x^2+y^2} = d\{\frac{1}{2}\ln(x^2+y^2)\}$, if $n = 1$

See Problems 13-19.

e) The equation $x^r y^s(my\,dx + nx\,dy) + x^\rho y^\sigma(\mu y\,dx + \nu x\,dy) = 0$, where $r, s, m, n, \rho, \sigma, \mu, \nu$ are constants and $m\nu - n\mu \neq 0$, has an integrating factor of the form $x^\alpha y^\beta$. The method of solution usually given consists of determining α and β by means of certain derived formulas. In Problems 20-22, a procedure, essentially that used in deriving the formulas, is followed.

SOLVED PROBLEMS

1. Show first by the use 2) and then by regrouping of terms that each equation is exact, and solve.

a) $(4x^3y^3 - 2xy)dx + (3x^4y^2 - x^2)dy = 0$

b) $(3e^{3x}y - 2x)dx + e^{3x}dy = 0$

c) $(\cos y + y\cos x)dx + (\sin x - x\sin y)dy = 0$

d) $2x(ye^{x^2} - 1)dx + e^{x^2}dy = 0$

e) $(6x^5y^3 + 4x^3y^5)dx + (3x^6y^2 + 5x^4y^4)dy = 0$

a) By 2): $\dfrac{\partial M}{\partial y} = 12x^3y^2 - 2x = \dfrac{\partial N}{\partial x}$ and the equation is exact.

By inspection: $(4x^3y^3dx + 3x^4y^2dy) - (2xy\,dx + x^2dy) = d(x^4y^3) - d(x^2y) = 0.$

The primitive is $x^4y^3 - x^2y = C.$

b) By 2): $\dfrac{\partial M}{\partial y} = 3e^{3x} = \dfrac{\partial N}{\partial x}$ and the equation is exact.

By inspection: $(3e^{3x}y\,dx + e^{3x}dy) - 2x\,dx = d(e^{3x}y) - d(x^2) = 0.$

The primitive is $e^{3x}y - x^2 = C.$

c) By 2): $\dfrac{\partial M}{\partial y} = -\sin y + \cos x = \dfrac{\partial N}{\partial x}$ and the equation is exact.

By inspection: $(\cos y\,dx - x\sin y\,dy) + (y\cos x\,dx + \sin x\,dy)$

$= d(x\cos y) + d(y\sin x) = 0.$ The primitive is $x\cos y + y\sin x = C.$

d) By 2): $\dfrac{\partial M}{\partial y} = 2xe^{x^2} = \dfrac{\partial N}{\partial x}$ and the equation is exact.

By inspection: $(2xye^{x^2}dx + e^{x^2}dy) - 2x\,dx = d(ye^{x^2}) - d(x^2) = 0.$

The primitive is $ye^{x^2} - x^2 = C.$

e) By 2): $\dfrac{\partial M}{\partial y} = 18x^5y^2 + 20x^3y^4 = \dfrac{\partial N}{\partial x}$ and the equation is exact.

By inspection: $(6x^5y^3dx + 3x^6y^2dy) + (4x^3y^5dx + 5x^4y^4dy) = d(x^6y^3) + d(x^4y^5) = 0.$

The primitive is $x^6y^3 + x^4y^5 = C.$

2. Solve $(2x^3 + 3y)dx + (3x + y - 1)dy = 0.$

$\dfrac{\partial M}{\partial y} = 3 = \dfrac{\partial N}{\partial x}$ and the equation is exact.

Solution 1. Set $\mu(x,y) = \int^x (2x^3 + 3y)dx = \frac{1}{2}x^4 + 3xy + \phi(y).$

Then $\dfrac{\partial \mu}{\partial y} = 3x + \phi'(y) = N(x,y) = 3x + y - 1, \quad \phi'(y) = y - 1, \quad \phi(y) = \frac{1}{2}y^2 - y,$

and the primitive is $\frac{1}{2}x^4 + 3xy + \frac{1}{2}y^2 - y = C_1$ or $x^4 + 6xy + y^2 - 2y = C.$

Solution 2. Grouping the terms thus $2x^3dx + y\,dy - dy + 3(y\,dx + x\,dy) = 0$

and recalling that $y\,dx + x\,dy = d(xy)$, we obtain, by integration, $\frac{1}{2}x^4 + \frac{1}{2}y^2 - y + 3xy = C_1$ as before.

3. Solve $(y^2e^{xy^2} + 4x^3)dx + (2xye^{xy^2} - 3y^2)dy = 0.$

$$\frac{\partial M}{\partial y} = 2ye^{xy^2} + 2xy^3e^{xy^2} = \frac{\partial N}{\partial x} \quad \text{and the equation is exact.}$$

Set $\mu(x,y) = \int^x (y^2e^{xy^2} + 4x^3)dx = e^{xy^2} + x^4 + \phi(y).$

Then $\dfrac{\partial \mu}{\partial y} = 2xye^{xy^2} + \phi'(y) = 2xye^{xy^2} - 3y^2, \quad \phi'(y) = -3y^2, \quad \phi(y) = -y^3,$

and the primitive is $e^{xy^2} + x^4 - y^3 = C.$

The equation may be solved by regrouping thus $4x^3dx - 3y^2dy + (y^2e^{xy^2}dx + 2xye^{xy^2}dy) = 0$ and noting that $y^2e^{xy^2}dx + 2xye^{xy^2}dy = d(e^{xy^2}).$

4. Solve $(x^2 + y^2 + x)dx + xy\,dy = 0.$

$\dfrac{\partial M}{\partial y} = 2y, \quad \dfrac{\partial N}{\partial x} = y;$ the equation is not exact.

However, $\dfrac{\dfrac{\partial M}{\partial y} - \dfrac{\partial N}{\partial x}}{N} = \dfrac{2y - y}{xy} = \dfrac{1}{x} = f(x)$ and $e^{\int f(x)dx} = e^{\int dx/x} = e^{\ln x} = x$

is an integrating factor. Introducing the integrating factor, we have

$$(x^3 + xy^2 + x^2)dx + x^2y\,dy = 0 \qquad \text{or} \qquad x^3dx + x^2dx + (xy^2dx + x^2y\,dy) = 0.$$

Then, noting that $xy^2dx + x^2y\,dy = d(\tfrac{1}{2}x^2y^2)$, we have for the primitive

$$\frac{x^4}{4} + \frac{x^3}{3} + \tfrac{1}{2}x^2y^2 = C_1 \qquad \text{or} \qquad 3x^4 + 4x^3 + 6x^2y^2 = C.$$

5. Solve $(2xy^4e^y + 2xy^3 + y)dx + (x^2y^4e^y - x^2y^2 - 3x)dy = 0.$

$\dfrac{\partial M}{\partial y} = 8xy^3e^y + 2xy^4e^y + 6xy^2 + 1, \quad \dfrac{\partial N}{\partial x} = 2xy^4e^y - 2xy^2 - 3;$ the equation is not exact.

However, $\dfrac{\partial M}{\partial y} - \dfrac{\partial N}{\partial x} = 8xy^3e^y + 8xy^2 + 4$ and $\dfrac{\dfrac{\partial M}{\partial y} - \dfrac{\partial N}{\partial x}}{M} = \dfrac{4}{y} = -g(y).$

Then $e^{\int g(y)dy} = e^{-4\int dy/y} = e^{-4\ln y} = 1/y^4$ is an integrating factor and, upon introducing it, the equation takes the form

$$(2xe^y + 2\frac{x}{y} + \frac{1}{y^3})dx + (x^2e^y - \frac{x^2}{y^2} - 3\frac{x}{y^4})dy = 0 \quad \text{and is exact.}$$

Set $\mu(x,y) = \int^x (2xe^y + 2\dfrac{x}{y} + \dfrac{1}{y^3})dx = x^2e^y + \dfrac{x^2}{y} + \dfrac{x}{y^3} + \phi(y).$

Then $\dfrac{\partial \mu}{\partial y} = x^2e^y - \dfrac{x^2}{y^2} - 3\dfrac{x}{y^4} + \phi'(y) = x^2e^y - \dfrac{x^2}{y^2} - 3\dfrac{x}{y^4}, \quad \phi'(y) = 0, \quad \phi(y) = \text{constant},$

and the primitive is $x^2e^y + \dfrac{x^2}{y} + \dfrac{x}{y^3} = C.$

6. Solve $(2x^3y^2 + 4x^2y + 2xy^2 + xy^4 + 2y)dx + 2(y^3 + x^2y + x)dy = 0.$

$\dfrac{\partial M}{\partial y} = 4x^3y + 4x^2 + 4xy + 4xy^3 + 2, \quad \dfrac{\partial N}{\partial x} = 2(2xy+1);$ the equation is not exact.

$\dfrac{\dfrac{\partial M}{\partial y} - \dfrac{\partial N}{\partial x}}{N} = 2x$ and the integrating factor is $e^{\int 2x\,dx} = e^{x^2}$. When it is introduced, the given equation becomes

$$(2x^3y^2 + 4x^2y + 2xy^2 + xy^4 + 2y)e^{x^2}dx + 2(y^3 + x^2y + x)e^{x^2}dy = 0 \quad \text{and is exact.}$$

Set
$$\mu(x,y) = \int^x (2x^3y^2 + 4x^2y + 2xy^2 + xy^4 + 2y)e^{x^2}dx$$
$$= \int^x (2xy^2 + 2x^3y^2)e^{x^2}dx + \int^x (2y + 4x^2y)e^{x^2}dx + \int^x xy^4e^{x^2}dx$$
$$= x^2y^2e^{x^2} + 2xye^{x^2} + \tfrac{1}{2}y^4e^{x^2} + \phi(y).$$

Then $\dfrac{\partial\mu}{\partial y} = 2x^2ye^{x^2} + 2xe^{x^2} + 2y^3e^{x^2} + \phi'(y) = 2(y^3 + x^2y + x)e^{x^2}, \quad \phi'(y) = 0,$

and the primitive is $(2x^2y^2 + 4xy + y^4)e^{x^2} = C.$

7. Show that $\dfrac{1}{Mx+Ny}$, where $Mx+Ny$ is not identically zero, is an integrating factor of the homogeneous equation $M(x,y)dx + N(x,y)dy = 0$ of degree n. Investigate the case $Mx+Ny=0$ identically.

We are to show that $\dfrac{M}{Mx+Ny}dx + \dfrac{N}{Mx+Ny}dy = 0$ is an exact equation, that is, that

$$\frac{\partial}{\partial y}\left(\frac{M}{Mx+Ny}\right) = \frac{\partial}{\partial x}\left(\frac{N}{Mx+Ny}\right).$$

$$\frac{\partial}{\partial y}\left(\frac{M}{Mx+Ny}\right) = \frac{(Mx+Ny)\dfrac{\partial M}{\partial y} - M\left(x\dfrac{\partial M}{\partial y} + N + y\dfrac{\partial N}{\partial y}\right)}{(Mx+Ny)^2} = \frac{Ny\dfrac{\partial M}{\partial y} - MN - My\dfrac{\partial N}{\partial y}}{(Mx+Ny)^2}$$

and

$$\frac{\partial}{\partial x}\left(\frac{N}{Mx+Ny}\right) = \frac{(Mx+Ny)\dfrac{\partial N}{\partial x} - N\left(x\dfrac{\partial M}{\partial x} + M + y\dfrac{\partial N}{\partial x}\right)}{(Mx+Ny)^2} = \frac{Mx\dfrac{\partial N}{\partial x} - MN - Nx\dfrac{\partial M}{\partial x}}{(Mx+Ny)^2}.$$

$$\frac{\partial}{\partial y}\left(\frac{M}{Mx+Ny}\right) - \frac{\partial}{\partial x}\left(\frac{N}{Mx+Ny}\right) = \frac{N\left(x\dfrac{\partial M}{\partial x} + y\dfrac{\partial M}{\partial y}\right) - M\left(x\dfrac{\partial N}{\partial x} + y\dfrac{\partial N}{\partial y}\right)}{(Mx+Ny)^2} = \frac{N(nM) - M(nN)}{(Mx+Ny)^2} = 0$$

(by Euler's Theorem on homogeneous functions).

If $Mx+Ny=0$ identically, then $\dfrac{M}{N} = -\dfrac{y}{x}$ and the differential equation reduces to $y\,dx - x\,dy = 0$ for which $1/xy$ is an integrating factor.

8. Solve $(x^4 + y^4)dx - xy^3dy = 0.$

The equation is homogeneous and $\dfrac{1}{Mx+Ny} = \dfrac{1}{x^5}$ is an integrating factor. Upon its introduction, the equation becomes $(\frac{1}{x} + \frac{y^4}{x^5})dx - \frac{y^3}{x^4}dy = 0$ and is exact.

Set $\mu(x,y) = \int^x (\frac{1}{x} + \frac{y^4}{x^5})dx = \ln x - \frac{1}{4}\frac{y^4}{x^4} + \phi(y)$.

Then $\frac{\partial\mu}{\partial y} = -\frac{y^3}{x^4} + \phi'(y) = -\frac{y^3}{x^4}$, $\phi'(y) = 0$, and the primitive is

$$\ln x - \frac{1}{4}\frac{y^4}{x^4} = C_1 \quad \text{or} \quad y^4 = 4x^4 \ln x + Cx^4.$$

Note. The same integrating factor is obtained by using the procedure of *a*) above. The equation may be solved by the method of Chapter 4.

9. Solve $y^2dx + (x^2 - xy - y^2)dy = 0$.

The equation is homogeneous and $\frac{1}{Mx + Ny} = \frac{1}{y(x^2 - y^2)}$ is an integrating factor.

Upon introducing it the given equation becomes $\frac{y}{x^2 - y^2}dx + \frac{x^2 - xy - y^2}{y(x^2 - y^2)}dy = 0$ which is exact.

Set $\mu(x,y) = \int^x \frac{y}{x^2 - y^2}dx = \frac{1}{2}\int^x (\frac{1}{x-y} - \frac{1}{x+y})dx = \frac{1}{2}\ln\frac{x-y}{x+y} + \phi(y)$.

Then $\frac{\partial\mu}{\partial y} = -\frac{x}{x^2 - y^2} + \phi'(y) = \frac{x^2 - xy - y^2}{y(x^2 - y^2)} = \frac{1}{y} - \frac{x}{x^2 - y^2}$, $\phi'(y) = \frac{1}{y}$, $\phi(y) = \ln y$,

and the primitive is $\frac{1}{2}\ln\frac{x-y}{x+y} + \ln y = \ln C_1$ or $(x-y)y^2 = C(x+y)$.

10. Show that $\frac{1}{Mx - Ny}$, when $Mx - Ny$ is not identically zero, is an integrating factor for the equation $M\,dx + N\,dy = yf_1(xy)dx + xf_2(xy)dy = 0$. Investigate the case $Mx - Ny = 0$ identically.

The equation $\frac{yf_1(xy)}{xy\{f_1(xy) - f_2(xy)\}}dx + \frac{xf_2(xy)}{xy\{f_1(xy) - f_2(xy)\}}dy = 0$ is exact

since

$$\frac{\partial}{\partial y}\{\frac{f_1}{x(f_1 - f_2)}\} = \frac{x(f_1 - f_2)\frac{\partial f_1}{\partial y} - f_1x(\frac{\partial f_1}{\partial y} - \frac{\partial f_2}{\partial y})}{x^2(f_1 - f_2)^2} = \frac{-f_2\frac{\partial f_1}{\partial y} + f_1\frac{\partial f_2}{\partial y}}{x(f_1 - f_2)^2},$$

$$\frac{\partial}{\partial x}\{\frac{f_2}{y(f_1 - f_2)}\} = \frac{y(f_1 - f_2)\frac{\partial f_2}{\partial x} - f_2y(\frac{\partial f_1}{\partial x} - \frac{\partial f_2}{\partial x})}{y^2(f_1 - f_2)^2} = \frac{f_1\frac{\partial f_2}{\partial x} - f_2\frac{\partial f_1}{\partial x}}{y(f_1 - f_2)^2},$$

and

$$\frac{\partial}{\partial y}\{\frac{f_1}{x(f_1 - f_2)}\} - \frac{\partial}{\partial x}\{\frac{f_2}{y(f_1 - f_2)}\} = \frac{f_2(-y\frac{\partial f_1}{\partial y} + x\frac{\partial f_1}{\partial x}) + f_1(y\frac{\partial f_2}{\partial y} - x\frac{\partial f_2}{\partial x})}{xy(f_1 - f_2)^2}.$$

This is identically zero since $y\frac{\partial f(xy)}{\partial y} = x\frac{\partial f(xy)}{\partial x}$.

If $Mx - Ny = 0$, then $\dfrac{M}{N} = \dfrac{y}{x}$ and the equation reduces to $x\,dy + y\,dx = 0$ with solution $xy = C$.

11. Solve $y(x^2y^2 + 2)dx + x(2 - 2x^2y^2)dy = 0$.

The equation is of the form $y f_1(xy)dx + x f_2(xy)dy = 0$ and $\dfrac{1}{Mx - Ny} = \dfrac{1}{3x^3y^3}$ is an integrating factor.

Upon introducing it, the equation becomes $\dfrac{x^2y^2 + 2}{3x^3y^2}dx + \dfrac{2 - 2x^2y^2}{3x^2y^3}dy = 0$ and is exact.

Set $\mu(x,y) = \int^x \left(\dfrac{x^2y^2 + 2}{3x^3y^2}\right)dx = \int^x \left(\dfrac{1}{3x} + \dfrac{2}{3x^3y^2}\right)dx = \dfrac{1}{3}\ln x - \dfrac{1}{3x^2y^2} + \phi(y)$.

Then $\dfrac{\partial\mu}{\partial y} = \dfrac{2}{3x^2y^3} + \phi'(y) = \dfrac{2 - 2x^2y^2}{3x^2y^3}$, $\phi'(y) = -\dfrac{2}{3y}$, $\phi(y) = -\dfrac{2}{3}\ln y$,

and the primitive is $\dfrac{1}{3}\ln x - \dfrac{1}{3x^2y^2} - \dfrac{2}{3}\ln y = \ln C_1$ or $x = Cy^2e^{1/x^2y^2}$.

The equation may be solved by the method of Chapter 4.

12. Solve $y(2xy + 1)dx + x(1 + 2xy - x^3y^3)dy = 0$.

The equation is of the form $y f_1(xy)dx + x f_2(xy)dy = 0$ and $\dfrac{1}{Mx - Ny} = \dfrac{1}{x^4y^4}$ is an integrating factor.

Upon introducing it, the equation becomes $\left(\dfrac{2}{x^3y^2} + \dfrac{1}{x^4y^3}\right)dx + \left(\dfrac{1}{x^3y^4} + \dfrac{2}{x^2y^3} - \dfrac{1}{y}\right)dy = 0$ and is exact.

Set $\mu(x,y) = \int^x \left(\dfrac{2}{x^3y^2} + \dfrac{1}{x^4y^3}\right)dx = -\dfrac{1}{x^2y^2} - \dfrac{1}{3x^3y^3} + \phi(y)$.

Then $\dfrac{\partial\mu}{\partial y} = \dfrac{2}{x^2y^3} + \dfrac{1}{x^3y^4} + \phi'(y) = \dfrac{1}{x^3y^4} + \dfrac{2}{x^2y^3} - \dfrac{1}{y}$, $\phi'(y) = -\dfrac{1}{y}$, $\phi(y) = -\ln y$,

and the primitive is $-\ln y - \dfrac{1}{x^2y^2} - \dfrac{1}{3x^3y^3} = C_1$ or $y = Ce^{-(3xy+1)/(3x^3y^3)}$.

13. Obtain an integrating factor by inspection for each of the following equations.

a) $(2xy^4e^y + 2xy^3 + y)dx + (x^2y^4e^y - x^2y^2 - 3x)dy = 0$ (Problem 5)

b) $(x^2y^3 + 2y)dx + (2x - 2x^3y^2)dy = 0$ (Problem 11)

c) $(2xy^2 + y)dx + (x + 2x^2y - x^4y^3)dy = 0$ (Problem 12)

a) When the equation is written in the form

$$y^4(2xe^y dx + x^2e^y dy) + 2xy^3dx - x^2y^2dy + y\,dx - 3x\,dy = 0$$

the term $y^4(2xe^ydx + x^2e^ydy) = y^4\cdot$(an exact differential) suggests that $1/y^4$ is a possible integrating factor. To show that it is an integrating factor, we verify that its introduction produces an exact equation.

b) When the equation is written in the form $2(y\,dx + x\,dy) + x^2y^3dx - 2x^3y^2dy = 0$, the term $(y\,dx + x\,dy)$ suggests $1/(xy)^k$ as a possible integrating factor. An examination of the remaining terms shows that each will be an exact differential if $k = 3$, i.e., $1/(xy)^3$ is an integrating factor.

c) When the equation is written in the form $(x\,dy + y\,dx) + 2xy(x\,dy + y\,dx) - x^4y^3dy = 0$ the first two terms suggest $1/(xy)^k$. The third term will be an exact differential if $k = 4$; thus, $1/(xy)^4$ is an integrating factor.

14. Solve $y\,dx + x(1-3x^2y^2)dy = 0$ or $x\,dy + y\,dx - 3x^3y^2dy = 0$.

The terms $x\,dy + y\,dx$ suggest $1/(xy)^k$ and the last term requires $k = 3$.

Upon introducing the integrating factor $\dfrac{1}{(xy)^3}$, the equation becomes $\dfrac{x\,dy + y\,dx}{x^3y^3} - \dfrac{3}{y}dy = 0$

whose primitive is $\dfrac{-1}{2x^2y^2} - 3\ln y = C_1$, $6\ln y = \ln C - \dfrac{1}{x^2y^2}$ or $y^6 = Ce^{-1/(x^2y^2)}$.

15. Solve $x\,dx + y\,dy + 4y^3(x^2+y^2)dy = 0$.

The last term suggests $1/(x^2+y^2)$ as an integrating factor.

Introducing it, the equation becomes $\dfrac{x\,dx + y\,dy}{x^2+y^2} + 4y^3dy = 0$ and is exact.

The primitive is $\frac{1}{2}\ln(x^2+y^2) + y^4 = \ln C_1$ or $(x^2+y^2)e^{2y^4} = C$.

16. Solve $x\,dy - y\,dx - (1-x^2)dx = 0$.

Here $1/x^2$ is the integrating factor, since all other possibilities suggested by $x\,dy - y\,dx$ render the last term inexact.

Upon introducing it, the equation becomes $\dfrac{x\,dy - y\,dx}{x^2} - (\dfrac{1}{x^2} - 1)dx = 0$ whose primitive is $\dfrac{y}{x} + \dfrac{1}{x} + x = C$ or $y + x^2 + 1 = Cx$.

17. Solve $(x + x^4 + 2x^2y^2 + y^4)dx + y\,dy = 0$ or $x\,dx + y\,dy + (x^2+y^2)^2dx = 0$.

An integrating factor suggested by the form of the equation is $\dfrac{1}{(x^2+y^2)^2}$. Using it, we have $\dfrac{x\,dx + y\,dy}{(x^2+y^2)^2} + dx = 0$ whose primitive is $-\dfrac{1}{2(x^2+y^2)} + x = C_1$ or $(C+2x)(x^2+y^2) = 1$.

18. Solve $x^2\dfrac{dy}{dx} + xy + \sqrt{1-x^2y^2} = 0$ or $x(x\,dy + y\,dx) + \sqrt{1-x^2y^2}\,dx = 0$.

The integrating factor $\dfrac{1}{x\sqrt{1-x^2y^2}}$ reduces the equation to the form $\dfrac{x\,dy + y\,dx}{\sqrt{1-x^2y^2}} + \dfrac{dx}{x} = 0$

whose primitive is $\arcsin(xy) + \ln x = C$.

19. Solve $\dfrac{dy}{dx} = \dfrac{y - xy^2 - x^3}{x + x^2y + y^3}$ or $(x^3 + xy^2 - y)dx + (y^3 + x^2y + x)dy = 0$.

When the equation is written thus $(x^2+y^2)(x\,dx + y\,dy) + x\,dy - y\,dx = 0$, the terms $x\,dy - y\,dx$ suggest several possible integrating factors. By trial, we determine $1/(x^2+y^2)$ which reduces the given equation to the form $x\,dx + y\,dy + \dfrac{x\,dy - y\,dx}{x^2+y^2} = x\,dx + y\,dy + \dfrac{\dfrac{x\,dy - y\,dx}{x^2}}{1+(\frac{y}{x})^2} = 0.$

The primitive is $\frac{1}{2}x^2 + \frac{1}{2}y^2 + \arctan\frac{y}{x} = C_1$ or $x^2 + y^2 + 2\arctan\frac{y}{x} = C.$

20. Solve $x(4y\,dx + 2x\,dy) + y^3(3y\,dx + 5x\,dy) = 0.$

Suppose that the effect of multiplying the given equation by $x^\alpha y^\beta$ is to produce an equation

A) $$(4x^{\alpha+1}y^{\beta+1}dx + 2x^{\alpha+2}y^\beta dy) + (3x^\alpha y^{\beta+4}dx + 5x^{\alpha+1}y^{\beta+3}dy) = 0$$

each of whose two terms is an exact differential. Then the first term of $A)$ is proportional to

B) $$d(x^{\alpha+2}y^{\beta+1}) = (\alpha+2)x^{\alpha+1}y^{\beta+1}dx + (\beta+1)x^{\alpha+2}y^\beta dy,$$

that is,

C) $$\frac{\alpha+2}{4} = \frac{\beta+1}{2} \quad \text{and} \quad \alpha - 2\beta = 0.$$

Also, the second term of $A)$ is proportional to

D) $$d(x^{\alpha+1}y^{\beta+4}) = (\alpha+1)x^\alpha y^{\beta+4}dx + (\beta+4)x^{\alpha+1}y^{\beta+3}dy,$$

that is,

E) $$\frac{\alpha+1}{3} = \frac{\beta+4}{5}. \quad \text{and} \quad 5\alpha - 3\beta = 7.$$

Solving $\alpha - 2\beta = 0$, $5\alpha - 3\beta = 7$ simultaneously, we find $\alpha = 2$, $\beta = 1$.

When these substitutions are made in $A)$, the equation becomes

$$(4x^3y^2dx + 2x^4y\,dy) + (3x^2y^5dx + 5x^3y^4dy) = 0.$$

The primitive is $$x^4y^2 + x^3y^5 = C.$$

21. Solve $(8y\,dx + 8x\,dy) + x^2y^3(4y\,dx + 5x\,dy) = 0.$

Suppose that the effect of multiplying the given equation by $x^\alpha y^\beta$ is to produce an equation

A) $$(8x^\alpha y^{\beta+1}dx + 8x^{\alpha+1}y^\beta dy) + (4x^{\alpha+2}y^{\beta+4}dx + 5x^{\alpha+3}y^{\beta+3}dy) = 0$$

each of whose two terms is an exact differential. Then the first term is proportional to

B) $$d(x^{\alpha+1}y^{\beta+1}) = (\alpha+1)x^\alpha y^{\beta+1}dx + (\beta+1)x^{\alpha+1}y^\beta dy,$$

that is,

C) $$\frac{\alpha+1}{8} = \frac{\beta+1}{8} \quad \text{and} \quad \alpha - \beta = 0.$$

Also, the second term is proportional to

D) $$d(x^{\alpha+3}y^{\beta+4}) = (\alpha+3)x^{\alpha+2}y^{\beta+4}dx + (\beta+4)x^{\alpha+3}y^{\beta+3}dy,$$

that is,

E) $$\frac{\alpha+3}{4} = \frac{\beta+4}{5} \quad \text{and} \quad 5\alpha - 4\beta = 1.$$

Solving $\alpha - \beta = 0$, $5\alpha - 4\beta = 1$ simultaneously, we find $\alpha = 1$, $\beta = 1$.

When these substitutions are made in A), the equation becomes

$$(8xy^2\,dx + 8x^2y\,dy) + (4x^3y^5\,dx + 5x^4y^4\,dy) = 0.$$

The primitive is $4x^2y^2 + x^4y^5 = C.$

Note. In this and the previous problem it was not necessary to write statements B) and D) since, after a little practice, the relations C) and E) may be obtained directly from A).

22. Solve $x^3y^3(2y\,dx + x\,dy) - (5y\,dx + 7x\,dy) = 0.$

Multiplying the given equation by $x^\alpha y^\beta$, we have

$$A)\qquad (2x^{\alpha+3}y^{\beta+4}dx + x^{\alpha+4}y^{\beta+3})dy - (5x^\alpha y^{\beta+1}dx + 7x^{\alpha+1}y^\beta dy) = 0.$$

If the first term of A) is to be exact, then $\dfrac{\alpha+4}{2} = \dfrac{\beta+4}{1}$ and $\alpha - 2\beta = 4.$

If the second term of A) is to be exact, then $\dfrac{\alpha+1}{5} = \dfrac{\beta+1}{7}$ and $7\alpha - 5\beta = -2.$

Solving $\alpha - 2\beta = 4$, $7\alpha - 5\beta = -2$ simultaneously, we find $\alpha = -8/3$, $\beta = -10/3$.

Then, from A), $(2x^{1/3}y^{2/3}dx + x^{4/3}y^{-1/3}dy) - (5x^{-8/3}y^{-7/3}dx + 7x^{-5/3}y^{-10/3}dy) = 0,$ each of the two terms is exact, and the primitive is

$$\frac{3}{2}x^{4/3}y^{2/3} + 3x^{-5/3}y^{-7/3} = C_1,\quad x^{4/3}y^{2/3} + 2x^{-5/3}y^{-7/3} = C \quad\text{or}\quad x^3y^3 + 2 = Cx^{5/3}y^{7/3}.$$

SUPPLEMENTARY PROBLEMS

23. Select from the following equations those which are exact and solve.

a) $(x^2 - y)dx - x\,dy = 0$ *Ans.* $xy = x^3/3 + C$

b) $y(x-2y)dx - x^2dy = 0$

c) $(x^2 + y^2)dx + xy\,dy = 0$

d) $(x^2 + y^2)dx + 2xy\,dy = 0$ *Ans.* $xy^2 + x^3/3 = C$

e) $(x + y\cos x)dx + \sin x\,dy = 0$ *Ans.* $x^2 + 2y\sin x = C$

f) $(1 + e^{2\theta})d\rho + 2\rho e^{2\theta}d\theta = 0$ *Ans.* $\rho(1 + e^{2\theta}) = C$

g) $dx - \sqrt{a^2 - x^2}\,dy = 0$

h) $(2x + 3y + 4)dx + (3x + 4y + 5)dy = 0$ *Ans.* $x^2 + 3xy + 2y^2 + 4x + 5y = C$

i) $(4x^3y^3 + \frac{1}{x})dx + (3x^4y^2 - \frac{1}{y})dy = 0$ *Ans.* $x^4y^3 + \ln(x/y) = C$

j) $2(u^2 + uv)du + (u^2 + v^2)dv = 0$ *Ans.* $2u^3 + 3u^2v + v^3 = C$

k) $(x\sqrt{x^2+y^2} - y)dx + (y\sqrt{x^2+y^2} - x)dy = 0$ *Ans.* $(x^2+y^2)^{3/2} - 3xy = C$

l) $(x+y+1)dx - (x-y-3)dy = 0$

m) $(x+y+1)dx - (y-x+3)dy = 0$ *Ans.* $x^2 + 2xy - y^2 + 2x - 6y = C$

n) $\csc\theta \tan\theta\, dr - (r\csc\theta + \tan^2\theta)d\theta = 0$ *Ans.* $r\csc\theta = \ln\sec\theta + C$

o) $(y^2 - \dfrac{y}{x(x+y)} + 2)dx + [\dfrac{1}{x+y} + 2y(x+1)]dy = 0$ *Ans.* $\ln\dfrac{x+y}{x} + (x+1)(y^2+2) = C$

p) $(2xye^{x^2y} + y^2e^{xy^2} + 1)dx + (x^2e^{x^2y} + 2xye^{xy^2} - 2y)dy = 0$ *Ans.* $e^{x^2y} + e^{xy^2} + x - y^2 = C$

24. Solve the remaining problems above [*b*), *c*), *g*), *l*)] using the appropriate procedure of Chap. 4.

Ans. *b*) $x/y = 2\ln x + C$ *g*) $y = \arcsin x/a + C$

c) $x^4 + 2x^2y^2 = C$ *l*) $\ln\sqrt{x^2+y^2-2x+4y+5} - \arctan\dfrac{y+2}{x-1} = C$

25. For each of the following, obtain an integrating factor by inspection and solve.

a) $x\,dx + y\,dy = (x^2+y^2)dx$ *Ans.* $1/(x^2+y^2)$; $x^2+y^2 = Ce^{2x}$

b) $(2y-3x)dx + x\,dy = 0$ *Ans.* x; $x^2y = x^3 + C$

c) $(x-y^2)dx + 2xy\,dy = 0$ *Ans.* $1/x^2$; $y^2 + x\ln x = Cx$

d) $x\,dy - y\,dx = 3x^2(x^2+y^2)dx$ *Ans.* $1/(x^2+y^2)$; $\arctan y/x = x^3 + C$

e) $y\,dx - x\,dy + \ln x\,dx = 0$ *Ans.* $1/x^2$; $y + \ln x + 1 = Cx$

f) $(3x^2+y^2)dx - 2xy\,dy = 0$ *Ans.* $1/x^2$; $3x^2 - y^2 = Cx$

g) $(xy-2y^2)dx - (x^2-3xy)dy = 0$ *Ans.* $1/xy^2$; $x/y + \ln(y^3/x^2) = C$

h) $(x+y)dx - (x-y)dy = 0$ *Ans.* $1/(x^2+y^2)$; $x^2+y^2 = Ce^{2\arctan y/x}$

i) $2y\,dx - 3xy^2dx - x\,dy = 0$ *Ans.* x/y^2; $x^2/y - x^3 = C$

j) $y\,dx + x(x^2y-1)dy = 0$ *Ans.* y/x^3; $3y^2 - 2x^2y^3 = Cx^2$

k) $(y + x^3y + 2x^2)dx + (x + 4xy^4 + 8y^3)dy = 0$ *Ans.* $1/(xy+2)$; $\ln(xy+2)^3 + x^3 + 3y^4 = C$

26. For each of the following, obtain an integrating factor and solve.

a) $x\,dy - y\,dx = x^2e^x dx$ *Ans.* $y = Cx + xe^x$

b) $(1+y^2)dx = (x+x^2)dy$ *Ans.* $\arctan y = \ln x/(x+1) + C$

c) $(2y-x^3)dx + x\,dy = 0$ *Ans.* $x^2y - x^5/5 = C$

d) $y^2dy + y\,dx - x\,dy = 0$ *Ans.* $y^2 + x = Cy$

e) $(3y^3-xy)dx - (x^2+6xy^2)dy = 0$ *Ans.* $3y^2 + x\ln(xy) = Cx$

f) $3x^2y^2dx + 4(x^3y-3)dy = 0$ *Ans.* $x^3y^4 - 4y^3 = C$

g) $y(x+y)dx - x^2dy = 0$ *Ans.* $x/y + \ln x = C$

h) $(2y+3xy^2)dx + (x+2x^2y)dy = 0$ *Ans.* $x^2y(1+xy) = C$

i) $y(y^2-2x^2)dx + x(2y^2-x^2)dy = 0$ *Ans.* $x^2y^2(y^2-x^2) = C$

27. Show that $\dfrac{1}{x^2}f(y/x)$ is an integrating factor of $x\,dy - y\,dx = 0$.

CHAPTER 6

Equations of First Order and First Degree

LINEAR EQUATIONS AND THOSE REDUCIBLE TO THAT FORM

THE EQUATION 1) $$\frac{dy}{dx} + yP(x) = Q(x),$$

whose left member is linear in both the dependent variable and its derivative, is called a *linear equation of the first order*. For example,
$\frac{dy}{dx} + 3xy = \sin x$ is called linear while $\frac{dy}{dx} + 3xy^2 = \sin x$ is not.

Since $\frac{d}{dx}(ye^{\int P(x)\,dx}) = \frac{dy}{dx}e^{\int P(x)\,dx} + y\,P(x)\,e^{\int P(x)\,dx} = e^{\int P(x)\,dx}(\frac{dy}{dx} + yP(x))$,

$e^{\int P(x)\,dx}$ is an integrating factor of 1) and its primitive is

$$ye^{\int P(x)\,dx} = \int Q(x)\cdot e^{\int P(x)\,dx}\,dx + C.$$ See Problems 1-7.

BERNOULLI'S EQUATION. An equation of the form

$$\frac{dy}{dx} + yP(x) = y^n Q(x) \qquad \text{or} \qquad y^{-n}\frac{dy}{dx} + y^{-n+1}P(x) = Q(x)$$

is reduced to the form 1), namely, $\frac{dv}{dx} + v\{(1-n)P(x)\} = (1-n)Q(x)$, by the transformation

$$y^{-n+1} = v, \qquad y^{-n}\frac{dy}{dx} = \frac{1}{1-n}\frac{dv}{dx}.$$ See Problems 8-12.

OTHER EQUATIONS may be reduced to the form 1) by means of appropriate transformations. As in previous chapters, no general rule can be stated; in each instance, the proper transformation is suggested by the form of the equation. See Problems 13-18.

SOLVED PROBLEMS

LINEAR EQUATIONS.

1. Solve $\frac{dy}{dx} + 2xy = 4x$.

$\int P(x)\,dx = \int 2x\,dx = x^2$ and $e^{\int P(x)\,dx} = e^{x^2}$ is an integrating factor.

Then $ye^{x^2} = \int 4xe^{x^2}\,dx = 2e^{x^2} + C$ or $y = 2 + Ce^{-x^2}$.

2. Solve $x\frac{dy}{dx} = y + x^3 + 3x^2 - 2x$ or $\frac{dy}{dx} - \frac{1}{x}y = x^2 + 3x - 2$.

$\int P(x)\,dx = -\int\frac{dx}{x} = -\ln x$ and $e^{-\ln x} = \frac{1}{x}$ is an integrating factor.

Then $y\frac{1}{x} = \int \frac{1}{x}(x^2+3x-2)\,dx = \int (x+3-\frac{2}{x})\,dx = \frac{1}{2}x^2+3x-2\ln x+C_1$ or

$$2y = x^3+6x^2-4x\ln x+Cx.$$

3. Solve $(x-2)\frac{dy}{dx} = y+2(x-2)^3$ or $\frac{dy}{dx} - \frac{1}{x-2}y = 2(x-2)^2$.

$\int P(x)\,dx = -\int \frac{dx}{x-2} = -\ln(x-2)$ and an integrating factor is $e^{-\ln(x-2)} = \frac{1}{x-2}$.

Then $y(\frac{1}{x-2}) = 2\int (x-2)^2\cdot\frac{1}{x-2}\,dx = 2\int (x-2)\,dx = (x-2)^2+C$ or $y = (x-2)^3+C(x-2)$.

4. Solve $\frac{dy}{dx} + y\cot x = 5e^{\cos x}$. Find the particular solution, given the initial conditions: $x = \frac{1}{2}\pi$, $y = -4$.

An integrating factor is $e^{\int \cot x\,dx} = e^{\ln\sin x} = \sin x$ and

$$y\sin x = 5\int e^{\cos x}\sin x\,dx = -5e^{\cos x}+C.$$

When $x = \frac{1}{2}\pi$, $y = -4$: $(-4)(1) = -5(1)+C$ and $C = 1$. The particular solution is

$$y\sin x + 5e^{\cos x} = 1.$$

5. Solve $x^3\frac{dy}{dx} + (2-3x^2)y = x^3$ or $\frac{dy}{dx} + \frac{2-3x^2}{x^3}y = 1$.

$\int \frac{2-3x^2}{x^3}\,dx = -\frac{1}{x^2} - 3\ln x$ and an integrating factor is $\frac{1}{x^3e^{1/x^2}}$.

Then $\frac{y}{x^3e^{1/x^2}} = \int \frac{dx}{x^3e^{1/x^2}} = \frac{1}{2e^{1/x^2}} + C_1$ or $2y = x^3+Cx^3e^{1/x^2}$.

6. Solve $\frac{dy}{dx} - 2y\cot 2x = 1 - 2x\cot 2x - 2\csc 2x$.

An integrating factor is $e^{-\int 2\cot 2x\,dx} = e^{-\ln\sin 2x} = \csc 2x$.

Then $y\csc 2x = \int (\csc 2x - 2x\cot 2x\csc 2x - 2\csc^2 2x)\,dx = x\csc 2x + \cot 2x + C$

or

$$y = x+\cos 2x + C\sin 2x.$$

7. Solve $y\ln y\,dx + (x-\ln y)dy = 0$.

The equation, with x taken as dependent variable, may be put in the form $\frac{dx}{dy} + \frac{1}{y\ln y}x = \frac{1}{y}$.

Then $e^{\int dy/(y\ln y)} = e^{\ln(\ln y)} = \ln y$ is an integrating factor.

Thus, $x\ln y = \int \ln y\frac{dy}{y} = \frac{1}{2}\ln^2 y + K$ and the solution is $2x\ln y = \ln^2 y + C$.

BERNOULLI'S EQUATION.

8. Solve $\dfrac{dy}{dx} - y = xy^5$ or $y^{-5}\dfrac{dy}{dx} - y^{-4} = x.$

The transformation $y^{-4} = v,\ y^{-5}\dfrac{dy}{dx} = -\dfrac{1}{4}\dfrac{dv}{dx}$ reduces the equation to

$-\dfrac{1}{4}\dfrac{dv}{dx} - v = x$ or $\dfrac{dv}{dx} + 4v = -4x.$ An integrating factor is $e^{4\int dx} = e^{4x}.$

Then $ve^{4x} = -4\int xe^{4x}dx = -xe^{4x} + \tfrac{1}{4}e^{4x} + C,$

$$y^{-4}e^{4x} = -xe^{4x} + \tfrac{1}{4}e^{4x} + C, \quad \text{or} \quad \frac{1}{y^4} = -x + \tfrac{1}{4} + Ce^{-4x}.$$

9. Solve $\dfrac{dy}{dx} + 2xy + xy^4 = 0$ or $y^{-4}\dfrac{dy}{dx} + 2xy^{-3} = -x.$

The transformation $y^{-3} = v,\ -3y^{-4}\dfrac{dy}{dx} = \dfrac{dv}{dx}$ reduces the equation to $\dfrac{dv}{dx} - 6xv = 3x.$

Using the integrating factor $e^{-\int 6x\,dx} = e^{-3x^2},$ we have

$$ve^{-3x^2} = \int 3xe^{-3x^2}dx = -\frac{1}{2}e^{-3x^2} + C \quad \text{or} \quad \frac{1}{y^3} = -\frac{1}{2} + Ce^{3x^2}.$$

10. Solve $\dfrac{dy}{dx} + \dfrac{1}{3}y = \dfrac{1}{3}(1-2x)y^4$ or $y^{-4}\dfrac{dy}{dx} + \dfrac{1}{3}y^{-3} = \dfrac{1}{3}(1-2x).$

The transformation $y^{-3} = v,\ -3y^{-4}\dfrac{dy}{dx} = \dfrac{dv}{dx}$ reduces the equation to $\dfrac{dv}{dx} - v = 2x - 1$

for which e^{-x} is an integrating factor. Then, integrating by parts,

$$ve^{-x} = \int(2x-1)e^{-x}dx = -2xe^{-x} - e^{-x} + C \quad \text{or} \quad \frac{1}{y^3} = -1 - 2x + Ce^{x}.$$

11. Solve $\dfrac{dy}{dx} + y = y^2(\cos x - \sin x)$ or $y^{-2}\dfrac{dy}{dx} + y^{-1} = \cos x - \sin x.$

The transformation $y^{-1} = v,\ -y^{-2}\dfrac{dy}{dx} = \dfrac{dv}{dx}$ reduces the equation to $\dfrac{dv}{dx} - v = \sin x - \cos x$

for which e^{-x} is an integrating factor. Then

$$ve^{-x} = \int(\sin x - \cos x)e^{-x}dx = -e^{-x}\sin x + C \quad \text{or} \quad \frac{1}{y} = -\sin x + Ce^{x}.$$

12. Solve $x\,dy - \{y + xy^3(1 + \ln x)\}dx = 0$ or $y^{-3}\dfrac{dy}{dx} - \dfrac{1}{x}y^{-2} = 1 + \ln x.$

The transformation $y^{-2} = v,\ -2y^{-3}\dfrac{dy}{dx} = \dfrac{dv}{dx}$ reduces the equation to $\dfrac{dv}{dx} + \dfrac{2}{x}v = -2(1+\ln x)$

for which $e^{\int 2\,dx/x} = x^2$ is an integrating factor. Then

$$vx^2 = -2\int(x^2 + x^2\ln x)dx = -\frac{4}{9}x^3 - \frac{2}{3}x^3\ln x + C \quad \text{or} \quad \frac{x^2}{y^2} = -\frac{2}{3}x^3(\frac{2}{3} + \ln x) + C.$$

MISCELLANEOUS SUBSTITUTIONS.

13. An equation of the form $f'(y)\dfrac{dy}{dx} + f(y)P(x) = Q(x)$ is a linear equation of the first order $\dfrac{dv}{dx} + vP(x) = Q(x)$ in the new variable $v = f(y)$. (Note that the Bernoulli equation $y^{-n}\dfrac{dy}{dx} + y^{-n+1}P(x) = Q(x)$ or $(-n+1)y^{-n}\dfrac{dy}{dx} + y^{-n+1}(-n+1)P(x) = (-n+1)Q(x)$ is an example.)

Solve $\dfrac{dy}{dx} + 1 = 4e^{-y}\sin x$ or $e^y\dfrac{dy}{dx} + e^y = 4\sin x$.

In the new variable $v = f(y) = e^y$, the equation becomes $\dfrac{dv}{dx} + v = 4\sin x$ for which e^x is an integrating factor. Then

$$ve^x = 4\int e^x \sin x\, dx = 2e^x(\sin x - \cos x) + C \quad\text{or}\quad e^y = 2(\sin x - \cos x) + Ce^{-x}.$$

14. Solve $\sin y\dfrac{dy}{dx} = \cos x(2\cos y - \sin^2 x)$ or $-\sin y\dfrac{dy}{dx} + \cos y(2\cos x) = \sin^2 x\cos x$.

In the new variable $v = \cos y$, the equation becomes $\dfrac{dv}{dx} + 2v\cos x = \sin^2 x\cos x$ for which $e^{2\int\cos x\,dx} = e^{2\sin x}$ is an integrating factor. Then

$$ve^{2\sin x} = \int e^{2\sin x}\sin^2 x\cos x\,dx = \tfrac{1}{2}e^{2\sin x}\sin^2 x - \tfrac{1}{2}e^{2\sin x}\sin x + \tfrac{1}{4}e^{2\sin x} + C$$

or

$$\cos y = \tfrac{1}{2}\sin^2 x - \tfrac{1}{2}\sin x + \tfrac{1}{4} + Ce^{-2\sin x}.$$

15. Solve $\sin y\dfrac{dy}{dx} = \cos y(1 - x\cos y)$ or $\dfrac{\sin y}{\cos^2 y}\dfrac{dy}{dx} - \dfrac{1}{\cos y} = -x$.

Since $\dfrac{d}{dy}\left(\dfrac{1}{\cos y}\right) = \dfrac{\sin y}{\cos^2 y}$, we take $v = \dfrac{1}{\cos y}$ and obtain the equation $\dfrac{dv}{dx} - v = -x$.

Using the integrating factor e^{-x}, we obtain

$$ve^{-x} = \int -xe^{-x}\,dx = xe^{-x} + e^{-x} + C \quad\text{or}\quad v = \frac{1}{\cos y} = \sec y = x + 1 + Ce^x.$$

16. Solve $x\dfrac{dy}{dx} - y + 3x^3y - x^2 = 0$ or $x\,dy - y\,dx + 3x^3y\,dx - x^2\,dx = 0$.

Here $(x\,dy - y\,dx)$ suggest the transformation $\dfrac{y}{x} = v$.

Then $\dfrac{x\,dy - y\,dx}{x^2} + 3x^2\dfrac{y}{x}dx - dx = 0$ is reduced to $\dfrac{dv}{dx} + 3x^2v = 1$ for which e^{x^3} is an integrating factor.

Thus $ve^{x^3} = \int e^{x^3}dx + C$ or $y = xe^{-x^3}\int e^{x^3}dx + Cxe^{-x^3}$.

The indefinite integral here cannot be evaluated in terms of elementary functions.

17. Solve $(4r^2 s - 6)dr + r^3 ds = 0$ or $(r\,ds + s\,dr) + 3s\,dr = \dfrac{6}{r^2}dr.$

The first term suggests the substitution $rs = t$ which reduces the equation to

$$dt + 3\frac{t}{r}dr = \frac{6}{r^2}dr \qquad \text{or} \qquad \frac{dt}{dr} + \frac{3}{r}t = \frac{6}{r^2}.$$

Then r^3 is an integrating factor and the solution is

$$tr^3 = r^4 s = 3r^2 + C \qquad \text{or} \qquad s = \frac{3}{r^2} + \frac{C}{r^4}.$$

18. Solve $x\sin\theta\,d\theta + (x^3 - 2x^2\cos\theta + \cos\theta)dx = 0$ or $-\dfrac{x\sin\theta\,d\theta + \cos\theta\,dx}{x^2} + 2\cos\theta\,dx = x\,dx.$

The substitution $xy = \cos\theta$, $dy = -\dfrac{x\sin\theta\,d\theta + \cos\theta\,dx}{x^2}$ reduces the equation to

$$dy + 2xy\,dx = x\,dx \qquad \text{or} \qquad \frac{dy}{dx} + 2xy = x.$$

An integrating factor is e^{x^2} and the solution is

$$ye^{x^2} = \frac{\cos\theta}{x}e^{x^2} = \int e^{x^2}x\,dx = \frac{1}{2}e^{x^2} + K \qquad \text{or} \qquad 2\cos\theta = x + Cxe^{-x^2}.$$

SUPPLEMENTARY PROBLEMS

19. From the following equations, select those which are linear, state the dependent variable, and solve.

a) $dy/dx + y = 2 + 2x$

b) $d\rho/d\theta + 3\rho = 2$

c) $dy/dx - y = xy^2$

d) $x\,dy - 2y\,dx = (x-2)e^x dx$

e) $di/dt - 6i = 10\sin 2t$

f) $dy/dx + y = y^2 e^x$

g) $y\,dx + (xy + x - 3y)dy = 0$

h) $(2s - e^{2t})ds = 2(se^{2t} - \cos 2t)dt$

i) $x\,dy + y\,dx = x^3 y^6 dx$

j) $dr + (2r\cot\theta + \sin 2\theta)d\theta = 0$

k) $y(1+y^2)dx = 2(1-2xy^2)dy$

l) $yy' - xy^2 + x = 0$

m) $x\,dy - y\,dx = x\sqrt{x^2 - y^2}\,dy$

n) $\phi_1(t)\,dx/dt + x\phi_2(t) = 1$

o) $2\,dx/dy - x/y + x^3\cos y = 0$

p) $xy' = y(1 - x\tan x) + x^2\cos x$

q) $(2+y^2)dx - (xy + 2y + y^3)dy = 0$

r) $(1+y^2)dx = (\arctan y - x)dy$

s) $(2xy^5 - y)dx + 2x\,dy = 0$

t) $(1+\sin y)dx = [2y\cos y - x(\sec y + \tan y)]dy$

Ans.

a) y; I.F., e^x; $y = 2x + Ce^{-x}$

b) ρ; I.F., $e^{3\theta}$; $3\rho = 2 + Ce^{-3\theta}$

d) y; I.F., $1/x^2$; $y = e^x + Cx^2$

e) i; I.F., e^{-6t}; $i = -\frac{1}{2}(3\sin 2t + \cos 2t) + Ce^{6t}$

g) x; I.F., ye^y; $xy = 3(y-1) + Ce^{-y}$

j) r; I.F., $\sin^2\theta$; $2r\sin^2\theta + \sin^4\theta = C$

k) x; I.F., $(1+y^2)^2$; $(1+y^2)^2 x = 2\ln y + y^2 + C$

n) x; I.F., $e^{\int \phi_2(t)\,dt/\phi_1(t)}$; $x\,e^{\int \phi_2(t)dt/\phi_1(t)} = \int \frac{1}{\phi_1(t)}\, e^{\int \phi_2(t)dt/\phi_1(t)}\,dt + C$

p) y; I.F., $\frac{1}{x\cos x}$; $y = x^2\cos x + Cx\cos x$

q) x; I.F., $1/\sqrt{2+y^2}$; $x = 2 + y^2 + C\sqrt{2+y^2}$

r) x; I.F., $e^{\arctan y}$; $x = \arctan y - 1 + Ce^{-\arctan y}$

t) x; I.F., $\sec y + \tan y$; $x(\sec y + \tan y) = y^2 + C$

20. From the remaining equations in Problem 19, solve those of the Bernoulli type.

Ans. c) $y^{-1} = v$; $1/y = 1 - x + Ce^{-x}$ *l*) $y^2 = v$; $y^2 = 1 + Ce^{x^2}$

f) $y^{-1} = v$; $(C+x)ye^x + 1 = 0$ *o*) $x^{-2} = v$; $x^{-2}y = \cos y + y\sin y + C$

i) $y^{-5} = v$; $2/y^5 = Cx^5 + 5x^3$ *s*) $y^{-4} = v$; $3x^2 = (4x^3 + C)y^4$

21. Solve the remaining equations, *h*) and *m*), of Problem 19.

Ans. *h*) $s^2 - se^{2t} + \sin 2t = C$ *m*) $y = x\sin(y + C)$

22. Solve:

a) $xy' = 2y + x^3e^x$ subject to $y = 0$ when $x = 1$.   *Ans.* $y = x^2(e^x - e)$

b) $L\frac{di}{dt} + Ri = E\sin 2t$, where L, R, E are constants, subject to the condition $i = 0$ when $t = 0$.

Ans. $i = \frac{E}{R^2 + 4L^2}(R\sin 2t - 2L\cos 2t + 2Le^{-Rt/L})$

23. Solve:

a) $x^2\cos y\,\frac{dy}{dx} = 2x\sin y - 1$, using $\sin y = z$.   *Ans.* $3x\sin y = Cx^3 + 1$

b) $4x^2yy' = 3x(3y^2 + 2) + 2(3y^2 + 2)^3$, using $3y^2 + 2 = z$.   *Ans.* $4x^9 = (C - 3x^8)(3y^2 + 2)^2$

c) $(xy^3 - y^3 - x^2e^x)dx + 3xy^2dy = 0$, using $y^3 = vx$.   *Ans.* $2y^3e^x = xe^{2x} + Cx$

d) $dy/dx + x(x+y) = x^3(x+y)^3 - 1$.   *Ans.* $1/(x+y)^2 = x^2 + 1 + Ce^{x^2}$

e) $(y + e^y - e^{-x})dx + (1 + e^y)dy = 0$.   *Ans.* $y + e^y = (x + C)e^{-x}$

CHAPTER 7

Geometric Applications

IN CHAPTER 1 it was shown how the differential equation

1) $$f(x,y,y') = 0$$

of a family of curves

2) $$g(x,y,C) = 0$$

could be obtained. The differential equation expresses analytically a certain property common to every curve of the family.

Conversely, if a property whose analytic representation involves the derivative is given, the solution of the resulting differential equation represents a one parameter family of curves, all possessing the given property. Each curve of the family is called an *integral curve* of 1) and particular integral curves may be singled out by giving additional properties, for example, a point through which the curve passes.

For convenience, the following properties of curves which involve the derivative, are listed.

RECTANGULAR COORDINATES. Let (x,y) be a general point of a curve $F(x,y) = 0$.

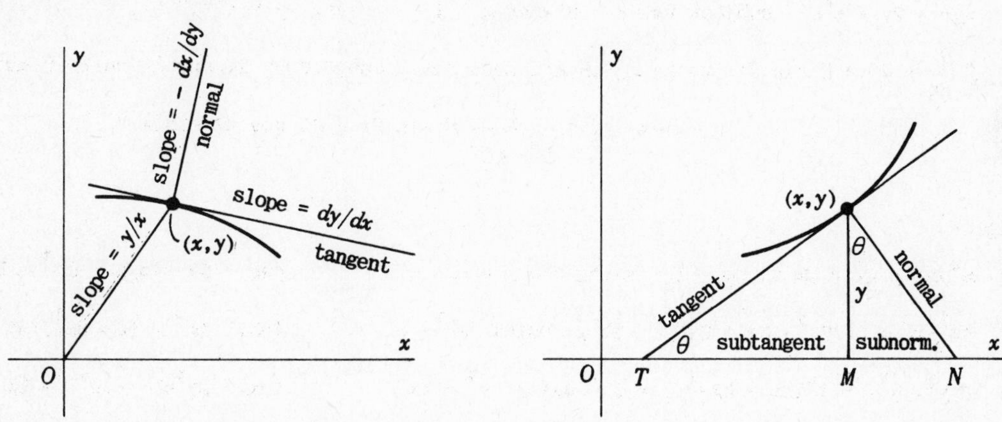

a) $\dfrac{dy}{dx}$ is the slope of the tangent to the curve at (x,y).

b) $-\dfrac{dx}{dy}$ is the slope of the normal to the curve at (x,y).

c) $Y-y = \dfrac{dy}{dx}(X-x)$ is the equation of the tangent at (x,y), where (X,Y) are the coordinates of any point on it.

d) $Y-y = -\dfrac{dx}{dy}(X-x)$ is the equation of the normal at (x,y), where (X,Y) are the coordinates of any point on it.

e) $x-y\dfrac{dx}{dy}$ and $y-x\dfrac{dy}{dx}$ are the x- and y-intercepts of the tangent.

f) $x+y\frac{dy}{dx}$ and $y+x\frac{dx}{dy}$ are the x- and y-intercepts of the normal.

g) $y\sqrt{1+(\frac{dx}{dy})^2}$ and $x\sqrt{1+(\frac{dy}{dx})^2}$ are the lengths of the tangent between (x,y) and the x- and y-axes.

h) $y\sqrt{1+(\frac{dy}{dx})^2}$ and $x\sqrt{1+(\frac{dx}{dy})^2}$ are the lengths of the normal between (x,y) and the x- and y-axes.

i) $y\frac{dx}{dy}$ and $y\frac{dy}{dx}$ are the lengths of the subtangent and subnormal.

j) $ds = \sqrt{(dx)^2+(dy)^2} = dx\sqrt{1+(\frac{dy}{dx})^2} = dy\sqrt{1+(\frac{dx}{dy})^2}$ is an element of length of arc.

k) $y\,dx$ or $x\,dy$ is an element of area.

POLAR COORDINATES. Let (ρ,θ) be a general point on a curve $\rho = f(\theta)$.

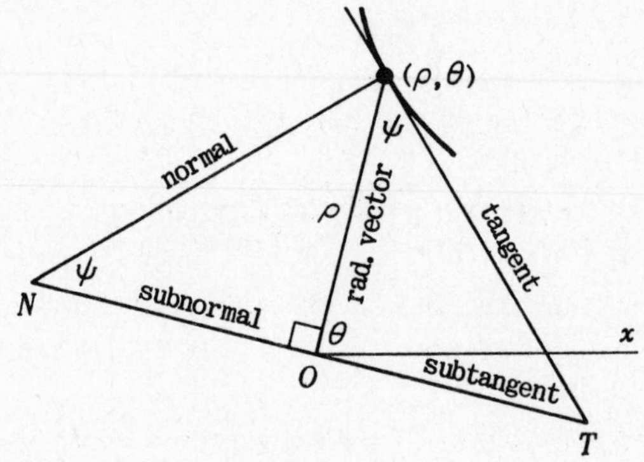

l) $\tan\psi = \rho\frac{d\theta}{d\rho}$, where ψ is the angle between the radius vector and the part of the tangent drawn toward the initial line.

m) $\rho\tan\psi = \rho^2\frac{d\theta}{d\rho}$ is the length of the polar subtangent.

n) $\rho\cot\psi = \frac{d\rho}{d\theta}$ is the length of the polar subnormal.

o) $\rho\sin\psi = \rho^2\frac{d\theta}{ds}$ is the length of the perpendicular from the pole to the tangent.

p) $ds = \sqrt{(d\rho)^2+\rho^2(d\theta)^2} = d\rho\sqrt{1+\rho^2(\frac{d\theta}{d\rho})^2} = d\theta\sqrt{(\frac{d\rho}{d\theta})^2+\rho^2}$ is an element of length of arc.

q) $\frac{1}{2}\rho^2\,d\theta$ is an element of area.

TRAJECTORIES. Any curve which cuts every member of a given family of curves at the constant angle ω is called an *ω-trajectory* of the family. A 90° trajectory of the family is commonly called an *orthogonal trajectory* of the family. For example, in Figure (*a*) below, the circles through the origin with centers on the y-axis are orthogonal trajectories of the family of circles through the origin with centers on the x-axis.

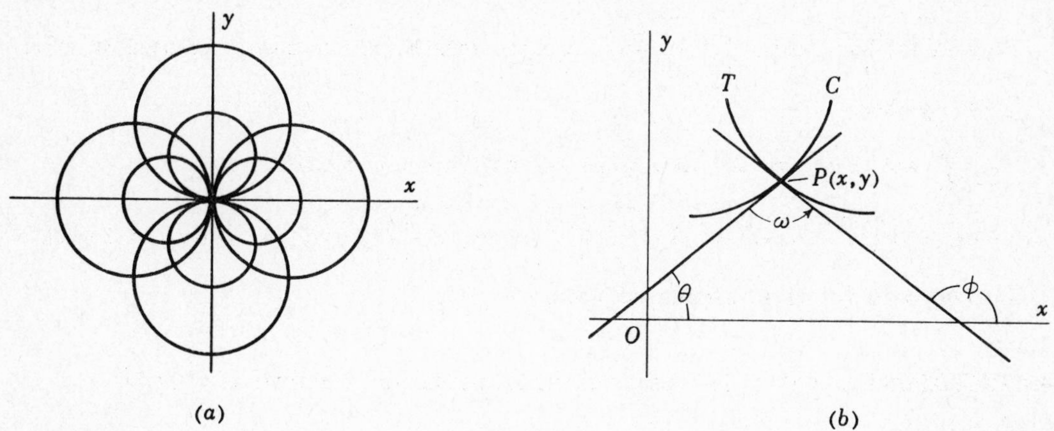

(*a*) (*b*)

In finding such trajectories, we shall use:

A) The integral curves of the differential equation

3) $$f(x,y,\ \frac{y' - \tan\omega}{1 + y'\tan\omega}) = 0$$

are the ω-trajectories of the family of integral curves of

1) $$f(x,y,y') = 0.$$

To prove this, consider the integral curve C of 1) and an ω-trajectory which intersect at $P(x,y)$, as shown in Figure (*b*) above. At each point of C for which 1) defines a value of y', we associate a triad of numbers $(x,y;y')$, the first two being the coordinates of the point and the third being the corresponding value of y' given by 1). Similarly, with each point of T for which there is a tangent line, we associate a triad $(x,y;y')$, the first two being the coordinates of the point and the third the slope of the tangent. To avoid confusion, since we are to consider the triads associated with P as a point on C and as a point on T, let us write the latter (associated with P on T) as $(\bar{x},\bar{y};\bar{y}')$. Now, from the figure, $x=\bar{x}$, $y=\bar{y}$ at P while $y'=\tan\theta$ and $\bar{y}'=\tan\phi$ are related by

$$y' = \tan\theta = \tan(\phi-\omega) = \frac{\tan\phi - \tan\omega}{1+\tan\phi\tan\omega} = \frac{\bar{y}' - \tan\omega}{1+\bar{y}'\tan\omega}.$$

Thus, at P (a general point in the plane) on an ω-trajectory, the relation

$$f(x,y,y') = f(\bar{x},\bar{y},\ \frac{\bar{y}' - \tan\omega}{1+\bar{y}'\tan\omega}) = 0$$

holds, or, dropping the dashes, $f(x,y,\ \dfrac{y' - \tan\omega}{1+y'\tan\omega}) = 0.$

B) The integral curves of the differential equation

4) $$f(x,y,-1/y') = 0$$

are the orthogonal trajectories of the family of integral curves of 1).

C) In polar coordinates, the integral curves of the differential equation

5) $$f(\rho, \theta, -\rho^2 \frac{d\theta}{d\rho}) = 0$$

are the orthogonal trajectories of the integral curves of

6) $$f(\rho, \theta, \frac{d\rho}{d\theta}) = 0.$$

SOLVED PROBLEMS

1. At each point (x,y) of a curve the intercept of the tangent on the y-axis is equal to $2xy^2$. Find the curve.

Using e), the differential equation of the curve is

$$y - x\frac{dy}{dx} = 2xy^2 \quad \text{or} \quad \frac{y\,dx - x\,dy}{y^2} = 2x\,dx.$$

Integrating, $\frac{x}{y} = x^2 + C$ or $x - x^2y = Cy$.

The differential equation may also be obtained directly from the adjoining figure as $\frac{dy}{dx} = \frac{y - 2xy^2}{x}$.

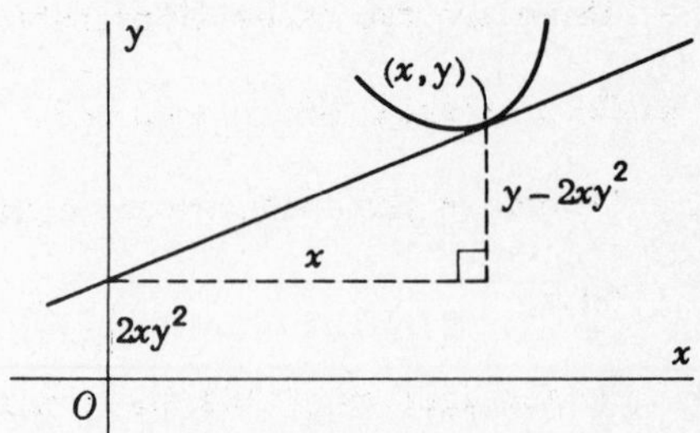

2. At each point (x,y) of a curve the subtangent is proportional to the square of the abscissa. Find the curve if it also passes through the point $(1,e)$.

Using i), the differential equation is $y\frac{dx}{dy} = kx^2$ or $\frac{dx}{x^2} = k\frac{dy}{y}$, where k is the proportionality factor.

Integrating, $k \ln y = -\frac{1}{x} + C$. When $x = 1$, $y = e$: $k = -1 + C$ and $C = k + 1$.

The required curve has equation $k \ln y = -\frac{1}{x} + k + 1$.

3. Find the family of curves for which the length of the part of the tangent between the point of contact (x,y) and the y-axis is equal to the y-intercept of the tangent.

From g) and e), we have $x\sqrt{1 + (\frac{dy}{dx})^2} = y - x\frac{dy}{dx}$ or A) $x^2 = y^2 - 2xy\frac{dy}{dx}$.

The transformation $y = vx$ reduces A) to

$$(1 + v^2)dx + 2vx\,dv = 0 \quad \text{or} \quad \frac{dx}{x} + \frac{2v\,dv}{1 + v^2} = 0.$$

Integrating, $\ln x + \ln(1 + v^2) = \ln C$.

Then $x(1 + \frac{y^2}{x^2}) = C$ or $x^2 + y^2 = Cx$ is the equation of the family.

4. Through any point (x,y) of a curve which passes through the origin, lines are drawn parallel to the coordinate axes. Find the curve given that it divides the rectangle formed by the two lines and the axes into two areas, one of which is three times the other.

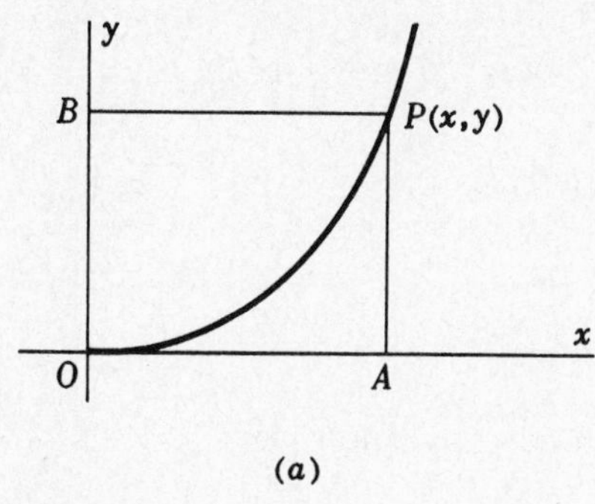

(a)

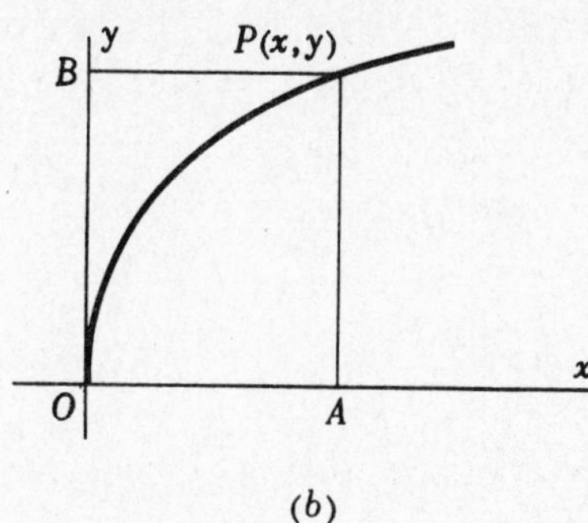

(b)

There are two cases illustrated in the figures.

a) Here 3(area OAP) = area OPB. Then $3\int_0^x y\,dx = xy - \int_0^x y\,dx$ or $4\int_0^x y\,dx = xy$.

To obtain the differential equation, we differentiate with respect to x.

Thus, $$4y = y + x\frac{dy}{dx} \quad \text{or} \quad \frac{dy}{dx} = \frac{3y}{x}.$$

An integration yields the family of curves $y = Cx^3$.

b) Here area OAP = 3(area OPB) and $4\int_0^x y\,dx = 3xy$.

The differential equation is $\dfrac{dy}{dx} = \dfrac{y}{3x}$, and the family of curves has equation $y^3 = Cx$.

Since the differential equation in each case was obtained by a differentiation, extraneous solutions may have been introduced. It is necessary therefore to compute the areas as a check. In each of the above cases, the curves found satisfy the conditions. However, see Problem 5.

5. The areas bounded by the x-axis, a fixed ordinate $x = a$, a variable ordinate, and the part of a curve intercepted by the ordinates is revolved about the x-axis. Find the curve if the volume generated is proportional to a) the sum of the two ordinates, b) the difference of the two ordinates.

a) Let A be the length of the fixed ordinate. The differential equation obtained by differentiating 1) $\pi\int_a^x y^2 dx = k(y + A)$ is $\pi y^2 = k\dfrac{dy}{dx}$. Integrating, we have 2) $y(C - \pi x) = k$.

When the value of y given by 2) is used in computing the left member of 1), we find

$$3)\quad \pi\int_a^x \frac{k^2\,dx}{(C - \pi x)^2} = \frac{k^2}{C - \pi x} - \frac{k^2}{C - \pi a} = k(y - A).$$

Thus, the solution is extraneous and no curve exists having the property a).

b) Repeating the above procedure with 1') $\pi\int_a^x y^2\,dx = k(y - A)$, we obtain the differential equation $\pi y^2 = k\dfrac{dy}{dx}$ whose solution is 2') $y(C - \pi x) = k$.

It is seen from 3) that this equation satisfies 1'). Thus, the family of curves 2') has the required property.

6. Find the curve such that at any point on it the angle between the radius vector and the tangent is equal to one-third the angle of inclination of the tangent.

Let θ denote the angle of inclination of the radius vector, τ the angle of inclination of the tangent, and ψ the angle between the radius vector and the tangent.

Since $\psi = \tau/3 = (\psi + \theta)/3$, then $\psi = \frac{1}{2}\theta$ and $\tan\psi = \tan\frac{1}{2}\theta$.

Using l), $\tan\psi = \rho\dfrac{d\theta}{d\rho} = \tan\frac{1}{2}\theta \cdot$ so that $\dfrac{d\rho}{\rho} = \cot\frac{1}{2}\theta\, d\theta$.

Integrating, $\ln\rho = 2\ln\sin\frac{1}{2}\theta + \ln C_1$ or $\rho = C_1\sin^2\frac{1}{2}\theta = C(1 - \cos\theta)$.

7. The area of the sector formed by an arc of a curve and the radii vectors to the end points is one-half the length of the arc. Find the curve.

Let the radii vectors be given by $\theta = \theta_1$ and $\theta = \theta$.

Using q) and p), $\frac{1}{2}\int_{\theta_1}^{\theta}\rho^2\,d\theta = \frac{1}{2}\int_{\theta_1}^{\theta}\sqrt{(\frac{d\rho}{d\theta})^2 + \rho^2}\;d\theta$.

Differentiating with respect to θ, we obtain the differential equation

$$\rho^2 = \sqrt{(\frac{d\rho}{d\theta})^2 + \rho^2} \quad \text{or} \quad 1)\;\; d\rho = \pm\rho\sqrt{\rho^2 - 1}\;d\theta.$$

If $\rho^2 = 1$, 1) reduces to $d\rho = 0$. It is easily verified that $\rho = 1$ satisfies the condition of the problem.

If $\rho^2 \neq 1$, we write the equation in the form $\dfrac{d\rho}{\rho\sqrt{\rho^2 - 1}} = \pm\, d\theta$ and obtain the solution $\rho = \sec(C \pm \theta)$. Thus, the conditions are satisfied by the circle $\rho = 1$ and the family of curves $\rho = \sec(C + \theta)$. Note that the families $\rho = \sec(C + \theta)$ and $\rho = \sec(C - \theta)$ are the same.

8. Find the curve for which the portion of the tangent between the point of contact and the foot of the perpendicular through the pole to the tangent is one-third the radius vector to the point of contact.

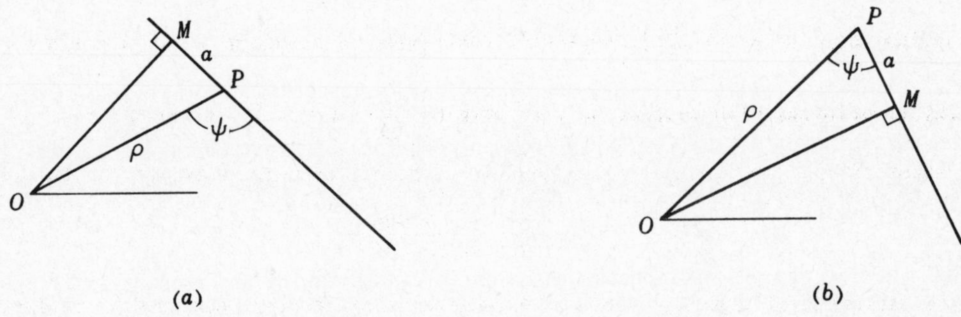

In Figure (a): $\rho = 3a = 3\rho\cos(\pi - \psi) = -3\rho\cos\psi$, $\cos\psi = -1/3$, and $\tan\psi = -2\sqrt{2}$.
In Figure (b): $\rho = 3a = 3\rho\cos\psi$ and $\tan\psi = 2\sqrt{2}$.

Using l) and combining the two cases, $\tan\psi = \rho\dfrac{d\theta}{d\rho} = \pm 2\sqrt{2}$ or $\dfrac{d\rho}{\rho} = \pm\dfrac{d\theta}{2\sqrt{2}}$.

The required curves are the families $\rho = Ce^{\theta/2\sqrt{2}}$ and $\rho = Ce^{-\theta/2\sqrt{2}}$.

9. Find the orthogonal trajectories of the hyperbolas $xy = C$.

The differential equation of the given family is $x\frac{dy}{dx} + y = 0$, obtained by differentiating $xy = C$. The differential equation of the orthogonal trajectories, obtained by replacing $\frac{dy}{dx}$ by $-\frac{dx}{dy}$, is $-x\frac{dx}{dy} + y = 0$ or $y\,dy - x\,dx = 0$.

Integrating, the orthogonal trajectories are the family of curves (hyperbolas) $y^2 - x^2 = C$.

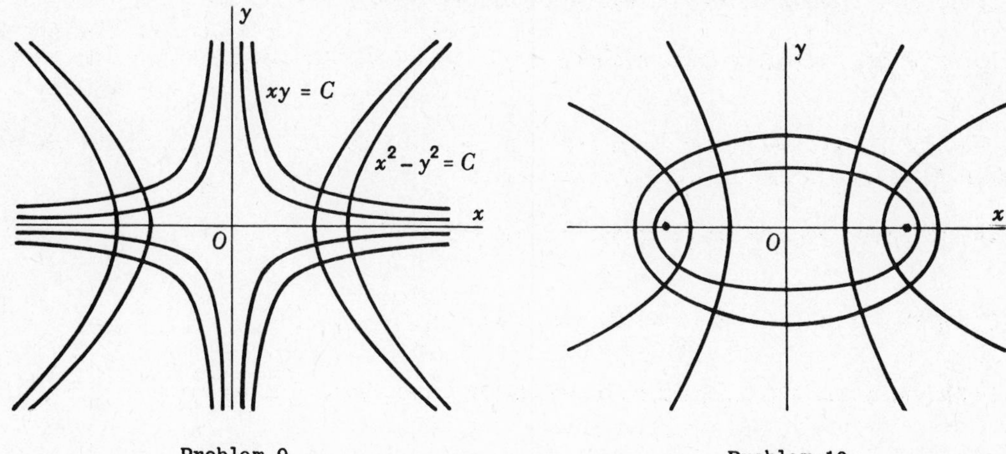

Problem 9

Problem 10

10. Show that the family of confocal conics $\frac{x^2}{C} + \frac{y^2}{C-\lambda} = 1$, where C is an arbitrary constant, is self-orthogonal.

Differentiating the equation of the family with respect to x yields $\frac{x}{C} + \frac{yp}{C-\lambda} = 0$, where $p = \frac{dy}{dx}$. Solving this for C, we find $C = \frac{\lambda x}{x+yp}$ so that $C - \lambda = \frac{-\lambda py}{x+yp}$. When these replacements are made in the equation of the family, the differential equation of the family is found to be

$$(x+yp)(px-y) - \lambda p = 0.$$

Since this equation is unchanged when p is replaced by $-1/p$, it is also the differential equation of the orthogonal trajectories of the given family.

11. Determine the orthogonal trajectories of the family of cardiods $\rho = C(1 + \sin\theta)$.

Differentiating with respect to θ to obtain $\frac{d\rho}{d\theta} = C\cos\theta$, solving for $C = \frac{1}{\cos\theta}\frac{d\rho}{d\theta}$, and substituting for C in the given equation, the differential equation of the given family is

$$\frac{d\rho}{d\theta} = \frac{\rho\cos\theta}{1+\sin\theta}.$$

The differential equation of the orthogonal trajectories, obtained by replacing $\frac{d\rho}{d\theta}$ by $-\rho^2\frac{d\theta}{d\rho}$ is

$$-\frac{d\theta}{d\rho} = \frac{\cos\theta}{\rho(1+\sin\theta)} \qquad \text{or} \qquad \frac{d\rho}{\rho} + (\sec\theta + \tan\theta)d\theta = 0.$$

Then $\ln\rho + \ln(\sec\theta + \tan\theta) - \ln\cos\theta = \ln C$ or $\rho = \frac{C\cos\theta}{\sec\theta + \tan\theta} = C(1 - \sin\theta)$.

12. Determine the 45° trajectories of the family of concentric circles $x^2 + y^2 = C$.

The differential equation of the family of circles is $x + yy' = 0$.

The differential equation of the 45° trajectories, obtained by replacing y' in the above equation by $\dfrac{y' - \tan 45^\circ}{1 + y'\tan 45^\circ} = \dfrac{y' - 1}{1 + y'}$, is $x + y\dfrac{y' - 1}{1 + y'} = 0$ or $(x + y)dy + (x - y)dx = 0$.

Using the transformation $y = vx$, this equation is reduced to

$$(v^2 + 1)dx + x(v + 1)dv = 0 \qquad \text{or} \qquad \frac{dx}{x} + \frac{v + 1}{v^2 + 1}dv = 0.$$

Integrating, $\ln x + \frac{1}{2}\ln(v^2 + 1) + \arctan v = \ln K_1$, $\ln x^2(1 + v^2) = \ln K - 2 \arctan v$, and $x^2 + y^2 = Ke^{-2 \arctan y/x}$.

In polar coordinates, the equation becomes $\rho^2 = Ke^{-2\theta}$ or $\rho e^{\theta} = C$.

SUPPLEMENTARY PROBLEMS

13. Find the equation of the curve for which

a) the normal at any point (x,y) passes through the origin. *Ans.* $x^2 + y^2 = C$

b) the slope of the tangent at any point (x,y) is $\frac{1}{2}$ the slope of the line from the origin to the point. *Ans.* $y^2 = Cx$

c) the normal at any point (x,y) and the line joining the origin to that point form an isosceles triangle having the x-axis as base. *Ans.* $y^2 - x^2 = C$

d) the part of the normal drawn at point (x,y) between this point and the x-axis is bisected by the y-axis. *Ans.* $y^2 + 2x^2 = C$

e) the perpendicular from the origin to a tangent line of the curve is equal to the abscissa of the point of contact (x,y). *Ans.* $x^2 + y^2 = Cx$

f) the arc length from the origin to the variable point (x,y) is equal to twice the square root of the abscissa of the point. *Ans.* $y = \pm(\arcsin\sqrt{x} + \sqrt{x - x^2}\,) + C$

g) the polar subnormal is twice the sine of the vectorial angle. *Ans.* $\rho = C - 2\cos\theta$

h) the angle between the radius vector and the tangent is $\frac{1}{2}$ the vectorial angle. *Ans.* $\rho = C(1 - \cos\theta)$

i) the polar subtangent is equal to the polar subnormal. *Ans.* $\rho = Ce^{\theta}$

14. Find the orthogonal trajectories of each of the following families of curves.

a) $x + 2y = C$ *Ans.* $y - 2x = K$

b) $xy = C$ $x^2 - y^2 = K$

c) $x^2 + 2y^2 = C$ $y = Kx^2$

d) $y = Ce^{-2x}$ $y^2 = x + K$

e) $y^2 = x^3/(C - x)$ $(x^2 + y^2)^2 = K(2x^2 + y^2)$

f) $y = x - 1 + Ce^{-x}$ *Ans.* $x = y - 1 + Ke^{-y}$

g) $y^2 = 2x^2(1 - Cx)$ $x^2 + 3y^2\ln(Ky) = 0$

h) $\rho = a\cos\theta$ $\rho = b\sin\theta$

i) $\rho = a(1 + \sin\theta)$ $\rho = b(1 - \sin\theta)$

j) $\rho = a(\sec\theta + \tan\theta)$ $\rho = be^{-\sin\theta}$

CHAPTER 8

Physical Applications

MANY OF THE APPLICATIONS of this and later chapters will be concerned with the motion of a body along a straight line. If the body moves with varying velocity v (that is, with accelerated motion) its acceleration, given by dv/dt, is due to one or more forces acting in the direction of motion or in the opposite direction. The net force on the mass is the (algebraic) sum of the several forces.

EXAMPLE 1. A boat is moving subject to a force of 20 pounds on its sail and a resisting force (lb) equal to 1/50 its velocity (ft/sec). If the direction of motion is taken as positive, the net force (lb) is $20 - v/50$.

EXAMPLE 2. To the free end of a spring of negligible mass, hanging vertically, a mass is attached and brought to rest. There are two forces acting on the mass – gravity acting downward and a restoring force, called the spring force, opposing gravity. The two forces, being opposite in direction, are equal in magnitude since the mass is at rest. Thus, the net force is zero.

Newton's Second Law of Motion states in part that the product of the mass and acceleration is proportional to the net force on the mass. When the system of units described below is used, the factor of proportionality is $k = 1$ and we have

$$\text{mass} \times \text{acceleration} = \text{net force.}$$

THE U. S. ENGINEERING SYSTEM is based on the fundamental units: the *pound of force* (the pound weight), the *foot of length*, and the *second of time*. The derived unit of *mass* is the *slug*, defined by

$$\text{mass in slugs} = \frac{\text{weight in pounds}}{g \text{ in ft/sec}^2}.$$

Hence,

$$\text{mass in slugs} \times \text{acceleration in ft/sec}^2 = \text{net force in pounds.}$$

The acceleration g of a freely falling body varies but slightly over the earth's surface. For convenience in computing, an approximate value $g = 32$ ft/sec^2 is used in the problems.

SOLVED PROBLEMS

1. If the population of a country doubles in 50 years, in how many years will it treble under the assumption that the rate of increase is proportional to the number of inhabitants?

Let y denote the population at time t years and y_0 the population at time $t = 0$. Then

1) $\dfrac{dy}{dt} = ky$ or $\dfrac{dy}{y} = k\,dt$, where k is the proportionality factor.

First Solution. Integrating 1), we have 2) $\ln y = kt + \ln C$ or $y = Ce^{kt}$.

At time $t = 0$, $y = y_0$ and, from 2), $C = y_0$. Thus, 3) $y = y_0 e^{kt}$.

At $t = 50$, $y = 2y_0$. From 3), $2y_0 = y_0 e^{50k}$ or $e^{50k} = 2$.

When $y = 3y_0$, 3) gives $3 = e^{kt}$. Then $3^{50} = e^{50kt} = (e^{50k})^t = 2^t$ and $t = 79$ years.

Second Solution. Integrating 1) between the limits $t = 0$, $y = y_0$ and $t = 50$, $y = 2y_0$,

$$\int_{y_0}^{2y_0} \frac{dy}{y} = k\int_0^{50} dt, \qquad \ln 2y_0 - \ln y_0 = 50k \quad \text{and} \quad 50k = \ln 2.$$

Integrating 1) between the limits $t = 0$, $y = y_0$ and $t = t$, $y = 3y_0$,

$$\int_{y_0}^{3y_0} \frac{dy}{y} = k\int_0^{t} dt, \quad \text{and} \quad \ln 3 = kt.$$

Then $50 \ln 3 = 50kt = t \ln 2$ and $t = \dfrac{50 \ln 3}{\ln 2} = 79$ years.

2. In a certain culture of bacteria the rate of increase is proportional to the number present. (*a*) If it is found that the number doubles in 4 hours, how many may be expected at the end of 12 hours? (*b*) If there are 10^4 at the end of 3 hours and $4\cdot10^4$ at the end of 5 hours, how many were there in the beginning?

Let x denote the number of bacteria at time t hours. Then

$$1) \quad \frac{dx}{dt} = kx \quad \text{or} \quad \frac{dx}{x} = k\,dt.$$

a) *First Solution.* Integrating 1), we have 2) $\ln x = kt + \ln C$ or $x = Ce^{kt}$.

Assuming that $x = x_0$ at time $t = 0$, $C = x_0$ and $x = x_0 e^{kt}$.

At time $t = 4$, $x = 2x_0$. Then $2x_0 = x_0 e^{4k}$ and $e^{4k} = 2$.

When $t = 12$, $x = x_0 e^{12k} = x_0(e^{4k})^3 = x_0(2^3) = 8x_0$, that is, there are 8 times the original number.

Second Solution. Integrating 1) between the limits $t = 0$, $x = x_0$ and $t = 4$, $x = 2x_0$,

$$\int_{x_0}^{2x_0} \frac{dx}{x} = k\int_0^{4} dt, \qquad \ln 2x_0 - \ln x_0 = 4k \quad \text{and} \quad 4k = \ln 2.$$

Integrating 1) between the limits $t = 0$, $x = x_0$ and $t = 12$, $x = x$,

$$\int_{x_0}^{x} \frac{dx}{x} = k\int_0^{12} dt, \quad \text{and} \quad \ln\frac{x}{x_0} = 12k = 3(4k) = 3\ln 2 = \ln 8.$$

Then $x = 8x_0$, as before.

b) *First Solution.* When $t = 3$, $x = 10^4$. Hence, from 2), $10^4 = Ce^{3k}$ and $C = \dfrac{10^4}{e^{3k}}$.

When $t = 5$, $x = 4\cdot10^4$. Hence, $4\cdot10^4 = Ce^{5k}$ and $C = \dfrac{4\cdot10^4}{e^{5k}}$.

Equating the values of C, $\dfrac{10^4}{e^{3k}} = \dfrac{4\cdot10^4}{e^{5k}}$. Then $e^{2k} = 4$ and $e^k = 2$.

Thus, the original number is $C = \dfrac{10^4}{e^{3k}} = \dfrac{10^4}{8}$ bacteria.

Second Solution. Integrating 1) between the limits $t=3$, $x=10^4$ and $t=5$, $x=4\cdot 10^4$,

$$\int_{10^4}^{4\cdot 10^4} \frac{dx}{x} = k\int_3^5 dt, \qquad \ln 4 = 2k \quad\text{and}\quad k = \ln 2.$$

Integrating 1) between the limits $t=0$, $x=x_0$ and $t=3$, $x=10^4$,

$$\int_{x_0}^{10^4} \frac{dx}{x} = k\int_0^3 dt, \qquad \ln\frac{10^4}{x_0} = 3k = 3\ln 2 = \ln 8 \quad\text{and}\quad x_0 = \frac{10^4}{8} \text{ as before.}$$

3. According to Newton's law of cooling, the rate at which a substance cools in moving air is proportional to the difference between the temperature of the substance and that of the air. If the temperature of the air is 30° and the substance cools from 100° to 70° in 15 minutes, find when the temperature will be 40°.

Let T be the temperature of the substance at time t minutes.

Then $$\frac{dT}{dt} = -k(T-30) \qquad\text{or}\qquad \frac{dT}{T-30} = -k\,dt.$$

(Note. The use of $-k$ here is optional. It will be found that k is positive, but if $+k$ is used it will be found that k is equally negative.)

Integrating between the limits $t=0$, $T=100$ and $t=15$, $T=70$,

$$\int_{100}^{70} \frac{dT}{T-30} = -k\int_0^{15} dt, \qquad \ln 40 - \ln 70 = -15k = \ln\frac{4}{7} \quad\text{and}\quad 15k = \ln\frac{7}{4} = 0.56.$$

Integrating between the limits $t=0$, $T=100$ and $t=t$, $T=40$,

$$\int_{100}^{40} \frac{dT}{T-30} = -k\int_0^{t} dt, \qquad \ln 10 - \ln 70 = -kt, \quad 15kt = 15\ln 7, \quad t = \frac{15\ln 7}{0.56} = 52 \text{ min.}$$

4. A certain chemical dissolves in water at a rate proportional to the product of the amount undissolved and the difference between the concentration in a saturated solution and the concentration in the actual solution. In 100 grams of a saturated solution it is known that 50 grams of the substance are dissolved. If when 30 grams of the chemical are agitated with 100 grams of water, 10 grams are dissolved in 2 hours, how much will be dissolved in 5 hours?

Let x denote the number of grams of the chemical undissolved after t hours. At this time the concentration of the actual solution is $\dfrac{30-x}{100}$ and that of a saturated solution is $\dfrac{50}{100}$. Then

$$\frac{dx}{dt} = kx\left(\frac{50}{100} - \frac{30-x}{100}\right) = kx\,\frac{x+20}{100} \qquad\text{or}\qquad \frac{dx}{x} - \frac{dx}{x+20} = \frac{k}{5}\,dt.$$

Integrating between $t=0$, $x=30$ and $t=2$, $x=30-10=20$,

$$\int_{30}^{20} \frac{dx}{x} - \int_{30}^{20} \frac{dx}{x+20} = \frac{k}{5}\int_0^2 dt, \qquad\text{and}\quad k = \frac{5}{2}\ln\frac{5}{6} = -0.46.$$

Integrating between $t=0$, $x=30$ and $t=5$, $x=x$,

$$\int_{30}^{x} \frac{dx}{x} - \int_{30}^{x} \frac{dx}{x+20} = \frac{k}{5}\int_0^5 dt, \qquad \ln\frac{5x}{3(x+20)} = k = -0.46, \quad \frac{x}{x+20} = \frac{3}{5}e^{-0.46}$$

$= 0.38$, and $x = 12$. Thus, the amount dissolved after 5 hours is $30-12 = 18$ grams.

5. A 100 gallon tank is filled with brine containing 60 pounds of dissolved salt. Water runs into the tank at the rate of 2 gallons per minute and the mixture, kept uniform by stirring, runs out at the same rate. How much salt is in the tank after 1 hour?

Let s be the number of pounds of salt in the tank after t minutes, the concentration then being $s/100$ lb/gal. During the interval dt, $2\,dt$ gallons of water flows into the tank and $2\,dt$ gallons of brine containing $\frac{2s}{100}\,dt = \frac{s}{50}\,dt$ pounds of salt flows out.

Thus, the change ds in the amount of salt in the tank is $ds = -\frac{s}{50}\,dt$.

Integrating, $s = Ce^{-t/50}$. At $t = 0$, $s = 60$; hence, $C = 60$ and $s = 60e^{-t/50}$.

When $t = 60$ minutes, $s = 60e^{-6/5} = 60(.301) = 18$ pounds.

6. The air in a certain room 150'×50'×12' tested 0.2% CO_2. Fresh air containing 0.05% CO_2 was then admitted by ventilators at the rate 9000 ft^3/min. Find the percentage CO_2 after 20 minutes.

Let x denote the number of cubic feet of CO_2 in the room at time t, the concentration of CO_2 then being $x/90{,}000$. During the interval dt, the amount of CO_2 entering the room is $9{,}000(.0005)dt$ ft^3 and the amount leaving is $9{,}000\,\frac{x}{90{,}000}\,dt$ ft^3.

Hence, the change dx in the interval is $dx = 9{,}000(.0005 - \frac{x}{90{,}000})dt = -\frac{x-45}{10}\,dt$.

Integrating, $10\ln(x-45) = -t + \ln C_1$ and $x = 45 + Ce^{-t/10}$.

At $t = 0$, $x = .002(90{,}000) = 180$. Then $C = 180 - 45 = 135$ and $x = 45 + 135e^{-t/10}$.

When $t = 20$, $x = 45 + 135e^{-2} = 63$. The percentage CO_2 is then $\frac{63}{90{,}000} = .0007 = 0.07\%$.

7. Under certain conditions the constant quantity Q calories/second of heat flowing through a wall is given by

$$Q = -kA\frac{dT}{dx},$$

where k is the conductivity of the material, $A(\text{cm}^2)$ is the area of a face of the wall perpendicular to the direction of flow, and T is the temperature x(cm) from that face such that T decreases as x increases. Find the number of calories of heat per hour flowing through 1 square meter of the wall of a refrigerator room 125 cm thick for which $k = 0.0025$, if the temperature of the inner face is -5°C and that of the outer face is 75°C.

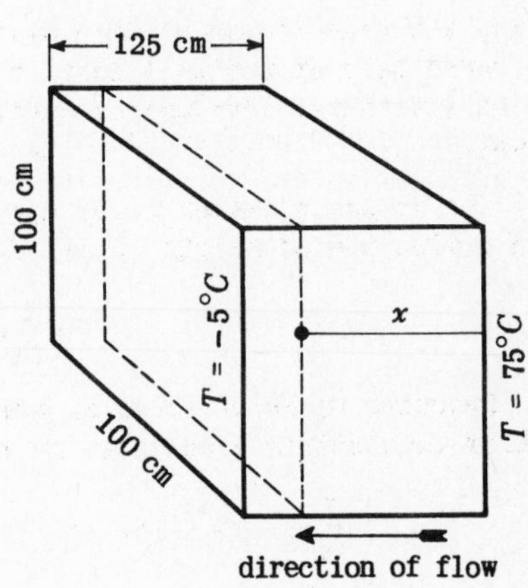

Let x denote the distance of a point within the wall from the outer face.

Integrating $dT = -\frac{Q}{kA}\,dx$ from $x = 0$, $T = 75$ to $x = 125$, $T = -5$,

$$\int_{75}^{-5} dT = -\frac{Q}{kA}\int_0^{125} dx, \quad 80 = \frac{Q}{kA}(125), \quad \text{and } Q = \frac{80kA}{125} = \frac{80(.0025)(100)^2}{125} = 16\,\frac{\text{cal}}{\text{sec}}.$$

Thus, the flow of heat per hour $= 3600Q = 57{,}600$ cal.

8. A steam pipe 20 cm in diameter is protected with a covering 6 cm thick for which k = 0.0003. (a) Find the heat loss per hour through a meter length of the pipe if the surface of the pipe is 200°C and that of the outer surface of the covering is 30°C. (b) Find the temperature at a distance $x > 10$ cm from the center of the pipe.

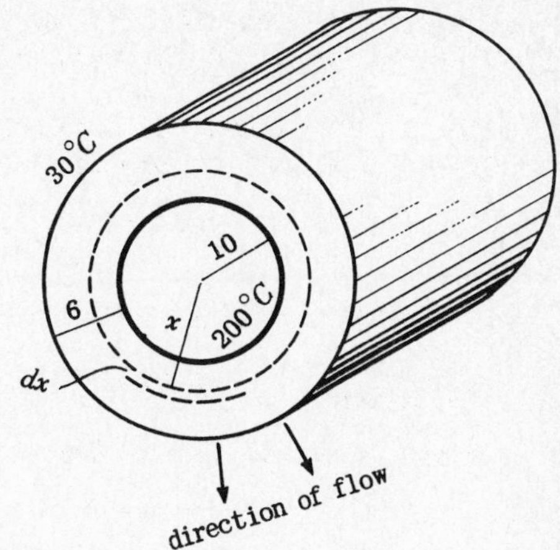

At a distance $x > 10$ cm from the center of the pipe, heat is flowing across a cylindrical shell of surface area $2\pi x$ cm^2 per cm of length of pipe. From Problem 7,

$$Q = -kA\frac{dT}{dx} = -2\pi kx\frac{dT}{dx} \quad \text{or} \quad 2\pi k\, dT = -Q\frac{dx}{x}.$$

a) Integrating between the limits $T = 30$, $x = 16$ and $T = 200$, $x = 10$,

$$2\pi k\int_{30}^{200} dT = -Q\int_{16}^{10}\frac{dx}{x}, \qquad 340\pi k = Q(\ln 16 - \ln 10) = Q\ln 1.6 \quad \text{and} \quad Q = \frac{340\pi k}{\ln 1.6}\ \text{cal/sec.}$$

Thus, the heat loss per hour through a meter length of pipe is $100(60)^2Q = 245{,}000$ cal.

b) Integrating $2\pi k\, dT = -\dfrac{340\pi k}{\ln 1.6}\dfrac{dx}{x}$ between the limits $T = 30$, $x = 16$ and $T = T$, $x = x$,

$$\int_{30}^{T} dT = -\frac{170}{\ln 1.6}\int_{16}^{x}\frac{dx}{x}, \qquad T - 30 = -\frac{170}{\ln 1.6}\ln\frac{x}{16} \quad \text{and} \quad T = (30 + \frac{170}{\ln 1.6}\ln\frac{16}{x})\,^\circ\text{C}.$$

Check. When $x = 10$, $T = 30 + \dfrac{170}{\ln 1.6}\ln 1.6 = 200$°C. When $x = 16$, $T = 30 + 0 = 30$°C.

9. Find the time required for a cylindrical tank of radius 8 ft and height 10 ft to empty through a round hole of radius 1 inch in the bottom of the tank, given that water will issue from such a hole with velocity approximately $v = 4.8\sqrt{h}$ ft/sec, h being the depth of the water in the tank.

The volume of water which runs out per second may be thought of as that of a cylinder 1 inch in radius and of height v. Hence, the volume which runs out in time dt sec is

$$\pi(\frac{1}{12})^2(4.8\sqrt{h})dt = \frac{\pi}{144}(4.8\sqrt{h})dt.$$

Denoting by dh the corresponding drop in the water level in the tank, the volume of water which runs out is also given by $64\pi dh$. Hence,

$$\frac{\pi}{144}(4.8\sqrt{h})\,dt = -64\pi\,dh \quad \text{or} \quad dt = -\frac{64(144)}{4.8}\frac{dh}{\sqrt{h}} = -1920\frac{dh}{\sqrt{h}}.$$

Integrating between $t = 0$, $h = 10$ and $t = t$, $h = 0$,

$$\int_0^t dt = -1920\int_{10}^{0}\frac{dh}{\sqrt{h}}, \qquad \text{and} \quad t = -3840\sqrt{h}\,\Big|_{10}^{0} = 3840\sqrt{10}\ \text{sec} = 3\ \text{hr}\ 22\ \text{min}.$$

10. A ship weighing 48,000 tons starts from rest under the force of a constant propeller thrust of 200,000 lb. a) Find its velocity as a function of time t, given that the resistance in pounds is $10{,}000v$, with v = velocity measured in ft/sec. b) Find the terminal velocity (i.e., v when $t \to \infty$) in miles per hour. (Take g = 32 ft/sec^2.)

Since mass (slugs) × acceleration (ft/sec^2) = net force (lb) = impetus of propeller − resistance,

then $\dfrac{48{,}000(2000)}{32}\dfrac{dv}{dt} = 200{,}000 - 10{,}000\,v$ or 1) $\dfrac{dv}{dt} + \dfrac{v}{300} = \dfrac{20}{300}$.

Integrating, $ve^{t/300} = \dfrac{20}{300}\int e^{t/300}\,dt = 20e^{t/300} + C$.

a) When $t = 0$, $v = 0$; $C = -20$ and $v = 20 - 20e^{-t/300} = 20(1 - e^{-t/300})$.

b) As $t \to \infty$, $v = 20$; the terminal velocity is 20 ft/sec = 13.6 mi/hr. This may also be obtained from 1) since, as v approaches a limiting value, $\dfrac{dv}{dt} \to 0$. Then $v = 20$, as before.

11. A boat is being towed at the rate 12 miles per hour. At the instant ($t = 0$) that the towing line is cast off, a man in the boat begins to row in the direction of motion, exerting a force of 20 lb. If the combined weight of the man and boat is 480 lb and the resistance (lb) is equal to $1.75v$, where v is measured in ft/sec, find the speed of the boat after $\frac{1}{2}$ minute.

Since mass (slugs) × acceleration (ft/sec^2) = net force (lb) = forward force − resistance,

then $\dfrac{480}{32}\dfrac{dv}{dt} = 20 - 1.75\,v$ or $\dfrac{dv}{dt} + \dfrac{7}{60}v = \dfrac{4}{3}$.

Integrating, $ve^{7t/60} = \dfrac{4}{3}\int e^{7t/60}\,dt = \dfrac{80}{7}e^{7t/60} + C$.

When $t = 0$, $v = \dfrac{12(5280)}{(60)^2} = \dfrac{88}{5}$; $C = \dfrac{216}{35}$ and $v = \dfrac{80}{7} + \dfrac{216}{35}e^{-7t/60}$.

When $t = 30$, $v = \dfrac{80}{7} + \dfrac{216}{35}e^{-3.5} = 11.6$ ft/sec.

12. A mass is being pulled across the ice on a sled, the total weight including the sled being 80 lb. Under the assumption that the resistance offered by the ice to the runners is negligible and that the air offers a resistance in pounds equal to 5 times the velocity (v ft/sec) of the sled, find
a) the constant force (pounds) exerted on the sled which will give it a terminal velocity of 10 miles per hour, and
b) the velocity and distance traveled at the end of 48 seconds.

Since mass (slugs) × acceleration (ft/sec^2) = net force (lb) = forward force − resistance,

then $\dfrac{80}{32}\dfrac{dv}{dt} = F - 5v$ or $\dfrac{dv}{dt} + 2v = \dfrac{2}{5}F$, where F (lb) is the forward force.

Integrating, $v = \dfrac{F}{5} + Ce^{-2t}$. When $t = 0$, $v = 0$; then $C = -\dfrac{F}{5}$ and *A*) $v = \dfrac{F}{5}(1 - e^{-2t})$.

a) As $t \to \infty$, $\dfrac{F}{5} = v = \dfrac{10(5280)}{(60)^2} = \dfrac{44}{3}$. The required force is $F = \dfrac{220}{3}$ lb.

b) Substituting from *a*) in *A*), $v = \dfrac{44}{3}(1 - e^{-2t})$.

When $t = 48$: $v = \dfrac{44}{3}(1 - e^{-96}) = \dfrac{44}{3}$ ft/sec, and $s = \int_0^{48} v\,dt = \dfrac{44}{3}\int_0^{48}(1 - e^{-2t})\,dt = 697$ ft.

13. A spring of negligible weight hangs vertically. A mass of m slugs is attached to the other end. If the mass is moving with velocity v_0 ft/sec when the spring is unstretched, find the velocity v as a function of the stretch x ft.

According to Hooke's law, the spring force (force opposing the stretch) is proportional to the stretch.

Net force on body = weight of body − spring force.

Then $m\dfrac{dv}{dt} = mg - kx$ or $m\dfrac{dv}{dx}\dfrac{dx}{dt} = mv\dfrac{dv}{dx} = mg - kx$, since $\dfrac{dx}{dt} = v$.

Integrating, $mv^2 = 2mgx - kx^2 + C$.

When $x = 0$, $v = v_0$. Then $C = mv_0^2$ and $mv^2 = 2mgx - kx^2 + mv_0^2$.

14. A parachutist is falling with speed 176 ft/sec when his parachute opens. If the air resistance is $Wv^2/256$ lb, where W is the total weight of the man and parachute, find his speed as a function of the time t after the parachute opened.

Net force on system = weight of system − air resistance.

Then $\dfrac{W}{g}\dfrac{dv}{dt} = W - \dfrac{Wv^2}{256}$ or $\dfrac{dv}{v^2 - 256} = -\dfrac{dt}{8}$.

Integrating between the limits $t = 0$, $v = 176$ and $t = t$, $v = v$,

$$\int_{176}^{v} \frac{dv}{v^2 - 256} = -\frac{1}{8}\int_0^t dt, \qquad \frac{1}{32}\ln\frac{v-16}{v+16}\bigg|_{176}^{v} = -\frac{t}{8}\bigg|_0^t,$$

$$\ln\frac{v-16}{v+16} - \ln\frac{5}{6} = -4t, \qquad \frac{v-16}{v+16} = \frac{5}{6}e^{-4t}, \quad \text{and} \quad v = 16\,\frac{6+5e^{-4t}}{6-5e^{-4t}}.$$

Note that the parachutist quickly attains an approximately constant speed, that is, the terminal speed of 16 ft/sec.

15. A body of mass m slugs falls from rest in a medium for which the resistance (lb) is proportional to the square of the velocity (ft/sec). If the terminal velocity is 150 ft/sec, find
a) the velocity at the end of 2 seconds, and
b) the time required for the velocity to become 100 ft/sec.

Let v denote the velocity of the body at time t seconds.

Net force on body = weight of body − resistance, and the equation of motion is

$$1) \quad m\frac{dv}{dt} = mg - Kv^2.$$

Taking $g = 32$ ft/sec^2, it is seen that some simplification is possible by choosing $K = 2mk^2$.

Then 1) reduces to $\dfrac{dv}{dt} = 2(16 - k^2v^2)$ or $\dfrac{dv}{k^2v^2 - 16} = -2\,dt$.

Integrating, $\ln\dfrac{kv-4}{kv+4} = -16kt + \ln C$ or $\dfrac{kv-4}{kv+4} = Ce^{-16kt}$.

When $t = 0$, $v = 0$. Then $C = -1$ and 2) $\dfrac{kv-4}{kv+4} = -e^{-16kt}$.

When $t \to \infty$, $v = 150$. Then $e^{-16kt} = 0$, $k = \dfrac{2}{75}$, and 2) becomes $\dfrac{v-150}{v+150} = -e^{-.43t}$.

$a)$ When $t = 2$, $\dfrac{v-150}{v+150} = -e^{-.86} = -.423$ and $v = 61$ ft/sec.

$b)$ When $v = 100$, $e^{-.43t} = .2 = e^{-1.6}$ and $t = 3.7$ sec.

16. A body of mass m falls from rest in a medium for which the resistance (lb) is proportional to the velocity (ft/sec). If the specific gravity of the medium is one-fourth that of the body and if the terminal velocity is 24 ft/sec, find (a) the velocity at the end of 3 sec and (b) the distance traveled in 3 sec.

Let v denote the velocity of the body at time t sec. In addition to the two forces acting as in Problem 15, there is a third force which results from the difference in specific gravities. This force is equal in magnitude to the weight of the medium which the body displaces and opposes gravity.

Net force on body = weight of body − buoyant force − resistance, and the equation of motion is

$$m\frac{dv}{dt} = mg - \frac{1}{4}mg - Kv = \frac{3}{4}mg - Kv.$$

Taking $g = 32$ ft/sec^2 and $K = 3mk$, the equation becomes $\dfrac{dv}{dt} = 3(8-kv)$ or $\dfrac{dv}{8-kv} = 3\,dt$.

Integrating from $t = 0$, $v = 0$ to $t = t$, $v = v$,

$$-\frac{1}{k}\ln(8-kv)\Big|_0^v = 3t\Big|_0^t, \quad -\ln(8-kv) + \ln 8 = 3kt, \quad \text{and} \quad kv = 8(1-e^{-3kt}).$$

When $t \to \infty$, $v = 24$. Then $k = 1/3$ and 1) $v = 24(1-e^{-t})$.

$a)$ When $t = 3$, $v = 24(1-e^{-3}) = 22.8$ ft/sec.

$b)$ Integrating $v = \dfrac{dx}{dt} = 24(1-e^{-t})$ between $t = 0$, $x = 0$ and $t = 3$, $x = x$,

$$x\Big|_0^x = 24(t + e^{-t})\Big|_0^3 \quad \text{and} \quad x = 24(2 + e^{-3}) = 49.2 \text{ ft.}$$

17. The gravitational pull on a mass m at a distance s feet from the center of the earth is proportional to m and inversely proportional to s^2. $a)$ Find the velocity attained by the mass in falling from rest at a distance $5R$ from the center to the earth's surface, where $R = 4000$ miles is taken as the radius of the earth. $b)$ What velocity would correspond to a fall from an infinite distance, that is, with what velocity must the mass be propelled vertically upward to escape the gravitational pull? (All other forces, including friction, are to be neglected.)

The gravitational force at a distance s from the earth's center is km/s^2. To determine k, note that the force is mg when $s = R$; thus $mg = km/R^2$ and $k = gR^2$. The equation of motion is

$$1) \quad m\frac{dv}{dt} = m\frac{ds}{dt}\frac{dv}{ds} = mv\frac{dv}{ds} = -\frac{mgR^2}{s^2} \quad \text{or} \quad v\,dv = -gR^2\frac{ds}{s^2},$$

the sign being negative since v increases as s decreases.

$a)$ Integrating 1) from $v = 0$, $s = 5R$ to $v = v$, $s = R$,

$$\int_0^v v\,dv = -gR^2\int_{5R}^R \frac{ds}{s^2}, \quad \tfrac{1}{2}v^2 = gR^2\left(\frac{1}{R}-\frac{1}{5R}\right) = \frac{4}{5}gR, \quad v^2 = \frac{8}{5}(32)(4000)(5280),$$

and $v = 2560\sqrt{165}$ ft/sec or approximately 6 mi/sec.

b) Integrating 1) from $v = 0,\ s \to \infty$ to $v = v,\ s = R$,

$$\int_0^v v\,dv = -gR^2 \int_\infty^R \frac{ds}{s^2}, \quad v^2 = 2gR, \quad v = 6400\sqrt{33} \text{ ft/sec or approximately 7 mi/sec.}$$

18. One of the basic equations in electric circuits is

1) $$L\frac{di}{dt} + Ri = E(t),$$

where L(henries) is called the inductance, R(ohms) the resistance, i(amperes) the current, and E(volts) the electromotive force or emf. (In this book, R and L will be constants.)

a) Solve 1) when $E(t) = E_0$ and the initial current is i_0.

b) Solve 1) when $L = 3$ henries, $R = 15$ ohms, $E(t)$ is the 60 cycle sine wave of amplitude 110 volts, and $i = 0$ when $t = 0$.

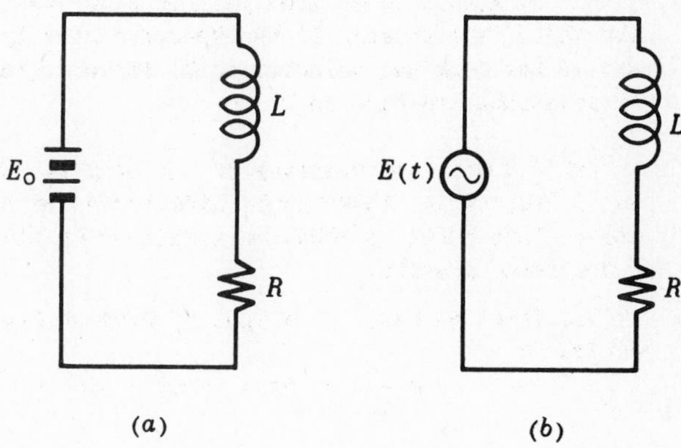

a) Integrating $L\frac{di}{dt} + Ri = E_0$, $ie^{Rt/L} = \frac{E_0}{L}\int e^{Rt/L}\,dt = \frac{E_0}{R}e^{Rt/L} + C$ or $i = \frac{E_0}{R} + Ce^{-Rt/L}$.

When $t = 0$, $i = i_0$. Then $C = i_0 - \frac{E_0}{R}$ and $i = \frac{E_0}{R}(1 - e^{-Rt/L}) + i_0e^{-Rt/L}$.

Note that as $t \to \infty$, $i = E_0/R$, a constant.

b) Integrating $3\frac{di}{dt} + 15i = E_0 \sin\omega t = 110 \sin 2\pi(60)t = 110 \sin 120\pi t$,

$$ie^{5t} = \frac{110}{3}\int e^{5t}\sin 120\pi t\,dt = \frac{110}{3}e^{5t}\frac{5\sin 120\pi t - 120\pi\cos 120\pi t}{25 + 14400\pi^2} + C$$

or $$i = \frac{22}{3}\,\frac{\sin 120\pi t - 24\pi\cos 120\pi t}{1 + 576\pi^2} + Ce^{-5t}.$$

When $t = 0$, $i = 0$. Then $C = \frac{22\cdot 24\pi}{3(1 + 576\pi^2)}$

and $$i = \frac{22}{3}\,\frac{\sin 120\pi t - 24\pi\cos 120\pi t + 24\pi e^{-5t}}{1 + 576\pi^2}.$$

A more useful form is obtained by noting that the sum of the squares of the coefficients of the sine and cosine terms is the denominator of the fraction above. Hence, we may define

$$\sin\phi = \frac{24\pi}{(1 + 576\pi^2)^{1/2}} \quad\text{and}\quad \cos\phi = \frac{1}{(1 + 576\pi^2)^{1/2}}$$

so that $$i = \frac{22}{3(1 + 576\pi^2)^{1/2}}(\cos\phi\sin 120\pi t - \sin\phi\cos 120\pi t) + \frac{176\pi e^{-5t}}{1 + 576\pi^2}.$$

$$= \frac{22}{3(1 + 576\pi^2)^{1/2}}\sin(120\pi t - \phi) + \frac{176\pi e^{-5t}}{1 + 576\pi^2}.$$

Note that after a short time the second term becomes very small; thus, the current quickly becomes a pure sine curve.

19. If an electric circuit contains a resistance R(ohms) and a condenser of capacitance C(farads) in series, and an emf E(volts), the charge q(coulombs) on the condenser is given by

$$R\frac{dq}{dt} + \frac{q}{C} = E.$$

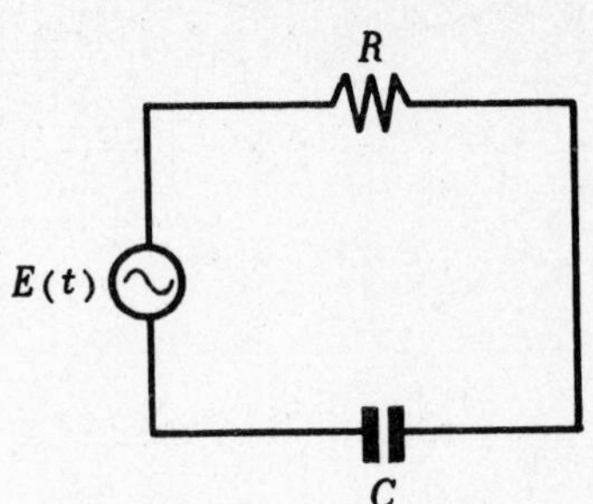

If $R = 10$ ohms, $C = 10^{-3}$ farad and $E(t) = 100 \sin 120\pi t$ volts,
a) find q, assuming that $q = 0$ when $t = 0$.
b) use $i = dq/dt$ to find i, assuming that $i = 5$ amperes when $t = 0$.

Integrating $\quad 10\frac{dq}{dt} + 10^3 q = 100 \sin 120\pi t,\quad$ we have

$$qe^{100t} = 10\int e^{100t}\sin 120\pi t\, dt = 10e^{100t}\frac{100 \sin 120\pi t - 120\pi \cos 120\pi t}{10{,}000 + 14{,}400\pi^2} + A$$

$$= e^{100t}\frac{10 \sin 120\pi t - 12\pi \cos 120\pi t}{100 + 144\pi^2} + A,$$

and $\quad$ 1) $\quad q = \dfrac{1}{(100 + 144\pi^2)^{1/2}}\sin(120\pi t - \phi) + Ae^{-100t}$

where $\sin\phi = \dfrac{12\pi}{(100 + 144\pi^2)^{1/2}}$ and $\cos\phi = \dfrac{10}{(100 + 144\pi^2)^{1/2}}$.

a) When $t = 0$, $q = 0$. Then $A = \dfrac{3\pi}{25 + 36\pi^2}$ and $q = \dfrac{1}{2(25 + 36\pi^2)^{1/2}}\sin(120\pi t - \phi) + \dfrac{3\pi e^{-100t}}{25 + 36\pi^2}$.

b) Differentiating 1) with respect to t, we obtain

$$i = \frac{dq}{dt} = \frac{60\pi}{(25 + 36\pi^2)^{1/2}}\cos(120\pi t - \phi) - 100Ae^{-100t}.$$

When $t = 0$, $i = 5$. Then $100A = \dfrac{60\pi}{(25 + 36\pi^2)^{1/2}}\cos\phi - 5 = \dfrac{300\pi}{25 + 36\pi^2} - 5$

and $\quad i = \dfrac{60\pi}{(25 + 36\pi^2)^{1/2}}\cos(120\pi t - \phi) - \left(\dfrac{300\pi}{25 + 36\pi^2} - 5\right)e^{-100t}.$

20. A boy, standing in corner A of a rectangular pool, has a boat in the adjacent corner B on the end of a string 20 feet long. He walks along the side of the pool toward C keeping the string taut. Locate the boy and boat when the latter is 12 feet from AC.

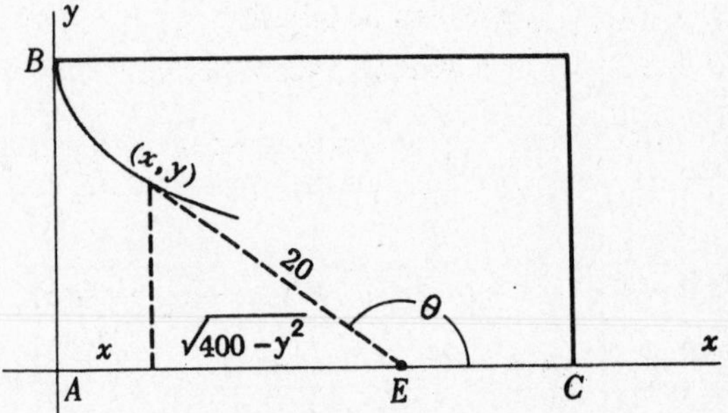

Choose the coordinate system so that AC is along the x-axis and AB is along the y-axis. Let (x,y) be the position of the boat when the boy has reached E, and let θ denote the angle of inclination of the string.

Then $\quad \tan\theta = \dfrac{dy}{dx} = \dfrac{-y}{\sqrt{400 - y^2}}\quad$ or $\quad dx = -\dfrac{\sqrt{400 - y^2}}{y}\,dy.$

Integrating, $x = -\sqrt{400-y^2} + 20 \ln \dfrac{20+\sqrt{400-y^2}}{y} + C.$

When the boat is at B, $x = 0$ and $y = 20$.

Then $C = 0$ and $x = -\sqrt{400-y^2} + 20 \ln \dfrac{20+\sqrt{400-y^2}}{y}$ is the equation of the boat's path.

Now $AE = x + \sqrt{400-y^2} = 20 \ln \dfrac{20+\sqrt{400-y^2}}{y}$. Hence, when the boat is 12 feet from AC (i.e., $y = 12$), $x + 16 = 20 \ln 3 = 22$.

The boy is 22 feet from A and the boat is 6 feet from AB.

21. A substance γ is being formed by the reaction of two substances α and β in which a grams of α and b grams of β form $(a+b)$ grams of γ. If initially there are x_0 grams of α, y_0 grams of β, and none of γ present and if the rate of formation of γ is proportional to the product of the quantities of α and β uncombined, express the amount (z grams) of γ formed as a function of time t.

The z grams of γ formed at time t consists of $\dfrac{az}{a+b}$ grams of α and $\dfrac{bz}{a+b}$ grams of β.

Hence, at time t there remain uncombined $(x_0 - \dfrac{az}{a+b})$ grams of α and $(y_0 - \dfrac{bz}{a+b})$ grams of β.

Then $$\frac{dz}{dt} = K(x_0 - \frac{az}{a+b})(y_0 - \frac{bz}{a+b}) = \frac{Kab}{(a+b)^2}(\frac{a+b}{a}x_0 - z)(\frac{a+b}{b}y_0 - z)$$

$$= k(A-z)(B-z), \quad \text{where } k = \frac{Kab}{(a+b)^2}, \quad A = \frac{(a+b)x_0}{a} \text{ and } B = \frac{(a+b)y_0}{b}.$$

There are two cases to be considered: 1) $A \neq B$, say $A > B$, and 2) $A = B$.

1) Here $$\frac{dz}{(A-z)(B-z)} = -\frac{1}{A-B}\cdot\frac{dz}{A-z} + \frac{1}{A-B}\cdot\frac{dz}{B-z} = k\,dt.$$

Integrating from $t = 0$, $z = 0$ to $t = t$, $z = z$, we obtain

$$\frac{1}{A-B}\ln\frac{A-z}{B-z}\Big|_0^z = kt\Big|_0^t, \quad \frac{1}{A-B}(\ln\frac{A-z}{B-z} - \ln\frac{A}{B}) = kt, \quad \frac{A-z}{B-z} = \frac{A}{B}e^{(A-B)kt},$$

and $$z = \frac{AB(1-e^{-(A-B)kt})}{A-Be^{-(A-B)kt}}.$$

2) Here $\dfrac{dz}{(A-z)^2} = k\,dt$. Integrating from $t = 0$, $z = 0$ to $t = t$, $z = z$, we obtain

$$\frac{1}{A-z}\Big|_0^z = kt\Big|_0^t, \quad \frac{1}{A-z} - \frac{1}{A} = kt, \quad \text{and} \quad z = \frac{A^2kt}{1+Akt}.$$

SUPPLEMENTARY PROBLEMS

22. A body moves in a straight line so that its velocity exceeds by 2 its distance from a fixed point of the line. If $v = 5$ when $t = 0$, find the equation of motion. *Ans.* $x = 5e^t - 2$

23. Find the time required for a sum of money to double itself at 5% per annum compounded continuously. Hint: $dx/dt = 0.05x$, where x is the amount after t years. *Ans.* 13.9 years

24. Radium decomposes at a rate proportional to the amount present. If half the original amount disappears in 1600 years, find the percentage lost in 100 years. *Ans.* 4.2 %

25. In a culture of yeast the amount of active ferment grows at a rate proportional to the amount present. If the amount doubles in 1 hour, how many times the original amount may be anticipated at the end of $2\frac{3}{4}$ hours? *Ans.* 6.73 times the original amount

26. If, when the temperature of the air is $20°C$, a certain substance cools from $100°C$ to $60°C$ in 10 minutes, find the temperature after 40 minutes. *Ans.* $25°C$

27. A tank contains 100 gal of brine made by dissolving 60 lb of salt in water. Salt water containing 1 lb of salt per gal runs in at the rate 2 gal/min and the mixture, kept uniform by stirring, runs out at the rate 3 gal/min. Find the amount of salt in the tank at the end of 1 hr. Hint: $dx/dt = 2 - 3x/(100-t)$. *Ans.* 37.4 lb

28. Find the time required for a square tank of side 6 ft and depth 9 ft to empty through a one inch circular hole in the bottom. (Assume, as in Prob. 9, $v = 4.8\sqrt{h}$ ft/sec.) *Ans.* 137 min

29. A brick wall ($k = 0.0012$) is 30 cm thick. If the inner surface is $20°C$ and the outer is $0°C$, find the temperature in the wall as a function of the distance from the outer surface and the heat loss per day through a square meter. *Ans.* $T = 2x/3$; 691,000 cal

30. A man and his boat weigh 320 lb. If the force exerted by the oars in the direction of motion is 16 lb and if the resistance (in lb) to the motion is equal to twice the speed (ft/sec), find the speed 15 sec after the boat starts from rest. *Ans.* 7.6 ft/sec

31. A tank contains 100 gal of brine made by dissolving 80 lb of salt in water. Pure water runs into the tank at the rate 4 gal/min and the mixture, kept uniform by stirring, runs out at the same rate. The outflow runs into a second tank which contains 100 gal of pure water initially and the mixture, kept uniform by stirring, runs out at the same rate. Find the amount of salt in the second tank after 1 hr.

    Hint: $\frac{dx}{dt} = 4(\frac{4}{5}e^{-0.04t}) - 4\frac{x}{100}$ for the second tank. *Ans.* 17.4 lb

32. A funnel 10 in. in diameter at the top and 1 in. in diameter at the bottom is 24 in. deep. If initially full of water, find the time required to empty. *Ans.* 13.7 sec

33. Water is flowing into a vertical cylindrical tank of radius 6 ft and height 9 ft at the rate 6π ft^3/min and is escaping through a hole 1 in. in diameter in the bottom. Find the time required to fill the tank. Hint: $(\frac{\pi}{10} - \frac{\pi}{(24)^2}4.8\sqrt{h})dt = 36\pi\, dh$. *Ans.* 65 min

34. A mass of 4 slugs slides on a table. The friction is equal to four times the velocity, and the mass is subjected to a force $12 \sin 2t$ lb. Find the velocity as a function of t if $v = 0$ when $t = 0$. *Ans.* $v = \frac{3}{5}(\sin 2t - 2\cos 2t + 2e^{-t})$

35. A steam pipe of diameter 1 ft has a jacket of insulating material ($k = 0.00022$) $\frac{1}{2}$ ft thick. The pipe is kept at $475°F$ and the outside of the jacket at $75°F$. Find the temperature in the jacket at a distance x ft from the center of the pipe and the heat loss per day per foot of pipe. *Ans.* $T = 75 - 400(\ln x)/(\ln 2)$; 69,000 B.T.U.

36. The differential equation of a circuit containing a resistance R, capacitance C, and emf $e = E \sin\omega t$ is $R\,di/dt + i/C = de/dt$. Assuming R, C, E, ω to be constants, find the current i at time t. *Ans.* $i = \frac{EC\omega}{1 + R^2C^2\omega^2}(\cos\omega t + RC\omega \sin\omega t) + C_1e^{-t/RC}$

CHAPTER 9

Equations of First Order and Higher Degree

A DIFFERENTIAL EQUATION of the first order has the form $f(x,y,y') = 0$ or $f(x,y,p)=0$, where for convenience $y' = \frac{dy}{dx}$ is replaced by p. If the degree of p is greater than one, as in $p^2 - 3px + 2y = 0$, the equation is of first order and higher (here, second) degree.

The general first order equation of degree n may be written in the form

1) $$p^n + P_1(x,y)p^{n-1} + \cdots\cdots + P_{n-1}(x,y)p + P_n(x,y) = 0.$$

It may be possible, at times, to solve such equations by one or more of the procedures outlined below. In each case the problem is reduced to that of solving one or more equations of the first order and first degree.

EQUATIONS SOLVABLE FOR p. Here the left member of 1), considered as a polynomial in p, can be resolved into n linear real factors, that is, 1) can be put in the form

$$(p-F_1)(p-F_2)\cdots\cdots(p-F_n) = 0,$$

where the F's are functions of x and y.

Set each factor equal to zero and solve the resulting n differential equations of first order and first degree

$$\frac{dy}{dx} = F_1(x,y), \quad \frac{dy}{dx} = F_2(x,y), \quad \cdots\cdots, \quad \frac{dy}{dx} = F_n(x,y)$$

to obtain

2) $$f_1(x,y,C) = 0, \quad f_2(x,y,C) = 0, \quad \cdots\cdots, \quad f_n(x,y,C) = 0.$$

The primitive of 1) is the product

3) $$f_1(x,y,C)\cdot f_2(x,y,C)\cdots\cdots\cdots f_n(x,y,C) = 0$$

of the n solutions 2).

Note. Each individual solution of 2) may be written in any one of its several possible forms before being combined into the product 3). See Prob. 1-3.

EQUATIONS SOLVABLE FOR y, i.e., $y = f(x,p)$.

Differentiate with respect to x to obtain

$$\frac{dy}{dx} = p = \frac{\partial f}{\partial x} + \frac{\partial f}{\partial p}\frac{dp}{dx} = F(x,p,\frac{dp}{dx}),$$

an equation of the first order and first degree.

Solve $p = F(x,p,\frac{dp}{dx})$ to obtain $\phi(x,p,C) = 0$.

Obtain the primitive by eliminating p between $y = f(x,p)$ and $\phi(x,p,C) = 0$, when possible, or express x and y separately as functions of the parameter p.

See Problems 4-7.

EQUATIONS SOLVABLE FOR x, i.e., $x = f(y,p)$.

Differentiate with respect to y to obtain

$$\frac{dx}{dy} = \frac{1}{p} = \frac{\partial f}{\partial y} + \frac{\partial f}{\partial p}\frac{dp}{dy} = F(y,p,\frac{dp}{dy}),$$

an equation of the first order and first degree.

Solve $\frac{1}{p} = F(y,p,\frac{dp}{dy})$ to obtain $\phi(y,p,C) = 0$.

Obtain the primitive by eliminating p between $x = f(y,p)$ and $\phi(y,p,C) = 0$, when possible, or express x and y separately as functions of the parameter p.
See Problems 8-10.

CLAIRAUT'S EQUATION. The differential equation of the form

$$y = px + f(p)$$

is called Clairaut's equation. Its primitive is

$$y = Cx + f(C)$$

and is obtained simply by replacing p by C in the given equation.
See Problems 11-16.

SOLVED PROBLEMS

1. Solve $p^4 - (x+2y+1)p^3 + (x+2y+2xy)p^2 - 2xyp = 0$ or $p(p-1)(p-x)(p-2y) = 0$.

The solutions of the component equations of first order and first degree

$$\frac{dy}{dx} = 0, \quad \frac{dy}{dx} = 1, \quad \frac{dy}{dx} - x = 0, \quad \frac{dy}{dx} - 2y = 0$$

are respectively $y - C = 0, \quad y - x - C = 0, \quad 2y - x^2 - C = 0, \quad y - Ce^{2x} = 0.$

The primitive of the given equation is $(y-C)(y-x-C)(2y-x^2-C)(y-Ce^{2x}) = 0.$

2. Solve $xyp^2 + (x^2+xy+y^2)p + x^2 + xy = 0$ or $(xp+x+y)(yp+x) = 0$.

The solutions of the component equations $x\frac{dy}{dx} + x + y = 0$ and $y\frac{dy}{dx} + x = 0$

are respectively $2xy + x^2 - C = 0$ and $x^2 + y^2 - C = 0$.

The primitive of the given equation is $(2xy + x^2 - C)(x^2+y^2-C) = 0.$

3. Solve $(x^2+x)p^2 + (x^2+x-2xy-y)p + y^2 - xy = 0$ or $[(x+1)p-y][xp+x-y] = 0$.

The solutions of the component equations $(x+1)\frac{dy}{dx} - y = 0$ and $x\frac{dy}{dx} + x - y = 0$

are respectively $y - C(x+1) = 0$ and $y + x\ln Cx = 0$.

The primitive of the given equation is $[y - C(x+1)][y + x\ln Cx] = 0.$

4. Solve $16x^2 + 2p^2y - p^3x = 0$ or $2y = px - 16\dfrac{x^2}{p^2}$.

Differentiating the latter form with respect to x, $2p = p + x\dfrac{dp}{dx} - \dfrac{32x}{p^2} + \dfrac{32x^2}{p^3}\dfrac{dp}{dx}$.

Clearing of fractions and combining, $p(p^3 + 32x) - x(p^3 + 32x)\dfrac{dp}{dx} = 0$

$$\text{or} \quad 1) \quad (p^3 + 32x)(p - x\frac{dp}{dx}) = 0.$$

This equation is satisfied when $p^3 + 32x = 0$ or $p - x\dfrac{dp}{dx} = 0$. From the latter, $\dfrac{dp}{p} = \dfrac{dx}{x}$ and $p = Kx$. When this replacement for p is made in the given equation, we have

$$16x^2 + 2K^2x^2y - K^3x^4 = 0 \quad \text{or} \quad 2 + C^2y - C^3x^2 = 0,$$

after replacing K by $2C$.

The factor $p^3 + 32x$ of 1) will not be considered here since it does not contain the derivative dp/dx. Its significance will be noted in Chapter 10.

5. Solve $y = 2px + p^4x^2$.

Differentiating with respect to x, $p = 2x\dfrac{dp}{dx} + 2p + 2p^4x + 4p^3x^2\dfrac{dp}{dx}$

$$\text{or} \quad (p + 2x\frac{dp}{dx})(1 + 2p^3x) = 0.$$

The factor $1 + 2p^3x$ is discarded as in Problem 4. From $p + 2x\dfrac{dp}{dx} = 0$, $xp^2 = C$.

In parametric form, we have $x = C/p^2$, $y = 2C/p + C^2$, the second relation being obtained by substituting $x = C/p^2$ in the differential equation.

Here p may be eliminated without difficulty between the relation $xp^2 = C$ or $p^2 = C/x$ and the given equation. The latter may be put in the form $y - p^4x^2 = 2px$ and squared to give $(y - p^4x^2)^2 = 4p^2x^2$. Then, substituting for p^2, we have $(y - C^2)^2 = 4Cx$.

6. Solve $x = yp + p^2$ or $y = \dfrac{x}{p} - p$.

Differentiating with respect to x, $p = \dfrac{1}{p} - \dfrac{x}{p^2}\dfrac{dp}{dx} - \dfrac{dp}{dx}$ or $p^3 - p + (x + p^2)\dfrac{dp}{dx} = 0$.

Then $(p^3 - p)\dfrac{dx}{dp} + x + p^2 = 0$ or $\dfrac{dx}{dp} + \dfrac{x}{p^3 - p} = -\dfrac{p}{p^2 - 1}$.

The latter is a linear equation for which $e^{\int dp/(p^3 - p)} = \dfrac{\sqrt{p^2 - 1}}{p}$ is an integrating factor. Using it,

$$\frac{x\sqrt{p^2 - 1}}{p} = -\int \frac{dp}{\sqrt{p^2 - 1}} = -\ln(p + \sqrt{p^2 - 1}) + C$$

and $x = -\dfrac{p}{\sqrt{p^2 - 1}}\ln(p + \sqrt{p^2 - 1}) + \dfrac{Cp}{\sqrt{p^2 - 1}}$, $y = -p - \dfrac{1}{\sqrt{p^2 - 1}}\ln(p + \sqrt{p^2 - 1}) + \dfrac{C}{\sqrt{p^2 - 1}}$.

7. Solve $y = (2+p)x + p^2$.

Differentiating with respect to x, $p = 2 + p + (x+2p)\dfrac{dp}{dx}$ or $\dfrac{dx}{dp} + \tfrac{1}{2}x = -p$.

This is a linear equation having $e^{\frac{1}{2}\int dp} = e^{\frac{1}{2}p}$ as an integrating factor.

Then $$xe^{\frac{1}{2}p} = -\int pe^{\frac{1}{2}p}\,dp = -2pe^{\frac{1}{2}p} + 4e^{\frac{1}{2}p} + C$$

and $$x = 2(2-p) + Ce^{-\frac{1}{2}p}, \qquad y = 8 - p^2 + (2+p)Ce^{-\frac{1}{2}p}.$$

8. Solve $y = 3px + 6p^2y^2$.

Solving for x, $3x = \dfrac{y}{p} - 6py^2$. Then, differentiating with respect to y,

$$\frac{3}{p} = \frac{1}{p} - \frac{y}{p^2}\frac{dp}{dy} - 6y^2\frac{dp}{dy} - 12py \qquad \text{and} \qquad (1 + 6p^2y)(2p + y\frac{dp}{dy}) = 0.$$

The second factor equated to zero yields $py^2 = C$. Solving for p and substituting in the original differential equation yields the primitive $y^3 = 3Cx + 6C^2$.

9. Solve $p^3 - 2xyp + 4y^2 = 0$ or $2x = \dfrac{p^2}{y} + \dfrac{4y}{p}$.

Differentiating with respect to y,

$$\frac{2}{p} = \frac{2p}{y}\frac{dp}{dy} - \frac{p^2}{y^2} + 4(\frac{1}{p} - \frac{y}{p^2}\frac{dp}{dy}) \qquad \text{or} \qquad (p - 2y\frac{dp}{dy})(2y^2 - p^3) = 0.$$

Integrating $p - 2y\dfrac{dp}{dy} = 0$ and eliminating p between the solution $p^2 = Ky$ and the original differential equation, we have $16y = K(K-2x)^2$. This may be put in the form $2y = C(C-x)^2$ by letting $K = 2C$.

10. Solve $4x = py(p^2-3)$.

Differentiating with respect to y,

$$\frac{4}{p} = p(p^2-3) + 3y(p^2-1)\frac{dp}{dy} \qquad \text{or} \qquad \frac{dy}{y} + \frac{3p(p^2-1)dp}{(p^2-4)(p^2+1)} = 0.$$

Integrating, by partial fractions, $\ln y + \dfrac{9}{10}\ln(p+2) + \dfrac{9}{10}\ln(p-2) + \dfrac{3}{5}\ln(p^2+1) = \ln C$.

Then $$y = \frac{C}{(p^2-4)^{9/10}(p^2+1)^{3/5}}, \qquad x = \frac{1}{4}\frac{Cp(p^2-3)}{(p^2-4)^{9/10}(p^2+1)^{3/5}}.$$

CLAIRAUT'S EQUATION.

11. Solve $y = px + \sqrt{4+p^2}$. The primitive is $y = Cx + \sqrt{4+C^2}$.

12. Solve $(y-px)^2 = 1 + p^2$.

Here $y = px \pm \sqrt{1+p^2}$.

The primitive is $(y - Cx - \sqrt{1+C^2})(y - Cx + \sqrt{1+C^2}) = 0$ or $(y-Cx)^2 = 1 + C^2$.

13. Solve $y = 3px + 6y^2p^2$. (See Problem 8.)

This may be reduced to the form of a Clairaut equation.

Multiply the equation by y^2 to obtain $y^3 = 3y^2px + 6y^4p^2$.

Using the transformation $y^3 = v$, $3y^2p = \frac{dv}{dx}$, this becomes $v = x\frac{dv}{dx} + \frac{2}{3}(\frac{dv}{dx})^2$.

The primitive is $v = Kx + \frac{2}{3}K^2$ or $y^3 = Kx + \frac{2}{3}K^2$ or $y^3 = 3Cx + 6C^2$.

14. Solve $\cos^2 y\, p^2 + \sin x \cos x \cos y\, p - \sin y \cos^2 x = 0$.

The transformation $\sin y = u$, $\sin x = v$, $p\frac{\cos y}{\cos x} = \frac{du}{dv}$ reduces the equation to $u = v\frac{du}{dv} + (\frac{du}{dv})^2$. Then $u = Cv + C^2$ or $\sin y = C\sin x + C^2$.

15. Solve $(px - y)(py + x) = 2p$.

The transformation $y^2 = u$, $x^2 = v$, $p = \frac{v^{1/2}}{u^{1/2}}\frac{du}{dv}$ reduces the equation to

$$(\frac{v}{u^{1/2}}\frac{du}{dv} - u^{1/2})(v^{1/2}\frac{du}{dv} + v^{1/2}) = 2\frac{v^{1/2}}{u^{1/2}}\frac{du}{dv} \qquad \text{or} \qquad (v\frac{du}{dv} - u)(\frac{du}{dv} + 1) = 2\frac{du}{dv}.$$

Then $u = v\frac{du}{dv} - \frac{2\frac{du}{dv}}{1 + \frac{du}{dv}}$, and $u = Cv - \frac{2C}{1+C}$ or $y^2 = Cx^2 - \frac{2C}{1+C}$.

16. Solve $p^2x(x-2) + p(2y - 2xy - x + 2) + y^2 + y = 0$.

The equation may be written as $(y - px + 2p)(y - px + 1) = 0$.

Each of $y = px - 2p$ and $y = px - 1$ is a Clairaut equation.

Thus the primitive is $(y - Cx + 2C)(y - Cx + 1) = 0$.

SUPPLEMENTARY PROBLEMS

Find the primitive of each of the following.

17. $x^2p^2 + xyp - 6y^2 = 0$ *Ans.* $(y - Cx^2)(y - Cx^{-3}) = 0$

18. $xp^2 + (y - 1 - x^2)p - x(y-1) = 0$ *Ans.* $(2y - x^2 + C)(xy - x + C) = 0$

19. $xp^2 - 2yp + 4x = 0$ *Ans.* $Cy = x^2 + C^2$

20. $3x^4p^2 - xp - y = 0$ *Ans.* $xy = C(3Cx - 1)$

21. $8yp^2 - 2xp + y = 0$ *Ans.* $y^2 - Cx + 2C^2 = 0$

22. $y^2p^2 + 3px - y = 0$ *Ans.* $y^3 - 3Cx - C^2 = 0$

23. $p^2 - xp + y = 0$ *Ans.* $y = Cx - C^2$

24. $16y^3p^2 - 4xp + y = 0$ *Ans.* $y^4 = C(x - C)$

25. $xp^5 - yp^4 + (x^2+1)p^3 - 2xyp^2 + (x+y^2)p - y = 0$ *Ans.* $(y - Cx - C^3)(C^2x - Cy + 1) = 0$

26. $xp^2 - yp - y = 0$ *Ans.* $x = C(p+1)e^p, \quad y = Cp^2e^p$

27. $y = 2px + y^2p^3$ (Use $y^2 = z$.) *Ans.* $y^2 = 2Cx + C^3$

28. $p^2 - xp - y = 0$ *Ans.* $3x = 2p + C/\sqrt{p}, \quad 3y = p^2 - C/\sqrt{p}$

29. $y = (1+p)x + p^2$ *Ans.* $x = 2(1-p) + Ce^{-p}, \quad y = 2 - p^2 + C(1+p)e^{-p}$

30. $y = 2p + \sqrt{1+p^2}$ *Ans.* $x = 2 \ln p + \ln(p + \sqrt{1+p^2}) + C, \quad y = 2p + \sqrt{1+p^2}$

31. $yp^2 - xp + 3y = 0$ *Ans.* $x = Cp^{1/2}(p^2+3)(p^2+2)^{-5/4}, \quad y = Cp^{3/2}(p^2+2)^{-5/4}$

CHAPTER 10

Singular Solutions—Extraneous Loci

THE DIFFERENTIAL EQUATION

1) $$y = px + 2p^2$$

has as primitive the family of straight lines of equation

2) $$y = Cx + 2C^2.$$

With each point (x,y) in the region of points for which $x^2 + 8y > 0$, equation 1) associates a pair of distinct real directions and equation 2) associates a pair of distinct real lines having the directions determined by 1). For example, when the coordinates $(-2,4)$ are substituted in 1), we have $4 = -2p + 2p^2$ or $p^2 - p - 2 = 0$ and then $p = 2, -1$. Similarly, when 2) is used, we obtain $C = 2, -1$. Thus, through the point $(-2,4)$ pass the lines $y = 2x + 8$ and $y = -x + 2$ of the family 2) whose slopes are given by 1). Points for which $x^2 + 8y < 0$ yield distinct imaginary p- and C-roots.

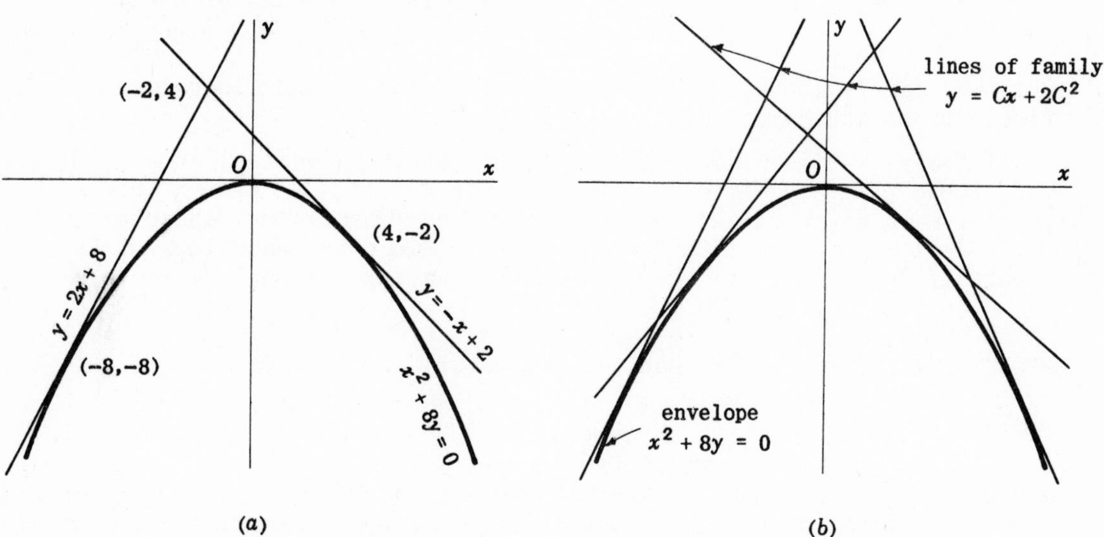

(*a*) (*b*)

Through each point of the parabola $x^2 + 8y = 0$ there passes but one line of the family, that is, the coordinates of any point on the parabola are so related that for them the two C-roots of 2) and the two p-roots of 1) are equal. For example, through the point $(-8,-8)$ there passes but one line, $y = 2x + 8$, and through the point $(4,-2)$ but one line, $y = -x + 2$. (See Fig. *a*.)

It is easily verified that the line of 2) through a point of $x^2 + 8y = 0$ is tangent to the parabola there, that is, the direction of the parabola at any one of its points is given by 1). Thus, $x^2 + 8y = 0$ is a solution of 1). It is called a *singular solution* since it cannot be obtained from 2) by a choice of the arbitrary constant, that is, since it is not a particular solution. The corresponding curve, the parabola, is called an *envelope* of the family of lines 2). (See Fig. *b* above.)

Summary and Extension:

A singular solution of a differential equation satisfies the differential equation but is not a particular solution of the equation.

At each point of its locus (envelope) the number of distinct directions given by the differential equation and the number of distinct curves given by the corresponding primitive are fewer than at points off the locus.

THE SINGULAR SOLUTIONS of a differential equation are to be found by expressing the conditions
a) that the differential equation (p-equation) have multiple roots, and
b) that the primitive (C-equation) have multiple roots.

In general, an equation of the first order does not have singular solutions; if it is of the first degree it cannot have singular solutions. Moreover, an equation $f(x,y,p) = 0$ cannot have singular solutions if $f(x,y,p)$ can be resolved into factors which are linear in p and rational in x and y.

The simplest expression, called the discriminant, involving the coefficients of an equation $F(X) = 0$ whose vanishing is the condition that the equation have multiple roots is obtained by eliminating X between $F(X) = 0$ and $F'(X) = 0$. The discriminant of

$$aX^2 + bX + c = 0 \qquad \text{is} \qquad b^2 - 4ac,$$

of $$aX^3 + bX^2 + cX + d = 0 \qquad \text{is} \qquad b^2c^2 + 18abcd - 4ac^3 - 4b^3d - 27a^2d^2.$$

See Problem 1.

For the example above, the discriminants of the p- and C-equations are identical, being $x^2 + 8y$.

If $E(x,y) = 0$ is a singular solution of the differential equation $f(x,y,p) = 0$, whose primitive is $g(x,y,C) = 0$, then $E(x,y)$ is a factor of both discriminants. Each discriminant, however, may have other factors which give rise to other loci associated with the primitive. Since the equations of these loci generally do not satisfy the differential equation, they are called *extraneous*.

EXTRANEOUS LOCI. (Differential equation, $f(x,y,p) = 0$; primitive, $g(x,y,C) = 0$.)

a) *Tac Locus.*

Let P be a point for which two or more of the n distinct curves of the family $g(x,y,C) = 0$ through it have a common tangent at P. Now the number of distinct directions at P is less than n so that the p-discriminant must vanish there. The locus, if there is one, of all such points is called a *tac locus*. If $T(x,y) = 0$ is the equation of the tac locus, then $T(x,y)$ is a factor of the p-discriminant. In general, $T(x,y)$ is not a factor of the C-discriminant and $T(x,y) = 0$ does not satisfy the differential equation.

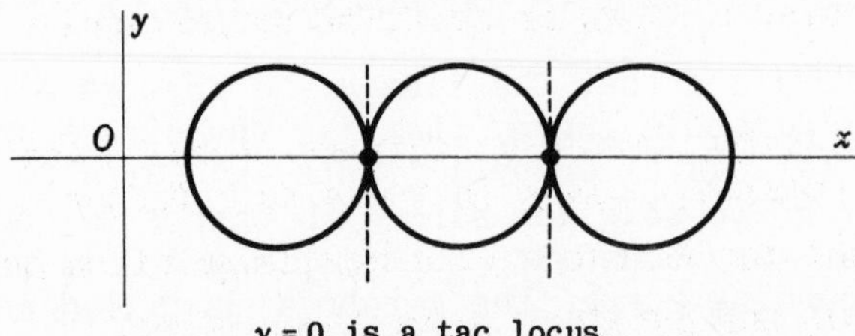

$y = 0$ is a tac locus.

b) Nodal Locus.

Let one of the curves of the family through P have a node (a double point with distinct tangents) there. Since two of the n values of p are thus accounted for, there can be no more than $n-1$ distinct curves through P; hence, the C-discriminant must vanish at P. The locus, if there is one, of all such points is called a *nodal locus*. If $N(x,y) = 0$ is the equation of the nodal locus, then $N(x,y)$ is a factor of the C-discriminant. In general, $N(x,y)$ is not a factor of the p-discriminant and $N(x,y) = 0$ does not satisfy the differential equation.

c) Cusp Locus.

Let one of the curves of the family through P have a cusp (a double point with coincident tangents) there. Since one of the p-roots is of multiplicity two, the p-discriminant must vanish at P. Moreover, as in the case of a node, there can be no more than $n-1$ curves through P and the C-discriminant must vanish at P. The locus, if there is one, of all such points is a cusp locus. If $C(x,y) = 0$ is the equation of the cusp locus, then $C(x,y)$ is a factor of both the p- and C-discriminants. In general, $C(x,y) = 0$ does not satisfy the differential equation.

$y = 0$ is a nodal locus.

$y = 0$ is a cusp locus.

If the curves of the family $g(x,y,C) = 0$ are straight lines, there are no extraneous loci.

If the curves of the family are conics, there can be neither a nodal nor cusp locus.

THE p-DISCRIMINANT RELATION. The discriminant of the differential equation $f(x,y,p) = 0$, the p-discriminant, equated to zero includes as a factor

1) the equation of the envelope (singular solution) once. See Problems 2-4. (The singular solution satisfies the differential equation.)
2) the equation of the cuspidal locus once. See Problem 7. (The equation of the cuspidal locus does not satisfy the differential equation unless it is also a singular solution or particular solution.)
3) the equation of the tac locus twice. See Problem 5. (The equation of the tac locus does not satisfy the differential equation unless it is also a singular solution or particular solution.)

THE C-DISCRIMINANT RELATION. The discriminant of the primitive $g(x,y,C) = 0$, the C-discriminant, equated to zero includes as a factor

1) the equation of the envelope or singular solution once.
2) the equation of the cuspidal locus three times.

3) the equation of the nodal locus twice. See Problem 6.
(The equation of the nodal locus does not satisfy the differential equation unless it is also a singular solution or particular solution.)

When any locus falls in two of the categories, the multiplicity of its equation in a discriminant relation is the sum of the multiplicities for each category; thus, a cuspidal locus which is also an envelope is included twice in the p-discriminant and four times in the C-discriminant relation.

The identification of extraneous loci is, however, more than a mere counting of multiplicities of factors.

SOLVED PROBLEMS

1. Find the discriminant relation for each of the following:

$a)$ $p^3 + px - y = 0$, $\quad b)$ $p^3x - 2p^2y - 16x^2 = 0$, $\quad c)$ $y = C(x-C)^2$.

Note. These discriminant relations may be written readily using the formula given above. We give here a procedure which may be preferred.

$a)$ We are to eliminate p between $f(x,y,p) = p^3 + px - y = 0$ and $\dfrac{\partial f}{\partial p} = 3p^2 + x = 0$. This is best done by eliminating p between

$3f - p\dfrac{\partial f}{\partial p} = 3p^3 + 3px - 3y - 3p^3 - px = 2px - 3y = 0$ and $\dfrac{\partial f}{\partial p} = 3p^2 + x = 0$. Solving the first for $p = \dfrac{3y}{2x}$ and substituting in the second, we find $3p^2 + x = \dfrac{27y^2}{4x^2} + x = 0$ or $4x^3 + 27y^2 = 0$.

Note. If $f(x,y,p) = 0$ is of degree n in p, we eliminate p between $nf - p\dfrac{\partial f}{\partial p} = 0$ and $\dfrac{\partial f}{\partial p} = 0$.

$b)$ We are to eliminate p between $3f - p\dfrac{\partial f}{\partial p} = 3p^3x - 6p^2y - 48x^2 - 3p^3x + 4p^2y = -2p^2y - 48x^2 = 0$ and $\dfrac{\partial f}{\partial p} = 3p^2x - 4py = 0$. From the latter we obtain $9p^4x^2 = 16p^2y^2$ or $9p^4x^2 - 16p^2y^2 = 0$ and from the former $p^2 = -24\dfrac{x^2}{y}$. Substituting for p^2, we obtain $x^2(2y^3 + 27x^4) = 0$.

$c)$ Here $g(x,y,C) = C^3 - 2C^2x + Cx^2 - y = 0$ and we are to eliminate C between

1) $3g - C\dfrac{\partial g}{\partial C} = 3C^3 - 6C^2x + 3Cx^2 - 3y - 3C^3 + 4C^2x - Cx^2 = -2C^2x + 2Cx^2 - 3y = 0$ and

2) $\dfrac{\partial g}{\partial C} = 3C^2 - 4Cx + x^2 = 0$.

Multiplying 1) by 3 and 2) by $2x$, and adding, we have $-2Cx^2 + 2x^3 - 9y = 0$.

Substituting $C = \dfrac{2x^3 - 9y}{2x^2}$ in 2) and simplifying, we obtain $y(4x^3 - 27y) = 0$.

2. Solve $y = 2xp - yp^2$ and examine for singular solutions.

Solving for $2x = \dfrac{y}{p} + yp$ and differentiating with respect to y, we have

$$\frac{2}{p} = \frac{1}{p} - \frac{y}{p^2}\frac{dp}{dy} + p + y\frac{dp}{dy} \qquad \text{or} \qquad (p^2-1)(p + y\frac{dp}{dy}) = 0.$$

Integrating $p + y\dfrac{dp}{dy} = 0$ to obtain $py = C$ and substituting for $p = \dfrac{C}{y}$ in the given differential equation, we obtain the primitive $y^2 = 2Cx - C^2$.

The p- and C-discriminant relations are $x^2 - y^2 = 0$. Since both $y = x$ and $y = -x$ satisfy the given differential equation, they are singular solutions.

If p is eliminated between the differential equation and the relation $p^2 - 1 = 0$, discarded in this solution, the equation of the envelope $x^2 - y^2 = 0$ is again obtained. The presence of such a factor implies the existence of a singular solution but not conversely. Hence, this procedure is not to be used in finding singular solutions.

The primitive represents a family of parabolas with principal axis along the x-axis. Each parabola is tangent to the line $y = x$ at the point (C,C) and to the line $y = -x$ at the point $(C,-C)$. See Figure (a) below.

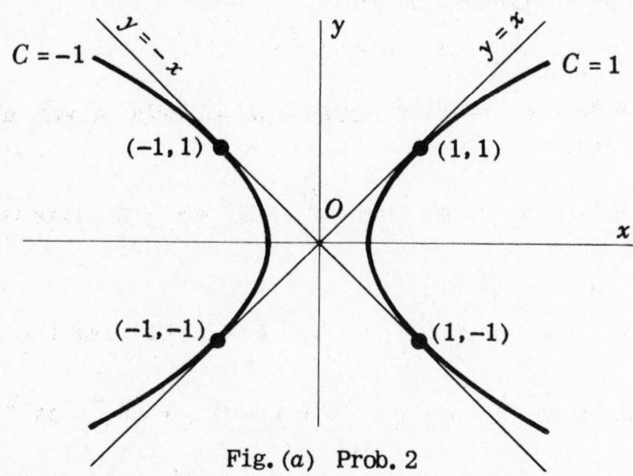

Fig. (a) Prob. 2

Family of parabolas $y^2 = 2Cx - C^2$, envelope $y = \pm x$.

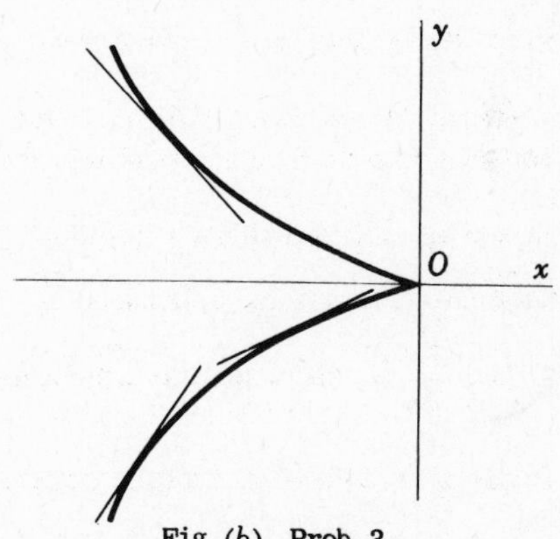

Fig. (b) Prob. 3

Family of straight lines $y = Cx + C^3$, envelope $4x^3 + 27y^2 = 0$.

3. Examine $p^3 + px - y = 0$ for singular solutions.

This is a Clairaut equation, the primitive being $y = Cx + C^3$.

The p- and C-discriminant relation $4x^3 + 27y^2 = 0$ is a singular solution since it satisfies the differential equation.

The primitive represents a family of straight lines tangent to the semi-cubical parabola $4x^3 + 27y^2 = 0$, the envelope. See Figure (b) above.

4. Examine $6p^2y^2 + 3px - y = 0$ for singular solutions.

From Problem 13, Chapter 9, the primitive is $y^3 = 3Cx + 6C^2$.

Both the p- and C-discriminant relations are $3x^2 + 8y^3 = 0$. Since this satisfies the differential equation, it is a singular solution.

5. Solve $(x^2-4)p^2 - 2xyp - x^2 = 0$ and examine for singular solutions and extraneous loci.

Solving for $2y = xp - \dfrac{4}{x}p - \dfrac{x}{p}$ and differentiating with respect to x, we have

$$2p = p + x\frac{dp}{dx} + \frac{4p}{x^2} - \frac{4}{x}\frac{dp}{dx} - \frac{1}{p} + \frac{x}{p^2}\frac{dp}{dx} \qquad \text{or} \qquad (p^2x^2 - 4p^2 + x^2)(p - x\frac{dp}{dx}) = 0.$$

From $p - x\dfrac{dp}{dx} = 0$, $p = Cx$ and the primitive is $C^2(x^2 - 4) - 2Cy - 1 = 0$. The p-discriminant relation is $x^2(x^2 + y^2 - 4) = 0$, and the C-discriminant relation is $x^2 + y^2 - 4 = 0$.

Now $x^2 + y^2 = 4$ occurs once in the p- and C-discriminant relations and satisfies the differential equation; it is a singular solution. Also $x = 0$ occurs twice in the p-discriminant relation, does not occur in the C-discriminant relation, and does not satisfy the differential equation; it is a tac locus.

The primitive represents a family of parabolas having the circle $x^2 + y^2 = 4$ as envelope. See Figure (c) below.

Note 1. The two parabolas through a point P of the tac locus $x = 0$ have at P a common tangent.

Note 2. A curve of the family meets the envelope in the points $(\pm\dfrac{\sqrt{4C^2 - 1}}{C}, -\dfrac{1}{C})$; hence, only those parabolas given by $C^2 \geqq \frac{1}{4}$ touch the circle.

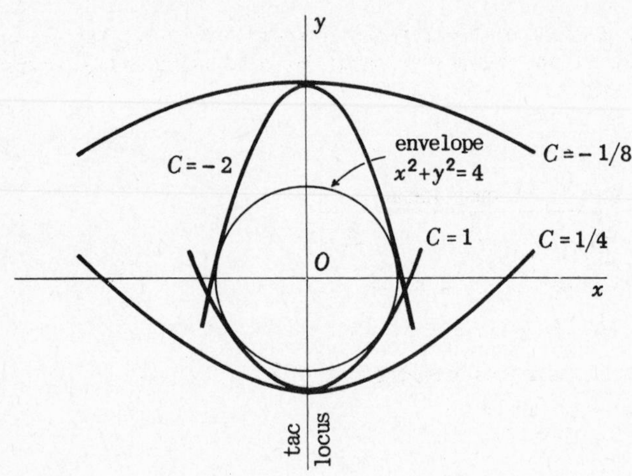

Family of parabolas
$C^2(x^2 - 4) - 2Cy - 1 = 0.$

Fig. (c) Prob. 5

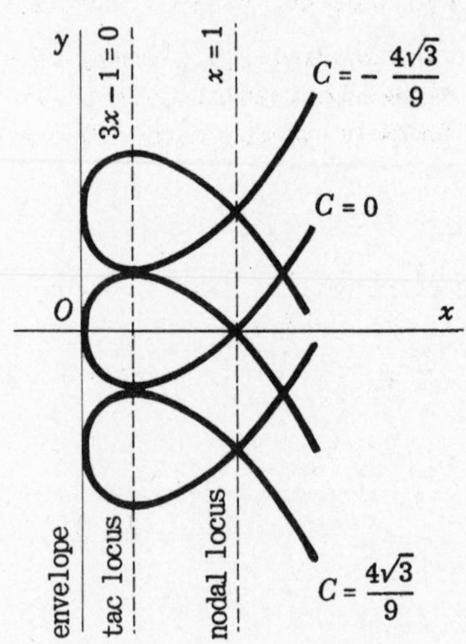

Family of cubic curves
$(y + C)^2 = x(x - 1)^2.$

Fig. (d) Prob. 6

6. Solve $4xp^2 - (3x - 1)^2 = 0$ and examine for singular solutions and extraneous loci.

Solving for $p = \pm(\frac{3}{2}x^{1/2} - \frac{1}{2}x^{-1/2})$, we obtain by integration $y = \pm(x^{3/2} - x^{1/2}) + C_1$ or $(y + C)^2 = x(x - 1)^2$. The p-discriminant relation is $x(3x - 1)^2 = 0$, and the C-discriminant relation is $x(x - 1)^2 = 0$.

Here $x = 0$ is common to the two relations and satisfies the differential equation, that is, $x = 0$, $\dfrac{dx}{dy} = 0$ satisfies the equation when written in the form $4x - (3x - 1)^2(\dfrac{dx}{dy})^2 = 0$. It is a singular solution.

$3x - 1 = 0$ is a tac locus since it occurs twice in the p-discriminant relation, does not occur in the C-discriminant relation, and does not satisfy the differential equation.

$x - 1 = 0$ is a nodal locus since it occurs twice in the C-discriminant relation, does not occur in the p-discriminant relation, and does not satisfy the differential equation.

The primitive represents a family of cubics obtained by moving $y^2 = x(x-1)^2$ along the y-axis. These curves are tangent to the y-axis and have a double point at $x = 1$. Moreover, through each point on $x = 1/3$ pass two curves of the family having a common tangent there. See Figure (d) above.

7. Solve $9yp^2 + 4 = 0$ and examine for singular solutions and extraneous loci.

Solving for $9y = -4/p^2$ and differentiating with respect to x, we have

$$dx = \frac{8}{9}\frac{dp}{p^4} \quad \text{and} \quad x + C = -\frac{8}{27p^3}.$$

Eliminating p between this latter relation and the differential equation, the primitive is $y^3 + (x+C)^2 = 0$.

The p-discriminant relation is $y = 0$, and the C-discriminant relation is $y^3 = 0$. Since $y = 0$ occurs once in the p-discriminant relation, three times in the C-discriminant relation, and does not satisfy the differential equation, it is a cusp locus.

The primitive represents the family of semi-cubical parabolas obtained by moving $y^3 + x^2 = 0$ along the x-axis. Each curve has a cusp at its intersection with the x-axis, and $y = 0$ is the locus of these cusps. See the figure below.

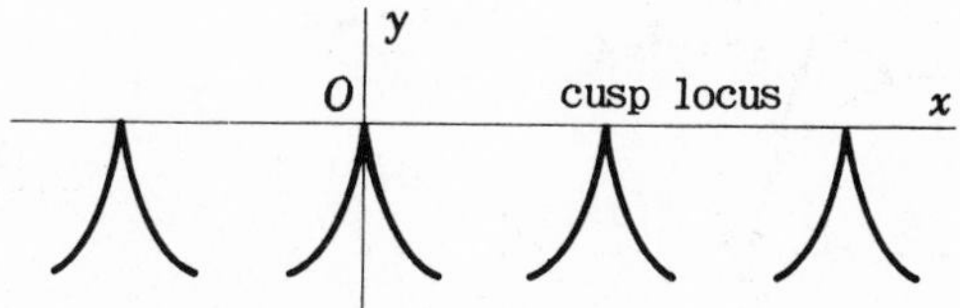

Family of semicubical parabolas
$y^3 + (x+C)^2 = 0$

8. Solve $x^3p^2 + x^2yp + 1 = 0$ and examine for singular solutions and extraneous loci.

Solving for $y = -\dfrac{1}{x^2p} - xp$ and differentiating with respect to x, we have

$$(1 - x^3p^2)(2p + x\frac{dp}{dx}) = 0.$$

From $2p + x\dfrac{dp}{dx} = 0$, $px^2 = C$ and, eliminating p between this and the differential equation, the primitive is $C^2 + Cxy + x = 0$.

The p-discriminant relation is $x^3(xy^2 - 4) = 0$, and the C-discriminant relation is $x(xy^2 - 4) = 0$.

$xy^2 - 4 = 0$ satisfies the differential equation and is a singular solution.

$x = 0$ is a particular solution ($C = 0$). Note that it occurs three times in the p-discriminant relation and once in the C-discriminant relation.

9. Examine $p^3x - 2p^2y - 16x^2 = 0$ for singular solutions and extraneous loci.

From Problem 4, Chapter 9, the primitive is $C^3x^2 - C^2y - 2 = 0$.

The p-discriminant relation is $x^2(2y^3 + 27x^4) = 0$, and the C-discriminant relation is $2y^3 + 27x^4 = 0$.

Since $2y^3 + 27x^4 = 0$ is common to the discriminant relations and satisfies the differential equation, it is a singular solution. At each point of the line $x = 0$, two parabolas of the family are tangent there (for $y < 0$, the parabolas are real). Thus, $x = 0$ is a tac locus. Also, $x = 0$ is a particular solution. Since it is obtained by letting $C \to \infty$, it is sometimes called an infinite solution. Note however that when the primitive is written as $x^2 - Ky - 2K^3 = 0$, this solution is obtained when $K = 0$.

SUPPLEMENTARY PROBLEMS

Investigate for singular solutions and extraneous loci.

10. $y = px - 2p^2$. *Ans.* primitive, $y = Cx - 2C^2$; singular solution, $x^2 = 8y$.

11. $y^2p^2 + 3xp - y = 0$. *Ans.* prim., $y^3 + 3Cx - C^2 = 0$; s.s., $9x^2 + 4y^3 = 0$.

12. $xp^2 - 2yp + 4x = 0$. *Ans.* prim., $C^2x^2 - Cy + 1 = 0$; s.s., $y^2 - 4x^2 = 0$.

13. $xp^2 - 2yp + x + 2y = 0$. *Ans.* prim., $2x^2 + 2C(x-y) + C^2 = 0$; s.s., $x^2 + 2xy - y^2 = 0$.

14. $(3y-1)^2p^2 = 4y$. *Ans.* prim., $(x+C)^2 = y(y-1)^2$; s.s., $y = 0$; t.l., $y = 1/3$; n.l., $y = 1$.

15. $y = -xp + x^4p^2$. *Ans.* prim., $xy = C + C^2x$; s.s., $1 + 4x^2y = 0$; t.l., $x = 0$.

16. $2y = p^2 + 4xp$. *Ans.* prim., $(4x^3 + 3xy + C)^2 = 2(2x^2 + y)^3$; no s.s.; c.l., $2x^2 + y = 0$.

17. $y(3-4y)^2p^2 = 4(1-y)$. *Ans.* prim., $(x-C)^2 = y^3(1-y)$; s.s., $y = 1$; c.l., $y = 0$; t.l., $y = 3/4$.

18. $p^3 - 4x^4p + 8x^3y = 0$. *Ans.* prim., $y = Cx^2 - C^3$; s.s., $4x^6 - 27y^2 = 0$; t.l., $x = 0$.

19. $(p^2+1)(x-y)^2 = (x+yp)^2$. *Ans.* prim., $(x-C)^2 + (y-C)^2 = C^2$; s.s., $xy = 0$; t.l., $y = x$.
Hint: Use $x = \rho\cos\theta$, $y = \rho\sin\theta$.

CHAPTER 11

Applications of First Order and Higher Degree Equations

IN FINDING THE EQUATION of a curve having a given property, (for example, that its slope at any point is twice the abscissa of the point), we obtained in Chapter 7 a *family* of curves ($y = x^2 + C$) having the property. In this chapter the family of curves will frequently be a family of straight lines. In such cases, the curve in which we are most interested is the envelope of the family.

SOLVED PROBLEMS

1. Find the curve for which:

   a) the sum of the intercepts of the tangent line on the coordinate axes is equal to k.

   b) the product of the intercepts of the tangent line on the coordinate axes is equal to k.

   c) the portion of the tangent line intercepted by the coordinate axes is of constant length k.

Let the equation of the tangent line be

$$y = px + f(p),$$

the x- and y-intercepts being $-f(p)/p$ and $f(p)$ respectively.

a) Since $f(p) - f(p)/p = k$, $f(p) = -kp/(1-p)$, and

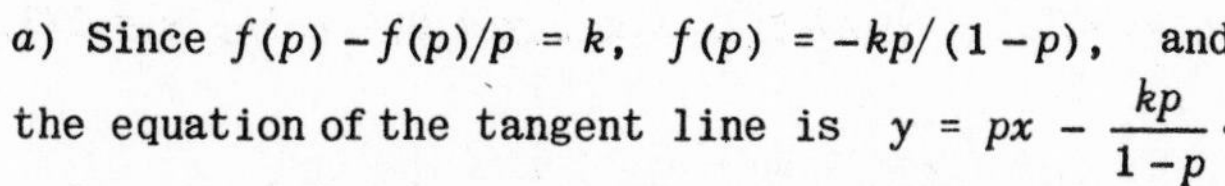

the equation of the tangent line is $y = px - \dfrac{kp}{1-p}$.

This is a Clairaut equation, the primitive being the family of lines $y = Cx - \dfrac{kC}{1-C}$ or $xC^2 - (x+y-k)C + y = 0$. The required curve, the envelope of the family, has equation $(x+y-k)^2 = 4xy$ or $x^{½} \pm y^{½} = k^{½}$. Note that this curve is an envelope (singular solution) since it satisfies the differential equation and cannot be obtained from the primitive by assigning a value to C.

b) Since $f(p)[-f(p)/p] = k$, $f(p) = \pm\sqrt{-kp}$, and the equation of the tangent line is $y = px \pm \sqrt{-kp}$. This is a Clairaut equation, the primitive being

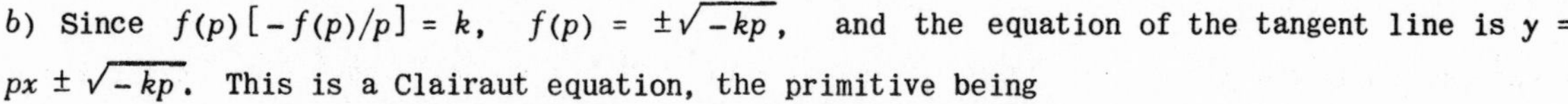

$$y - Cx = \pm\sqrt{-Ck} \qquad \text{or} \qquad x^2C^2 + (k-2xy)C + y^2 = 0.$$

The required curve, the envelope of the family, has equation $4xy = k$.

c) Since $[\{f(p)\}^2 + \{-f(p)/p\}^2]^{½} = k$, $f(p) = \pm\, kp/\sqrt{1+p^2}$, and the equation of the tangent line is $y = px \pm kp/\sqrt{1+p^2}$. The primitive of this equation is $y = Cx \pm kC/\sqrt{1+C^2}$.

Differentiating with respect to C, we have $0 = x \pm k/(1+C^2)^{3/2}$.

Then $x = \mp\, k/(1+C^2)^{3/2}$, $y = Cx \pm kC/(1+C^2)^{1/2} = \pm\, kC^3/(1+C^2)^{3/2}$, and the equation of the envelope is $\quad x^{2/3} + y^{2/3} = k^{2/3}/(1+C^2) + k^{2/3}C^2/(1+C^2) = k^{2/3}$.

2. Find the curve for which:
   a) the sum of the distances of the points $(a,0)$ and $(-a,0)$ from the tangent line is equal to k.
   b) the sum of the distances of the points $(a,0)$ and $(0,a)$ from the tangent line is equal to k.

Take $\dfrac{px - y + f(p)}{\sqrt{1+p^2}} = 0$ as the normal form of the equation of a tangent line.

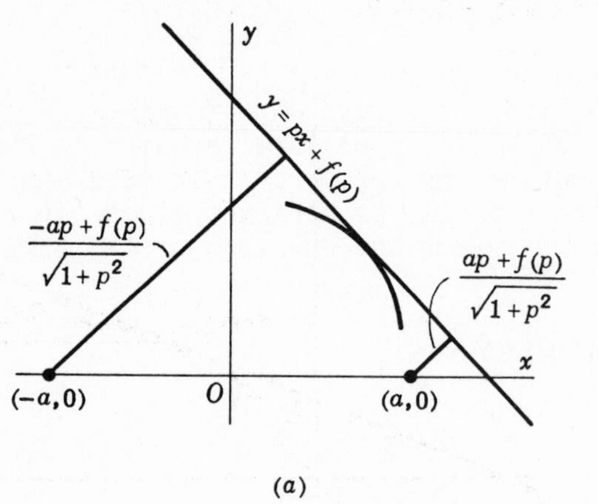

(a)

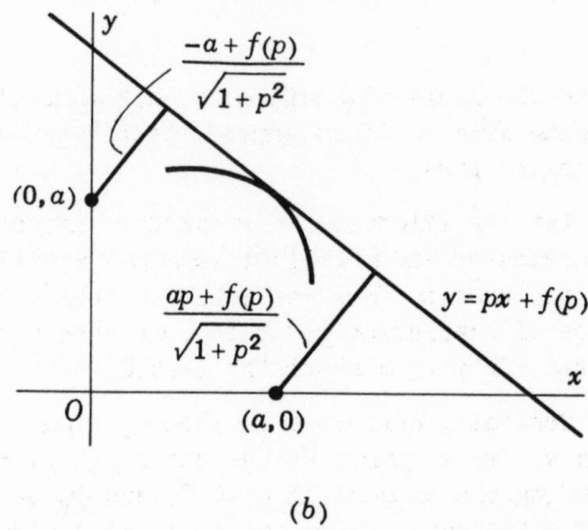

(b)

a) The distances of the points $(a,0)$ and $(-a,0)$ from the line are $\dfrac{ap+f(p)}{\sqrt{1+p^2}}$ and $\dfrac{-ap+f(p)}{\sqrt{1+p^2}}$ respectively. Thus, $\dfrac{2f(p)}{\sqrt{1+p^2}} = k$, $f(p) = \frac{1}{2}k\sqrt{1+p^2}$, and the equation of the tangent line is $y = px + \frac{1}{2}k\sqrt{1+p^2}$. The primitive of this Clairaut equation is

$$y = Cx + \tfrac{1}{2}k\sqrt{1+C^2} \quad \text{or} \quad (4x^2-k^2)C^2 - 8xyC + 4y^2 - k^2 = 0.$$

The required curve, the envelope of this family of lines, has as equation $x^2 + y^2 = \frac{1}{4}k^2$.

b) The distances of the points $(a,0)$ and $(0,a)$ from the line are $\dfrac{ap+f(p)}{\sqrt{1+p^2}}$ and $\dfrac{-a+f(p)}{\sqrt{1+p^2}}$ respectively. Thus, $\dfrac{-a+ap+2f(p)}{\sqrt{1+p^2}} = k$, $f(p) = \frac{1}{2}[k\sqrt{1+p^2} - ap + a]$, and the equation of the tangent line is $y = px + \frac{1}{2}[k\sqrt{1+p^2} - ap + a]$. The primitive is $y = Cx + \frac{1}{2}[k\sqrt{1+C^2} - aC + a]$.

Differentiating with respect to C, we have $0 = x + \frac{1}{2}[kC/\sqrt{1+C^2} - a]$.

Then $x = -\frac{1}{2}[kC/\sqrt{1+C^2} - a]$, $y = \frac{1}{2}[k/\sqrt{1+C^2} + a]$, and the envelope of the family of lines has equation $x^2 + y^2 - ax - ay = \frac{1}{4}(k^2 - 2a^2)$.

3. Find the curve such that the tangent line at any of its points P bisects the angle between the ordinate at P and the line joining P and the origin.

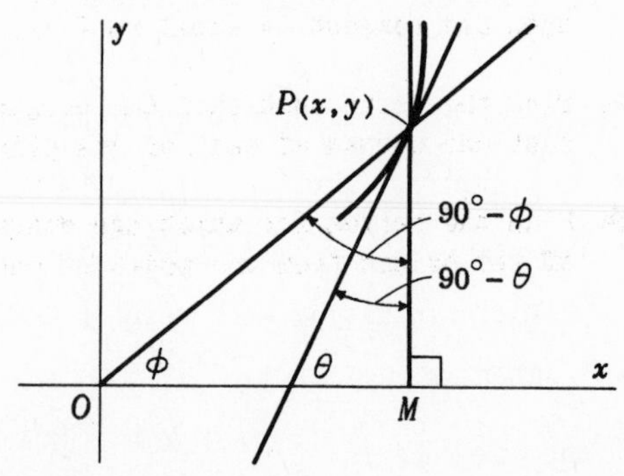

Let θ be the angle of inclination of a tangent line and ϕ be the angle of inclination of OP. Then, if M is the foot of the ordinate through P,
angle $OPM = 90° - \phi = 2(90° - \theta) = 180° - 2\theta$.

Now $\tan(90° - \phi) = \cot\phi = \tan(180° - 2\theta) = -\tan 2\theta$ and $\tan\phi \tan 2\theta = -1$.

Since $\tan\phi = y/x$ and $\tan\theta = y' = p$, we obtain the differential equation of the curve $\frac{y}{x}\cdot\frac{2p}{1-p^2} = -1$ or $2y = xp - x/p$. Differentiating with respect to x, $2p = p - \frac{1}{p} + (x + \frac{x}{p^2})\frac{dp}{dx}$, $p(p^2+1) = x(p^2+1)\frac{dp}{dx}$, and $x\,dp - p\,dx = 0$.

Integrating, $\ln p = \ln x + \ln C$ or $p = Cx$. Substituting for p in the differential equation, we obtain the family of parabolas $C^2x^2 - 2Cy - 1 = 0$.

4. Find the shape of a reflector such that light coming from a fixed source is reflected in parallel rays.

Let the fixed point be at the origin of coordinates and the reflected rays be parallel to the x-axis. The reflector is then a surface of revolution generated by revolving a curve $f(x,y) = 0$ about the x-axis.

Confining ourselves to the xOy plane, let $P(x,y)$ be a point on the curve $f(x,y) = 0$, TPT' be the tangent line at P, and PQ be the reflected ray. Since the angle of incidence is equal to the angle of reflection, it follows that $\angle OPT = \phi = \angle QPT'$.

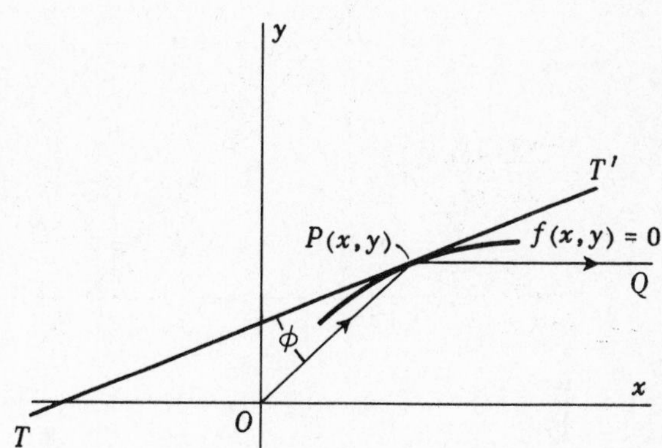

Now $p = \frac{dy}{dx} = \tan\angle OTP = \tan\phi$ and $\tan\angle TOP = \tan(\pi - 2\phi) = -\tan 2\phi = \frac{-2\tan\phi}{1-\tan^2\phi} = -\frac{y}{x}$; hence, $\frac{y}{x} = \frac{2p}{1-p^2}$ or $2x = \frac{y}{p} - yp$.

Differentiating with respect to y, $\frac{2}{p} = \frac{1}{p} - \frac{y}{p^2}\frac{dp}{dy} - p - y\frac{dp}{dy}$ and $\frac{dp}{p} = -\frac{dy}{y}$. Then, $p = \frac{C}{y}$.

Eliminating p between this relation and the original differential equation, we have the family of curves $y^2 = 2Cx + C^2$. Thus, the reflector is a member of the family of paraboloids of revolution $y^2 + z^2 = 2Cx + C^2$.

SUPPLEMENTARY PROBLEMS

5. Find the curve for which each of its tangent lines forms with the coordinate axes a triangle of constant area a^2. *Ans.* $2xy = a^2$

6. Find the curve for which the product of the distances of the points $(a,0)$ and $(-a,0)$ from the tangent lines is equal to k. *Ans.* $kx^2 = (k+a^2)(k-y^2)$

7. Find the curve for which the projection upon the y-axis of the perpendicular from the origin upon any tangent is equal to k. *Ans.* $x^2 = 4k(k-y)$

8. Find the curve such that the origin bisects the portion of the y-axis intercepted by the tangent and normal at each of its points. *Ans.* $x^2 + 2Cy = C^2$

9. Find the curves for which the distance of the tangent from the origin varies as the distance of the origin from the point of contact.

Hint: $\frac{\rho^2}{\sqrt{\rho^2 + (d\rho/d\theta)^2}} = k\rho$. *Ans.* $\rho = Ce^{\theta\frac{\sqrt{1-k^2}}{k}}$

CHAPTER 12

Linear Equations of Order n

A LINEAR DIFFERENTIAL EQUATION of order n has the form

$$1)\quad P_0 \frac{d^n y}{dx^n} + P_1 \frac{d^{n-1} y}{dx^{n-1}} + P_2 \frac{d^{n-2} y}{dx^{n-2}} + \cdots\cdots + P_{n-1} \frac{dy}{dx} + P_n y = Q,$$

where $P_0 \neq 0$, P_1, P_2, ..., P_n, Q are functions of x or constants.

If $Q = 0$, 1) has the form

$$2)\quad P_0 \frac{d^n y}{dx^n} + P_1 \frac{d^{n-1} y}{dx^{n-1}} + P_2 \frac{d^{n-2} y}{dx^{n-2}} + \cdots\cdots + P_{n-1} \frac{dy}{dx} + P_n y = 0$$

and is called *homogeneous* to indicate that all of the terms are of the same (first) degree in y and its derivatives.

Examples. $A)$ $x^3 \dfrac{d^3 y}{dx^3} + 2x \dfrac{d^2 y}{dx^2} - 5 \dfrac{dy}{dx} - xy = \sin x$, of order 3.

$B)$ $\dfrac{d^2 y}{dx^2} - 3 \dfrac{dy}{dx} + 2y = 0$, of order 2.

Equation $B)$ is an example of a homogeneous linear equation.

SOLUTIONS. If $y = y_1(x)$ is a solution of 2), then $y = C_1 y_1(x)$, where C_1 is an arbitrary constant, is also a solution. If $y = y_1(x)$, $y = y_2(x)$, $y = y_3(x)$, ... are solutions of 2), then $y = C_1 y_1(x) + C_2 y_2(x) + C_3 y_3(x) + \cdots\cdots$ is also a solution.

A set of solutions $y = y_1(x)$, $y = y_2(x)$, ..., $y = y_n(x)$ of 2) is said to be *linearly independent* if the equality

$$c_1 y_1 + c_2 y_2 + c_3 y_3 \quad \cdots\cdots\cdots + c_n y_n = 0,$$

where the c's are constants, holds only when $c_1 = c_2 = c_3 = \cdots\cdot = c_n = 0$.

Example 1. The functions e^x and e^{-x} are linearly independent. To show this, form $c_1 e^x + c_2 e^{-x} = 0$, where c_1 and c_2 are constants, and differentiate to obtain $c_1 e^x - c_2 e^{-x} = 0$. When the two relations are solved simultaneously for c_1 and c_2, we find $c_1 = c_2 = 0$.

Example 2. The functions e^x, $2e^x$, and e^{-x} are linearly dependent, since $c_1 e^x + 2c_2 e^x + c_3 e^{-x} = 0$ when $c_1 = 2$, $c_2 = -1$, $c_3 = 0$.

A necessary and sufficient condition that the set of n solutions be linearly independent is that:

$$W = \begin{vmatrix} y_1 & y_2 & y_3 & \cdots & y_n \\ y_1' & y_2' & y_3' & \cdots & y_n' \\ y_1'' & y_2'' & y_3'' & \cdots & y_n'' \\ \cdots & \cdots & \cdots & \cdots & \cdots \\ y_1^{(n-1)} & y_2^{(n-1)} & y_3^{(n-1)} & \cdots & y_n^{(n-1)} \end{vmatrix} \neq 0.$$

If $y = y_1(x)$, $y = y_2(x)$, ..., $y = y_n(x)$ are n linearly independent solutions of 2), then

3) $$y = C_1 y_1(x) + C_2 y_2(x) + \cdots\cdots + C_n y_n(x)$$

is the primitive of 2).

If $y = R(x)$ is a particular solution, also called *particular integral*, of 1), then

4) $$y = C_1 y_1(x) + C_2 y_2(x) + \cdots\cdots + C_n y_n(x) + R(x)$$

is the primitive of 1). Note that 4) contains all of 3). This part of 4) is called the *complementary function*. Thus the primitive of 1) consists of the sum of the complementary function and a particular integral.

Attention has been called to the fact that the primitive of a differential equation is not necessarily the complete solution of the equation. However, when the equation is linear, the primitive is its complete solution. Thus 3) and 4) may be called complete solutions of 2) and 1) respectively.

LINEAR DIFFERENTIAL EQUATIONS with constant coefficients (equation B) above) will be treated in Chapters 13-16. Those with variable coefficients (equation A) above) will be considered in Chapters 17-19.

SOLVED PROBLEMS

1. Show that the equation $\dfrac{d^2y}{dx^2} - \dfrac{dy}{dx} - 2y = 0$ has two distinct solutions of the form $y = e^{ax}$.

If $y = e^{ax}$, for some value of a, is a solution then the given equation is satisfied when the replacements $y = e^{ax}$, $\dfrac{dy}{dx} = ae^{ax}$, $\dfrac{d^2y}{dx^2} = a^2e^{ax}$ are made in it.

We obtain $\dfrac{d^2y}{dx^2} - \dfrac{dy}{dx} - 2y = e^{ax}(a^2 - a - 2) = 0$ which is satisfied when $a = -1, 2$.

Thus $y = e^{-x}$ and $y = e^{2x}$ are solutions.

2. Show that $y = C_1e^{-x} + C_2e^{2x}$ is the primitive of the equation of Problem 1.

Substituting for y and its derivatives in the differential equation, it is readily checked that $y = C_1e^{-x} + C_2e^{2x}$ is a solution. To show that it is the primitive, we note first that the number (2) of arbitrary constants and the order (2) of the equation agree and second that since $W = \begin{vmatrix} e^{-x} & e^{2x} \\ -e^{-x} & 2e^{2x} \end{vmatrix} = 3e^{x} \neq 0$, $y = e^{-x}$ and $y = e^{2x}$ are linearly independent.

3. Show that the differential equation $x^3 \dfrac{d^3y}{dx^3} - 6x\dfrac{dy}{dx} + 12y = 0$ has three linearly independent solutions of the form $y = x^r$.

After making the replacements

$$y = x^r, \quad \frac{dy}{dx} = rx^{r-1}, \quad \frac{d^2y}{dx^2} = r(r-1)x^{r-2}, \quad \frac{d^3y}{dx^3} = r(r-1)(r-2)x^{r-3}$$

in the left member of the given equation, we have $x^r(r^3 - 3r^2 - 4r + 12) = 0$ which is satisfied when $r = 2, 3, -2$. The corresponding solutions $y = x^2$, $y = x^3$, $y = x^{-2}$ are linearly independent

since $W = \begin{vmatrix} x^2 & x^3 & x^{-2} \\ 2x & 3x^2 & -2x^{-3} \\ 2 & 6x & 6x^{-4} \end{vmatrix} = 20 \neq 0$. The primitive is $y = C_1x^2 + C_2x^3 + C_3x^{-2}$.

4. Verify that $y = -\sin x$ is a particular integral of $\dfrac{d^2y}{dx^2} - \dfrac{dy}{dx} - 2y = \cos x + 3\sin x$ and write the primitive.

Substituting for y and its derivatives in the differential equation, it is found that the equation is satisfied. From Problem 2, the complementary function is $y = C_1e^{-x} + C_2e^{2x}$.

Hence, the primitive is $y = C_1e^{-x} + C_2e^{2x} - \sin x$.

5. Verify that $y = \ln x$ is a particular integral of $x^3\dfrac{d^3y}{dx^3} - 6x\dfrac{dy}{dx} + 12y = 12\ln x - 4$ and write the primitive.

Substituting for y and its derivatives in the given equation, it is found that the equation is satisfied. From Problem 3, the complementary function is $y = C_1x^2 + C_2x^3 + C_3x^{-2}$.

Hence, the primitive is $y = C_1x^2 + C_2x^3 + C_3x^{-2} + \ln x$.

6. Show that $\dfrac{d^4y}{dx^4} - \dfrac{d^3y}{dx^3} - 3\dfrac{d^2y}{dx^2} + 5\dfrac{dy}{dx} - 2y = 0$ has only two linearly independent solutions of the form $y = e^{ax}$.

Substituting for y and its derivatives in the given equation, we have $e^{ax}(a^4 - a^3 - 3a^2 + 5a - 2) = 0$ which is satisfied when $a = 1, 1, 1, -2$.

Since $\begin{vmatrix} e^x & e^{-2x} \\ e^x & -2e^{-2x} \end{vmatrix} \neq 0$, but $\begin{vmatrix} e^x & e^x & e^x & e^{-2x} \\ e^x & e^x & e^x & -2e^{-2x} \\ e^x & e^x & e^x & 4e^{-2x} \\ e^x & e^x & e^x & -8e^{-2x} \end{vmatrix} = 0$, the linearly independent solutions are $y = e^x$ and $y = e^{-2x}$.

7. Verify that $y = e^x$, $y = xe^x$, $y = x^2e^x$, and $y = e^{-2x}$ are four linearly independent solutions of the equation of Problem 6 and write the primitive.

By Problem 6, $y = e^x$ and $y = e^{-2x}$ are solutions. By direct substitution in the given equation it is found that the others are solutions.

Since
$$W = \begin{vmatrix} e^x & xe^x & x^2e^x & e^{-2x} \\ e^x & xe^x + e^x & x^2e^x + 2xe^x & -2e^{-2x} \\ e^x & xe^x + 2e^x & x^2e^x + 4xe^x + 2e^x & 4e^{-2x} \\ e^x & xe^x + 3e^x & x^2e^x + 6xe^x + 6e^x & -8e^{-2x} \end{vmatrix} = e^x \begin{vmatrix} 1 & 0 & 0 & 1 \\ 1 & 1 & 0 & -2 \\ 1 & 2 & 2 & 4 \\ 1 & 3 & 6 & -8 \end{vmatrix} = -54e^x \neq 0,$$

these solutions are linearly independent and the primitive is

$$y = C_1e^x + C_2xe^x + C_3x^2e^x + C_4e^{-2x}.$$

8. Verify that $y = e^{-2x}\cos 3x$ and $y = e^{-2x}\sin 3x$ are solutions of $\dfrac{d^2y}{dx^2} + 4\dfrac{dy}{dx} + 13y = 0$ and write the primitive.

Substituting for y and its derivatives, it is found that the equation is satisfied.

Since $W = 3e^{-4x} \neq 0$, the solutions are linearly independent.

Hence, the primitive is $y = e^{-2x}(C_1 \cos 3x + C_2 \sin 3x)$.

SUPPLEMENTARY PROBLEMS

9. Show that each of the following sets of functions are linearly independent.

a) $\sin ax$, $\cos ax$ c) $1, x, x^2$ e) $\ln x$, $x \ln x$, $x^2 \ln x$

b) $e^{ax}\sin bx$, $e^{ax}\cos bx$ d) e^{ax}, e^{bx}, e^{cx} $(a \neq b \neq c)$

Form the differential equation having the given primitive.

10. $y = C_1e^{2x} + C_2e^{-3x}$ *Ans.* $y'' + y' - 6y = 0$

11. $y = C_1e^{2x} + C_2xe^{2x} + C_3x^2e^{2x}$ $y''' - 6y'' + 12y' - 8y = 0$

12. $y = C_1e^x + C_2e^{2x} + e^{5x}/12$ $y'' - 3y' + 2y = e^{5x}$

13. $y = C_1\cos 3x + C_2\sin 3x + (4x\cos x + \sin x)/32$ $y'' + 9y = x\cos x$

14. $y = C_1x^2 + C_2x^2 \ln x$ $x^2y'' - 3xy' + 4y = 0$

15. $y = C_1x + C_2x\ln x + C_3x\ln^2 x + x^4/9$ $x^3y''' + xy' - y = 3x^4$

16. $y = C_1 \sin x^2 + C_2 \cos x^2$ $xy'' - y' + 4x^3y = 0$

17. $y = \ln \sin(x - C_1) + C_2$ $y'' + (y')^2 + 1 = 0$

18. $y^2 = C_1x + C_2 + 2x^2$ $yy'' + (y')^2 = 2$

19. $x = C_1 + C_2y + y\ln y$ $yy'' + (y')^3 = 0$

CHAPTER 13

Homogeneous Linear Equations with Constant Coefficients

THE HOMOGENEOUS LINEAR EQUATION with constant coefficients has the form

$$1)\qquad P_0\frac{d^ny}{dx^n} + P_1\frac{d^{n-1}y}{dx^{n-1}} + P_2\frac{d^{n-2}y}{dx^{n-2}} + \cdots\cdots + P_{n-1}\frac{dy}{dx} + P_n y = 0$$

in which $P_0 \neq 0$, P_1, P_2, ⋯⋯, P_n are constants.

By a convenient change of notation, writing $\frac{dy}{dx} = Dy$, $\frac{d^2y}{dx^2} = \frac{d}{dx}(\frac{dy}{dx}) = D\cdot Dy = D^2y$, etc., 1) becomes

$$2)\qquad (P_0D^n + P_1D^{n-1} + P_2D^{n-2} + \cdots\cdots\cdots + P_{n-1}D + P_n)y = 0.$$

Now $D = \frac{d}{dx}$ is an operator which acts on y, and

$$3)\qquad P_0D^n + P_1D^{n-1} + P_2D^{n-2} + \cdots\cdots\cdots + P_{n-1}D + P_n$$

is simply a much more complex operator. However, we shall find it very convenient to consider 3) at times as a polynomial in the variable D and to denote it by $F(D)$. Thus, 1) may be written briefly as

$$4)\qquad F(D)y = 0.$$

It can be shown in general and will be indicated by an example that when 3) is treated as a polynomial and factored as

$$5)\qquad F(D) = P_0(D-m_1)(D-m_2)(D-m_3)\cdots\cdots\cdots(D-m_{n-1})(D-m_n),$$

then

$$6)\qquad F(D)y = P_0(D-m_1)(D-m_2)(D-m_3)\cdots\cdots\cdots(D-m_{n-1})(D-m_n)y = 0$$

remains valid, i.e., is equivalent to 1) when D is treated as an operator.

EXAMPLE. In the D notation $\frac{d^3y}{dx^3} - \frac{d^2y}{dx^2} - 4\frac{dy}{dx} + 4y = 0$ becomes $(D^3 - D^2 - 4D + 4)y = 0$ and, in factored form, $(D-1)(D-2)(D+2)y = 0$. Now

$$(D-1)(D-2)(D+2)y = (D-1)(D-2)(\frac{d}{dx} + 2)y = (D-1)(D-2)(\frac{dy}{dx} + 2y)$$

$$= (D-1)\{\frac{d}{dx}(\frac{dy}{dx} + 2y) - 2(\frac{dy}{dx} + 2y)\} = (D-1)(\frac{d^2y}{dx^2} - 4y)$$

$$= \frac{d}{dx}(\frac{d^2y}{dx^2} - 4y) - 1(\frac{d^2y}{dx^2} - 4y)$$

$$= \frac{d^3y}{dx^3} - 4\frac{dy}{dx} - \frac{d^2y}{dx^2} + 4y = \frac{d^3y}{dx^3} - \frac{d^2y}{dx^2} - 4\frac{dy}{dx} + 4y = 0.$$

In Problem 1 below, it will be indicated that the order of the factors here is immaterial.

THE EQUATION $F(D) = (D-m_1)(D-m_2)(D-m_3)\cdots\cdots(D-m_{n-1})(D-m_n) = 0$ is sometimes called the characteristic equation of 1) and the roots $m_1, m_2, m_3, \cdots, m_n$ are called the characteristic roots. Note that it is never necessary to write the characteristic equation since its roots can be read directly from 6).

TO OBTAIN THE PRIMITIVE of 1) we first write the equation in the form 6).

a) Suppose $m_1 \neq m_2 \neq m_3 \neq \cdots\cdots\cdots \neq m_{n-1} \neq m_n$. Then

$$y = C_1 e^{m_1 x} + C_2 e^{m_2 x} + C_3 e^{m_3 x} + \cdots\cdots\cdots + C_n e^{m_n x},$$

involving n linearly independent solutions of 1) with n arbitrary constants, is the primitive.

Thus in the example above, where $\dfrac{d^3y}{dx^3} - \dfrac{d^2y}{dx^2} - 4\dfrac{dy}{dx} + 4y = 0$ or $(D-1)(D-2)(D+2)y = 0$, the characteristic roots are $1, 2, -2$ and the primitive is $y = C_1 e^x + C_2 e^{2x} + C_3 e^{-2x}$. See also Problems 5-7.

b) Suppose $m_1 = m_2 \neq m_3 \neq \cdots\cdots\cdots \neq m_{n-1} \neq m_n$. Then

$$y = C_1 e^{m_1 x} + C_2 x e^{m_1 x} + C_3 e^{m_3 x} + \cdots\cdots\cdots + C_n e^{m_n x}$$

is the primitive.

In general, to a root m occurring r times there corresponds

$$C_1 e^{mx} + C_2 x e^{mx} + C_3 x^2 e^{mx} + \cdots\cdots + C_r x^{r-1} e^{mx}$$

in the primitive.

Thus to solve $\dfrac{d^3y}{dx^3} - 2\dfrac{d^2y}{dx^2} - 4\dfrac{dy}{dx} + 8y = 0$, write the equations as $(D^3 - 2D^2 - 4D + 8)y = (D-2)^2(D+2)y = 0$. The characteristic roots are $2, 2, -2$ and the primitive is $y = C_1 e^{2x} + C_2 x e^{2x} + C_3 e^{-2x}$. See also Problems 8-10.

c) If the coefficients of 1) are real and if $a+bi$ is a complex root of 6), so also is $a-bi$. The corresponding terms in the primitive are

$$\begin{aligned} Ae^{(a+bi)x} + Be^{(a-bi)x} &= e^{ax}(Ae^{bix} + Be^{-bix}) \\ &= e^{ax}(C_1 \cos bx + C_2 \sin bx) \\ &= Pe^{ax} \sin(bx + Q) = Pe^{ax} \cos(bx + R), \end{aligned}$$

where A, B, C_1, C_2, P, Q, R are arbitrary constants.

Thus the characteristic roots of $\dfrac{d^2y}{dx^2} - 4\dfrac{dy}{dx} + 5y = 0$ or $(D^2 - 4D + 5)y = 0$ are $2 \pm i$. Here $a = 2$, $b = 1$ and the primitive is $y = e^{2x}(C_1 \cos x + C_2 \sin x)$.

See also Problems 11-15.

SOLVED PROBLEMS

1. Show that $(D-a)(D-b)(D-c)y = (D-b)(D-c)(D-a)y$.

$$(D-a)(D-b)(D-c)y = (D-a)(D-b)(\frac{dy}{dx} - cy) = (D-a)(\frac{d^2y}{dx^2} - (b+c)\frac{dy}{dx} + bcy)$$

$$= \frac{d^3y}{dx^3} - (a+b+c)\frac{d^2y}{dx^2} + (ab+bc+ac)\frac{dy}{dx} - abcy.$$

$$(D-b)(D-c)(D-a)y = (D-b)(D-c)(\frac{dy}{dx} - ay) = (D-b)(\frac{d^2y}{dx^2} - (a+c)\frac{dy}{dx} + acy)$$

$$= \frac{d^3y}{dx^3} - (a+b+c)\frac{d^2y}{dx^2} + (ab+ac+bc)\frac{dy}{dx} - abcy.$$

2. Verify that $y = C_1e^{ax} + C_2e^{bx} + C_3e^{cx}$ satisfies the differential equation $(D-a)(D-b)(D-c)y = 0$.

We are to show that $(D-a)(D-b)(D-c)(C_1e^{ax} + C_2e^{bx} + C_3e^{cx}) = 0$.

$(D-a)(D-b)(D-c)C_1e^{ax} = (D-b)(D-c)(D-a)C_1e^{ax} = (D-b)(D-c)0 = 0$, and similarly for the other two terms.

3. Verify that $y = C_1e^{mx} + C_2xe^{mx} + C_3x^2e^{mx}$ satisfies the differential equation $(D-m)^3y = 0$.

This follows since: a) $(D-m)^3C_1e^{mx} = (D-m)^2(D-m)C_1e^{mx} = (D-m)^2 0 = 0$,

b) $(D-m)^3C_2xe^{mx} = (D-m)^2C_2e^{mx} = (D-m)0 = 0$, and

c) $(D-m)^3C_3x^2e^{mx} = 2(D-m)^2C_3xe^{mx} = 2(D-m)0 = 0$.

4. Find the primitive of $(D-m)^2y = 0$ (a) by assuming a solution of the form $y = x^re^{mx}$ and (b) by solving the equivalent pair of equations $(D-m)y = v$, $(D-m)v = 0$.

a) $(D-m)^2y = (D-m)(D-m)x^re^{mx} = (D-m)rx^{r-1}e^{mx} = r(r-1)x^{r-2}e^{mx} = 0$ when $r = 0,1$.

Thus the equation has two linearly independent solutions $y = e^{mx}$ and $y = xe^{mx}$.

The primitive is $y = C_1e^{mx} + C_2xe^{mx}$.

b) If we write $(D-m)y = v$, then $(D-m)^2y = (D-m)(D-m)y = (D-m)v = 0$.

Solving $(D-m)v = 0$, we obtain $v = C_2e^{mx}$. Since $(D-m)y = \frac{dy}{dx} - my = C_2e^{mx}$ is linear of the first order, its solution by the method of Chapter 6 is

$$ye^{-mx} = \int e^{-mx}(C_2e^{mx})dx = C_1 + C_2x \quad \text{or} \quad y = C_1e^{mx} + C_2xe^{mx}.$$

DISTINCT REAL ROOTS.

5. Solve $\frac{d^2y}{dx^2} + \frac{dy}{dx} - 6y = 0$.

We write the equation as $(D^2+D-6)y = (D-2)(D+3)y = 0$.

The characteristic roots are 2,−3, and the primitive is $y = C_1e^{2x} + C_2e^{-3x}$.

6. Solve $\dfrac{d^3y}{dx^3} - \dfrac{d^2y}{dx^2} - 12\dfrac{dy}{dx} = 0.$

We write the equation as $(D^3 - D^2 - 12D)y = 0$ or $D(D-4)(D+3)y = 0$.

The characteristic roots are 0,4,-3, and the primitive is $y = C_1 + C_2e^{4x} + C_3e^{-3x}$.

7. Solve $\dfrac{d^3y}{dx^3} + 2\dfrac{d^2y}{dx^2} - 5\dfrac{dy}{dx} - 6y = 0.$

We write the equation as $(D^3 + 2D^2 - 5D - 6)y$ or $(D-2)(D+1)(D+3)y = 0$.

The characteristic roots are 2,-1,-3, and the primitive is $y = C_1e^{2x} + C_2e^{-x} + C_3e^{-3x}$.

REPEATED ROOTS.

8. Solve $(D^3 - 3D^2 + 3D - 1)y = 0$ or $(D-1)^3y = 0$.

The characteristic roots are 1,1,1, and the primitive is $y = C_1e^x + C_2xe^x + C_3x^2e^x$.

9. Solve $(D^4 + 6D^3 + 5D^2 - 24D - 36)y = 0$ or $(D-2)(D+2)(D+3)^2y = 0$.

The characteristic roots are 2,-2,-3,-3. The primitive is $y = C_1e^{2x} + C_2e^{-2x} + C_3e^{-3x} + C_4xe^{-3x}$.

10. Solve $(D^4 - D^3 - 9D^2 - 11D - 4)y = 0$ or $(D+1)^3(D-4)y = 0$.

The characteristic roots are -1,-1,-1,4. The primitive is $y = e^{-x}(C_1 + C_2x + C_3x^2) + C_4e^{4x}$.

COMPLEX ROOTS.

11. Solve $(D^2 - 2D + 10)y = 0$.

The characteristic roots are $1 \pm 3i$, and the primitive is

$$y = e^x(C_1\cos 3x + C_2\sin 3x) \quad\text{or}\quad C_3e^x\sin(3x + C_4) \quad\text{or}\quad C_3e^x\cos(3x + C_5).$$

12. Solve $(D^3 + 4D)y = 0$ or $D(D^2 + 4)y = 0$.

The characteristic roots are $0, \pm 2i$, and the primitive is $y = C_1 + C_2\cos 2x + C_3\sin 2x$.

13. Solve $(D^4 + D^3 + 2D^2 - D + 3)y = 0$ or $(D^2 + 2D + 3)(D^2 - D + 1)y = 0$.

The characteristic roots are $-1 \pm i\sqrt{2}$, $\frac{1}{2} \pm \frac{1}{2}i\sqrt{3}$. and the primitive is

$$y = e^{-x}(C_1\cos\sqrt{2}\,x + C_2\sin\sqrt{2}\,x) + e^{\frac{1}{2}x}(C_3\cos\tfrac{1}{2}\sqrt{3}\,x + C_4\sin\tfrac{1}{2}\sqrt{3}\,x).$$

14. Solve $(D^4 + 5D^2 - 36)y = 0$ or $(D^2 - 4)(D^2 + 9)y = 0$.

The characteristic roots are ± 2, $\pm 3i$, and the primitive is

$$y = Ae^{2x} + Be^{-2x} + C_3\cos 3x + C_4\sin 3x$$
$$= C_1\cosh 2x + C_2\sinh 2x + C_3\cos 3x + C_4\sin 3x$$

since $\cosh 2x = \frac{1}{2}(e^{2x} + e^{-2x})$ and $\sinh 2x = \frac{1}{2}(e^{2x} - e^{-2x})$.

15. Solve $(D^2 - 2D + 5)^2y = 0$. The characteristic roots are $1 \pm 2i$, $1 \pm 2i$, and the primitive is

$$y = e^x(C_1\cos 2x + C_2\sin 2x) + xe^x(C_3\cos 2x + C_4\sin 2x)$$
$$= e^x\{(C_1 + C_3x)\cos 2x + (C_2 + C_4x)\sin 2x\}.$$

SUPPLEMENTARY PROBLEMS

Solve.

16. $(D^2 + 2D - 15)y = 0$ *Ans.* $y = C_1 e^{3x} + C_2 e^{-5x}$

17. $(D^3 + D^2 - 2D)y = 0$ $y = C_1 + C_2 e^{x} + C_3 e^{-2x}$

18. $(D^2 + 6D + 9)y = 0$ $y = C_1 e^{-3x} + C_2 x e^{-3x}$

19. $(D^4 - 6D^3 + 12D^2 - 8D)y = 0$ $y = C_1 + C_2 e^{2x} + C_3 x e^{2x} + C_4 x^2 e^{2x}$

20. $(D^2 - 4D + 13)y = 0$ $y = e^{2x}(C_1 \cos 3x + C_2 \sin 3x)$

21. $(D^2 + 25)y = 0$ $y = C_1 \cos 5x + C_2 \sin 5x$

22. $(D^3 - D^2 + 9D - 9)y = 0$ $y = C_1 e^{x} + C_2 \cos 3x + C_3 \sin 3x$

23. $(D^4 + 4D^2)y = 0$ $y = C_1 + C_2 x + C_3 \cos 2x + C_4 \sin 2x$

24. $(D^4 - 6D^3 + 13D^2 - 12D + 4)y = 0$ $y = (C_1 + C_2 x)e^{x} + (C_3 + C_4 x)e^{2x}$

25. $(D^6 + 9D^4 + 24D^2 + 16)y = 0$ $y = C_1 \cos x + C_2 \sin x + (C_3 + C_4 x)\cos 2x + (C_5 + C_6 x)\sin 2x$

CHAPTER 14

Linear Equations with Constant Coefficients

THE PRIMITIVE OF

1) $$F(D)y = (P_0 D^n + P_1 D^{n-1} + \cdots\cdots\cdots + P_{n-1}D + P_n)y = Q(x),$$

where $P_0 \neq 0$, P_1, P_2, $\cdots\cdots$, P_n are constants and $Q \equiv Q(x) \neq 0$, is the sum of the complementary function (primitive of $F(D)y = 0$ obtained in the preceding chapter) and any particular integral of 1). (See Chapter 12.)

At times a particular integral may be found by inspection. For example, $y = \frac{1}{2}x$ is a particular integral of $(D^3 - 3D^2 + 2)y = x$, since $D^3y = D^2y = 0$. Such equations occur infrequently, however, and we proceed to consider in this chapter two general procedures for obtaining a *particular integral*. Other procedures will be given in the next two chapters.

In each of the procedures below, use will be made of an operator $\dfrac{1}{F(D)}$ defined by the relation $\dfrac{1}{F(D)}\cdot F(D)y = y$. When the operator is applied to 1) we obtain

$$\frac{1}{F(D)}\cdot F(D)y = y = \frac{1}{F(D)}Q$$

or

2) $$y = \frac{1}{D-m_1}\cdot\frac{1}{D-m_2}\cdot\frac{1}{D-m_3}\cdots\cdots\cdots\cdots\frac{1}{D-m_n}Q.$$

FIRST METHOD. This consists of solving a succession of linear differential equations of order one, as follows:

SET	SOLVE	TO OBTAIN
$u = \dfrac{1}{D-m_n}Q$	$\dfrac{du}{dx} - m_n u = Q$	$u = e^{m_n x}\int Qe^{-m_n x}\,dx$
$v = \dfrac{1}{D-m_{n-1}}u$	$\dfrac{dv}{dx} - m_{n-1}v = u$	$v = e^{m_{n-1}x}\int ue^{-m_{n-1}x}\,dx$
$\cdots\cdots$	$\cdots\cdots$	$\cdots\cdots$
$\cdots\cdots$	$\cdots\cdots$	$\cdots\cdots$
$y = \dfrac{1}{D-m_1}w$	$\dfrac{dy}{dx} - m_1 y = w$	$y = e^{m_1 x}\int we^{-m_1 x}\,dx.$

As is indicated in Problem 3 below, the following formula may be established:

A) $$y = e^{m_1x}\int e^{(m_2-m_1)x}\int e^{(m_3-m_2)x}\int\cdots\cdots\int e^{(m_n-m_{n-1})x}\int Qe^{-m_nx}(dx)^n.$$

See Problems 1-6.

SECOND METHOD. This consists of expressing $\dfrac{1}{F(D)}$ as the sum of n partial fractions: $\dfrac{N_1}{D-m_1} + \dfrac{N_2}{D-m_2} + \cdots\cdots\cdots + \dfrac{N_n}{D-m_n}$. Then

$$B)\quad y = N_1 e^{m_1 x}\int Q e^{-m_1 x}\,dx + N_2 e^{m_2 x}\int Q e^{-m_2 x}\,dx + \cdots\cdots + N_n e^{m_n x}\int Q e^{-m_n x}\,dx.$$

See Problems 4-5.

In evaluating both A) and B), it is customary to discard the constants of integration as they appear; otherwise, one obtains the primitive rather than a particular integral of the differential equation. The complementary function is then obtained by inspection and added to the particular solution to form the primitive.

THE FOLLOWING FORMULAS will be found useful.

$$e^{ibx} = \cos bx + i\sin bx \qquad e^{-ibx} = \cos bx - i\sin bx$$

$$\sin bx = \frac{e^{ibx} - e^{-ibx}}{2i} \qquad \cos bx = \frac{e^{ibx} + e^{-ibx}}{2}$$

$$e^{bx} = \cosh bx + \sinh bx \qquad e^{-bx} = \cosh bx - \sinh bx$$

$$\sinh bx = \tfrac{1}{2}(e^{bx} - e^{-bx}) \qquad \cosh bx = \tfrac{1}{2}(e^{bx} + e^{-bx})$$

SOLVED PROBLEMS

1. Solve $(D^2 - 3D + 2)y = e^x$ or $(D-1)(D-2)y = e^x$.

The complementary function is $y = C_1 e^x + C_2 e^{2x}$, and a particular integral is $y = \dfrac{1}{D-1}\cdot\dfrac{1}{D-2}e^x$.

Let $u = \dfrac{1}{D-2}e^x$. Then $(D-2)u = e^x$ or $\dfrac{du}{dx} - 2u = e^x$, $ue^{-2x} = \int e^x e^{-2x}\,dx = \int e^{-x}\,dx = -e^{-x}$, and $u = -e^x$.

Now $y = \dfrac{1}{D-1}u$, $(D-1)y = u$ or $\dfrac{dy}{dx} - y = -e^x$, and $y = e^x\int -e^x e^{-x}\,dx = -xe^x$.

The primitive is $y = C_1 e^x + C_2 e^{2x} - xe^x$.

2. Solve $(D^3 + 3D^2 - 4)y = xe^{-2x}$ or $(D-1)(D+2)^2 y = xe^{-2x}$.

The complementary function is $y = C_1 e^x + C_2 e^{-2x} + C_3 xe^{-2x}$, and a particular integral is

$$y = \frac{1}{D-1}\cdot\frac{1}{D+2}\cdot\frac{1}{D+2}xe^{-2x}.$$

Let $u = \dfrac{1}{D+2}xe^{-2x}$. Then $\dfrac{du}{dx} + 2u = xe^{-2x}$ and $u = e^{-2x}\int xe^{-2x}\cdot e^{2x}\,dx = \dfrac{1}{2}x^2 e^{-2x}$.

Let $v = \dfrac{1}{D+2}u$. Then $\dfrac{dv}{dx} + 2v = \dfrac{1}{2}x^2 e^{-2x}$ and $v = e^{-2x}\int \dfrac{1}{2}x^2 e^{-2x}\cdot e^{2x}\,dx = \dfrac{1}{6}x^3 e^{-2x}$.

Now $y = \frac{1}{D-1}v$. Then $\frac{dy}{dx} - y = \frac{1}{6}x^3e^{-2x}$ and $y = e^x\int\frac{1}{6}x^3e^{-2x}\cdot e^{-x}\,dx = \frac{1}{6}e^x\int x^3e^{-3x}\,dx$

$$= -\frac{1}{18}e^{-2x}(x^3 + x^2 + \frac{2}{3}x + \frac{2}{9}).$$

The primitive is $y = C_1e^x + C_2e^{-2x} + C_3xe^{-2x} - \frac{1}{18}(x^3 + x^2)e^{-2x}$, the remaining terms of the particular integral being absorbed by the complementary function.

3. Find a particular integral of $(D-a)(D-b)y = Q$.

A particular integral is given by $y = \frac{1}{D-a}\cdot\frac{1}{D-b}Q$.

Let $\frac{1}{D-b}Q = u$. Then $\frac{du}{dx} - bu = Q$ and $u = e^{bx}\int Qe^{-bx}\,dx$.

Now $y = \frac{1}{D-a}u$. Then $\frac{dy}{dx} - ay = u = e^{bx}\int Qe^{-bx}\,dx$ and

$$y = e^{ax}\int e^{bx}e^{-ax}\int Qe^{-bx}dx\,dx = e^{ax}\int e^{(b-a)x}\int Qe^{-bx}(dx)^2.$$

4. Solve $(D^2 - 3D + 2)y = e^{5x}$ or $(D-1)(D-2)y = e^{5x}$.

The complementary function is $y = C_1e^x + C_2e^{2x}$, and a particular integral is $y = \frac{1}{D-1}\cdot\frac{1}{D-2}e^{5x}$.

First Method. $y = \frac{1}{D-1}\cdot\frac{1}{D-2}e^{5x} = e^x\int e^{(2-1)x}\int e^{5x}\cdot e^{-2x}(dx)^2$

$$= e^x\int e^x\int e^{3x}(dx)^2 = e^x\int e^x\frac{1}{3}e^{3x}\,dx = \frac{1}{3}e^x\int e^{4x}\,dx = \frac{1}{12}e^{5x}.$$

Second Method. $y = \frac{1}{(D-1)(D-2)}e^{5x} = (-\frac{1}{D-1} + \frac{1}{D-2})e^{5x}$

$$= -e^x\int e^{5x}\cdot e^{-x}\,dx + e^{2x}\int e^{5x}\cdot e^{-2x}\,dx$$

$$= -\frac{1}{4}e^xe^{4x} + \frac{1}{3}e^{2x}e^{3x} = \frac{1}{12}e^{5x}.$$

The primitive is $y = C_1e^x + C_2e^{2x} + \frac{1}{12}e^{5x}$.

5. Solve $(D^2 + 5D + 4)y = 3 - 2x$ or $(D+1)(D+4)y = 3 - 2x$.

The complementary function is $y = C_1e^{-x} + C_2e^{-4x}$, and a particular integral is

$$y = \frac{1}{(D+1)(D+4)}(3 - 2x).$$

First Method. $y = \frac{1}{D+1}\cdot\frac{1}{D+4}(3-2x) = e^{-x}\int e^{(-4+1)x}\int(3-2x)e^{4x}(dx)^2$

$$= e^{-x}\int e^{-3x}(\frac{3}{4}e^{4x} - \frac{1}{2}xe^{4x} + \frac{1}{8}e^{4x})dx = e^{-x}\int(\frac{7}{8}e^x - \frac{1}{2}xe^x)dx = e^{-x}(\frac{7}{8}e^x - \frac{1}{2}xe^x + \frac{1}{2}e^x) = \frac{11}{8} - \frac{1}{2}x.$$

Second Method. $y = \frac{1}{D+1}\cdot\frac{1}{D+4}(3-2x) = (\frac{1/3}{D+1} - \frac{1/3}{D+4})(3-2x)$

$$= \frac{1}{3}e^{-x}\int(3-2x)e^{x}\,dx - \frac{1}{3}e^{-4x}\int(3-2x)e^{4x}\,dx$$

$$= \frac{1}{3}e^{-x}(3e^{x} - 2xe^{x} + 2e^{x}) - \frac{1}{3}e^{-4x}(\frac{3}{4}e^{4x} - \frac{1}{2}xe^{4x} + \frac{1}{8}e^{4x}) = \frac{11}{8} - \frac{1}{2}x.$$

The primitive is $y = C_1e^{-x} + C_2e^{-4x} - \frac{1}{2}x + \frac{11}{8}$.

6. Solve $(D^3 - 5D^2 + 8D - 4)y = e^{2x}$ or $(D-1)(D-2)^2y = e^{2x}$.

The complementary function is $y = C_1e^{x} + C_2e^{2x} + C_3xe^{2x}$, and a particular integral is

$$y = \frac{1}{(D-1)(D-2)^2}e^{2x}.$$

$$y = \frac{1}{D-1}\cdot\frac{1}{D-2}\cdot\frac{1}{D-2}e^{2x}$$

$$= e^{x}\int e^{(2-1)x}\int e^{(2-2)x}\int e^{2x}e^{-2x}(dx)^3$$

$$= e^{x}\int e^{x}\iint(dx)^3 = e^{x}\int e^{x}\int x(dx)^2 = e^{x}\int e^{x}\tfrac{1}{2}x^2\,dx$$

$$= \tfrac{1}{2}e^{x}\int x^2e^{x}\,dx = \tfrac{1}{2}e^{x}(x^2e^{x} - 2xe^{x} + 2e^{x}) = \tfrac{1}{2}e^{2x}(x^2 - 2x + 2).$$

The primitive is $y = C_1e^{x} + C_2e^{2x} + C_3xe^{2x} + \frac{1}{2}x^2e^{2x}$, the remaining terms of the particular integral being absorbed in the complementary function.

7. Solve $(D^2 + 9)y = x\cos x$.

The complementary function is $y = C_1\cos 3x + C_2\sin 3x$, and a particular integral is

$$y = \frac{1}{D^2+9}x\cos x = e^{-3ix}\int e^{(3i+3i)x}\int x\cos x\, e^{-3ix}(dx)^2.$$

It will be simpler here to use $\cos x = \frac{1}{2}(e^{ix} + e^{-ix})$, so that:

$$y = \tfrac{1}{2}e^{-3ix}\int e^{6ix}\int x(e^{-2ix} + e^{-4ix})(dx)^2$$

$$= \frac{1}{2}e^{-3ix}\int e^{6ix}(\frac{1}{2}ixe^{-2ix} + \frac{1}{4}e^{-2ix} + \frac{1}{4}ixe^{-4ix} + \frac{1}{16}e^{-4ix})\,dx$$

$$= \frac{1}{2}e^{-3ix}\int(\frac{1}{2}ixe^{4ix} + \frac{1}{4}e^{4ix} + \frac{1}{4}ixe^{2ix} + \frac{1}{16}e^{2ix})\,dx$$

$$= \frac{1}{2}e^{-3ix}(\frac{1}{8}xe^{4ix} + \frac{1}{32}ie^{4ix} - \frac{1}{16}ie^{4ix} + \frac{1}{8}xe^{2ix} + \frac{1}{16}ie^{2ix} - \frac{1}{32}ie^{2ix})$$

$$= \frac{1}{16}x(e^{ix} + e^{-ix}) - \frac{1}{64}i(e^{ix} - e^{-ix}) = \frac{1}{8}x(\frac{e^{ix} + e^{-ix}}{2}) + \frac{1}{32}(\frac{e^{ix} - e^{-ix}}{2i})$$

$= \frac{1}{8}x\cos x + \frac{1}{32}\sin x$. The primitive is $y = C_1\cos 3x + C_2\sin 3x + \frac{1}{8}x\cos x + \frac{1}{32}\sin x$.

8. Solve $(D^2+4)y = 2\cos x\cos 3x = \cos 2x + \cos 4x.$

The complementary function is $y = C_1\cos 2x + C_2\sin 2x$, and a particular integral is

$$y = \frac{1}{D^2+4}(\cos 2x + \cos 4x) = \frac{1}{4}i\left(\frac{1}{D+2i} - \frac{1}{D-2i}\right)(\cos 2x + \cos 4x)$$

$$= \tfrac{1}{4}i\{ e^{-2ix}\int(\cos 2x + \cos 4x)e^{2ix}\,dx - e^{2ix}\int(\cos 2x + \cos 4x)e^{-2ix}\,dx \}$$

$$= \tfrac{1}{4}i\{ e^{-2ix}\int e^{2ix}[\tfrac{1}{2}(e^{2ix}+e^{-2ix}) + \tfrac{1}{2}(e^{4ix}+e^{-4ix})]\,dx$$

$$- e^{2ix}\int e^{-2ix}[\tfrac{1}{2}(e^{2ix}+e^{-2ix}) + \tfrac{1}{2}(e^{4ix}+e^{-4ix})]\,dx \}$$

$$= \frac{1}{8}\{e^{-2ix}\int(e^{4ix}+1+e^{6ix}+e^{-2ix})i\,dx - e^{2ix}\int(1+e^{-4ix}+e^{2ix}+e^{-6ix})i\,dx\}$$

$$= \frac{1}{8}\{e^{-2ix}(\frac{1}{4}e^{4ix} + ix + \frac{1}{6}e^{6ix} - \frac{1}{2}e^{-2ix}) - e^{2ix}(ix - \frac{1}{4}e^{-4ix} + \frac{1}{2}e^{2ix} - \frac{1}{6}e^{-6ix})\}$$

$$= \frac{1}{8}\{-ix(e^{2ix}-e^{-2ix}) + \frac{1}{4}(e^{2ix}+e^{-2ix}) - \frac{1}{3}(e^{4ix}+e^{-4ix})\}$$

$$= \frac{1}{4}x\left(\frac{e^{2ix}-e^{-2ix}}{2i}\right) + \frac{1}{16}\left(\frac{e^{2ix}+e^{-2ix}}{2}\right) - \frac{1}{12}\left(\frac{e^{4ix}+e^{-4ix}}{2}\right)$$

$$= \frac{1}{4}x\sin 2x + \frac{1}{16}\cos 2x - \frac{1}{12}\cos 4x.$$

The primitive is $y = C_1\cos 2x + C_2\sin 2x + \frac{1}{4}x\sin 2x - \frac{1}{12}\cos 4x.$

9. Solve $(D^2 - 9D + 18)y = e^{e^{-3x}}.$

The complementary function is $y = C_1e^{3x} + C_2e^{6x}$, and a particular integral is

$$y = \frac{1}{(D-6)(D-3)}e^{e^{-3x}} = e^{6x}\int e^{-3x}\int e^{e^{-3x}}\cdot e^{-3x}(dx)^2$$

$$= e^{6x}\int e^{-3x}(-\frac{1}{3}e^{e^{-3x}})dx = \frac{1}{3}e^{6x}\int e^{e^{-3x}}(-e^{-3x})\,dx = \frac{1}{9}e^{e^{-3x}}\cdot e^{6x}.$$

The complete solution is $y = C_1e^{3x} + (C_2 + \frac{1}{9}e^{e^{-3x}})e^{6x}.$

Note. When the factors are reversed, a particular integral is $y = e^{3x}\int e^{3x}\int e^{e^{-3x}}\cdot e^{-6x}(dx)^2$. Using the substitution $e^{-3x} = v$, we obtain

$$y = \frac{1}{9v}\int\frac{1}{v^2}\int e^v\,v\,(dv)^2 = \frac{1}{9v}\int e^v(\frac{1}{v} - \frac{1}{v^2})dv = \frac{1}{9v^2}e^v$$

or $y = \frac{1}{9}e^{e^{-3x}}\cdot e^{6x}$, as before.

SUPPLEMENTARY PROBLEMS

10. Evaluate, omitting the arbitrary constant.

a) $\frac{1}{D+1}e^x$ Ans. $\frac{1}{2}e^x$

b) $\frac{1}{D-1}e^x$ Ans. xe^x

c) $\frac{1}{D+1}(x+1)$ Ans. x

d) $\frac{1}{D+1}(x^2+1)$ Ans. x^2-2x+3

e) $\frac{1}{D+2}\sin 3x$ Ans. $\frac{1}{13}(2\sin 3x - 3\cos 3x)$

f) $\frac{1}{D+2}e^{-2x}\sin 3x$ Ans. $-\frac{1}{3}e^{-2x}\cos 3x$

Solve.

11. $(D^2-4D+3)y = 1$ Ans. $y = C_1e^x + C_2e^{3x} + 1/3$

12. $(D^2-4D)y = 5$ $y = C_1 + C_2e^{4x} - 5x/4$

13. $(D^3-4D^2)y = 5$ $y = C_1 + C_2x + C_3e^{4x} - 5x^2/8$

14. $(D^5-4D^3)y = 5$ $y = C_1 + C_2x + C_3x^2 + C_4e^{2x} + C_5e^{-2x} - 5x^3/24$

15. $(D^3-4D)y = x$ $y = C_1 + C_2e^{2x} + C_3e^{-2x} - x^2/8$

16. $(D^2-6D+9)y = e^{2x}$ $y = C_1e^{3x} + C_2xe^{3x} + e^{2x}$

17. $(D^2+D-2)y = 2(1+x-x^2)$ $y = C_1e^x + C_2e^{-2x} + x^2$

18. $(D^2-1)y = 4xe^x$ $y = C_1e^x + C_2e^{-x} + e^x(x^2-x)$

19. $(D^2-1)y = \sin^2 x = \frac{1}{2}(1-\cos 2x)$ $y = C_1e^x + C_2e^{-x} - \frac{1}{2} + \frac{1}{10}\cos 2x$

20. $(D^2-1)y = (1+e^{-x})^{-2}$ $y = C_1e^x + C_2e^{-x} - 1 + e^{-x}\ln(1+e^x)$

21. $(D^2+1)y = \csc x$ $y = C_1\cos x + C_2\sin x + \sin x\ln\sin x - x\cos x$

22. $(D^2-3D+2)y = \sin e^{-x}$ $y = C_1e^x + C_2e^{2x} - e^{2x}\sin e^{-x}$

CHAPTER 15

Linear Equations with Constant Coefficients

VARIATION OF PARAMETERS, UNDETERMINED COEFFICIENTS

TWO OTHER METHODS for determining a *particular integral* of a linear differential equation with constant coefficients

1) $F(D)y = (D^n + P_1D^{n-1} + P_2D^{n-2} + \cdots\cdots + P_{n-1}D + P_n)y = Q$

will be exhibited by means of examples.

VARIATION OF PARAMETERS. From the complementary function of 1),

$$y = C_1y_1(x) + C_2y_2(x) + \cdots\cdots + C_ny_n(x),$$

we obtain a basic relation

2) $y = L_1(x)\,y_1(x) + L_2(x)\,y_2(x) + \cdots\cdots + L_n(x)\,y_n(x)$

by replacing the C's by unknown functions of x, the L's. The method consists of a procedure for determining the L's so that 2) satisfies 1).
See Problems 1-4.

UNDETERMINED COEFFICIENTS. The basic relation here is

3) $y = A\,r_1(x) + B\,r_2(x) + C\,r_3(x) + \cdots\cdots + G\,r_t(x),$

where the functions $r_1(x), \cdots\cdots, r_t(x)$ are the terms of Q and those arising from these terms by differentiation, and $A,B,C,\cdots,G$ are constants.

For example, if the equation is $F(D)y = x^3$, we take for 3)

$$y = Ax^3 + Bx^2 + Cx + D;$$

if the equation is $f(D)y = e^x + e^{3x}$, we take for 3)

$$y = Ae^x + Be^{3x},$$

since no new terms are obtained by differentiating e^x and e^{3x};

if the equation is $F(D)y = \sin ax$, we take for 3)

$$y = A\sin ax + B\cos ax;$$

if the equation is $F(D)y = \sec x$, the method fails since the number of new terms obtained by differentiating $Q = \sec x$ is infinite.

Substituting 3) in 1), the coefficients $A,B,C,\cdots\cdots$ are found from the resulting identity.
See Problems 5-6.

The procedure must be modified in case:

a) A term of Q is also a term of the complementary function. If a term of Q, say u, is also a term of the complementary function corresponding to an s-fold root m, then in 3) we introduce a term x^su plus terms arising from it by differentiation.

For example, in finding a particular integral of $(D-2)^2(D+3)y = e^{2x} + x^2$, the basic relation is $y = Ax^2e^{2x} + Bxe^{2x} + Ce^{2x} + Dx^2 + Ex + F$, the first three terms arising from the fact that the term e^{2x} of Q is also a term of the complementary function corresponding to a double root $m=2$; hence, use is made of x^2e^{2x} and all terms arising by differentiation. See Problems 7-8.

b) A term of Q is $x^r u$ and u is a term of the complementary function. If u corresponds to an s-fold root m, 3) must contain the term $x^{r+s}u$ plus terms arising from it by differentiation.

For example, in finding a particular integral of $(D-2)^3(D+3)y = x^2e^{2x} + x^2$, the basic relation is

$$y = Ax^5e^{2x} + Bx^4e^{2x} + Cx^3e^{2x} + Dx^2e^{2x} + Exe^{2x} + Fe^{2x} + Gx^2 + Hx + J,$$

the first six terms arising from the fact that e^{2x} is a part of the complementary function corresponding to the triple root $m=2$. See Problem 9.

SOLVED PROBLEMS

VARIATION OF PARAMETERS.

1. Show that if $y = C_1y_1 + C_2y_2 + C_3y_3$ is the complementary function of

$$F(D)y = (D^3 + P_1D^2 + P_2D + P_3)y = Q$$

then

1) $$y = L_1y_1 + L_2y_2 + L_3y_3,$$

where L_1, L_2, L_3 satisfy the conditions

A) $$\begin{aligned} L_1'y_1 + L_2'y_2 + L_3'y_3 &= 0 \\ L_1'y_1' + L_2'y_2' + L_3'y_3' &= 0 \\ L_1'y_1'' + L_2'y_2'' + L_3'y_3'' &= Q, \end{aligned}$$

is a particular solution of the differential equation.

We obtain, in view of A), by successively differentiating $y = L_1y_1 + L_2y_2 + L_3y_3$:

B) $$\begin{aligned} Dy &= L_1y_1' + L_2y_2' + L_3y_3' + (L_1'y_1 + L_2'y_2 + L_3'y_3) = L_1y_1' + L_2y_2' + L_3y_3' \\ D^2y &= L_1y_1'' + L_2y_2'' + L_3y_3'' + (L_1'y_1' + L_2'y_2' + L_3'y_3') = L_1y_1'' + L_2y_2'' + L_3y_3'' \\ D^3y &= L_1y_1''' + L_2y_2''' + L_3y_3''' + (L_1'y_1'' + L_2'y_2'' + L_3'y_3'') = L_1y_1''' + L_2y_2''' + L_3y_3''' + Q. \end{aligned}$$

Then $$\begin{aligned} F(D)y &= L_1\{y_1''' + P_1y_1'' + P_2y_1' + P_3y_1\} + L_2\{y_2''' + P_1y_2'' + P_2y_2' + P_3y_2\} \\ &\quad + L_3\{y_3''' + P_1y_3'' + P_2y_3' + P_3y_3\} + Q \\ &= L_1F(D)y_1 + L_2F(D)y_2 + L_3F(D)y_3 + Q = 0+0+0+Q = Q, \end{aligned}$$

since y_1, y_2, y_3 are solutions of $F(D)y = 0$.

In using this method:

a) Write the complementary function.

b) Form the L function 1), which is to be a particular integral, by replacing the C's of the complementary function with L's.

c) Obtain equations B) by differentiating 1) as many times as the degree of the differential equation. After each differentiation, set the sum of all terms containing derivatives of the L's equal to zero, except in the case of the last differentiation when the sum is set equal to Q. The equations obtained by setting the sums equal to zero and Q are the equations A).

d) Solve these equations for $L_1', L_2', \cdots\cdots$.

e) Obtain $L_1, L_2, \cdots\cdots$ by integration.

2. Solve $(D^2 - 2D)y = e^x \sin x$.

The complementary function is $y = C_1 + C_2 e^{2x}$.

We form the relation $y = L_1 + L_2 e^{2x}$,

obtain, by differentiation $Dy = 2L_2 e^{2x} + (L_1' + L_2' e^{2x})$,

and set 1) $L_1' + L_2' e^{2x} = 0$.

Since now $Dy = 2L_2 e^{2x}$, $D^2 y = 4L_2 e^{2x} + 2L_2' e^{2x}$ and we set $2L_2' e^{2x} = Q = e^x \sin x$.

Thus, $L_2' = \frac{1}{2}e^{-x}\sin x$ and $L_2 = -\frac{1}{4}e^{-x}(\sin x + \cos x)$.

From 1), $L_1' = -L_2' e^{2x} = -\frac{1}{2}e^x \sin x$ and $L_1 = -\frac{1}{4}e^x(\sin x - \cos x)$.

A particular integral of the given equation is

$$y = L_1 + L_2 e^{2x} = -\tfrac{1}{4}e^x(\sin x - \cos x) - \tfrac{1}{4}e^x(\sin x + \cos x) = -\tfrac{1}{2}e^x \sin x.$$

and the primitive is $y = C_1 + C_2 e^{2x} - \frac{1}{2}e^x \sin x$.

3. Solve $(D^3 + D)y = \csc x$.

The complementary function is $y = C_1 + C_2 \cos x + C_3 \sin x$.

From the relation $y = L_1 + L_2 \cos x + L_3 \sin x$

we obtain $Dy = (-L_2 \sin x + L_3 \cos x) + (L_1' + L_2' \cos x + L_3' \sin x)$

and set 1) $L_1' + L_2' \cos x + L_3' \sin x = 0$.

Then $Dy = -L_2 \sin x + L_3 \cos x$,

$D^2 y = (-L_2 \cos x - L_3 \sin x) + (-L_2' \sin x + L_3' \cos x)$,

and we set 2) $-L_2' \sin x + L_3' \cos x = 0$.

Then $D^2 y = -L_2 \cos x - L_3 \sin x$,

$D^3 y = (L_2 \sin x - L_3 \cos x) + (-L_2' \cos x - L_3' \sin x)$,

and we set 3) $-L_2' \cos x - L_3' \sin x = Q = \csc x$.

Adding 1) and 3), $L_1' = \csc x$ and $L_1 = -\ln(\csc x + \cot x)$.

Solving 2) and 3), $L_3' = -1$ and $L_2' = -\cot x$, so that $L_3 = -x$ and $L_2 = -\ln \sin x$.

Thus, a particular integral of the differential equation is

$$y = L_1 + L_2 \cos x + L_3 \sin x = -\ln(\csc x + \cot x) - \cos x \ln \sin x - x \sin x,$$

and the primitive is

$$y = C_1 + C_2 \cos x + C_3 \sin x - \ln(\csc x + \cot x) - \cos x \ln \sin x - x \sin x.$$

4. Solve $(D^2 - 6D + 9)y = e^{3x}/x^2$.

The complementary function is $y = C_1 e^{3x} + C_2 x e^{3x}$.

From the relation $y = L_1 e^{3x} + L_2 x e^{3x}$

we obtain $Dy = (3L_1 + L_2)e^{3x} + 3L_2 x e^{3x} + (L_1' e^{3x} + L_2' x e^{3x})$

and set 1) $L_1' e^{3x} + L_2' x e^{3x} = 0$.

Then $$D^2y = (9L_1 + 6L_2)e^{3x} + 9L_2xe^{3x} + (3L_1' + L_2')e^{3x} + 3L_2'xe^{3x},$$

and we set 2) $$(3L_1' + L_2')e^{3x} + 3L_2'xe^{3x} = e^{3x}/x^2.$$

Solving 1) and 2), $L_1' = -1/x$ and $L_2' = 1/x^2$, so that $L_1 = -\ln x$ and $L_2 = -1/x$.

Thus, a particular integral of the differential equation is

$$y = L_1e^{3x} + L_2xe^{3x} = -e^{3x}\ln x - e^{3x},$$

and the primitive is $y = C_1e^{3x} + C_2xe^{3x} - e^{3x}\ln x$.

UNDETERMINED COEFFICIENTS.

5. Solve $(D^2 - 2D)y = e^x \sin x$.

The complementary function is $y = C_1 + C_2e^x$. As a particular integral, we take

$$y = Ae^x \sin x + Be^x \cos x.$$

Then
$$Dy = (A-B)e^x \sin x + (A+B)e^x \cos x,$$
$$D^2y = -2Be^x \sin x + 2Ae^x \cos x,$$
and
$$(D^2 - 2D)y = -2Ae^x \sin x - 2Be^x \cos x = e^x \sin x = Q.$$

Equating coefficients of like terms, $-2A = 1$ and $-2B = 0$, so that $A = -\frac{1}{2}$ and $B = 0$.

Hence, a particular integral of the differential equation is

$$y = Ae^x \sin x + Be^x \cos x = -\tfrac{1}{2}e^x \sin x,$$

and the primitive is $y = C_1 + C_2e^{2x} - \frac{1}{2}e^x \sin x$.

This was solved above as Problem 2.

6. Solve $(D^2 - 2D + 3)y = x^3 + \sin x$.

The complementary function is $y = e^x(C_1\cos\sqrt{2}\,x + C_2\sin\sqrt{2}\,x)$. As a particular integral, we take

$$y = Ax^3 + Bx^2 + Cx + E + F\sin x + G\cos x$$

Then
$$Dy = 3Ax^2 + 2Bx + C - G\sin x + F\cos x,$$
$$D^2y = 6Ax + 2B - F\sin x - G\cos x,$$
and
$$(D^2 - 2D + 3)y = 3Ax^3 + 3(B-2A)x^2 + (3C-4B+6A)x + (3E-2C+2B) + 2(F+G)\sin x + 2(G-F)\cos x$$
$$= x^3 + \sin x.$$

Equating coefficients of like terms, $3A = 1$ and $A = 1/3$; $B - 2A = 0$ and $B = 2/3$; $3C - 4B + 6A = 0$ and $C = 2/9$; $3E - 2C + 2B = 0$ and $E = -8/27$; $2(F+G) = 1$, $G - F = 0$ and $F = G = \frac{1}{4}$.

Thus, a particular integral of the differential equation is

$$y = \frac{1}{3}x^3 + \frac{2}{3}x^2 + \frac{2}{9}x - \frac{8}{27} + \frac{1}{4}(\sin x + \cos x),$$

and the primitive is

$$y = e^x(C_1\cos\sqrt{2}\,x + C_2\sin\sqrt{2}\,x) + \frac{1}{27}(9x^3 + 18x^2 + 6x - 8) + \frac{1}{4}(\sin x + \cos x).$$

7. Solve $(D^3 + 2D^2 - D - 2)y = e^x + x^2$.

The complementary function is $y = C_1e^x + C_2e^{-x} + C_3e^{-2x}$. Since e^x occurs in Q and also in the complementary function corresponding to a root of multiplicity one, we take as a particular integral

$$1) \qquad y = Ax^2 + Bx + C + Exe^x + Fe^x$$

Then

$$Dy = 2Ax + B + Exe^x + (E+F)e^x,$$
$$D^2y = 2A + Exe^x + (2E+F)e^x,$$
$$D^3y = Exe^x + (3E+F)e^x,$$

and $(D^3 + 2D^2 - D - 2)y = -2Ax^2 - 2(B+A)x + (4A - B - 2C) + 6Ee^x = e^x + x^2$.

Equating coefficients of like terms, $-2A = 1$, $B + A = 0$, $4A - B - 2C = 0$, $6E = 1$; hence, $A = -\frac{1}{2}$, $B = \frac{1}{2}$, $C = -\frac{5}{4}$, $E = \frac{1}{6}$, and F is arbitrary. Now F should be arbitrary here, since C_1e^x is a term of the complementary function. Thus, in writing 1), the inclusion of Fe^x was unnecessary.

Hence, a particular integral is $y = -\frac{1}{2}x^2 + \frac{1}{2}x - \frac{5}{4} + \frac{1}{6}xe^x$,

and the primitive is $y = C_1e^x + C_2e^{-x} + C_3e^{-2x} - \frac{1}{2}x^2 + \frac{1}{2}x - \frac{5}{4} + \frac{1}{6}xe^x$.

8. Solve $(D^2 - 4D + 4)y = x^3e^{2x} + xe^{2x}$.

The complementary function is $y = C_1e^{2x} + C_2xe^{2x}$ Now e^{2x} is a part of Q and also occurs in the complementary function corresponding to a root of multiplicity two. As a particular integral, we take

$$y = Ax^5e^{2x} + Bx^4e^{2x} + Cx^3e^{2x} + Ex^2e^{2x}.$$

Note that terms involving xe^{2x} and e^{2x} are not included, since they appear in the complementary function with arbitrary coefficients. Then

$$Dy = 2Ax^5e^{2x} + (5A + 2B)x^4e^{2x} + (4B + 2C)x^3e^{2x} + (3C + 2E)x^2e^{2x} + 2Exe^{2x},$$
$$D^2y = 4Ax^5e^{2x} + (20A + 4B)x^4e^{2x} + (20A + 16B + 4C)x^3e^{2x} + (12B + 12C + 4E)x^2e^{2x} + (6C + 8E)xe^{2x} + 2Ee^{2x},$$

and $(D^2 - 4D + 4)y = 20Ax^3e^{2x} + 12Bx^2e^{2x} + 6Cxe^{2x} + 2Ee^{2x} = x^3e^{2x} + xe^{2x}$.

Equating coefficients of like terms, $20A = 1$, $12B = 0$, $6C = 1$, $2E = 0$; hence, $A = 1/20$, $B = 0$, $C = 1/6$, $E = 0$.

Thus, a particular integral is $y = \frac{1}{20}x^5e^{2x} + \frac{1}{6}x^3e^{2x}$,

and the primitive is $y = C_1e^{2x} + C_2xe^x + \frac{1}{20}x^5e^{2x} + \frac{1}{6}x^3e^{2x}$.

9. Solve $(D^2 + 4)y = x^2 \sin 2x$.

The complementary function is $y = C_1\cos 2x + C_2\sin 2x$.

Since $x^2 \sin 2x$ occurs in Q and $\sin 2x$ is a part of the complementary function corresponding to a root of multiplicity one, we take as a particular integral

$$y = Ax^3 \cos 2x + Bx^3 \sin 2x + Cx^2 \cos 2x + Ex^2 \sin 2x + Fx \cos 2x + Gx \sin 2x.$$

Note that $H \cos 2x + K \sin 2x$ is not included, since these terms are in the complementary function. Then

$$Dy = 2Bx^3 \cos 2x - 2Ax^3 \sin 2x + (3A + 2E)x^2 \cos 2x + (3B - 2C)x^2 \sin 2x$$
$$+ (2C + 2G)x \cos 2x + (2E - 2F)x \sin 2x + F \cos 2x + G \sin 2x,$$

$$D^2y = -4Ax^3\cos 2x - 4Bx^3\sin 2x + (12B-4C)x^2\cos 2x + (-12A-4E)x^2\sin 2x$$
$$+ (6A+8E-4F)x\cos 2x + (6B-8C-4G)x\sin 2x + (2C+4G)\cos 2x + (2E-4F)\sin 2x,$$

and

$$(D^2+4)y = 12Bx^2\cos 2x - 12Ax^2\sin 2x + (6A+8E)x\cos 2x + (6B-8C)x\sin 2x$$
$$+ (2C+4G)\cos 2x + (2E-4F)\sin 2x = x^2\sin 2x.$$

Equating coefficients of like terms, $-12A = 1$, $12B = 0$, $6A + 8E = 0$, $6B - 8C = 0$, $2C + 4G = 0$, $2E - 4F = 0$; hence, $A = -1/12$, $B = 0$, $C = 0$, $E = 1/16$, $F = 1/32$, $G = 0$.

A particular integral is $y = -\dfrac{1}{12}x^3\cos 2x + \dfrac{1}{16}x^2\sin 2x + \dfrac{1}{32}x\cos 2x$,

and the primitive is $y = C_1\cos 2x + C_2\sin 2x - \dfrac{1}{12}x^3\cos 2x + \dfrac{1}{16}x^2\sin 2x + \dfrac{1}{32}x\cos 2x$.

SUPPLEMENTARY PROBLEMS

Solve, using the method of variation of parameters.

10. $(D^2 + 1)y = \csc x$ *Ans.* $y = C_1\cos x + C_2\sin x + \sin x \ln\sin x - x\cos x$

11. $(D^2 + 4)y = 4\sec^2 2x$ *Ans.* $y = C_1\cos 2x + C_2\sin 2x - 1 + \sin 2x \ln(\sec 2x + \tan 2x)$

12. $(D^2 - 4D + 3)y = (1+e^{-x})^{-1}$ *Ans.* $y = C_1e^x + C_2e^{3x} + \frac{1}{2}e^{2x} + \frac{1}{2}(e^x - e^{3x})\ln(1+e^{-x})$

13. $(D^2 - 1)y = e^{-x}\sin e^{-x} + \cos e^{-x}$ *Ans.* $y = C_1e^x + C_2e^{-x} - e^x\sin e^{-x}$

14. $(D^2 - 1)y = (1+e^{-x})^{-2}$ *Ans.* $y = C_1e^x + C_2e^{-x} - 1 + e^{-x}\ln(1+e^x)$

Solve, using the method of undetermined coefficients.

15. $(D^2 + 2)y = e^x + 2$ *Ans.* $y = C_1\cos\sqrt{2}\,x + C_2\sin\sqrt{2}\,x + e^x/3 + 1$

16. $(D^2 - 1)y = e^x\sin 2x$ *Ans.* $y = C_1e^x + C_2e^{-x} - e^x(\sin 2x + \cos 2x)/8$

17. $(D^2 + 2D + 2)y = x^2 + \sin x$ *Ans.* $y = e^{-x}(C_1\cos x + C_2\sin x) + \frac{1}{2}(x-1)^2 + \frac{1}{5}(\sin x - 2\cos x)$

18. $(D^2 - 9)y = x + e^{2x} - \sin 2x$ *Ans.* $y = C_1e^{3x} + C_2e^{-3x} - x/9 - e^{2x}/5 + \frac{1}{13}\sin 2x$

19. $(D^3 + 3D^2 + 2D)y = x^2 + 4x + 8$ (Use $Ax^3 + Bx^2 + Cx$.)
    Ans. $y = C_1 + C_2e^{-x} + C_3e^{-2x} + \frac{1}{6}x^3 + \frac{1}{4}x^2 + \frac{11}{4}x$

20. $(D^2 + 1)y = -2\sin x + 4x\cos x$ *Ans.* $y = C_1\cos x + C_2\sin x + 2x\cos x + x^2\sin x$

21. $(D^3 - D^2 - 4D + 4)y = 2x^2 - 4x - 1 + 2x^2e^{2x} + 5xe^{2x} + e^{2x}$
    Ans. $y = C_1e^x + C_2e^{2x} + C_3e^{-2x} + \frac{1}{2}x^2 + \frac{1}{6}x^3e^{2x}$

CHAPTER 16

Linear Equations with Constant Coefficients

SHORT METHODS

A PARTICULAR INTEGRAL of a linear differential equation $F(D)y = Q$ with constant coefficients is given by $y = \frac{1}{F(D)}Q$. For certain forms of Q the labor involved in evaluating this symbol may be considerably shortened, as follows:

a) If Q is of the form e^{ax},

$$y = \frac{1}{F(D)}e^{ax} = \frac{1}{F(a)}e^{ax}, \qquad F(a) \neq 0.$$

See Problems 2-3 when $F(a) \neq 0$, and Problems 4-5 when $F(a) = 0$.

b) If Q is of the form $\sin(ax+b)$ or $\cos(ax+b)$,

$$y = \frac{1}{F(D^2)}\sin(ax+b) = \frac{1}{F(-a^2)}\sin(ax+b), \qquad F(-a^2) \neq 0,$$

$$y = \frac{1}{F(D^2)}\cos(ax+b) = \frac{1}{F(-a^2)}\cos(ax+b), \qquad F(-a^2) \neq 0.$$

See Problems 7-11 when $F(-a^2) \neq 0$, and Problem 12 when $F(-a^2) = 0$.

c) If Q is of the form x^m,

$$y = \frac{1}{F(D)}x^m = (a_0 + a_1D + a_2D^2 + \cdots\cdots + a_mD^m)x^m, \qquad a_0 \neq 0,$$

obtained by expanding $\frac{1}{F(D)}$ in ascending powers of D and suppressing all terms beyond D^m, since $D^nx^m = 0$ when $n > m$. See Problems 13-15.

d) If Q is of the form $e^{ax}V(x)$, $\quad y = \frac{1}{F(D)}e^{ax}V = e^{ax}\frac{1}{F(D+a)}V.$

See Problems 17-20.

e) If Q is of the form $xV(x)$, $\quad y = \frac{1}{F(D)}xV = x\frac{1}{F(D)}V - \frac{F'(D)}{\{F(D)\}^2}V.$

See Problems 21-23.

SOLVED PROBLEMS

1. Establish the rule in *a*) above.

Since when $y = e^{ax}$, $Dy = ae^{ax}$, $D^2y = a^2e^{ax}$, ..., $D^re^{ax} = a^re^{ax}$,

$F(D)e^{ax} = \sum_r P_rD^re^{ax} = \sum_r P_ra^re^{ax} = F(a)e^{ax}$. Hence, $\frac{1}{F(D)}e^{ax} = \frac{1}{F(a)}e^{ax}$.

2. Solve $(D^3 - 2D^2 - 5D + 6)y = e^{4x}$ or $(D-1)(D-3)(D+2)y = e^{4x}$.

The complementary function is $y = C_1 e^{x} + C_2 e^{3x} + C_3 e^{-2x}$.

A particular integral is
$$y = \frac{1}{(D-1)(D-3)(D+2)} e^{4x}$$
$$= \frac{1}{(4-1)(4-3)(4+2)} e^{4x} = \frac{1}{3\cdot 1\cdot 6} e^{4x} = \frac{1}{18} e^{4x}.$$

Hence, the primitive is $y = C_1 e^{x} + C_2 e^{3x} + C_3 e^{-2x} + \frac{1}{18} e^{4x}$.

3. Solve $(D^3 - 2D^2 - 5D + 6)y = (e^{2x} + 3)^2$.

The complementary function is, from Problem 2, $y = C_1 e^{x} + C_2 e^{3x} + C_3 e^{-2x}$.

A particular integral is
$$y = \frac{1}{(D-1)(D-3)(D+2)} (e^{2x} + 3)^2$$
$$= \frac{1}{(D-1)(D-3)(D+2)} e^{4x} + \frac{6}{(D-1)(D-3)(D+2)} e^{2x} + \frac{9}{(D-1)(D-3)(D+2)} e^{0x}$$
$$= \frac{1}{3(1)6} e^{4x} + \frac{6}{1(-1)4} e^{2x} + \frac{9}{(-1)(-3)2} = \frac{e^{4x}}{18} - \frac{3e^{2x}}{2} + \frac{3}{2}.$$

The primitive is $y = C_1 e^{x} + C_2 e^{3x} + C_3 e^{-2x} + \dfrac{e^{4x}}{18} - \dfrac{3e^{2x}}{2} + \dfrac{3}{2}$.

4. Solve $(D^3 - 2D^2 - 5D + 6)y = e^{3x}$.

The complementary function is $y = C_1 e^{x} + C_2 e^{3x} + C_3 e^{-2x}$.

A particular integral is $y = \dfrac{1}{(D-1)(D-3)(D+2)} e^{3x}$. Now $F(a) = F(3) = 0$, and the short method does not apply. However, we may write

$$y = \frac{1}{(D-1)(D-3)(D+2)} e^{3x} = \frac{1}{D-3}\left(\frac{1}{(D-1)(D+2)} e^{3x}\right) = \frac{1}{D-3}\left(\frac{1}{2\cdot 5} e^{3x}\right) = \frac{1}{10}\frac{1}{D-3} e^{3x}$$
$$= \frac{1}{10} e^{3x} \int e^{3x} e^{-3x}\, dx = \frac{1}{10} e^{3x} \int dx = \frac{1}{10} x e^{3x}.$$

The primitive is $y = C_1 e^{x} + C_2 e^{3x} + C_3 e^{-2x} + xe^{3x}/10$.

5. Solve $(D^3 - 5D^2 + 8D - 4)y = e^{2x} + 2e^{x} + 3e^{-x}$.

The complementary function is $y = C_1 e^{x} + C_2 e^{2x} + C_3 x e^{2x}$, and a particular integral is

$$y = \frac{1}{(D-1)(D-2)^2} e^{2x} + \frac{2}{(D-1)(D-2)^2} e^{x} + \frac{3}{(D-1)(D-2)^2} e^{-x}$$
$$= \frac{1}{(D-2)^2}\left(\frac{1}{D-1} e^{2x}\right) + \frac{2}{D-1}\left(\frac{1}{(D-2)^2} e^{x}\right) + \frac{3}{(D-1)(D-2)^2} e^{-x}$$
$$= \frac{1}{(D-2)^2} e^{2x} + \frac{2}{D-1} e^{x} + \frac{3}{(-2)(-3)^2} e^{-x}$$
$$= e^{2x} \iint (dx)^2 + 2e^{x} \int dx - \frac{1}{6} e^{-x} = \frac{1}{2} x^2 e^{2x} + 2xe^{x} - \frac{1}{6} e^{-x}.$$

The primitive is $y = C_1e^x + C_2e^{2x} + C_3xe^{2x} + \frac{1}{2}x^2e^{2x} + 2xe^x - \frac{1}{6}e^{-x}$.

6. Establish the rule in b) above for $\cos(ax+b)$.

Since, when $y = \cos(ax+b)$, $D^2y = -a^2\cos(ax+b)$, $D^4y = (-a^2)^2\cos(ax+b)$, ..., $D^{2r}y = (-a^2)^r\cos(ax+b)$, then

$F(D^2)\cos(ax+b) = \sum_r P_rD^{2r}\cos(ax+b) = \sum_r P_r(-a^2)^r\cos(ax+b) = F(-a^2)\cos(ax+b)$.

Hence, $\dfrac{1}{F(D^2)}\cos(ax+b) = \dfrac{1}{F(-a^2)}\cos(ax+b)$.

7. Solve $(D^2+4)y = \sin 3x$.

The complementary function is $y = C_1\cos 2x + C_2\sin 2x$, and a particular solution is

$$y = \frac{1}{D^2+4}\sin 3x = \frac{1}{-(3)^2+4}\sin 3x = -\frac{1}{5}\sin 3x.$$

The primitive is $y = C_1\cos 2x + C_2\sin 2x - \frac{1}{5}\sin 3x$.

8. Solve $(D^4+10D^2+9)y = \cos(2x+3)$.

The complementary function is $y = C_1\cos x + C_2\sin x + C_3\cos 3x + C_4\sin 3x$, and a particular integral is

$$y = \frac{1}{(D^2+1)(D^2+9)}\cos(2x+3) = \frac{1}{(-3)(5)}\cos(2x+3) = -\frac{1}{15}\cos(2x+3).$$

The primitive is $y = C_1\cos x + C_2\sin x + C_3\cos 3x + C_4\sin 3x - \frac{1}{15}\cos(2x+3)$.

9. Solve $(D^2+3D-4)y = \sin 2x$.

The complementary function is $y = C_1e^x + C_2e^{-4x}$, and a particular integral is

$$y = \frac{1}{D^2+3D-4}\sin 2x = \frac{1}{(D-1)(D+4)}\sin 2x.$$

The operator here is not of the form $\dfrac{1}{F(D^2)}$, and the short method does not apply. However, we may use either of the following procedures to shorten the work.

a) $y = \dfrac{1}{(D-1)(D+4)}\sin 2x = \dfrac{(D+1)(D-4)}{(D^2-1)(D^2-16)}\sin 2x = \dfrac{1}{100}(D^2-3D-4)\sin 2x$

$= \dfrac{1}{100}(-4\sin 2x - 6\cos 2x - 4\sin 2x) = -\dfrac{1}{50}(4\sin 2x + 3\cos 2x)$.

b) $y = \dfrac{1}{D^2+3D-4}\sin 2x = \dfrac{1}{(-4)+3D-4}\sin 2x = \dfrac{1}{3D-8}\sin 2x = \dfrac{3D+8}{9D^2-64}\sin 2x$

$= -\dfrac{1}{100}(3D+8)\sin 2x = -\dfrac{1}{100}(6\cos 2x + 8\sin 2x) = -\dfrac{1}{50}(4\sin 2x + 3\cos 2x)$.

The primitive is $y = C_1e^x + C_2e^{-4x} - \frac{1}{50}(4\sin 2x + 3\cos 2x)$.

10. Solve $(D^3+D^2+D+1)y = \sin 2x + \cos 3x$.

The complementary function is $y = C_1\cos x + C_2\sin x + C_3e^{-x}$, and a particular integral is

$$y = \frac{1}{(D^2+1)(D+1)}(\sin 2x + \cos 3x) = \frac{1}{(D^2+1)(D+1)}\sin 2x + \frac{1}{(D^2+1)(D+1)}\cos 3x$$

$$= -\frac{1}{3}\frac{1}{D+1}\sin 2x - \frac{1}{8}\frac{1}{D+1}\cos 3x = -\frac{1}{3}\frac{D-1}{D^2-1}\sin 2x - \frac{1}{8}\frac{D-1}{D^2-1}\cos 3x$$

$$= \frac{1}{15}(D-1)\sin 2x + \frac{1}{80}(D-1)\cos 3x = \frac{1}{15}(2\cos 2x - \sin 2x) - \frac{1}{80}(3\sin 3x + \cos 3x).$$

The primitive is

$$y = C_1\cos x + C_2\sin x + C_3e^{-x} + \frac{1}{15}(2\cos 2x - \sin 2x) - \frac{1}{80}(3\sin 3x + \cos 3x).$$

11. Solve $(D^2-D+1)y = \sin 2x$.

The complementary function is $y = e^{\frac{1}{2}x}(C_1\cos\frac{1}{2}\sqrt{3}\,x + C_2\sin\frac{1}{2}\sqrt{3}\,x)$, and a particular integral is

$$y = \frac{1}{D^2-D+1}\sin 2x = \frac{1}{(-4)-D+1}\sin 2x = -\frac{1}{D+3}\sin 2x = -\frac{D-3}{D^2-9}\sin 2x$$

$$= \frac{1}{13}(D-3)\sin 2x = \frac{1}{13}(2\cos 2x - 3\sin 2x).$$

The primitive is $y = e^{\frac{1}{2}x}(C_1\cos\frac{1}{2}\sqrt{3}\,x + C_2\sin\frac{1}{2}\sqrt{3}\,x) + \frac{1}{13}(2\cos 2x - 3\sin 2x)$.

12. Solve $(D^2+4)y = \cos 2x + \cos 4x$.

The complementary function is $y = C_1\cos 2x + C_2\sin 2x$, and a particular integral is

$$y = \frac{1}{D^2+4}(\cos 2x + \cos 4x) = \frac{1}{D^2+4}\cos 2x + \frac{1}{D^2+4}\cos 4x.$$

The method of this chapter cannot be used to evaluate $\frac{1}{D^2+4}\cos 2x$ since, when D^2 is replaced by -4, $D^2+4=0$. However, the following procedure may be used.

Consider $\frac{1}{D^2+4}\cos(2+h)x = \frac{1}{-(2+h)^2+4}\cos(2+h)x = -\frac{1}{4h+h^2}\cos(2+h)x$

$$= -\frac{1}{h(4+h)}(\cos 2x - hx\sin 2x - \tfrac{1}{2}(hx)^2\cos 2x + \cdots\cdots\cdots)$$

by Taylor's theorem. The first term, $\cos 2x$, is part of the complementary function and need not be considered here. Hence, a particular integral is

$$\frac{1}{D^2+4}\cos(2+h)x = \frac{1}{h(4+h)}(hx\sin 2x + \tfrac{1}{2}(hx)^2\cos 2x - \cdots\cdots\cdots)$$

$$= \frac{1}{4+h}(x\sin 2x + \tfrac{1}{2}hx^2\cos 2x - \cdots\cdots\cdots).$$

Letting $h\to 0$, we obtain $\frac{1}{D^2+4}\cos 2x = \frac{1}{4}x\sin 2x$. Since $\frac{1}{D^2+4}\cos 4x = -\frac{1}{12}\cos 4x$, the primitive is $y = C_1\cos 2x + C_2\sin 2x + \frac{1}{4}x\sin 2x - \frac{1}{12}\cos 4x$. (Compare this solution with that given in Problem 8, Chapter 14.)

13. Solve $(2D^2+2D+3)y = x^2+2x-1.$

The complementary function is $y = e^{-\frac{1}{2}x}(C_1\cos\frac{1}{2}\sqrt{5}\,x + C_2\sin\frac{1}{2}\sqrt{5}\,x)$, and a particular integral is

$$y = \frac{1}{2D^2+2D+3}(x^2+2x-1) = (\frac{1}{3} - \frac{2}{9}D - \frac{2}{27}D^2)(x^2+2x-1)$$

$$= \frac{1}{3}(x^2+2x-1) - \frac{2}{9}(2x+2) - \frac{2}{27}(2) = \frac{1}{3}x^2 + \frac{2}{9}x - \frac{25}{27}.$$

Note: $\dfrac{1}{2D^2+2D+3} = (\frac{1}{3} - \frac{2}{9}D - \frac{2}{27}D^2 + \cdots\cdots\cdots)$ by direct division.

The primitive is $y = e^{-\frac{1}{2}x}(C_1\cos\frac{1}{2}\sqrt{5}\,x + C_2\sin\frac{1}{2}\sqrt{5}\,x) + \frac{1}{3}x^2 + \frac{2}{9}x - \frac{25}{27}.$

14. Solve $(D^3-2D+4)y = x^4+3x^2-5x+2.$

The complementary function is $y = C_1e^{-2x} + e^x(C_2\cos x + C_3\sin x)$, and a particular integral is

$$y = \frac{1}{D^3-2D+4}(x^4+3x^2-5x+2) = (\frac{1}{4} + \frac{1}{8}D + \frac{1}{16}D^2 - \frac{1}{32}D^3 - \frac{3}{64}D^4)(x^4+3x^2-5x+2)$$

$$= \frac{1}{4}x^4 + \frac{1}{2}x^3 + \frac{3}{2}x^2 - \frac{5}{4}x - \frac{7}{8}.$$

The primitive is $y = C_1e^{-2x} + e^x(C_2\cos x + C_3\sin x) + \frac{1}{4}x^4 + \frac{1}{2}x^3 + \frac{3}{2}x^2 - \frac{5}{4}x - \frac{7}{8}.$

15. Solve $(D^3-4D^2+3D)y = x^2.$

The complementary function is $y = C_1 + C_2e^x + C_3e^{3x}$, and a particular integral is

$$y = \frac{1}{D(D^2-4D+3)}x^2 = \frac{1}{D}(\frac{1}{D^2-4D+3})x^2 = \frac{1}{D}(\frac{1}{3} + \frac{4}{9}D + \frac{13}{27}D^2)x^2$$

$$= \frac{1}{D}(\frac{1}{3}x^2 + \frac{8}{9}x + \frac{26}{27}) = \frac{1}{9}x^3 + \frac{4}{9}x^2 + \frac{26}{27}x, \quad \text{since } \frac{1}{D}\{f(x)\} = \int f(x)\,dx.$$

The primitive is $y = C_1 + C_2e^x + C_3e^{3x} + \frac{1}{9}x^3 + \frac{4}{9}x^2 + \frac{26}{27}x.$

16. Solve $(D^4+2D^3-3D^2)y = x^2+3e^{2x}+4\sin x.$

The complementary function is $y = C_1 + C_2x + C_3e^x + C_4e^{-3x}$, and a particular integral is

$$y = \frac{1}{D^2(D^2+2D-3)}(x^2+3e^{2x}+4\sin x)$$

$$= \frac{1}{D^2(D^2+2D-3)}x^2 + 3\frac{1}{D^2(D^2+2D-3)}e^{2x} + 4\frac{1}{D^2(D^2+2D-3)}\sin x$$

$$= \frac{1}{D^2}\{\frac{1}{D^2+2D-3}x^2\} + \frac{3}{4(4+4-3)}e^{2x} + \frac{4}{(-1)(-1+2D-3)}\sin x$$

$$= \frac{1}{D^2}(-\frac{1}{3} - \frac{2}{9}D - \frac{7}{27}D^2)x^2 + \frac{3}{20}e^{2x} - \frac{2}{D-2}\sin x$$

$$= -\frac{1}{27}\frac{1}{D^2}(9x^2 + 12x + 14) + \frac{3}{20}e^{2x} - 2\frac{D+2}{D^2-4}\sin x$$

$$= -\frac{1}{27}(\frac{3}{4}x^4 + 2x^3 + 7x^2) + \frac{3}{20}e^{2x} + \frac{2}{5}(\cos x + 2\sin x).$$

The primitive is

$$y = C_1 + C_2x + C_3e^x + C_4e^{-3x} - \frac{x^2}{108}(3x^2 + 8x + 28) + \frac{3}{20}e^{2x} + \frac{2}{5}(\cos x + 2\sin x).$$

17. Establish the rule in $d)$ above by first showing that $F(D)e^{ax}U = e^{ax}F(D+a)U$.

Since when $y = e^{ax}U$, $Dy = ae^{ax}U + e^{ax}DU = e^{ax}(D+a)U$,

$D^2y = ae^{ax}(D+a)U + e^{ax}D(D+a)U = e^{ax}(D^2 + 2aD + a^2)U = e^{ax}(D+a)^2U$, ...,

$D^ry = e^{ax}(D+a)^rU$, and

1) $F(D)e^{ax}U = \sum_r P_rD^r(e^{ax}U) = \sum_r P_re^{ax}(D+a)^rU = e^{ax}\sum_r P_r(D+a)^rU = e^{ax}F(D+a)U.$

Let $V = F(D+a)U$ so that $U = \dfrac{1}{F(D+a)}V$. Then, from 1),

$$F(D)e^{ax}\frac{1}{F(D+a)}V = e^{ax}V \quad\text{and}\quad \frac{1}{F(D)}e^{ax}V = \frac{1}{F(D)}\{F(D)e^{ax}\frac{1}{F(D+a)}V\} = e^{ax}\frac{1}{F(D+a)}V.$$

18. Solve $(D^2-4)y = x^2e^{3x}$.

The complementary function is $y = C_1e^{2x} + C_2e^{-2x}$, and a particular integral is

$$y = \frac{1}{D^2-4}x^2e^{3x} = e^{3x}\frac{1}{(D+3)^2-4}x^2 = e^{3x}\frac{1}{D^2+6D+5}x^2$$

$$= e^{3x}(\frac{1}{5} - \frac{6}{25}D + \frac{31}{125}D^2)x^2 = e^{3x}(\frac{x^2}{5} - \frac{12}{25}x + \frac{62}{125}).$$

The primitive is $y = C_1e^{2x} + C_2e^{-2x} + \dfrac{1}{125}e^{3x}(25x^2 - 60x + 62)$.

19. Solve $(D^2 + 2D + 4)y = e^x\sin 2x$.

The complementary function is $y = e^{-x}(C_1\cos\sqrt{3}\,x + C_2\sin\sqrt{3}\,x)$, and a particular integral is

$$y = \frac{1}{D^2+2D+4}e^x\sin 2x = e^x\frac{1}{(D+1)^2+2(D+1)+4}\sin 2x = e^x\frac{1}{D^2+4D+7}\sin 2x$$

$$= e^x\frac{1}{4D+3}\sin 2x = e^x\frac{4D-3}{16D^2-9}\sin 2x = -\frac{e^x}{73}(4D-3)\sin 2x = -\frac{e^x}{73}(8\cos 2x - 3\sin 2x).$$

The primitive is $y = e^{-x}(C_1\cos\sqrt{3}\,x + C_2\sin\sqrt{3}\,x) - \dfrac{e^x}{73}(8\cos 2x - 3\sin 2x)$.

20. Solve $(D^2-4D+3)y = 2xe^{3x} + 3e^x \cos 2x$.

The complementary function is $y = C_1e^x + C_2e^{3x}$, and a particular integral is

$$y = \frac{1}{D^2-4D+3}(2xe^{3x} + 3e^x \cos 2x) = 2\frac{1}{D^2-4D+3}xe^{3x} + 3\frac{1}{D^2-4D+3}e^x \cos 2x$$

$$= 2e^{3x}\frac{1}{D^2+2D}x + 3e^x\frac{1}{D^2-2D}\cos 2x = 2e^{3x}\frac{1}{D}\cdot\frac{1}{D+2}x + 3e^x\frac{1}{-4-2D}\cos 2x$$

$$= 2e^{3x}\frac{1}{D}(\frac{1}{2}-\frac{1}{4}D)x - \frac{3}{2}e^x\frac{D-2}{D^2-4}\cos 2x = \frac{1}{2}e^{3x}\frac{1}{D}(2x-1) + \frac{3}{16}e^x(D-2)\cos 2x$$

$$= \frac{1}{2}e^{3x}(x^2-x) - \frac{3}{8}e^x(\cos 2x + \sin 2x).$$

The primitive is $y = C_1e^x + C_2e^{3x} + \frac{1}{2}e^{3x}(x^2-x) - \frac{3}{8}e^x(\cos 2x + \sin 2x)$.

21. Establish the rule in *e*) above by first showing that $F(D)xU = xF(D)U + F'(D)U$.

Since when $y = xU$, $Dy = xDU + U$, $D^2y = xD^2U + 2DU$, ..., $D^ry = xD^rU + rD^{r-1}U = xD^rU + (\frac{d}{dD}D^r)U$, then

1) $F(D)xU = \sum_r P_rD^r(xU) = \sum_r P_rxD^rU + \sum_r P_r(\frac{d}{dD}D^r)U = xF(D)U + F'(D)U.$

Let $V = F(D)U$ so that $U = \frac{1}{F(D)}V$. Then, substituting in 1),

$$F(D)x\frac{1}{F(D)}V = xF(D)\frac{1}{F(D)}V + F'(D)\frac{1}{F(D)}V, \quad xV = F(D)x\frac{1}{F(D)}V - F'(D)\frac{1}{F(D)}V,$$

and $$\frac{1}{F(D)}xV = x\frac{1}{F(D)}V - \frac{1}{F(D)}F'(D)\frac{1}{F(D)}V = x\frac{1}{F(D)}V - \frac{F'(D)}{\{F(D)\}^2}V.$$

22. Solve $(D^2+3D+2)y = x \sin 2x$.

The complementary function is $y = C_1e^{-x} + C_2e^{-2x}$, and a particular integral is

$$y = \frac{1}{D^2+3D+2}x\sin 2x = x\frac{1}{D^2+3D+2}\sin 2x - \frac{2D+3}{(D^2+3D+2)^2}\sin 2x$$

$$= x\frac{1}{3D-2}\sin 2x - \frac{2D+3}{D^4+6D^3+13D^2+12D+4}\sin 2x$$

$$= x\frac{1}{3D-2}\sin 2x - \frac{2D+3}{(-4)^2+6(-4)D+13(-4)+12D+4}\sin 2x, \text{ replacing } D^2 \text{ by } -4,$$

$$= x\frac{3D+2}{9D^2-4}\sin 2x + \frac{1}{4}\frac{(2D+3)(3D-8)}{9D^2-64}\sin 2x$$

$$= \frac{-x(3\cos 2x + \sin 2x)}{20} + \frac{24\sin 2x + 7\cos 2x}{200}.$$

The primitive is $y = C_1e^{-x} + C_2e^{-2x} - \frac{30x-7}{200}\cos 2x - \frac{5x-12}{100}\sin 2x.$

23. Solve $(D^2-1)y = x^2 \sin 3x$.

The complementary function is $y = C_1 e^x + C_2 e^{-x}$, and a particular integral is

$$y = \frac{1}{D^2-1}x^2 \sin 3x = x\frac{1}{D^2-1}x \sin 3x - \frac{2D}{(D^2-1)^2}x \sin 3x$$

$$= x^2\frac{1}{D^2-1}\sin 3x - x\frac{2D}{(D^2-1)^2}\sin 3x - 2D\{x\frac{1}{D^4-2D^2+1}\sin 3x - \frac{4D^3-4D}{(D^4-2D^2+1)^2}\sin 3x\}$$

$$= x^2\frac{1}{D^2-1}\sin 3x - x\frac{2D}{(D^2-1)^2}\sin 3x - 2D\{x\frac{1}{(D^2-1)^2}\sin 3x\} + \frac{8D^2}{(D^2-1)^3}\sin 3x$$

$$= -\frac{1}{10}x^2 \sin 3x - \frac{3}{50}x\cos 3x - \frac{1}{50}D(x \sin 3x) + \frac{9}{125}\sin 3x$$

$$= -\frac{1}{10}x^2 \sin 3x - \frac{3}{25}x\cos 3x + \frac{13}{250}\sin 3x.$$

The primitive is $y = C_1 e^x + C_2 e^{-x} - \dfrac{25x^2-13}{250}\sin 3x - \dfrac{3}{25}x\cos 3x$.

24. Solve $(D^3-3D^2-6D+8)y = xe^{-3x}$.

The complementary function is $y = C_1 e^x + C_2 e^{4x} + C_3 e^{-2x}$, and a particular integral is

$$y = \frac{1}{D^3-3D^2-6D+8}xe^{-3x}.$$

By a): $$y = e^{-3x}\frac{1}{(D-3)^3-3(D-3)^2-6(D-3)+8}x = e^{-3x}\frac{1}{D^3-12D^2+39D-28}x$$

$$= e^{-3x}(-\frac{1}{28}-\frac{39}{784}D)x = e^{-3x}(-\frac{1}{28}x-\frac{39}{784}).$$

By e): $$y = x\frac{1}{D^3-3D^2-6D+8}e^{-3x} - \frac{3D^2-6D-6}{(D^3-3D^2-6D+8)^2}e^{-3x}$$

$$= -\frac{1}{28}xe^{-3x} - \frac{3D^2-6D-6}{(-28)^2}e^{-3x} = -\frac{1}{28}xe^{-3x} - \frac{39}{784}e^{-3x}.$$

The primitive is $y = C_1 e^x + C_2 e^{4x} + C_3 e^{-2x} - \dfrac{e^{-3x}}{784}(28x+39)$.

25. Denote the real and imaginary parts of a complex number z by Re[z] and Im[z] respectively. An alternate short method for Problems 9-11 makes use of $\sin bx = \text{Im}[e^{ibx}]$ and $\cos bx = \text{Re}[e^{ibx}]$.

Consider, for example, $F(D)z = (D^3+D^2+D+1)z = e^{2ix}+e^{3ix}$ for which

$$z = \frac{e^{2ix}}{F(2i)} + \frac{e^{3ix}}{F(3i)} = -\frac{e^{2ix}}{3+6i} - \frac{e^{3ix}}{8+24i}$$

$$= \frac{2i-1}{15}(\cos 2x + i\sin 2x) + \frac{3i-1}{80}(\cos 3x + i\sin 3x) = z_1 + z_2$$

is a particular integral. Then

$$\text{Re}[z_1] + \text{Re}[z_2] = -\frac{1}{15}(2\sin 2x + \cos 2x) - \frac{1}{80}(3\sin 3x + \cos 3x)$$

is a particular integral of $F(D)y = \cos 2x + \cos 3x$,

$$\text{Im}[z_1] + \text{Im}[z_2] = \frac{1}{15}(2\cos 2x - \sin 2x) + \frac{1}{80}(3\cos 3x - \sin 3x)$$

is a particular integral of $F(D)y = \sin 2x + \sin 3x$,

$$\text{Re}[z_1] + \text{Im}[z_2]$$

is a particular integral of $F(D)y = \cos 2x + \sin 3x$, and

$$\text{Im}[z_1] + \text{Re}[z_2]$$

is the particular integral in Problem 10.

SUPPLEMENTARY PROBLEMS

Find a particular integral.

26. $(D^2 + D + 1)y = e^{3x} + 6e^x - 3e^{-2x} + 5$ *Ans.* $y = e^{3x}/13 + 2e^x - e^{-2x} + 5$

27. $(D^2 - 1)y = e^x$ $y = xe^x/2$

28. $(D - 2)^2 y = e^x + xe^{2x}$ $y = e^x + x^3e^{2x}/6$

29. $(D^4 - 1)y = \sin 2x$ $y = \frac{1}{15}\sin 2x$

30. $(D^3 + 1)y = \cos x$ $y = \frac{1}{2}(\cos x - \sin x)$

31. $(D^2 + 4)y = \sin 2x$ $y = -\frac{1}{4}x\cos 2x$

32. $(D^2 + 5)y = \cos\sqrt{5}\,x$ $y = \frac{\sqrt{5}}{10}x\sin\sqrt{5}\,x$

33. $(D^3 + D^2 + D + 1)y = e^x + e^{-x} + \sin x$ $y = \frac{1}{4}(e^x + 2xe^{-x}) - \frac{1}{4}x(\sin x + \cos x)$

34. $(D^2 - 1)y = x^2$ $y = -x^2 - 2$

35. $D^4(D^2 - 1)y = x^2$ $y = -\frac{1}{360}(x^6 + 30x^4)$

36. $(D^2 + 2)y = x^3 + x^2 + e^{-2x} + \cos 3x$ $y = \frac{1}{2}(x^3 + x^2 - 3x - 1) + \frac{1}{6}e^{-2x} - \frac{1}{7}\cos 3x$

37. $(D^2 - 2D - 1)y = e^x\cos x$ $y = -\frac{1}{3}e^x\cos x$

38. $(D - 2)^2 y = e^{2x}/x^2$ $y = -e^{2x}\ln x$

39. $(D^2 - 1)y = xe^{3x}$ $y = \frac{1}{32}e^{3x}(4x - 3)$

40. $(D^2 + 5D + 6)y = e^{-2x}(\sec^2 x)(1 + 2\tan x)$ $y = e^{-2x}\tan x$

CHAPTER 17

Linear Equations with Variable Coefficients

THE CAUCHY AND LEGENDRE LINEAR EQUATIONS

THE CAUCHY LINEAR EQUATION

1) $$p_0 x^n \frac{d^n y}{dx^n} + p_1 x^{n-1} \frac{d^{n-1}y}{dx^{n-1}} + \cdots\cdots + p_{n-1} x \frac{dy}{dx} + p_n y = Q(x),$$

in which $p_0, p_1, \cdots\cdots, p_n$ are constants, and the *Legendre linear equation*

2) $$p_0 (ax+b)^n \frac{d^n y}{dx^n} + p_1 (ax+b)^{n-1} \frac{d^{n-1}y}{dx^{n-1}} + \cdots\cdots + p_{n-1}(ax+b)\frac{dy}{dx} + p_n y = Q(x),$$

of which 1) is the special case $(a=1,\ b=0)$, may be reduced to linear equations with constant coefficients by properly chosen transformations of the independent variable.

THE CAUCHY LINEAR EQUATION. Let $x = e^z$; then if $\mathcal{D}$ is defined by $\mathcal{D} = \dfrac{d}{dz}$,

$$Dy = \frac{dy}{dx} = \frac{dy}{dz}\frac{dz}{dx} = \frac{1}{x}\frac{dy}{dz} \qquad \text{and} \qquad xDy = \frac{dy}{dz} = \mathcal{D}y,$$

$$D^2 y = \frac{d}{dx}\left(\frac{1}{x}\frac{dy}{dz}\right) = \frac{1}{x^2}\left(\frac{d^2y}{dz^2} - \frac{dy}{dz}\right) \qquad \text{and} \qquad x^2 D^2 y = \mathcal{D}(\mathcal{D}-1)y,$$

$$D^3 y = -\frac{2}{x^3}\left(\frac{d^2y}{dz^2} - \frac{dy}{dz}\right) + \frac{1}{x^3}\left(\frac{d^3y}{dz^3} - \frac{d^2y}{dz^2}\right)$$

$$= \frac{1}{x^3}\left(\frac{d^3y}{dz^3} - 3\frac{d^2y}{dz^2} + 2\frac{dy}{dz}\right) \qquad \text{and} \qquad x^3 D^3 y = \mathcal{D}(\mathcal{D}-1)(\mathcal{D}-2)y,$$

....

....

$$x^r D^r y = \mathcal{D}(\mathcal{D}-1)(\mathcal{D}-2)\cdots\cdots(\mathcal{D}-r+1)y.$$

After making these replacements, 1) becomes

$$\{p_0\mathcal{D}(\mathcal{D}-1)(\mathcal{D}-2)\cdots\cdots(\mathcal{D}-n+1) + p_1\mathcal{D}(\mathcal{D}-1)(\mathcal{D}-2)\cdots\cdots(\mathcal{D}-n+2) + \cdots\cdots\cdots$$
$$+ p_{n-1}\mathcal{D} + p_n\}y = Q(e^z),$$

a linear equation with constant coefficients.

See Problems 1-3.

THE LEGENDRE LINEAR EQUATION. Let $ax+b = e^z$; then

$$Dy = \frac{dy}{dz}\frac{dz}{dx} = \frac{a}{ax+b}\frac{dy}{dz} \qquad \text{and} \qquad (ax+b)Dy = a\frac{dy}{dz} = a\mathcal{D}y,$$

$$D^2 y = \frac{a^2}{(ax+b)^2}\left(\frac{d^2y}{dz^2} - \frac{dy}{dz}\right) \qquad \text{and} \qquad (ax+b)^2 D^2 y = a^2\mathcal{D}(\mathcal{D}-1)y,$$

....

....

$$(ax+b)^r D^r y = a^r \mathcal{D}(\mathcal{D}-1)\cdots\cdots(\mathcal{D}-r+1)y.$$

After making these replacements, 2) becomes

$$\{p_0 a^n \mathcal{D}(\mathcal{D}-1)(\mathcal{D}-2)\cdots\cdots(\mathcal{D}-n+1) + p_1 a^{n-1}\mathcal{D}(\mathcal{D}-1)(\mathcal{D}-2)\cdots\cdots(\mathcal{D}-n+2) + \cdots\cdots$$
$$+ p_{n-1} a\mathcal{D} + p_n\}y = Q(\frac{e^z - b}{a}),$$

a linear equation with constant coefficients.

See Problems 4-5.

SOLVED PROBLEMS

1. Solve $(x^3D^3 + 3x^2D^2 - 2xD + 2)y = 0$.

The transformation $x = e^z$ reduces the equation to

$$\{\mathcal{D}(\mathcal{D}-1)(\mathcal{D}-2) + 3\mathcal{D}(\mathcal{D}-1) - 2\mathcal{D} + 2\}y = (\mathcal{D}^3 - 3\mathcal{D} + 2)y = 0$$

whose solution is $y = C_1e^z + C_2ze^z + C_3e^{-2z}$.

Since $z = \ln x$, the complete solution of the given equation is $y = C_1x + C_2x\ln x + C_3/x^2$.

2. Solve $(x^3D^3 + 2xD - 2)y = x^2\ln x + 3x$.

The transformation $x = e^z$ reduces the equation to

$$\{\mathcal{D}(\mathcal{D}-1)(\mathcal{D}-2) + 2\mathcal{D} - 2\}y = (\mathcal{D}-1)(\mathcal{D}^2 - 2\mathcal{D}+2)y = ze^{2z} + 3e^z.$$

The complementary function is $y = C_1e^z + e^z(C_2\cos z + C_3\sin z)$, and a particular integral is

$$y = \frac{1}{\mathcal{D}^3 - 3\mathcal{D}^2 + 4\mathcal{D} - 2}(ze^{2z} + 3e^z) = e^{2z}\frac{1}{(\mathcal{D}+2)^3 - 3(\mathcal{D}+2)^2 + 4(\mathcal{D}+2) - 2}z + 3\frac{1}{(\mathcal{D}-1)(\mathcal{D}^2-2\mathcal{D}+2)}e^z$$

$$= e^{2z}\frac{1}{\mathcal{D}^3 + 3\mathcal{D}^2 + 4\mathcal{D} + 2}z + 3\frac{1}{(\mathcal{D}-1)(1)}e^z$$

$$= e^{2z}(\tfrac{1}{2} - \mathcal{D})z + 3e^z\int e^z\cdot e^{-z}\,dz = e^{2z}(\tfrac{1}{2}z - 1) + 3ze^z.$$

Thus, the solution is
$$y = C_1e^z + e^z(C_2\cos z + C_3\sin z) + \tfrac{1}{2}e^{2z}(z-2) + 3ze^z$$
$$= C_1x + x(C_2\cos\ln x + C_3\sin\ln x) + \tfrac{1}{2}x^2(\ln x - 2) + 3x\ln x.$$

3. Solve $(x^2D^2 - xD + 4)y = \cos\ln x + x\sin\ln x$.

The transformation $x = e^z$ reduces the equation to

$$\{\mathcal{D}(\mathcal{D}-1) - \mathcal{D} + 4\}y = (\mathcal{D}^2 - 2\mathcal{D} + 4)y = \cos z + e^z\sin z.$$

The complementary function is $y = e^z(C_1\cos\sqrt{3}\,z + C_2\sin\sqrt{3}\,z)$, and a particular solution is

$$y = \frac{1}{\mathcal{D}^2 - 2\mathcal{D} + 4}\cos z + \frac{1}{\mathcal{D}^2 - 2\mathcal{D} + 4}e^z\sin z$$

$$= \frac{1}{3 - 2\mathcal{D}}\cos z + e^z\frac{1}{\mathcal{D}^2 + 3}\sin z = \frac{1}{13}(3\cos z - 2\sin z) + \frac{1}{2}e^z\sin z.$$

Thus, the solution is

$$y = e^z(C_1\cos\sqrt{3}\,z + C_2\sin\sqrt{3}\,z) + \frac{1}{13}(3\cos z - 2\sin z) + \frac{1}{2}e^z\sin z$$

$$= x(C_1\cos\sqrt{3}\cdot\ln x + C_2\sin\sqrt{3}\cdot\ln x) + \frac{1}{13}(3\cos\ln x - 2\sin\ln x) + \frac{1}{2}x\sin\ln x.$$

4. Solve $(x+2)^2\dfrac{d^2y}{dx^2} - (x+2)\dfrac{dy}{dx} + y = 3x + 4.$

Put $x+2 = e^z$; then the given equation becomes

$$\{\mathcal{D}(\mathcal{D}-1) - \mathcal{D} + 1\}y = (\mathcal{D}-1)^2 y = 3e^z - 2.$$

The complementary function is $y = C_1e^z + C_2ze^z$, and a particular integral is

$$y = \frac{1}{(\mathcal{D}-1)^2}(3e^z - 2) = 3e^z\iint(dz)^2 - 2\frac{1}{(\mathcal{D}-1)^2}e^{0z} = \frac{3}{2}z^2e^z - 2.$$

The solution is $y = C_1e^z + C_2ze^z + \frac{3}{2}z^2e^z - 2$ or, since $z = \ln(x+2)$,

$$y = (x+2)[C_1 + C_2\ln(x+2) + \frac{3}{2}\ln^2(x+2)] - 2.$$

5. Solve $\{(3x+2)^2D^2 + 3(3x+2)D - 36\}y = 3x^2 + 4x + 1.$

The transformation $3x + 2 = e^z$ reduces the equation to

$$\{9\mathcal{D}(\mathcal{D}-1) + 9\mathcal{D} - 36\}y = 9(\mathcal{D}^2-4)y = \frac{1}{3}(9x^2 + 12x + 3) = \frac{1}{3}(e^{2z} - 1) \quad\text{or}\quad (\mathcal{D}^2-4)y = \frac{1}{27}(e^{2z} - 1).$$

The complete solution is $y = C_1e^{2z} + C_2e^{-2z} + \frac{1}{27}\left(\frac{1}{\mathcal{D}^2-4}e^{2z} - \frac{1}{\mathcal{D}^2-4}e^{0z}\right)$

$$= C_1e^{2z} + C_2e^{-2z} + \frac{1}{108}(ze^{2z} + 1)$$

or $$y = C_1(3x+2)^2 + C_2(3x+2)^{-2} + \frac{1}{108}[(3x+2)^2\ln(3x+2) + 1].$$

SUPPLEMENTARY PROBLEMS

Solve.

6. $(x^2D^2 - 3xD + 4)y = x + x^2\ln x$ *Ans.* $y = C_1x^2 + C_2x^2\ln x + x + \frac{1}{6}x^2\ln^3 x$

7. $(x^2D^2 - 2xD + 2)y = \ln^2 x - \ln x^2$ *Ans.* $y = C_1x + C_2x^2 + \frac{1}{2}(\ln^2 x + \ln x) + \frac{1}{4}$

8. $(x^3D^3 + 2x^2D^2)y = x + \sin(\ln x)$ *Ans.* $y = C_1 + C_2x + C_3\ln x + x\ln x + \frac{1}{2}(\cos\ln x + \sin\ln x)$

9. $x^3y''' + xy' - y = 3x^4$ *Ans.* $y = C_1x + C_2x\ln x + C_3x\ln^2 x + x^4/9$

10. $[(x+1)^2D^2 + (x+1)D - 1]y = \ln(x+1)^2 + x - 1$
Ans. $y = C_1(x+1) + C_2(x+1)^{-1} - \ln(x+1)^2 + \frac{1}{2}(x+1)\cdot\ln(x+1) + 2$

11. $(2x+1)^2y'' - 2(2x+1)y' - 12y = 6x$ *Ans.* $y = C_1(2x+1)^{-1} + C_2(2x+1)^3 - 3x/8 + 1/16$

CHAPTER 18

Linear Equations with Variable Coefficients

EQUATIONS OF THE SECOND ORDER

A LINEAR DIFFERENTIAL EQUATION of the second order has the form

1) $$\frac{d^2y}{dx^2} + R(x)\frac{dy}{dx} + S(x)\,y = Q(x).$$

If the coefficients R and S are constants, the equation can be solved by the methods of the preceding chapter; otherwise, no general method is known. In this chapter certain procedures are given which at times will yield a solution.

CHANGE OF DEPENDENT VARIABLE. Under the transformation

$$y = uv, \qquad u = u(x) \quad \text{and} \quad v = v(x),$$

$$\frac{dy}{dx} = u\frac{dv}{dx} + v\frac{du}{dx}, \qquad \frac{d^2y}{dx^2} = u\frac{d^2v}{dx^2} + 2\frac{dv}{dx}\frac{du}{dx} + v\frac{d^2u}{dx^2},$$

1) becomes

2) $$\frac{d^2v}{dx^2} + R_1(x)\frac{dv}{dx} + S_1(x)\,v = Q_1(x)$$

with $R_1(x) = \dfrac{2}{u}\dfrac{du}{dx} + R(x)$, $S_1(x) = \dfrac{1}{u}\{\dfrac{d^2u}{dx^2} + R(x)\dfrac{du}{dx} + S(x)\cdot u\}$, $Q_1(x) = \dfrac{Q(x)}{u}$.

a) If u is a particular integral of $\dfrac{d^2y}{dx^2} + R(x)\dfrac{dy}{dx} + S(x)\,y = 0$, then $S_1 = 0$ and 2) becomes

3) $$\frac{d^2v}{dx^2} + R_1(x)\frac{dv}{dx} = Q_1(x).$$

The further substitution $\dfrac{dv}{dx} = p$, $\dfrac{d^2v}{dx^2} = \dfrac{dp}{dx}$ reduces 3) to

4) $$\frac{dp}{dx} + R_1(x)\,p = Q_1(x),$$

a linear equation of the first order. See Problems 1-6.

b) If u is chosen so that $R_1(x) = \dfrac{2}{u}\dfrac{du}{dx} + R(x) = 0$ or $\dfrac{du}{u} = -\tfrac{1}{2}R(x)\,dx$, then

$$u = e^{-\frac{1}{2}\int R(x)\,dx}$$

Now $\dfrac{du}{dx} = -\tfrac{1}{2}u\,R(x)$ and $\dfrac{d^2u}{dx^2} = -\tfrac{1}{2}R(x)\dfrac{du}{dx} - \tfrac{1}{2}u\dfrac{dR}{dx}$ so that

$$S_1(x) = S(x) + \frac{R(x)}{u}\frac{du}{dx} + \frac{1}{u}\frac{d^2u}{dx^2} = S(x) + \tfrac{1}{2}\frac{R(x)}{u}\frac{du}{dx} - \tfrac{1}{2}\frac{dR}{dx} = S - \tfrac{1}{4}R^2 - \tfrac{1}{2}\frac{dR}{dx}$$

and $Q_1 = Q/u$.

If $S_1(x) = S - \tfrac{1}{4}R^2 - \tfrac{1}{2}\dfrac{dR}{dx} = A$, a constant, 2) becomes $\dfrac{d^2v}{dx^2} + Av = Q/u$, a linear equation with constant coefficients.

If $S_1(x) = A/x^2$, 2) becomes $x^2\dfrac{d^2v}{dx^2} + Av = Qx^2/u$, a Cauchy equation, and the substitution $x = e^z$ will reduce it to one with constant coefficients.

See Problems 7-10.

CHANGE OF INDEPENDENT VARIABLE. Let the transformation be $z = \theta(x)$. Then

$$\frac{dy}{dx} = \frac{dy}{dz}\frac{dz}{dx}, \qquad \frac{d^2y}{dx^2} = \frac{d^2y}{dz^2}\left(\frac{dz}{dx}\right)^2 + \frac{dy}{dz}\frac{d^2z}{dx^2},$$

and 1) becomes

$$\frac{d^2y}{dz^2}\left(\frac{dz}{dx}\right)^2 + \left(\frac{d^2z}{dx^2} + R\frac{dz}{dx}\right)\frac{dy}{dz} + Sy = Q$$

or

5) $$\frac{d^2y}{dz^2} + \frac{\dfrac{d^2z}{dx^2} + R\dfrac{dz}{dx}}{\left(\dfrac{dz}{dx}\right)^2}\frac{dy}{dz} + \frac{Sy}{\left(\dfrac{dz}{dx}\right)^2} = \frac{Q}{\left(\dfrac{dz}{dx}\right)^2}.$$

Let $z = \theta(x)$ be chosen so that $\dfrac{dz}{dx} = \sqrt{\dfrac{\pm S}{a^2}}$, the sign being that which makes $\dfrac{dz}{dx}$ real and a^2 being any positive constant. (One may consistently take $a^2 = 1$.)

If now $\dfrac{\dfrac{d^2z}{dx^2} + R\dfrac{dz}{dx}}{\left(\dfrac{dz}{dx}\right)^2} = A$, a constant, then 5) becomes $\dfrac{d^2y}{dz^2} + A\dfrac{dy}{dz} \pm a^2y = \dfrac{Q}{\left(\dfrac{dz}{dx}\right)^2}$, a linear equation with constant coefficients.

See Problems 11-14.

OPERATIONAL FACTORING. It may be possible to separate the left member of

$$\{P(x)D^2 + R(x)D + S(x)\}y = Q(x)$$

into two linear operators $F_1(D)$ and $F_2(D)$ so that

6) $\{F_1(D)\cdot F_2(D)\}y = F_1(D)\{F_2(D)y\} = \{P(x)D^2 + R(x)D + S(x)\}y = Q(x)$.

Then, setting $F_2(D)y = v$, 6) becomes $F_1(D)v = Q(x)$, a linear equation of order one.

The factorization in this section differs from that of Chapter 13. With possible exceptions, the factors here contain the independent variable x, they are not commutative, and the factorization differs from that when D is treated as a variable. For example,

$$\{xD^2 - (x^2+2)D + x\}y = \{(xD-2)(D-x)\}y,$$

since

$$\{(xD-2)(D-x)\}y = (xD-2)(\frac{d}{dx}-x)y = (xD-2)(y'-xy)$$

$$= (x\frac{d}{dx}-2)(y'-xy) = x(y''-y-xy') - 2(y'-xy)$$

$$= xy'' - (x^2+2)y' + xy = \{(xD^2-(x^2+2)D+x\}y.$$

The factors are not commutative, since

$$\{(D-x)(xD-2)\}y = (D-x)(xy'-2y) = xy''+y'-2y'-x^2y'+2xy$$

$$= xy''-(x^2+1)y'+2xy = \{xD^2-(x^2+1)D+2x\}y.$$

Finally, when D is treated as a *variable* rather than an operator,

$$\{(xD-2)(D-x)\}y = \{xD^2-(x^2+2)D+2x\}y.$$ See Problems 15-17.

IN SUMMARY, the following procedure is suggested for solving

$$\frac{d^2y}{dx^2} + R(x)\frac{dy}{dx} + S(x)y = Q(x).$$

1) Find by inspection, or otherwise, a particular integral $u = u(x)$ of the equation when $Q(x) = 0$. The substitution $y = uv$ will yield a linear equation in which the dependent variable v does not appear. This equation is of the first order in $dv/dx = p$.

2) If a particular integral cannot be found, compute $S - \frac{1}{4}R^2 - \frac{1}{2}\frac{dR}{dx}$. If this is a constant K or K/x^2, the transformation $y = ve^{-\frac{1}{2}\int R\,dx}$ reduces the given equation to a linear equation with constant coefficients or to a Cauchy equation.

3) If the above procedure does not apply, put $\frac{dz}{dx} = \sqrt{\frac{\pm S}{a^2}}$ (choosing the sign so that the square root is real) and substitute in $\dfrac{\frac{d^2z}{dx^2} + R\frac{dz}{dx}}{(\frac{dz}{dx})^2}$. If this is a constant, the transformation $z = \int\sqrt{\frac{\pm S}{a^2}}\,dx$ yields a linear equation with constant coefficients.

4) If the left member of the equation is operationally factorable, the problem is then reduced to that of solving two linear equations of order one.

Note. As a partial check on the work, it is desirable to know the type of equation which results when the transformations in 1)-3) are made.

SOLVED PROBLEMS

1. For the equation $(D^2 + RD + S)y = 0$, show that

a) $y = x$ is a particular integral if $R + xS = 0$,

b) $y = e^x$ is a particular integral if $1 + R + S = 0$,

c) $y = e^{-x}$ is a particular integral if $1 - R + S = 0$,

d) $y = e^{mx}$ is a particular integral if $m^2 + mR + S = 0$.

a) If $y = x$ is a particular integral of $(D^2 + RD + S)y = 0$ then, since $Dy = 1$ and $D^2y = 0$, $R + Sx = 0$.

d) If $y = e^{mx}$ is a particular integral of $(D^2 + RD + S)y = 0$ then, since $Dy = my$ and $D^2y = m^2y$, $(m^2 + mR + S)y = 0$ and $m^2 + mR + S = 0$. b) and c) are special cases ($m = 1$, $m = -1$) of d).

2. Solve $(D^2 - \dfrac{3}{x}D + \dfrac{3}{x^2})y = 2x - 1$.

Here $R + Sx = 0$ and $y = x$ is a particular integral of $(D^2 - \dfrac{3}{x}D + \dfrac{3}{x^2})y = 0$.

The transformation $y = xv$, $Dy = x\dfrac{dv}{dx} + v$, $D^2y = x\dfrac{d^2v}{dx^2} + 2\dfrac{dv}{dx}$ reduces the given equation to

$$x\frac{d^2v}{dx^2} + 2\frac{dv}{dx} - 3\frac{dv}{dx} - \frac{3}{x}v + \frac{3}{x}v = x\frac{d^2v}{dx^2} - \frac{dv}{dx} = 2x - 1 \qquad \text{or} \qquad \frac{d^2v}{dx^2} - \frac{1}{x}\frac{dv}{dx} = 2 - \frac{1}{x}.$$

Putting $\dfrac{dv}{dx} = p$, $\dfrac{d^2v}{dx^2} = \dfrac{dp}{dx}$, this becomes $\dfrac{dp}{dx} - \dfrac{1}{x}p = 2 - \dfrac{1}{x}$ for which $e^{\int -dx/x} = 1/x$ is an integrating factor. Then

$$\frac{p}{x} = \int(\frac{2}{x} - \frac{1}{x^2})dx = 2\ln x + \frac{1}{x} + K, \qquad p = \frac{dv}{dx} = 2x\ln x + 1 + Kx,$$

$$v = \frac{y}{x} = \int(2x\ln x + 1 + Kx)dx = x^2\ln x + x + C_1x^2 + C_2, \quad \text{and} \quad y = C_1x^3 + C_2x + x^3\ln x + x^2.$$

3. Solve $x^2(x+1)\dfrac{d^2y}{dx^2} - x(2 + 4x + x^2)\dfrac{dy}{dx} + (2 + 4x + x^2)y = -x^4 - 2x^3$.

Here, $R + Sx = -\dfrac{x(2 + 4x + x^2)}{x^2(x+1)} + x\dfrac{2 + 4x + x^2}{x^2(x+1)} = 0$ and $y = x$ is a particular integral of the equation with its right member replaced by zero.

The transformation $y = xv$, $\dfrac{dy}{dx} = x\dfrac{dv}{dx} + v$, $\dfrac{d^2y}{dx^2} = x\dfrac{d^2v}{dx^2} + 2\dfrac{dv}{dx}$ reduces the given equation

to $$x^2(x+1)(x\frac{d^2v}{dx^2} + 2\frac{dv}{dx}) - x(2 + 4x + x^2)(x\frac{dv}{dx} + v) + (2 + 4x + x^2)xv = -x^4 - 2x^3$$

or $$\frac{d^2v}{dx^2} - \frac{x+2}{x+1}\frac{dv}{dx} = -\frac{x+2}{x+1}.$$

Putting $\dfrac{dv}{dx} = p$, this becomes $\dfrac{dp}{dx} - \dfrac{x+2}{x+1}p = -\dfrac{x+2}{x+1}$ for which $e^{-\int(1 + \frac{1}{x+1})dx} = \dfrac{e^{-x}}{x+1}$ is an integrating factor. Then

$$\frac{e^{-x}}{x+1}p \;=\; -\int \frac{(x+2)e^{-x}}{(x+1)^2}\,dx \;=\; \frac{e^{-x}}{x+1} + C_1, \qquad p \;=\; \frac{dv}{dx} \;=\; 1 + C_1(x+1)e^x,$$

$$v \;=\; \frac{y}{x} \;=\; x + C_1 x e^x + C_2, \qquad \text{and} \qquad y \;=\; C_1 x^2 e^x + C_2 x + x^2.$$

4. Solve $x\dfrac{d^2y}{dx^2} - (2x+1)\dfrac{dy}{dx} + (x+1)y \;=\; (x^2+x-1)e^{2x}.$

Here $1+R+S = 1 - \dfrac{2x+1}{x} + \dfrac{x+1}{x} = 0$ and $y = e^x$ is a particular integral of the equation with its right member replaced by zero.

The transformation $y = e^x v$, $\dfrac{dy}{dx} = e^x(\dfrac{dv}{dx} + v)$, $\dfrac{d^2y}{dx^2} = e^x(\dfrac{d^2v}{dx^2} + 2\dfrac{dv}{dx} + v)$ reduces the given equation to $\dfrac{d^2v}{dx^2} - \dfrac{1}{x}\dfrac{dv}{dx} \;=\; (x + 1 - \dfrac{1}{x})e^x.$

Putting $\dfrac{dv}{dx} = p$, this becomes $\dfrac{dp}{dx} - \dfrac{1}{x}p \;=\; (x+1-\dfrac{1}{x})e^x$ for which $\dfrac{1}{x}$ is an integrating factor. Then

$$\frac{p}{x} \;=\; \int (e^x + \frac{xe^x - e^x}{x^2})\,dx \;=\; e^x + \frac{e^x}{x} + K, \qquad p \;=\; \frac{dv}{dx} \;=\; xe^x + e^x + Kx,$$

$$v \;=\; \frac{y}{e^x} \;=\; xe^x + C_1 x^2 + C_2; \qquad \text{and} \qquad y \;=\; C_1 x^2 e^x + C_2 e^x + xe^{2x}.$$

5. Solve $(x-2)\dfrac{d^2y}{dx^2} - (4x-7)\dfrac{dy}{dx} + (4x-6)y \;=\; 0.$

Here $m^2 + mR + S \;=\; m^2 - m\dfrac{4x-7}{x-2} + \dfrac{4x-6}{x-2} = 0$ when $m = 2$, and $y = e^{2x}$ is a particular integral.

The transformation $y = e^{2x}v$, $\dfrac{dy}{dx} = e^{2x}\dfrac{dv}{dx} + 2e^{2x}v$, $\dfrac{d^2y}{dx^2} = e^{2x}\dfrac{d^2v}{dx^2} + 4e^{2x}\dfrac{dv}{dx} + 4e^{2x}v$ reduces the given equation to

$$(x-2)(\frac{d^2v}{dx^2} + 4\frac{dv}{dx} + 4v) - (4x-7)(\frac{dv}{dx} + 2v) + (4x-6)v = 0 \quad \text{or} \quad \frac{d^2v}{dx^2} - \frac{1}{x-2}\frac{dv}{dx} = 0.$$

Putting $\dfrac{dv}{dx} = p$, this becomes $\dfrac{dp}{dx} - \dfrac{1}{x-2}p \;=\; 0.$ Then

$$p \;=\; \frac{dv}{dx} \;=\; K(x-2), \qquad v \;=\; \frac{y}{e^{2x}} \;=\; C_1(x-2)^2 + C_2, \quad \text{and} \quad y = C_1 e^{2x}(x-2)^2 + C_2 e^{2x}.$$

6. Solve $\dfrac{d^2y}{dx^2} - 2\tan x\dfrac{dy}{dx} + 3y \;=\; 2\sec x.$

By inspection, it is seen that $y = \sin x$ is a particular integral of $(D^2 - 2\tan x\, D + 3)y = 0$. The transformation $y = v \sin x$ reduces the given equation to

$$\sin x\,\frac{d^2v}{dx^2} + 2(\cos x - \frac{\sin^2 x}{\cos x})\frac{dv}{dx} = 2\sec x, \quad\text{or}\quad \frac{d^2v}{dx^2} + 2(\cot x - \tan x)\frac{dv}{dx} = 4\csc 2x.$$

The substitution $\frac{dv}{dx} = p$ reduces this to $\frac{dp}{dx} + 2(\cot x - \tan x)p = 4\csc 2x$ for which an integrating factor is $\frac{1}{4}\sin^2 2x$. Then

$$\tfrac{1}{4}p\sin^2 2x = \int \sin 2x\,dx = -\tfrac{1}{2}\cos 2x + \tfrac{1}{4}K_1, \qquad p = \frac{dv}{dx} = -2\csc 2x\cot 2x + K_1\csc^2 2x,$$

$$v = \frac{y}{\sin x} = \csc 2x + K\cot 2x + C_2, \quad\text{and}\quad y = \tfrac{1}{2}\sec x + C_1(\cos x - \tfrac{1}{2}\sec x) + C_2\sin x.$$

7. Solve $\frac{d^2y}{dx^2} - \frac{2}{x}\frac{dy}{dx} + (1 + \frac{2}{x^2})y = xe^x$.

Here $R = -\frac{2}{x}$, $S = 1 + \frac{2}{x^2}$, $S - \frac{1}{4}R^2 - \frac{1}{2}\frac{dR}{dx} = 1$ and $u = e^{-\frac{1}{2}\int R\,dx} = e^{\int dx/x} = x$.

The transformation $y = uv = xv$, $\frac{dy}{dx} = x\frac{dv}{dx} + v$, $\frac{d^2y}{dx^2} = x\frac{d^2v}{dx^2} + 2\frac{dv}{dx}$ reduces the given equation to $\frac{d^2v}{dx^2} + v = e^x$, a linear equation with constant coefficients, whose complete solution is $v = \frac{y}{x} = C_1\cos x + C_2\sin x + \frac{1}{D^2+1}e^x = C_1\cos x + C_2\sin x + \frac{1}{2}e^x$.

Thus, $y = C_1 x\cos x + C_2 x\sin x + \frac{1}{2}xe^x$.

8. Solve $\frac{d^2y}{dx^2} - 2x\frac{dy}{dx} + (x^2+2)y = e^{\frac{1}{2}(x^2+2x)}$.

Here $R = -2x$, $S = x^2+2$, $S - \frac{1}{4}R^2 - \frac{1}{2}\frac{dR}{dx} = 3$, and $u = e^{-\frac{1}{2}\int R\,dx} = e^{\frac{1}{2}x^2}$.

The transformation $y = e^{\frac{1}{2}x^2}v$ reduces the equation to $\frac{d^2v}{dx^2} + 3v = e^x$ whose complete solution is $v = y/e^{\frac{1}{2}x^2} = C_1\cos\sqrt{3}\,x + C_2\sin\sqrt{3}\,x + \frac{1}{D^2+3}e^x = C_1\cos\sqrt{3}x + C_2\sin\sqrt{3}x + \frac{1}{4}e^x$.

Thus, $y = e^{\frac{1}{2}x^2}(C_1\cos\sqrt{3}\,x + C_2\sin\sqrt{3}\,x) + \frac{1}{4}e^{\frac{1}{2}(x^2+2x)}$.

9. Solve $(D^2 - \frac{3}{x}D + \frac{3}{x^2})y = 2x - 1$. (Problem 2.)

Here $S - \frac{1}{4}R^2 - \frac{1}{2}\frac{dR}{dx} = \frac{3}{x^2} - \frac{9}{4x^2} - \frac{3}{2x^2} = -\frac{3}{4x^2}$ and $u = e^{-\frac{1}{2}\int R\,dx} = e^{-\frac{1}{2}\int -3\,dx/x} = x^{3/2}$.

The transformation $y = uv = x^{3/2}v$ reduces the equation to

$$\frac{d^2v}{dx^2} - \frac{3}{4x^2}v = \frac{2x-1}{x^{3/2}} \quad\text{or}\quad x^2\frac{d^2v}{dx^2} - \frac{3}{4}v = 2x^{3/2} - x^{1/2},\ \text{a Cauchy equation.}$$

Putting $x = e^z$, we have $(\mathcal{D}^2 - \mathcal{D} - \frac{3}{4})v = 2e^{3z/2} - e^{z/2}$.

The complementary function is $v = C_1 e^{-z/2} + C_2 e^{3z/2}$, and a particular integral is

$$v = \frac{1}{\mathcal{D}^2 - \mathcal{D} - 3/4}(2e^{3z/2} - e^{z/2}) = \frac{1}{\mathcal{D} - 3/2}e^{3z/2} + e^{z/2} = ze^{3z/2} + e^{z/2}.$$

The complete solution is $v = y/x^{3/2} = C_1 x^{-1/2} + C_2 x^{3/2} + x^{3/2}\ln x + x^{1/2}$

and $$y = C_1 x + C_2 x^3 + x^3 \ln x + x^2.$$

10. Solve $\dfrac{d^2y}{dx^2} - 4x\dfrac{dy}{dx} + 4x^2 y = xe^{x^2}$.

Here $S - \frac{1}{4}R^2 - \frac{1}{2}\dfrac{dR}{dx} = 2$ and $u = e^{2\int x\,dx} = e^{x^2}$.

The transformation $y = ve^{x^2}$ reduces the equation to $\dfrac{d^2v}{dx^2} + 2v = x$ whose complete solution is

$$v = C_1\cos\sqrt{2}\,x + C_2\sin\sqrt{2}\,x + \tfrac{1}{2}x.$$

Then $y = ve^{x^2} = e^{x^2}(C_1\cos\sqrt{2}\,x + C_2\sin\sqrt{2}\,x) + \frac{1}{2}xe^{x^2}$.

11. Solve $\dfrac{d^2y}{dx^2} - (1 + 4e^x)\dfrac{dy}{dx} + 3e^{2x}y = e^{2(x+e^x)}$.

When $\dfrac{dz}{dx} = \sqrt{\dfrac{S}{a^2}} = \sqrt{\dfrac{3e^{2x}}{3}} = e^x$, $\dfrac{d^2z/dx^2 + R(dz/dx)}{(dz/dx)^2} = \dfrac{e^x - (1+4e^x)e^x}{(e^x)^2} = -4 = A$.

The introduction of $z = e^x$ as new independent variable leads to

$$\frac{d^2y}{dz^2} + A\frac{dy}{dz} + a^2y = \frac{Q}{(dz/dx)^2} \quad \text{or} \quad \frac{d^2y}{dz^2} - 4\frac{dy}{dz} + 3y = \frac{e^{2(x+e^x)}}{e^{2x}} = e^{2e^x} = e^{2z}$$

whose complete solution is $y = C_1e^z + C_2e^{3z} + \dfrac{1}{\mathcal{D}^2 - 4\mathcal{D} + 3}e^{2z} = C_1e^z + C_2e^{3z} - e^{2z}$.

Replacing z by e^x, we have $y = C_1e^{e^x} + C_2e^{3e^x} - e^{2e^x}$.

Note. The choice of $a^2 = 3$ is one of convenience only. Taking $a^2 = 1$, $\dfrac{dz}{dx} = \sqrt{3}\,e^x$ and $A = \dfrac{-4}{\sqrt{3}}$.

The transformation $z = \sqrt{3}\,e^x$ yields $\dfrac{d^2y}{dz^2} - \dfrac{4}{\sqrt{3}}\dfrac{dy}{dz} + y = \dfrac{1}{3}e^{2z/\sqrt{3}}$ whose solution is

$y = C_1e^{z/\sqrt{3}} + C_2e^{\sqrt{3}\,z} - e^{2z/\sqrt{3}}$. Then $y = C_1e^{e^x} + C_2e^{3e^x} - e^{2e^x}$, as before.

12. Solve $\dfrac{d^2y}{dx^2} - \cot x\dfrac{dy}{dx} - \sin^2 x\, y = \cos x - \cos^3 x$.

Here $S = -\sin^2 x$ and when $\dfrac{dz}{dx} = \sqrt{\dfrac{-S}{1}} = \sin x$, $\dfrac{d^2z/dx^2 + R(dz/dx)}{(dz/dx)^2} = \dfrac{\cos x + (-\cot x)(\sin x)}{\sin^2 x} = 0$.

Thus the introduction of $z = -\cos x$ as new independent variable leads to

$\frac{d^2y}{dz^2} - y = \cos x = -z$ whose complete solution is $y = C_1e^z + C_2e^{-z} + z$.

Upon replacing z by $-\cos x$, we have $y = C_1e^{-\cos x} + C_2e^{\cos x} - \cos x$.

13. Solve $\frac{d^2y}{dx^2} + \frac{2}{x}\frac{dy}{dx} + \frac{1}{x^4}y = \frac{2x^2+1}{x^6}$.

When $\frac{dz}{dx} = \sqrt{S} = \sqrt{\frac{1}{x^4}} = \frac{1}{x^2}$, $\frac{d^2z/dx^2 + R(dz/dx)}{(dz/dx)^2} = 0$. Thus the introduction of $z = -\frac{1}{x}$ as new independent variable leads to $\frac{d^2y}{dz^2} + y = 2 + z^2$ whose complete solution is $y = C_1\cos z + K\sin z + z^2$.

Upon replacing z by $-1/x$, we have $y = C_1\cos(-1/x) + K\sin(-1/x) + 1/x^2$

$= C_1\cos(1/x) + C_2\sin(1/x) + 1/x^2$.

14. Solve $\frac{d^2y}{dx^2} + (4x - \frac{1}{x})\frac{dy}{dx} + 4x^2y = 3xe^{-x^2}$.

When $\frac{dz}{dx} = \sqrt{S} = \sqrt{4x^2} = 2x$, $\frac{d^2z/dx^2 + R(dz/dx)}{(dz/dx)^2} = \frac{2 + (4x - 1/x)2x}{(2x)^2} = 2$. Thus the introduction of $z = x^2$ as new independent variable leads to $\frac{d^2y}{dz^2} + 2\frac{dy}{dz} + y = \frac{3e^{-z}}{4\sqrt{z}}$ whose complete solution is $y = C_1e^{-z} + C_2ze^{-z} + \frac{3/4}{(D+1)^2}e^{-z}z^{-1/2} = C_1e^{-z} + C_2ze^{-z} + z^{3/2}e^{-z}$.

Upon replacing z by x^2, we have $y = C_1e^{-x^2} + C_2x^2e^{-x^2} + x^3e^{-x^2}$.

15. Solve $(D^2 - \frac{3}{x}D + \frac{3}{x^2})y = 2x - 1$. (Problem 2.)

a. The equation is equivalent to $D^2y - D(\frac{3}{x}y) = D(D - \frac{3}{x})y = 2x - 1$.

Putting $(D - \frac{3}{x})y = v$, we have $Dv = 2x - 1$ and $v = x^2 - x + K$.

Now $(D - \frac{3}{x})y = x^2 - x + K$ for which $\frac{1}{x^3}$ is an integrating factor. Then

$\frac{y}{x^3} = \int(\frac{1}{x} - \frac{1}{x^2} + \frac{K}{x^3})dx = \ln x + \frac{1}{x} + \frac{C_1}{x^2} + C_2$ and $y = C_1x + C_2x^3 + x^2(1 + x\ln x)$.

b. The equation $(xD^2 - 3D + \frac{3}{x})y = 2x^2 - x$ is equivalent to $(D - \frac{3}{x})(xD - 1)y = 2x^2 - x$.

Putting $(xD-1)y = v$, we have $(D - \frac{3}{x})v = 2x^2 - x$ for which $\frac{1}{x^3}$ is an integrating factor.

Then $\frac{v}{x^3} = \int(\frac{2}{x} - \frac{1}{x^2})dx = 2\ln x + \frac{1}{x} + K$ and

$$(xD-1)y = v = 2x^3 \ln x + x^2 + Kx^3 \quad \text{or} \quad (D-1/x)y = 2x^2 \ln x + x + Kx^2.$$

Here $1/x$ is an integrating factor so that

$$y/x = \int(2x \ln x + 1 + Kx)dx = x^2 \ln x - \tfrac{1}{2}x^2 + x + K_1 x^2 + C_2 = x^2 \ln x + x + C_1 x^2 + C_2$$

and
$$y = C_1 x^3 + C_2 x + x^2(1 + x \ln x).$$

16. Solve $[xD^2 + (1-x)D - 2(1+x)]y = e^{-x}(1-6x)$.

The equation is equivalent to $[xD + (1+x)][D-2]y = e^{-x}(1-6x)$.

Putting $(D-2)y = v$, we have $[xD+1+x]v = e^{-x}(1-6x)$ or $(D + \frac{1}{x} + 1)v = e^{-x}(\frac{1}{x} - 6)$.

Now xe^x is an integrating factor so that $vxe^x = \int(1-6x)dx = x - 3x^2 + K$

and
$$(D-2)y = v = (1-3x)e^{-x} + Ke^{-x}/x.$$

Here e^{-2x} is an integrating factor so that

$$ye^{-2x} = \int[(1-3x)e^{-3x} + Ke^{-3x}/x]dx = xe^{-3x} + C_1\int \frac{e^{-3x}}{x}dx + C_2$$

and
$$y = xe^{-x} + C_1 e^{2x}\int\frac{e^{-3x}}{x}dx + C_2 e^{2x}.$$

17. Solve $[(x+3)D^2 - (2x+7)D + 2]y = (x+3)^2 e^x$.

The equation may be written as $[(x+3)D - 1][D-2]y = (x+3)^2 e^x$.

Putting $(D-2)y = v$, we have $[(x+3)D-1]v = (x+3)^2 e^x$ or $(D - \frac{1}{x+3})v = (x+3)e^x$.

Using the integrating factor $1/(x+3)$, we have $v/(x+3) = \int e^x\,dx = e^x + K$

so that
$$(D-2)y = v = (x+3)e^x + K(x+3).$$

Using the integrating factor e^{-2x}, we have

$$ye^{-2x} = \int[(x+3)e^{-x} + K(x+3)e^{-2x}]dx = -xe^{-x} - 4e^{-x} + K(-\frac{1}{2}xe^{-2x} - \frac{7}{4}e^{-2x}) + C_2$$

and
$$y = -xe^x - 4e^x + C_1(2x+7) + C_2 e^{2x}.$$

18. Show that the Riccati equation $\frac{dy}{dx} + yP(x) + y^2 Q(x) = R(x)$, $Q(x) \neq 0$, is reduced to a linear equation of the second order by the substitution $y = \frac{1}{Qu}\frac{du}{dx}$.

Since $\frac{dy}{dx} = \frac{1}{Qu}\frac{d^2u}{dx^2} - \frac{1}{Qu^2}(\frac{du}{dx})^2 - \frac{1}{Q^2u}\frac{dQ}{dx}\frac{du}{dx}$, the substitution yields

$$\frac{1}{Qu}\frac{d^2u}{dx^2} - \frac{1}{Qu^2}(\frac{du}{dx})^2 - \frac{1}{Q^2u}\frac{dQ}{dx}\frac{du}{dx} + \frac{P}{Qu}\frac{du}{dx} + \frac{1}{Qu^2}(\frac{du}{dx})^2 - R = 0 \quad \text{or} \quad \frac{d^2u}{dx^2} + (P - \frac{1}{Q}\frac{dQ}{dx})\frac{du}{dx} - RQu = 0.$$

19. Use the procedure outlined in Problem 18 to solve $\frac{dy}{dx} + \frac{2}{x}y + \frac{1}{2}x^3y^2 = \frac{1}{2x}$.

The substitution $y = \frac{1}{Qu}\frac{du}{dx} = \frac{2}{x^3u}\frac{du}{dx}$ reduces the equation to

$$\frac{d^2u}{dx^2} + \left(\frac{2}{x} - \frac{3x^2/2}{x^3/2}\right)\frac{du}{dx} - \frac{1}{2x}\cdot\frac{x^3}{2}u = 0 \quad \text{or} \quad \frac{d^2u}{dx^2} - \frac{1}{x}\frac{du}{dx} - \frac{1}{4}x^2u = 0.$$

In turn, the substitution $\dfrac{dz}{dx} = \sqrt{\dfrac{-S}{\frac{1}{4}}} = \sqrt{\dfrac{\frac{1}{4}x^2}{\frac{1}{4}}} = x$ reduces this equation to

$\dfrac{d^2u}{dz^2} - \dfrac{1}{4}u = 0$ whose solution is $u = C_1e^{\frac{1}{2}z} + C_2e^{-\frac{1}{2}z}$.

Then $y = \dfrac{1}{Qu}\dfrac{du}{dx} = \dfrac{2}{x^3}\cdot\dfrac{\frac{1}{2}(C_1e^{\frac{1}{2}z} - C_2e^{-\frac{1}{2}z})}{C_1e^{\frac{1}{2}z} + C_2e^{-\frac{1}{2}z}}x = \dfrac{1}{x^2}\cdot\dfrac{e^{\frac{1}{4}x^2} - ke^{-\frac{1}{4}x^2}}{e^{\frac{1}{4}x^2} + ke^{-\frac{1}{4}x^2}}$, where $k = \dfrac{C_2}{C_1}$.

20. Solve $\dfrac{dy}{dx} - (\tan x + 3\cos x)y + y^2\cos^2 x = -2$.

The substitution $y = \dfrac{1}{Qu}\dfrac{du}{dx} = \dfrac{\sec^2 x}{u}\dfrac{du}{dx}$ reduces the equation to

$$\frac{d^2u}{dx^2} + (\tan x - 3\cos x)\frac{du}{dx} + 2u\cos^2 x = 0.$$

In turn, the substitution $\dfrac{dz}{dx} = \sqrt{\dfrac{2\cos^2 x}{2}} = \cos x$, or $z = \sin x$, reduces this equation to $\dfrac{d^2u}{dz^2} - 3\dfrac{du}{dz} + 2u = 0$ whose solution is $u = C_1e^{z} + C_2e^{2z}$.

Then $y = \dfrac{1}{Qu}\dfrac{du}{dx} = \dfrac{\sec^2 x(C_1e^{z} + 2C_2e^{2z})}{C_1e^{z} + C_2e^{2z}}\cos x = \sec x\,\dfrac{e^{\sin x} + 2ke^{2\sin x}}{e^{\sin x} + ke^{2\sin x}}$.

SUPPLEMENTARY PROBLEMS

Solve.

21. $xy'' - (x+2)y' + 2y = 0$ — *Ans.* $y = C_1e^{x} + C_2(x^2 + 2x + 2)$

22. $(1+x^2)y'' - 2xy' + 2y = 2$ — $y = C_1x + C_2(x^2-1) + x^2$

23. $(x^2+4)y'' - 2xy' + 2y = 8$ — $y = C_1(x^2-4) + C_2x + x^2$

24. $(x+1)y'' - (2x+3)y' + (x+2)y = (x^2+2x+1)e^{2x}$ — $y = C_1e^{x} + C_2e^{x}(x+1)^2 + xe^{2x}$

25. $y'' - 2\tan x\, y' - 10y = 0$ — $y = (C_1e^{3x} + C_2e^{-3x})\sec x$

26. $x^2y'' - x(2x+3)y' + (x^2+3x+3)y = (6-x^2)e^{x}$ — $y = C_1x^3e^{x} + C_2xe^{x} + e^{x}(x^2+2)$

27. $4x^2y'' + 4x^3y' + (x^2+1)^2y = 0$ — $y = \sqrt{x}\,e^{-x^2/4}(C_1 + C_2\ln x)$

28. $x^2y'' + (x-4x^2)y' + (1-2x+4x^2)y = (x^2-x+1)e^{x}$ — $y = e^{2x}(C_1\cos\ln x + C_2\sin\ln x) + e^{x}$

29. $xy'' - y' + 4x^3y = 0$ — $y = C_1\sin x^2 + C_2\cos x^2$

30. $x^4y'' + 2x^3y' + y = (1+x)/x$ — $y = C_1\cos(1/x) + C_2\sin(1/x) + (1+x)/x$

31. $x^8y'' + 4x^7y' + y = 1/x^3$ *Ans.* $y = C_1\cos(1/3x^3) + C_2\sin(1/3x^3) + 1/x^3$

32. $(x\sin x + \cos x)y'' - x\cos x\, y' + y\cos x = x$ $y = C_1x + C_2\cos x - \sin x$

33. $xy'' - 3y' + 3y/x = x + 2$ $y = C_1x + C_2x^3 - x^2 - x\ln x$

34. Solve Problem 21 by factoring.

35. $[(x+1)D^2 - (3x+4)D + 3]y = (3x+2)e^{3x}$ $y = C_1(3x+4) + C_2e^{3x} + xe^{3x}$

36. $x^2y'' - 4xy' + (6+9x^2)y = 0$ $y = x^2(C_1\cos 3x + C_2\sin 3x)$

37. $xy'' + 2y' + 4xy = 4$ $y = (C_1\cos 2x + C_2\sin 2x + 1)/x$

38. $(1+x^2)y'' - 2xy' + 2y = (1-x^2)/x$ $y = C_1(x^2-1) + C_2x + x\ln x$

CHAPTER 19

Linear Equations with Variable Coefficients

MISCELLANEOUS TYPES

IN THIS CHAPTER various types of differential equations of order higher than the first and with variable coefficients will be considered. There is no general procedure comparable to that for linear equations. However, for the types treated here, the procedure consists in obtaining from the given equation another of lower order. For example, if the given equation is of order three and if, by some means, an equation of order two, which is solvable by one of the methods of the previous chapters can be obtained from it, the given equation can be solved.

DEPENDENT VARIABLE ABSENT. If the equation is free of y, that is, is of the form

$$1) \qquad f\left(\frac{d^n y}{dx^n}, \frac{d^{n-1} y}{dx^{n-1}}, \ldots\ldots, \frac{dy}{dx}, x\right) = 0,$$

the substitution $\dfrac{dy}{dx} = p$, $\dfrac{d^2y}{dx^2} = \dfrac{dp}{dx}$, ... will reduce the order by one.

EXAMPLE. The equation $x^2 \dfrac{d^3y}{dx^3} + 2\dfrac{d^2y}{dx^2}\dfrac{dy}{dx} - 3x\left(\dfrac{dy}{dx}\right)^2 + x^3 = 0$, of order three, is reduced to $x^2 \dfrac{d^2p}{dx^2} + 2p\dfrac{dp}{dx} - 3xp^2 + x^3 = 0$, of order two, by the substitution $\dfrac{dy}{dx} = p$, $\dfrac{d^2y}{dx^2} = \dfrac{dp}{dx}$, $\dfrac{d^3y}{dx^3} = \dfrac{d^2p}{dx^2}$.

If the given equation is of the form

$$2) \qquad f\left(\frac{d^n y}{dx^n}, \frac{d^{n-1} y}{dx^{n-1}}, \ldots\ldots, \frac{d^k y}{dx^k}, x\right) = 0,$$

the substitution $\dfrac{d^ky}{dx^k} = q$, $\dfrac{d^{k+1}y}{dx^{k+1}} = \dfrac{dq}{dx}$, ... will reduce the order by k.

See Problems 1-5.

INDEPENDENT VARIABLE ABSENT. If the equation is free of x, that is, is of the form

$$3) \qquad f\left(\frac{d^n y}{dx^n}, \frac{d^{n-1} y}{dx^{n-1}}, \ldots\ldots, \frac{dy}{dx}, y\right) = 0,$$

the substitution $\dfrac{dy}{dx} = p$, $\dfrac{d^2y}{dx^2} = \dfrac{dp}{dy}\dfrac{dy}{dx} = p\dfrac{dp}{dy}$,

$$\frac{d^3y}{dx^3} = \frac{d}{dy}\left(p\frac{dp}{dy}\right)\frac{dy}{dx} = \left\{p\frac{d^2p}{dy^2} + \left(\frac{dp}{dy}\right)^2\right\}\frac{dy}{dx} = p^2\frac{d^2p}{dy^2} + p\left(\frac{dp}{dy}\right)^2, \text{ etc.,}$$

will reduce the order of the differential equation by one.

EXAMPLE. The substitution $\frac{dy}{dx} = p$, $\frac{d^2y}{dx^2} = p\frac{dp}{dy}$, $\frac{d^3y}{dx^3} = p^2\frac{d^2p}{dy^2} + p(\frac{dp}{dy})^2$ reduces the equation $yy''' - y''(y')^2 = 1$, of order three, to $yp^2\frac{d^2p}{dy^2} + py(\frac{dp}{dy})^2 - p^3\frac{dp}{dy} = 1$, of order two.

See Problems 6-10.

LINEAR EQUATIONS WITH KNOWN PARTICULAR INTEGRAL. If a particular integral $y = u(x)$ of the equation

$$4)\qquad (P_0D^n + P_1D^{n-1} + \cdots\cdots + P_{n-1}D + P_n)y = 0$$

is known, then the substitution $y = uv$ will transform

$$5)\qquad (P_0D^n + P_1D^{n-1} + \cdots\cdots + P_{n-1}D + P_n)y = Q(x)$$

into an equation of the same order but with the dependent variable absent. In turn, the order of this equation may be reduced by the procedure of the first section of this chapter. Equation 4) is called the reduced equation of 5).

EXAMPLE. Since $y = x$ is a solution of $(D^2 - xD + 1)y = 0$, the substitution $y = vx$, $\frac{dy}{dx} = x\frac{dv}{dx} + v$, $\frac{d^2y}{dx^2} = x\frac{d^2v}{dx^2} + 2\frac{dv}{dx}$ reduces $(D^2 - xD + 1)y = e^{2x}$ to $\frac{d^2v}{dx^2} + \frac{2-x^2}{x}\frac{dv}{dx} = \frac{e^{2x}}{x}$. Here, the dependent variable v is missing and the procedure of the first section above applies.

See Problems 11-14.

EXACT EQUATIONS. The differential equation

$$6)\qquad f(\frac{d^ny}{dx^n}, \frac{d^{n-1}y}{dx^{n-1}}, \cdots\cdots, \frac{dy}{dx}, y, x) = Q(x)$$

is called an exact equation if it can be obtained by differentiating once an equation

$$7)\qquad g(\frac{d^{n-1}y}{dx^{n-1}}, \frac{d^{n-2}y}{dx^{n-2}}, \cdots\cdots, \frac{dy}{dx}, y, x) = Q_1(x) + C$$

of one lower order. For example, the equation

$$3y^2y''' + 14yy'y'' + 4(y')^3 + 12y'y'' = 2x$$

is an exact equation since it may be obtained by differentiating once the equation

$$3y^2y'' + 4y(y')^2 + 6(y')^2 = x^2 + C.$$

The linear equation 4) is exact provided

$P_n - P'_{n-1} + P''_{n-2} + \cdots\cdots + (-1)^nP_0^{(n)} = 0$, identically.

EXAMPLE. Consider the equation $(x^3 - 2x)y''' + (8x^2 - 5)y'' + 15xy' + 5y = 0$ in which $P_3 = 5$, $P_2 = 15x$ and $P_2' = 15$, $P_1 = 8x^2 - 5$ and $P_1'' = 16$, and $P_0 = x^3 - 2x$ and $P_0''' = 6$. The equation is exact since $P_3 - P_2' + P_1'' - P_0''' = 5-15+16-6 = 0$. The given equation is the exact derivative of $(x^3 - 2x)y'' + (5x^2 - 3)y' + 5xy = C$.

If equation 6) is not linear no simple test for exactness can be stated. In this case, we show that 6) is exact by producing the equation of one lower order from which it may be obtained by a differentiation.

If 6) is not exact, it may be possible to find an integrating factor. Again, no general rule can be stated for determining an integrating factor.

See Problems 15-21.

SOLVED PROBLEMS

DEPENDENT VARIABLE ABSENT.

1. Solve $2\dfrac{d^2y}{dx^2} - (\dfrac{dy}{dx})^2 + 4 = 0.$

The substitution $\dfrac{dy}{dx} = p$ reduces the equation to $2\dfrac{dp}{dx} = p^2 - 4$ or $\dfrac{2\,dp}{p^2-4} = dx.$

Integrating, $\frac{1}{2}\ln\dfrac{p-2}{p+2} = x + \ln K;\quad \dfrac{p-2}{p+2} = C_1e^{2x},\quad p = \dfrac{2(1+C_1e^{2x})}{1-C_1e^{2x}} = 2(1 + \dfrac{2C_1e^{2x}}{1-C_1e^{2x}}),$

and $y = 2x - 2\ln(1-C_1e^{2x}) + C_2.$

2. Solve $x\dfrac{d^3y}{dx^3} - 2\dfrac{d^2y}{dx^2} = 0.$

The substitution $\dfrac{d^2y}{dx^2} = q$ reduces the equation to $x\dfrac{dq}{dx} - 2q = 0.$

Then $\ln q = \ln x^2 + \ln K,\quad q = \dfrac{d^2y}{dx^2} = Kx^2,$ and $y = C_1x^4 + C_2x + C_3.$

3. Solve $\dfrac{d^4y}{dx^4}\cdot\dfrac{d^3y}{dx^3} = 1.$

The substitution $\dfrac{d^3y}{dx^3} = q$ reduces the equation to $q\dfrac{dq}{dx} = 1$ and $q^2 = 2x + C_1.$

Then $q = \dfrac{d^3y}{dx^3} = \pm(2x+C_1)^{1/2},\quad \dfrac{d^2y}{dx^2} = \pm\dfrac{1}{3}(2x+C_1)^{3/2} + K,\quad \dfrac{dy}{dx} = \pm\dfrac{1}{15}(2x+C_1)^{5/2} + Kx + K_3,$

$y = \pm\dfrac{1}{105}(2x+C_1)^{7/2} + K_2x^2 + K_3x + K_4$ or $105y = \pm(2x+C_1)^{7/2} + C_2x^2 + C_3x + C_4.$

4. Solve $(\dfrac{d^3y}{dx^3})^2 + x\dfrac{d^3y}{dx^3} - \dfrac{d^2y}{dx^2} = 0.$

The substitution $\dfrac{d^2y}{dx^2} = q$ reduces the equation to $(\dfrac{dq}{dx})^2 + x\dfrac{dq}{dx} - q = 0$ or $q = x\dfrac{dq}{dx} + (\dfrac{dq}{dx})^2,$ a Clairaut equation.

Then $q = \dfrac{d^2y}{dx^2} = Kx + K^2,\quad \dfrac{dy}{dx} = \dfrac{1}{2}Kx^2 + K^2x + C_2 = Cx^2 + 4C^2x + C_2,$ and

$$y = \frac{1}{3}Cx^3 + 2C^2x^2 + C_2x + C_3 = C_1x^3 + 18C_1^2x^2 + C_2x + C_3.$$

5. Solve $(1+2x)\dfrac{d^3y}{dx^3} + 4x\dfrac{d^2y}{dx^2} - (1-2x)\dfrac{dy}{dx} = e^{-x}$.

The transformation $p = \dfrac{dy}{dx}$ reduces the equation to

$$(1+2x)p'' + 4xp' - (1-2x)p = e^{-x} \qquad \text{or} \qquad p'' + \frac{4x}{1+2x}p' - \frac{1-2x}{1+2x}p = \frac{e^{-x}}{1+2x}.$$

Since $1 - R + S = 0$, we use the substitution

$$p = e^{-x}v, \quad p' = e^{-x}(v'-v), \quad p'' = e^{-x}(v''-2v'+v)$$

to obtain $(1+2x)v'' - 2v' = 1$ or $(1+2x)^2v'' - 2(1+2x)v' = (1+2x)$, a Legendre linear equation. The substitution $1+2x = e^t$ reduces the equation to $[4\mathcal{D}(\mathcal{D}-1) - 4\mathcal{D}]v = e^t$ or $\mathcal{D}(\mathcal{D}-2)v = \frac{1}{4}e^t$.

Then $$v = K_1 + K_2e^{2t} - \tfrac{1}{4}e^t = K_1 + K_2(1+2x)^2 - \tfrac{1}{4}(1+2x),$$

$$p = \frac{dy}{dx} = e^{-x}v = K_1e^{-x} + K_2(1+2x)^2e^{-x} - \tfrac{1}{4}(1+2x)e^{-x},$$

and $$y = C_1e^{-x} + C_2(4x^2+12x+13)e^{-x} + C_3 + \tfrac{1}{4}(2x+3)e^{-x}$$

or $$y = Ae^{-x} + B(x^2+3x)e^{-x} + C + \tfrac{1}{2}xe^{-x}.$$

INDEPENDENT VARIABLE ABSENT.

6. Solve $y'' = (y')^3 + y'$.

The substitution $y' = p$, $y'' = p\dfrac{dp}{dy}$ reduces the equation to $p\dfrac{dp}{dy} = p^3 + p$ or $\dfrac{dp}{dy} = p^2 + 1$.

Then $\dfrac{dp}{p^2+1} = dy$, $\quad$ arc tan $p = y + K_1$, $\quad$ and $p = \dfrac{dy}{dx} = \tan(y+K_1)$.

Now $\cot(y+K_1)dy = dx$, $\ln\sin(y+K_1) = x + K_2$, $\sin(y+K_1) = C_2e^x$, and $y = \arcsin C_2e^x + C_1$.

7. Solve $yy'' = 2(y')^2 - 2y'$.

The substitution $y' = p$, $y'' = p\dfrac{dp}{dy}$ reduces the equation to $p(y\dfrac{dp}{dy} - 2p + 2) = 0$.

Here $p = 0$ and $y = C$ is a solution, or

$$\frac{dp}{p-1} = 2\frac{dy}{y}, \qquad \ln(p-1) = \ln A^2y^2, \qquad p = A^2y^2 + 1, \qquad \text{or} \qquad \frac{dy}{1+A^2y^2} = dx.$$

Then $\dfrac{1}{A}\arctan Ay = x + K$, $\quad \arctan Ay = Ax + B$, $\quad$ and $\quad Ay = \tan(Ax+B)$.

8. Solve $yy'' - (y')^2 = y^2\ln y$.

The substitution $y' = p$, $y'' = p\dfrac{dp}{dy}$ reduces the equation to

$$yp\frac{dp}{dy} - p^2 = y^2\ln y \qquad \text{or} \qquad \frac{2y^2p\,dp - 2p^2y\,dy}{y^4} = 2\ln y\frac{dy}{y}.$$

Then $\dfrac{p^2}{y^2} = \ln^2 y + C$, $\quad \dfrac{dy}{y\sqrt{\ln^2 y + C}} = \pm dx$, $\quad$ and $\quad \ln(\ln y + \sqrt{\ln^2 y + C}) = \pm x + \ln K$.

Now $\ln y + \sqrt{\ln^2 y + C} = Ke^{\pm x}$, $\sqrt{\ln^2 y + C} = Ke^{\pm x} - \ln y$, and $C = K^2 e^{\pm 2x} - 2Ke^{\pm x} \ln y$.

This may be written as $\ln y = C_1 e^{\pm x} + C_2 e^{\mp x}$ or, finally, $\ln y = C_1 e^{x} + C_2 e^{-x}$ since C_1 and C_2 are arbitrary constants.

9. Solve $yy'' + (y')^2 = y^2$.

The substitution $y' = p$, $y'' = p\dfrac{dp}{dy}$ reduces the equation to $py\dfrac{dp}{dy} + p^2 = y^2$ for which y is an integrating factor. The solution of $py^2\,dp + p^2 y\,dy = y^3\,dy$ is $2p^2y^2 = y^4 + C^2$.

Now $\sqrt{2}\,p = \sqrt{2}\dfrac{dy}{dx} = \pm\dfrac{\sqrt{y^4 + C^2}}{y}$ whose solution is $\sqrt{2}\sinh^{-1}\dfrac{y^2}{C} = \pm 2x + K\sqrt{2}$. Then

$\sinh^{-1}\dfrac{y^2}{C} = \pm\sqrt{2}\,x + K$, $\dfrac{y^2}{C} = \sinh(\pm\sqrt{2}\,x + K) = \pm\sinh(\sqrt{2}\,x + K_1)$, and $y^2 = C_1 \sinh(\sqrt{2}\,x + C_2)$.

10. Solve $\dfrac{d^2y}{dx^2} = e^{2y}$ given that $y = y' = 0$ when $x = 0$.

Putting $y'' = p\dfrac{dp}{dy}$, we have $2p\,dp = 2e^{2y}dy$ whose solution is $p^2 = e^{2y} + K$.

Using the initial conditions, $0 = 1 + K$ and $K = -1$. Now $p = \dfrac{dy}{dx} = \pm\sqrt{e^{2y} - 1}$ which, by the substitution $e^{2y} = z$, becomes $\dfrac{dz}{2z\sqrt{z-1}} = \pm\,dx$. The solution of this equation is $\arctan\sqrt{z-1} = \pm x + C$ or, in the original variables, $\arctan\sqrt{e^{2y} - 1} = \pm x + C$. Here, the initial conditions require $C = 0$ so that $\sqrt{e^{2y} - 1} = \tan(\pm x) = \pm\tan x$ and, finally, $e^{2y} = \sec^2 x$.

It should be noted that the form of the solution of the given equation depends on the sign of the first constant of integration. If in $p^2 = e^{2y} + K$, K is positive and $= A^2$, we solve $\dfrac{dz}{2z\sqrt{z + A^2}} = \pm\,dx$. and obtain $\dfrac{1}{2A}\ln\dfrac{\sqrt{z + A^2} - A}{\sqrt{z + A^2} + A} = \pm x + C$. Then $\dfrac{\sqrt{z + A^2} - A}{\sqrt{z + A^2} + A} = Be^{\pm 2Ax}$ and

$\dfrac{A(1 + Be^{\pm 2Ax})}{1 - Be^{\pm 2Ax}} = \sqrt{z + A^2}$. Since A is arbitrary, we may write $z + A^2 = \dfrac{A^2(1 + Be^{2Ax})^2}{(1 - Be^{2Ax})^2}$ and

obtain $z = e^{2y} = \dfrac{4A^2Be^{2Ax}}{(1 - Be^{2Ax})^2}$ or $e^{y} = \dfrac{2ACe^{Ax}}{1 - C^2e^{2Ax}}$.

LINEAR EQUATIONS WITH KNOWN PARTICULAR INTEGRAL.

11. Solve $x^3(\sin x)y''' - (3x^2\sin x + x^3\cos x)y'' + (6x\sin x + 2x^2\cos x)y' - (6\sin x + 2x\cos x)y = 0$.

By inspection it is seen that $y = x$ is a particular integral.

By means of the substitution $y = xv$, $y' = xv' + v$, $y'' = xv'' + 2v'$, $y''' = xv''' + 3v''$, the equation is reduced to $\sin x\dfrac{d^3v}{dx^3} - \cos x\dfrac{d^2v}{dx^2} = 0$. In turn, the substitution $\dfrac{d^2v}{dx^2} = q$

reduces this equation to $\sin x \frac{dq}{dx} - q\cos x = 0$ or $\frac{dq}{q} = \cot x\, dx$.

Then $\ln q = \ln\sin x + \ln C$, $q = \frac{d^2v}{dx^2} = C\sin x$, and $v = \frac{y}{x} = C_1\sin x + C_2x + C_3$.

Thus, the solution is $y = C_1x\sin x + C_2x^2 + C_3x$.

12. Solve $(x^3 - 3x^2 + 6x - 6)y^{iv} - x^3y''' + 3x^2y'' - 6xy' + 6y = 0$.

By inspection it is seen that $y = x$ is a particular integral.

The substitution $y = xv$, $y' = xv' + v$, $y'' = xv'' + 2v'$, $y''' = xv''' + 3v''$, $y^{iv} = xv^{iv} + 4v'''$ reduces the equation to $(x^4 - 3x^3 + 6x^2 - 6x)v^{iv} + (-x^4 + 4x^3 - 12x^2 + 24x - 24)v''' = 0$.

Putting $\frac{d^3v}{dx^3} = q$, this equation becomes

$$x(x^3 - 3x^3 + 6x - 6)\frac{dq}{dx} + (-x^4 + 4x^3 - 12x^2 + 24x - 24)q = 0 \quad\text{or}\quad \frac{dq}{q} + \left(-1 + \frac{4}{x} - \frac{3x^2 - 6x + 6}{x^3 - 3x^2 + 6x - 6}\right)dx = 0.$$

Integrating, $\ln q = x - 4\ln x + \ln(x^3 - 3x^2 + 6x - 6) + \ln A$ or $q = \frac{d^3v}{dx^3} = A\frac{x^3 - 3x^2 + 6x - 6}{x^4}e^x$.

Then
$$\frac{d^2v}{dx^2} = A\int \frac{x^3 - 3x^2 + 6x - 6}{x^4}e^x\,dx = A\cdot\frac{1}{D}\left(\frac{x^3 - 3x^2 + 6x - 6}{x^4}e^x\right)$$

$$= Ae^x\frac{1}{D+1}\left(\frac{x^3 - 3x^2 + 6x - 6}{x^4}\right) = Ae^x\frac{1}{D+1}\left(\frac{1}{x} - \frac{3}{x^2} + \frac{6}{x^3} - \frac{6}{x^4}\right).$$

Now $D(\frac{1}{x}) = -\frac{1}{x^2}$, $D^2(\frac{1}{x}) = \frac{2}{x^3}$, and $D^3(\frac{1}{x}) = -\frac{6}{x^4}$,

so that
$$\frac{1}{D+1}\left(\frac{1}{x} - \frac{3}{x^2} + \frac{6}{x^3} - \frac{6}{x^4}\right) = \frac{1}{D+1}\left[\frac{1}{x} + 3D(\tfrac{1}{x}) + 3D^2(\tfrac{1}{x}) + D^3(\tfrac{1}{x})\right] = \frac{1}{D+1}(D+1)^3(\tfrac{1}{x})$$

$$= (D^2 + 2D + 1)(\tfrac{1}{x}) = \frac{x^2 - 2x + 2}{x^3}.$$

Thus,
$$\frac{d^2v}{dx^2} = A\frac{x^2 - 2x + 2}{x^3}e^x + B, \qquad \frac{dv}{dx} = A\frac{(x-1)e^x}{x^2} + Bx + C,$$

$$v = \frac{y}{x} = C_1\frac{e^x}{x} + C_2x^2 + C_3x + C_4, \quad\text{and}\quad y = C_1e^x + C_2x^3 + C_3x^2 + C_4x.$$

In this example, it is fairly easy to see that $y = x$, $y = x^2$, $y = x^3$, and $y = e^x$ are particular integrals. Thus the complete solution could have been written down immediately.

13. Solve $(2\sin x - x\sin x - x\cos x)y''' + (2x\cos x - \sin x - \cos x)y'' + x(\sin x - \cos x)y'$
$+ (\cos x - \sin x)y = 2\sin x - x\cos x - x\sin x$.

By inspection it is seen that $y = x$, $y = e^x$, and $y = \sin x$ are particular integrals of the reduced equation. We shall obtain a particular integral of the given equation using the *method of variation of parameters*.

We take $$y = L_1 x + L_2 e^x + L_3 \sin x .$$

Then $$y' = L_1 + L_2 e^x + L_3 \cos x + (L_1' x + L_2' e^x + L_3' \sin x)$$
and we set
$$A) \qquad L_1' x + L_2' e^x + L_3' \sin x = 0.$$

Now $$y'' = L_2 e^x - L_3 \sin x + (L_1' + L_2' e^x + L_3' \cos x)$$
and we set
$$B) \qquad L_1' + L_2' e^x + L_3' \cos x = 0.$$

Then $$y''' = L_2 e^x - L_3 \cos x + (L_2' e^x - L_3' \sin x)$$
and we set
$$C) \qquad L_2' e^x - L_3' \sin x = 2 \sin x - x \cos x - x \sin x.$$

Solving $A)$, $B)$, $C)$ simultaneously, we obtain

$L_1' = -\sin x + \cos x$ and $L_1 = \cos x + \sin x$,

$L_2' = -e^{-x}(x \cos x - \sin x)$ and $L_2 = \frac{1}{2} x e^{-x}(-\sin x + \cos x) - e^{-x} \sin x - \frac{1}{2} e^{-x} \cos x$,

$L_3' = -1 + x$ and $L_3 = -x + \frac{1}{2} x^2$.

Thus, the complete solution is
$$y = C_1 x + C_2 e^x + C_3 \sin x + \frac{1}{2} x^2 \sin x + \frac{3}{2} x \cos x - \frac{1}{2} x \sin x - \frac{1}{2} \cos x .$$

14. Solve $(x^2 + x)y''' - (x^2 + 3x + 1)y'' + (x + 4 + \frac{2}{x})y' - (1 + \frac{4}{x} + \frac{2}{x^2})y = 3x^2(x+1)^2$.

By inspection it is seen that $y = x$ is a particular integral of the reduced equation. The substitution $y = xv$ reduces the given equation to
$$(x^2 + x)v''' - (x^2 - 2)v'' - (x + 2)v' = 3x(x+1)^2$$
and, in turn, the substitution $v' = u$ reduces this to
$$A) \qquad (x^2 + x)u'' - (x^2 - 2)u' - (x + 2)u = 3x(x+1)^2 .$$

Since the sum of the coefficients of the reduced equation of $A)$ is identically zero, $u = e^x$ is a particular integral and we use the substitution
$$u = e^x w, \quad u' = e^x w' + e^x w, \quad u'' = e^x w'' + 2e^x w' + e^x w$$
to reduce $A)$ to
$$(x^2 + x)w'' + (x^2 + 2x + 2)w' = 3xe^{-x}(x+1)^2 .$$
Using the substitution $w' = z$, this becomes
$$(x^2 + x)z' + (x^2 + 2x + 2)z = 3xe^{-x}(x+1)^2$$

or $\dfrac{dz}{dx} + (1 + \dfrac{2}{x} - \dfrac{1}{x+1})z = 3e^{-x}(x+1)$ for which $\dfrac{x^2 e^x}{x+1}$ is an integrating factor.

Then $z \dfrac{x^2 e^x}{x+1} = \displaystyle\int 3x^2 dx = x^3 + K_1$, $\qquad \dfrac{dw}{dx} = z = x(x+1)e^{-x} + K_1 \dfrac{x+1}{x^2} e^{-x}$,

$\dfrac{u}{e^x} = w = -x^2 e^{-x} - 3xe^{-x} - 3e^{-x} + C_1 \dfrac{e^{-x}}{x} + C_2$, $\qquad \dfrac{dv}{dx} = u = -x^2 - 3x - 3 + \dfrac{C_1}{x} + C_2 e^x$,

and $$y = xv = -\frac{x^4}{3} - \frac{3}{2}x^3 - 3x^2 + C_1 x \ln x + C_2 x e^x + C_3 x.$$

EXACT EQUATIONS.

15. Show that $P_0(x)y^{iv} + P_1(x)y''' + P_2(x)y'' + P_3(x)y' + P_4(x)y = 0$ is exact if and only if $P_4 - P_3' + P_2'' - P_1''' + P_0^{iv} = 0$.

Let the given differential equation be obtained by differentiating

$$R_0(x)y''' + R_1(x)y'' + R_2(x)y' + R_3(x)y = C_1.$$

Since this differentiation yields $R_0y^{iv} + (R_0'+R_1)y''' + (R_1'+R_2)y'' + (R_2'+R_3)y' + R_3'y = 0$, we have $P_0 = R_0$, $P_1 = R_0' + R_1$, $P_2 = R_1' + R_2$, $P_3 = R_2' + R_3$, and $P_4 = R_3'$.
Now $P_4 - P_3' + P_2'' - P_1''' + P_0^{iv} = R_3' - (R_2''+R_3') + (R_1''' + R_2'') - (R_0^{iv} + R_1''') + R_0^{iv} = 0$.

Conversely, suppose $P_4 - P_3' + P_2'' - P_1''' + P_0^{iv} = 0$. Since

$$\frac{d}{dx}[P_0y''' + (P_1 - P_0')y'' + (P_2 - P_1' + P_0'')y' + (P_3 - P_2' + P_1'' - P_0''')y]$$

$$= P_0y^{iv} + P_1y''' + P_2y'' + P_3y' - (-P_3' + P_2'' - P_1''' + P_0^{iv})y$$

$$= P_0y^{iv} + P_1y''' + P_2y'' + P_3y' + P_4y,$$ the given differential equation is exact.

16. Solve $xy''' + (x^2+x+3)y'' + (4x+2)y' + 2y = 0$.

The equation is exact since $P_3 - P_2' + P_1'' - P_0''' = 2 - 4 + 2 - 0 = 0$.

Consider the left member $xy''' + (x^2+x+3)y'' + (4x+2)y' + 2y$.

To obtain the first term we must differentiate xy''. Now $\frac{d}{dx}(xy'') = xy''' + y''$ and when this is removed, we have $(x^2+x+2)y'' + (4x+2)y' + 2y$. To obtain the first term of the resulting relation, we must differentiate $(x^2+x+2)y'$. When $\frac{d}{dx}(x^2+x+2)y' = (x^2+x+2)y'' + (2x+1)y'$ is removed, we have $(2x+1)y' + 2y = \frac{d}{dx}(2x+1)y$. Thus the given equation is the exact derivative of

$$A)\qquad xy'' + (x^2+x+2)y' + (2x+1)y = C_1.$$

Since $P_2 - P_1' + P_0'' = (2x+1) - (2x+1) + 0 = 0$, we now treat the left member of *A*) precisely as we did the corresponding member of the original equation.

We remove $\frac{d}{dx}(xy') = xy'' + y'$ and have $(x^2+x+1)y' + (2x+1)y = \frac{d}{dx}(x^2+x+1)y$.

Hence *A*) is the exact derivative of

$$B)\qquad xy' + (x^2+x+1)y = C_1x + C_2,$$

a linear equation for which $xe^{\frac{1}{2}x(x+2)}$ is an integrating factor.

Thus, the complete solution of the given equation is

$$xye^{\frac{1}{2}x(x+2)} = C_1\int xe^{\frac{1}{2}x(x+2)}dx + C_2\int e^{\frac{1}{2}x(x+2)}dx + C_3.$$

The following scheme will be found convenient.

$$xy''' + (x^2+x+3)y'' + (4x+2)y' + 2y = 0$$

xy'': $\quad xy''' + y''$

$$(x^2+x+2)y'' + (4x+2)y' + 2y$$

$(x^2+x+2)y'$: $\quad (x^2+x+2)y'' + (2x+1)y'$

$$(2x+1)y' + 2y$$

$(2x+1)y$: $\quad (2x+1)y' + 2y$

A) $\quad xy'' + (x^2+x+2)y' + (2x+1)y = C_1$

xy': $\quad xy'' + y'$

$$(x^2+x+1)y' + (2x+1)y$$

$(x^2+x+1)y$: $\quad (x^2+x+1)y' + (2x+1)y$

B) $\quad xy' + (x^2+x+1)y = C_1x + C_2.$

17. Solve $\quad 2y\dfrac{d^3y}{dx^3} + 6\dfrac{d^2y}{dx^2}\dfrac{dy}{dx} = -\dfrac{1}{x^2}.$

We write

$$2y\frac{d^3y}{dx^3} + 6\frac{d^2y}{dx^2}\frac{dy}{dx}$$

$2y\dfrac{d^2y}{dx^2}$: $\quad 2y\dfrac{d^3y}{dx^3} + 2\dfrac{d^2y}{dx^2}\dfrac{dy}{dx}$

$2(\dfrac{dy}{dx})^2$: $\quad 4\dfrac{d^2y}{dx^2}\dfrac{dy}{dx}$

Thus, the given equation is exact, being obtained by differentiating

$$2y\frac{d^2y}{dx^2} + 2(\frac{dy}{dx})^2 = -\int\frac{1}{x^2}\,dx = \frac{1}{x} + K_1.$$

A second integration yields $\quad 2y\dfrac{dy}{dx} = \ln x + K_1x + K_2 \quad$ whose solution is

$$y^2 = x\ln x + C_1x^2 + C_2x + C_3.$$

18. Solve $\quad (1+3xy^2)y''' + 9(y^2+2xyy')y'' + 18y(y')^2 + 6x(y')^3 = 6.$

We write

$$(1+3xy^2)y''' + 9y^2y'' + 18xyy'y'' + 18y(y')^2 + 6x(y')^3$$

$(1+3xy^2)y''$: $\quad (1+3xy^2)y''' + 3y^2y'' + 6xyy'y''$

$$6y^2y'' + 12xyy'y'' + 18y(y')^2 + 6x(y')^3$$

$6y^2y'$: $\quad 6y^2y'' + 12y(y')^2$

$$12xyy'y'' + 6y(y')^2 + 6x(y')^3$$

$6xy(y')^2$: $\quad 12xyy'y'' + 6y(y')^2 + 6x(y')^3$

The given equation is exact, being obtained by differentiating

$$(1+3xy^2)y'' + 6y^2y' + 6xy(y')^2 = 6x + K$$

$(1+3xy^2)y'$: $\quad (1+3xy^2)y'' + 3y^2y' + 6xy(y')^2$

$$3y^2y'$$

y^3: $\quad 3y^2y'$

and this equation is obtained by differentiating $(1+3xy^2)y' + y^3 = 3x^2 + Kx + C_2$.

In turn, this equation is exact and we have $xy^3 + y = x^3 + C_1x^2 + C_2x + C_3$.

19. Solve $x^3y''' + 5x^2y'' + (2x - x^3)y' - (2+x^2)y = 40x^3 - 4x^5$.

It is readily verified that this linear equation is not exact. To test whether or not it has an integrating factor of the form x^m, we multiply by x^m to get

$$x^{m+3}y''' + 5x^{m+2}y'' + (2x^{m+1} - x^{m+3})y' - (2x^m + x^{m+2})y = (40x^3 - 4x^5)x^m$$

and write the condition

$$-(2x^m + x^{m+2}) - 2(m+1)x^m + (m+3)x^{m+2} + 5(m+2)(m+1)x^m - (m+3)(m+2)(m+1)x^m$$

$$= (m+2)x^{m+2} + (m+2)(m-m^2)x^m = 0, \quad \text{for all values of } x.$$

Then $m = -2$, and x^{-2} is an integrating factor. Using it, we have

$$xy''' + 5y'' + (\frac{2}{x} - x)y' - (\frac{2}{x^2} + 1)y = 40x - 4x^3$$

$$xy'' \qquad \underline{xy''' + y''}$$

$$4y'' + (\frac{2}{x} - x)y' - (\frac{2}{x^2} + 1)y$$

$$4y' + (\frac{2}{x} - x)y \qquad \underline{4y'' + (\frac{2}{x} - x)y' - (\frac{2}{x^2} + 1)y}$$

and

$$xy'' + 4y' + (\frac{2}{x} - x)y = 20x^2 - x^4 + K.$$

The transformation $y = \dfrac{v}{x^2}$ reduces this equation to $v'' - v = (D^2-1)v = 20x^3 - x^5 + Kx$,

and the complete solution is $v = x^2y = C_1e^x + C_2e^{-x} - (1+D^2+D^4+D^6+\cdots)(20x^3 - x^5 + Kx)$

$$= C_1e^x + C_2e^{-x} + C_3x + x^5.$$

20. Solve $2yy''' + 2(y+3y')y'' + 2(y')^2 = 2$.

We write

$$2yy''' + 2yy'' + 6y'y'' + 2(y')^2 = 2$$

$$2yy'' \qquad \underline{2yy''' \quad + 2y'y''}$$

$$2yy'' + 4y'y'' + 2(y')^2$$

$$2yy' + 2(y')^2 \qquad \underline{2yy'' + 4y'y'' + 2(y')^2}$$

and thus obtain by integration

$$2yy'' + 2(y')^2 + 2yy' = 2x + K_1$$

$$2yy' \qquad \underline{2yy'' + 2(y')^2}$$

$$2yy'$$

$$y^2 \qquad \underline{2yy'}$$

and obtain $2yy' + y^2 = x^2 + K_1x + K_2$. By inspection, e^x is an integrating factor; then

$$y^2e^x = x^2e^x - 2xe^x + 2e^x + K_1(xe^x - e^x) + K_2e^x + C_3 \qquad \text{or} \qquad y^2 = x^2 + C_1 + C_2x + C_3e^{-x}.$$

21. Solve $x\cos y\, y''' - 3x\sin y\, y'y'' - \cos y\, y'' - x\cos y\,(y')^3 + \sin y\,(y')^2 + x\cos y\, y' - \sin y = 0.$

Since $\frac{d}{dx}\left(\frac{\sin y}{x}\right) = \frac{x\cos y\, y' - \sin y}{x^2}$, the last two terms of the given equation suggest $\frac{1}{x^2}$ as a possible integrating factor. Using it and integrating,

$$\frac{\cos y\, y'' - \sin y\,(y')^2}{x} + \frac{\sin y}{x} = C_1 \quad \text{or} \quad \cos y\, y'' - \sin y\,(y')^2 + \sin y = C_1 x.$$

The substitution $\sin y = z$ reduces this equation to $z'' + z = C_1 x$ whose complete solution is

$$z = \sin y = C_1 x + C_2\cos x + C_3\sin x.$$

SUPPLEMENTARY PROBLEMS

Solve.

22. $y'' + (y')^2 + 1 = 0$ *Ans.* $y = \ln\cos(x - C_1) + C_2$

23. $(1+x^2)y'' + 2xy' = 2x^{-3}$ $y = C_1 + C_2\arctan x + 1/x$

24. $xy'' - y' = -2/x - \ln x$ $y = C_1x^2 + C_2 + (x+1)\ln x$

25. $y''' + y'' = x^2$ $y = C_1e^{-x} + C_2x + C_3 + x^2(x^2 - 4x + 12)/12$

26. $yy'' + (y')^3 = 0$ $x = C_1 + C_2y + y\ln y$

27. $yy'' + (y')^2 = 2$ $y^2 = 2x^2 + C_1x + C_2$

28. $yy'' = (y')^2(1 - y'\cos y + yy'\sin y)$ $x = C_1 + C_2\ln y + \sin y$

29. $(2x-3)y''' - (6x-7)y'' + 4xy' - 4y = 8$ $y = C_1x + C_2e^x + C_3e^{2x} - 2$
Hint: $y = x$ is a particular integral of the reduced equation.

30. $(2x^3 - 1)y''' - 6x^2y'' + 6xy' = 0$ $y = C_1(x^4 + 4x) + C_2x^2 + C_3$

31. $yy'' - (y')^2 = y^2\ln y$ $\ln y = C_1e^x + C_2e^{-x}$
Hint: Use $\ln y = z$.

32. $(x+2y)y'' + 2(y')^2 + 2y' = 2$ $y(x+y) = x^2 + C_1x + C_2$

33. $(1 + 2y + 3y^2)y''' + 6y'[y'' + (y')^2 + 3yy''] = x$ $y + y^2 + y^3 = C_1x^2 + C_2x + C_3 + x^4/24$

34. $3x[y^2y''' + 6yy'y'' + 2(y')^3] - 3y[yy'' + 2(y')^2] = -2/x$
Hint: $1/x^2$ is an integrating factor. *Ans.* $y^3 = C_1x^3 + C_2x + C_3 + x\ln x$

35. $yy''' + 3y'y'' - 2yy'' - 2(y')^2 + yy' = e^{2x}$ *Ans.* $y^2 = C_1 + C_2e^x + C_3xe^x + e^{2x}$
Hint: e^{-x} is an integrating factor. Solve also using $y^2 = v$.

36. $2(y+1)y'' + 2(y')^2 + y^2 + 2y = 0$ *Ans.* $y^2 + 2y = C_1\cos x + C_2\sin x$
Hint: Use $y^2 + 2y = v$.

CHAPTER 20

Applications of Linear Equations

GEOMETRICAL APPLICATIONS. In rectangular coordinates the radius of curvature R of a curve $y = f(x)$ at a general point on it is given by

$$R = \frac{\left[1 + \left(\frac{dy}{dx}\right)^2\right]^{3/2}}{\frac{d^2y}{dx^2}}.$$

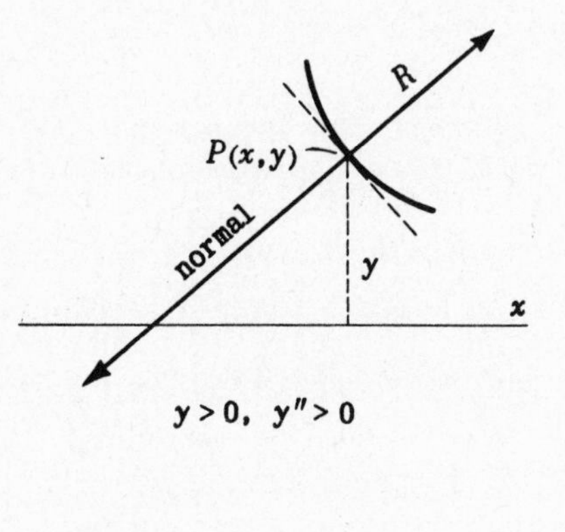

$y > 0,\ y'' > 0$

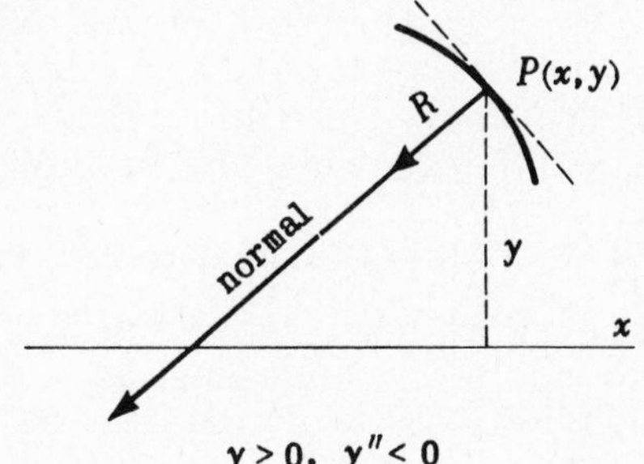

$y > 0,\ y'' < 0$

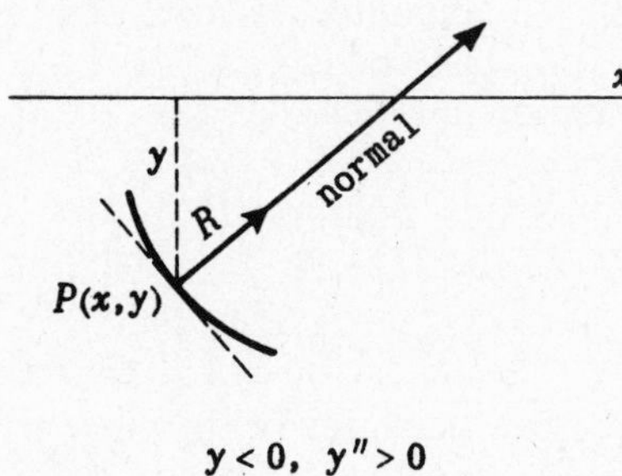

$y < 0,\ y'' > 0$

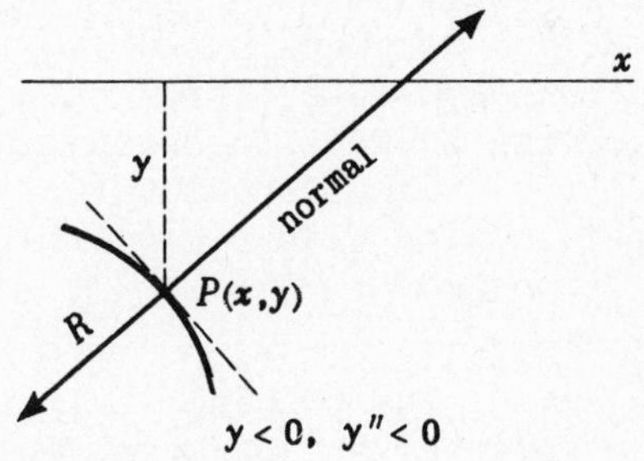

$y < 0,\ y'' < 0$

Let the normal at the point be drawn toward the x-axis. It is clear from the figures that the normal and radius of curvature at any point have the same direction when y and d^2y/dx^2 have opposite signs and have opposite directions when y and d^2y/dx^2 have the same signs.

PHYSICAL APPLICATIONS. OSCILLATORY MOTION. Consider a ball bobbing up and down at the end of a rubber string.

If the other end of the string is held fixed and no external force is applied to the ball to keep it moving once it has been started, and if the mass of the string and the resistance offered by the air are such that they may be neglected, the ball will move with *simple harmonic motion*

$$x = A \cos \omega t + B \sin \omega t$$

where x is the displacement of the ball at time t from its position of rest or equilibrium.

For simple harmonic motion:

a) The *amplitude* or maximum displacement from equilibrium position is $\sqrt{A^2+B^2}$, since when $dx/dt = 0$, $\tan \omega t = A/B$, and $x = \sqrt{A^2 + B^2}$.

b) The *period* or number of units (sec) of time for a complete oscillation is $2\pi/\omega$ sec, since when t is changed by $2\pi/\omega$ sec the values of x and dx/dt are unchanged, while for any change of t less than this amount one (or both) of x and dx/dt is changed.

c) The *frequency* or number of oscillations (cycles) per sec is $\omega/2\pi$ cycles/sec.

d) The *differential equation* of simple harmonic motion is $m\dfrac{d^2x}{dt^2} = -kx$, where k is a positive quantity. In the above illustration

$$m\frac{d^2x}{dt^2} = -m\omega^2(A\cos\omega t + B\sin\omega t) = -kx$$

where m is the mass of the ball and $k = m\omega^2$.

If the above assumptions are modified so that the resistance of the air cannot be neglected, the ball will move with *free damped motion*

$$x = e^{-st}(A\cos\omega t + B\sin\omega t).$$

The motion is oscillatory as before but never repeats itself. Since the *damping factor* e^{-st} decreases as t increases, the amplitude of each oscillation is less than that of the preceding one. The frequency is $\omega/2\pi$ cycles/sec. See Problem 8*a*.

If the resistance offered to the motion is sufficiently great, other cases will arise. See Problem 8*b*.

If in addition to a resistance, there is an external force acting on the ball or the complete system is given a motion, the motion of the ball is said to be *forced*. If the forcing function is harmonic with period $2\pi/\lambda$, the motion of the ball is the result of two motions — a free damping motion which dies out as time increases (called the *transient phenomenon*) and a simple harmonic motion with period that of the forcing function (called the *steady-state phonomenon*). See Problem 9.

HORIZONTAL BEAMS. The problem is that of determining the deflection (bending) of a beam under given loadings. Only beams which are uniform in material and shape will be considered. It is convenient to think of the beam as consisting of fibers running lengthwise. In the bent beam shown, the fibers of the upper half are compressed and those of the lower half stretched, the two halves being separated by a neutral surface whose fibers are neither compressed nor stretched. The fiber which originally coincided with the horizontal axis of the beam now lies in the neutral surface along a curve (the elastic curve or curve of deflection). We seek the equation of this curve.

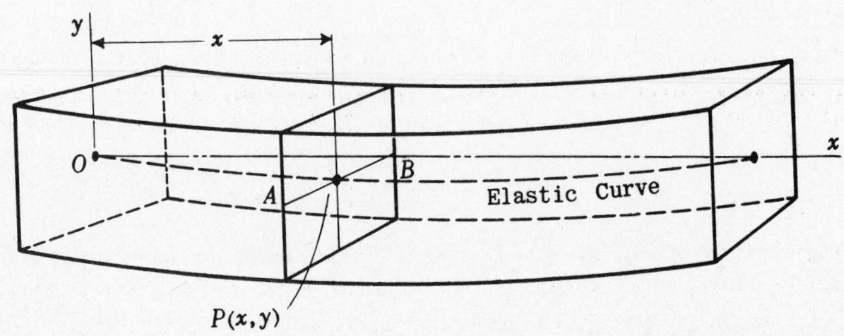

Consider a cross section of the beam at a distance x from one end. Let AB be its intersection with the neutral surface and P its intersection with the elastic curve. It is shown in Mechanics that the moment M with respect to AB of all external forces acting on either of the two segments into which the beam is separated by the cross section (*a*) is independent of the segment considered and (*b*) is given by

$$A) \qquad EI/R = M.$$

Here, E = the modulus of elasticity of the beam and I = the moment of inertia of the cross section with respect to AB are constants associated with the beam, and R is the radius of curvature of the elastic curve at P.

For convenience, think of the beam as replaced by its elastic curve and the cross section by the point P. Take the origin at the left end of the beam with the x-axis horizontal and let P have coordinates (x,y). Since the slope dy/dx of the elastic curve at all of its points is numerically small,

$$R = \frac{\left[1 + \left(\frac{dy}{dx}\right)^2\right]^{3/2}}{\frac{d^2y}{dx^2}} = \frac{1}{\frac{d^2y}{dx^2}}, \text{ approximately,}$$

and $A)$ reduces to

$$B) \qquad EI\frac{d^2y}{dx^2} = M.$$

The bending moment M at the cross section (point P of the elastic curve) is the algebraic sum of the moments of the external forces acting on the segment of the beam (segment of the elastic curve) about the line AB in the cross section (about the point P of the elastic curve). We shall assume here that upward forces give positive moments and downward forces give negative moments.

EXAMPLE. Consider a 30 foot beam resting on two vertical supports, as in the figure below. Suppose the beam carries a uniform load of 200 lb/ft of length and a load of 2000 lb at its middle.

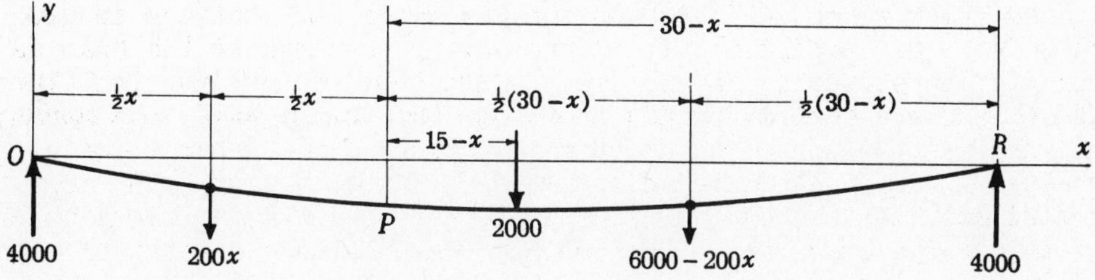

The external forces acting on OP are (*a*) an upward thrust at O, x feet from P, equal to one-half the total load, i.e., $\frac{1}{2}(2000 + 30\cdot 200) = 4000$ lb, and (*b*) a downward force of $200x$ lb thought of as concentrated at the middle of OP and thus $\frac{1}{2}x$ feet from P. The bending moment at P is

$$M = 4000x - 200x(\tfrac{1}{2}x) = 4000x - 100x^2.$$

To show that the bending moment at P is independent of the segment used, consider the forces acting on PR: (*a*) an upward thrust of 4000 lb at R, $30 - x$ ft from P, (*b*) the load of 2000 lb acting downward at the middle of the beam, $15-x$ ft from P, and (c) $200(30-x)$ lb downward thought of as concentrated at the middle of PR, $\frac{1}{2}(30-x)$ ft from P. Then

$$M = 4000(30-x) - 2000(15-x) - 200(30-x)\cdot\tfrac{1}{2}(30-x)$$
$$= 4000x - 100x^2, \qquad \text{as before.}$$

A beam is said to be *fixed* at one end if it is held horizontal there by the masonry. In the example above the beam is not horizontal at O and is said to be freely *supported* there.

SIMPLE ELECTRIC CIRCUITS. The sum of the voltage drops across the elements of a closed circuit is equal to the total electromotive force E in the circuit. The voltage drop across a resistance R ohms is Ri, across a coil of inductance L henries is $L\,di/dt$, and across a condenser of capacitance (capacity) C farads is q/C. Here, the current i amperes and the charge q coulombs are related by $i = dq/dt$. We will consider R, L, and C as constants.

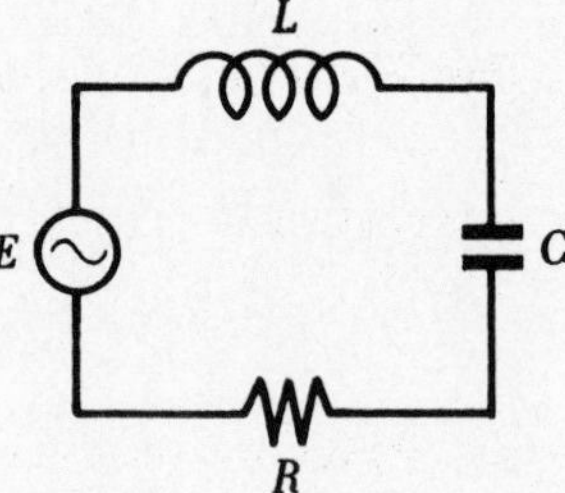

The differential equation of an electric circuit containing an inductance L, a resistance R, a condenser of capacitance C, and an electromotive force $E(t)$ is therefore

C')
$$L\frac{di}{dt} + Ri + \frac{q}{C} = E(t)$$

or, since $i = dq/dt$, $di/dt = d^2q/dt^2$,

C)
$$L\frac{d^2q}{dt^2} + R\frac{dq}{dt} + \frac{q}{C} = E(t)$$

from which $q = q(t)$ may be found.

By differentiating C') and using $\frac{dq}{dt} = i$, we have

D)
$$L\frac{d^2i}{dt^2} + R\frac{di}{dt} + \frac{i}{C} = E'(t)$$

from which $i = i(t)$ may be found.

SOLVED PROBLEMS

GEOMETRIC APPLICATIONS.

1. Determine the curve whose radius of curvature at any point $P(x,y)$ is equal to the normal at P and (a) in the same direction, (b) in the opposite direction.

a) Here $\dfrac{[1+(y')^2]^{3/2}}{y''} = -y[1+(y')^2]^{1/2}$ or $yy'' + (y')^2 + 1 = 0$.

The equation is exact and an integration yields $yy' + x - C_1 = 0$ or $y\,dy + (x-C_1)dx = 0$.

Integrating again, $\frac{1}{2}y^2 + \frac{1}{2}(x-C_1)^2 = K$ or $y^2 + (x-C_1)^2 = C_2$, a family of circles with centers on the x-axis.

b) Here $\dfrac{[1+(y')^2]^{3/2}}{y''} = y[1+(y')^2]^{1/2}$ or $yy'' - (y')^2 - 1 = 0$.

The substitution $y' = p$, $y'' = p\dfrac{dp}{dy}$ of Chapter 19 reduces the equation to

$$yp\frac{dp}{dy} - p^2 - 1 = 0 \qquad \text{or} \qquad \frac{p\,dp}{1+p^2} = \frac{dy}{y}.$$

Then $\ln(1+p^2) = \ln y^2 + \ln C_1^2$, $1+p^2 = C_1^2y^2$, or $\dfrac{dy}{\sqrt{C_1^2y^2-1}} = \pm\,dx$.

Integrating, $\cosh^{-1} C_1y = \pm C_1x + C_2$, $C_1y = \cosh(\pm C_1x + C_2)$, or

$$y = \frac{1}{2C_1}[e^{(\pm C_1x + C_2)} + e^{-(\pm C_1x + C_2)}].$$

The curves are catenaries and the equation may be written in the form

$$y = \tfrac{1}{2}A[e^{(B\pm x)/A} + e^{-(B\pm x)/A}], \quad \text{where } A = \frac{1}{C_1} \text{ and } B = \frac{C_2}{C_1}.$$

PHYSICAL APPLICATIONS

MOTION OF A PENDULUM.

2. A pendulum, of length l and mass m, suspended at P (see figure) moves in a vertical plane through P. Disregarding all forces except that of gravity, find its motion.

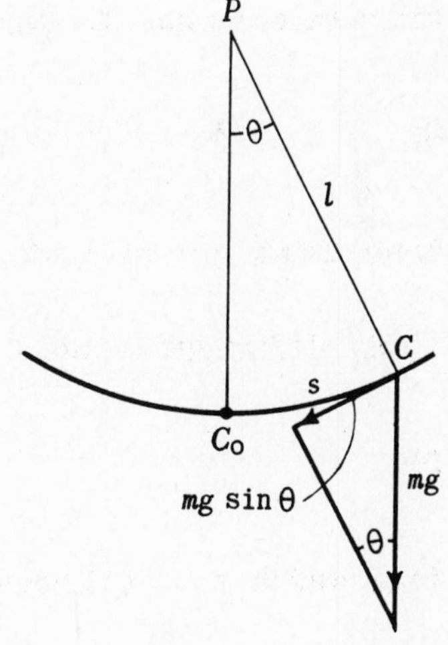

Under the assumptions, the center of gravity C of the bob moves on a circle with center P and radius l. Let θ, positive when measured counterclockwise, be the angle which the string makes with the vertical at time t. The only force is gravity, positive when measured downward, and its component along the tangent to the path of the bob is $mg\sin\theta$. If s denotes the length of arc C_0C, then $s = l\theta$ and the acceleration along the arc is

$$\frac{d^2s}{dt^2} = l\frac{d^2\theta}{dt^2}.$$

Thus $m\cdot l\dfrac{d^2\theta}{dt^2} = -mg\sin\theta$ or $l\dfrac{d^2\theta}{dt^2} = -g\sin\theta$.

Multiplying by $2\dfrac{d\theta}{dt}$ and integrating, $l(\dfrac{d\theta}{dt})^2 = 2g\cos\theta + C_1$ or $\dfrac{d\theta}{\sqrt{2g\cos\theta + C_1}} = \pm\dfrac{dt}{\sqrt{l}}$.

This integral cannot be expressed in terms of elementary functions.

When θ is small, $\sin\theta = \theta$, approximately. When this replacement is made in the original differential equation, we have $\dfrac{d^2\theta}{dt^2} + \dfrac{g}{l}\theta = 0$ whose solution is $\theta = C_1\cos\sqrt{\dfrac{g}{l}}\,t + C_2\sin\sqrt{\dfrac{g}{l}}\,t$.

This is an example of simple harmonic motion. The amplitude is $\sqrt{C_1^2 + C_2^2}$ and the period is $2\pi\sqrt{\dfrac{l}{g}}$.

MOTION ALONG A STRAIGHT LINE.

3. A mass m is projected vertically upward from O with initial velocity v_0. Find the maximum height reached, assuming that the resistance of the air is proportional to the velocity.

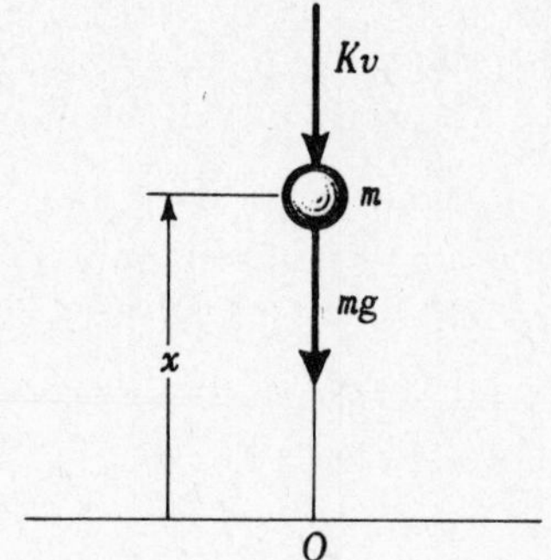

Take upward direction from O as positive, and let x denote the distance of the mass from O at time t. The mass is acted upon by two forces, the gravitational force of magnitude mg and the resistance of magnitude $Kv = K\dfrac{dx}{dt}$ each directed down. Hence, using

$$\text{mass} \times \text{acceleration} = \text{net force},$$

$$m\frac{d^2x}{dt^2} = -mg - K\frac{dx}{dt} \quad \text{or} \quad \frac{d^2x}{dt^2} + k\frac{dx}{dt} = -g, \quad \text{where } K = mk.$$

Integrating, 1) $x = C_1 + C_2e^{-kt} - \dfrac{g}{k}t$, and then differentiating once with respect to t,

2) $v = \dfrac{dx}{dt} = -kC_2e^{-kt} - \dfrac{g}{k}$.

When $t = 0$, $x = 0$ and $v = v_0$. Then $C_1 + C_2 = 0$, $v_0 = -kC_2 - \dfrac{g}{k}$, and $C_1 = -C_2 = \dfrac{v_0}{k} + \dfrac{g}{k^2}$.

Making these replacements in 1), we have $x = \dfrac{1}{k^2}(g + kv_0)(1 - e^{-kt}) - \dfrac{g}{k}t$.

The maximum height is reached when $v = 0$. From 2), $e^{-kt} = \dfrac{-g}{k^2C_2} = \dfrac{g}{g + kv_0}$ and $t = \dfrac{1}{k}\ln\dfrac{g + kv_0}{g}$.

Then the maximum height is $x = \dfrac{1}{k^2}(g + kv_0)(1 - \dfrac{g}{g + kv_0}) - \dfrac{g}{k}(\dfrac{1}{k}\ln\dfrac{g + kv_0}{g}) = \dfrac{1}{k}(v_0 - \dfrac{g}{k}\ln\dfrac{g + kv_0}{g})$.

4. A mass m, free to move along the x-axis, is attracted toward the origin with a force proportional to its distance from the origin. Find the motion (a) if it starts from rest at $x = x_0$ and (b) if it starts at $x = x_0$ with initial velocity v_0 moving away from the origin.

Let x denote the distance from the origin to the mass at time t.

Then $m\dfrac{d^2x}{dt^2} = -Kx$ or $\dfrac{d^2x}{dt^2} + k^2x = 0$, where $K = mk^2$.

Integrating, 1) $x = C_1 \sin kt + C_2 \cos kt$, and differentiating once with respect to t,

2) $v = -kC_2 \sin kt + kC_1 \cos kt$.

a) When $t = 0$, $x = x_0$ and $v = 0$. Then $C_1 = 0$ from 2), $C_2 = x_0$ from 1), and

$$x = x_0 \cos kt.$$

b) When $t = 0$, $x = x_0$ and $v = v_0$. Then $C_2 = x_0$, $C_1 = v_0/k$, and

$$x = \frac{v_0}{k}\sin kt + x_0 \cos kt.$$

In a) the motion is simple harmonic motion of amplitude x_0 and period $2\pi/k$.

In b) the motion is simple harmonic of amplitude $\dfrac{\sqrt{v_0^2 + k^2x_0^2}}{k}$ and period $2\pi/k$.

MOTION OF A COMPLEX SYSTEM.

5. A chain hangs over a smooth peg, 8 feet being on one side and 12 feet on the other. Find the time required for it to slide off (*a*) neglecting friction and (*b*) if the friction is equal to the weight of 1 foot of the chain.

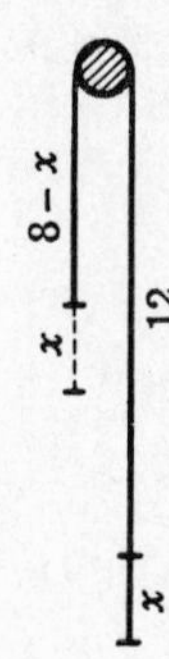

a) Denote the total mass of the chain by m and the length (feet) of the chain which has moved over the peg at time t by x. At time t there are $(8-x)$ feet of chain on one side and $(12+x)$ feet on the other. The excess $(4+2x)$ feet on one side produces an unbalanced force of $(4+2x)\frac{mg}{20}$ pounds. Thus,

$$m\frac{d^2x}{dt^2} = (4+2x)\frac{mg}{20} \qquad \text{or} \qquad 10\frac{d^2x}{dt^2} = gx + 2g.$$

Solution 1.

Integrating $\dfrac{d^2x}{dt^2} - \dfrac{g}{10}x = \dfrac{g}{5}$, we have $x = C_1e^{\sqrt{g/10}\,t} + C_2e^{-\sqrt{g/10}\,t} - 2.$

Differentiating once with respect to t, $v = \sqrt{\dfrac{g}{10}}\,(C_1e^{\sqrt{g/10}\,t} - C_2e^{-\sqrt{g/10}\,t}).$

When $t=0$, $x=0$ and $v=0$. Then $C_1 = C_2 = 1$ and $x = e^{\sqrt{g/10}\,t} + e^{-\sqrt{g/10}\,t} - 2 = 2\cosh\sqrt{\dfrac{g}{10}}\,t - 2.$

Hence $t = \sqrt{\dfrac{10}{g}}\cosh^{-1}\tfrac{1}{2}(x+2) = \sqrt{\dfrac{10}{g}}\ln\dfrac{x+2+\sqrt{x^2+4x}}{2}.$

When $x = 8$ ft has moved over the peg, $t = \sqrt{\dfrac{10}{g}}\ln(5+2\sqrt{6})$ sec.

Solution 2. Multiplying the equation by $\dfrac{dx}{dt}$ and integrating, we have

$$10\frac{dx}{dt}\frac{d^2x}{dt^2} = gx\frac{dx}{dt} + 2g\frac{dx}{dt} \qquad \text{and} \qquad 5\left(\frac{dx}{dt}\right)^2 = \tfrac{1}{2}gx^2 + 2gx + C_1.$$

When $t=0$, $x=0$ and $dx/dt=0$. Then $C_1 = 0$ and

$$5\left(\frac{dx}{dt}\right)^2 = \tfrac{1}{2}gx^2 + 2gx \qquad \text{or} \qquad dt = \sqrt{\frac{10}{g}}\,\frac{dx}{\sqrt{x^2+4x}}.$$

(The positive square root is used here since x increases with t.)

Integrating, $t = \sqrt{\dfrac{10}{g}}\displaystyle\int\frac{dx}{\sqrt{(x+2)^2-4}} = \sqrt{\dfrac{10}{g}}\ln(x+2+\sqrt{x^2+4x}) + C_2.$

When $t=0$, $x=0$. Then $C_2 = -\sqrt{\dfrac{10}{g}}\ln 2$, and $t = \sqrt{\dfrac{10}{g}}\ln\dfrac{x+2+\sqrt{x^2+4x}}{2}$ as before.

b) Here $m\dfrac{d^2x}{dt^2} = (4+2x)\dfrac{mg}{20} - \dfrac{mg}{20}$ or $20\dfrac{d^2x}{dt^2} = (2x+3)g.$

Multiplying by $\dfrac{dx}{dt}$ and integrating, we have $10\left(\dfrac{dx}{dt}\right)^2 = gx^2 + 3gx + C_1.$

When $t=0$, $x=0$ and $v=0$. Then $C_1=0$, and $\dfrac{dx}{dt} = \sqrt{\dfrac{g}{10}(x^2+3x)}$ or $dt = \sqrt{\dfrac{10}{g}}\,\dfrac{dx}{\sqrt{x^2+3x}}.$

Then $t = \sqrt{\dfrac{10}{g}}\ln(x+\dfrac{3}{2}+\sqrt{x^2+3x}) + C_2.$

When $t=0$, $x=0$. Then $C_2 = -\sqrt{\dfrac{10}{g}}\ln\dfrac{3}{2}$ and $t = \sqrt{\dfrac{10}{g}}\ln\dfrac{2}{3}(x+\dfrac{3}{2}+\sqrt{x^2+3x})$.

When $x=8$, $t = \sqrt{\dfrac{10}{g}}\ln\dfrac{19+4\sqrt{22}}{3} = 1.4$ sec.

6. A bead slides without friction along a straight rod of negligible mass as the rod rotates with constant angular velocity ω about its midpoint O. Determine the motion (a) if the bead is initially at rest at O and (b) if the bead is initially at O moving with velocity $g/2\omega$.

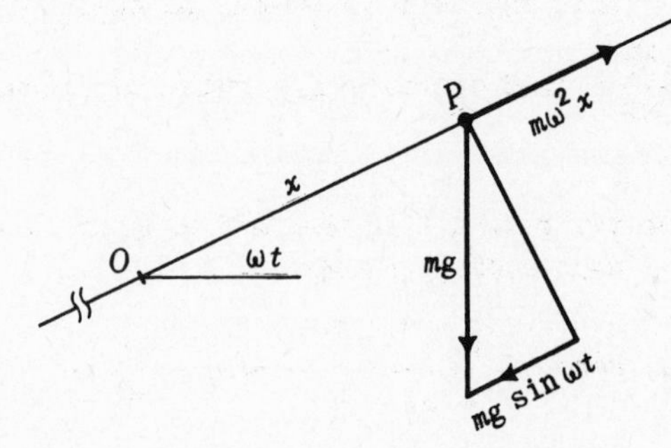

Let the bead be x units from O at time t. It is being acted upon by two forces, (i) gravity and (ii) the centrifugal force $m\omega^2 x$ acting along the rod and directed away from O. Since the rod has rotated through an angle ωt, the component of the gravitational force along the rod has magnitude $mg\sin\omega t$; its direction is toward O. Hence

$$m\frac{d^2x}{dt^2} = m\omega^2 x - mg\sin\omega t \qquad \text{or} \qquad \frac{d^2x}{dt^2} - \omega^2 x = -g\sin\omega t.$$

Integrating, 1) $x = C_1e^{\omega t} + C_2e^{-\omega t} + \dfrac{g}{2\omega^2}\sin\omega t$. Differentiating once with respect to t,

$$2)\quad v = \omega C_1e^{\omega t} - \omega C_2e^{-\omega t} + \frac{g}{2\omega}\cos\omega t.$$

$a)$ When $t=0$, $x=0$ and $v=0$.

Then $C_1+C_2=0$ from 1), $C_1-C_2+\dfrac{g}{2\omega^2}=0$ from 2), $C_1=-C_2=-\dfrac{g}{4\omega^2}$, and

$$x = \frac{g}{4\omega^2}(e^{-\omega t}-e^{\omega t}) + \frac{g}{2\omega^2}\sin\omega t = -\frac{g}{2\omega^2}\sinh\omega t + \frac{g}{2\omega^2}\sin\omega t.$$

$b)$ When $t=0$, $x=0$ and $v=g/2\omega$.

Then $C_1+C_2=0$, $C_1-C_2=0$, $C_1=C_2=0$, and $x = \dfrac{g}{2\omega^2}\sin\omega t$.

SPRINGS.

7. A spring, for which k = 48 lb/ft, hangs in a vertical position with its upper end fixed. A mass, weighing 16 lb, is attached to the lower end. After coming to rest, the mass is pulled down 2 inches and released. Discuss the resulting motion of the mass, neglecting air resistance.

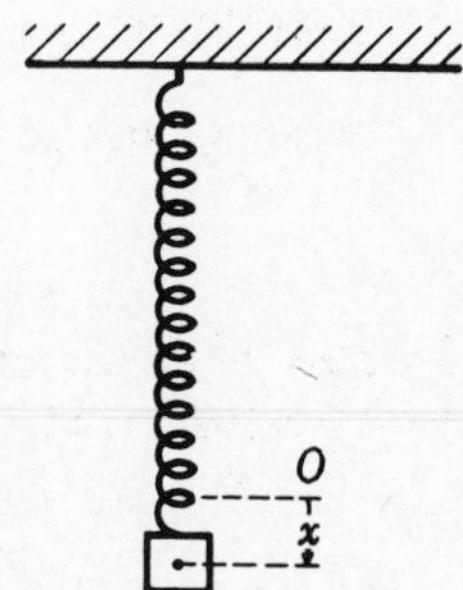

Take the origin at the center of gravity of the mass, after coming to rest, and let x, positive when measured downward, be the change in position of the mass at time t. When the mass is at rest the spring force is equal to but opposite in direction to the gravitational force. The net force at time t is the spring force $-kx$ corresponding to the change x in the position of the mass. Then

$$\frac{16}{g}\frac{d^2x}{dt^2} = -48x \qquad \text{or} \qquad \frac{d^2x}{dt^2} + 96x = 0, \quad \text{taking } g = 32 \text{ ft/sec}^2.$$

Integrating, $x = C_1 \sin\sqrt{96}\,t + C_2 \cos\sqrt{96}\,t$.

Differentiating once with respect to t, $v = \dfrac{dx}{dt} = \sqrt{96}\,(C_1 \cos\sqrt{96}\,t - C_2 \sin\sqrt{96}\,t)$.

When $t = 0$, $x = \dfrac{1}{6}$ and $v = 0$. Then $C_2 = \dfrac{1}{6}$, $C_1 = 0$, and $x = \dfrac{1}{6}\cos\sqrt{96}\,t$.

This represents a simple harmonic motion. The period is $\dfrac{2\pi}{\sqrt{96}} = 0.641$ sec, the frequency is $\dfrac{\sqrt{96}}{2\pi} = 1.56$ cycles/sec, and the amplitude is $\dfrac{1}{6}$ ft.

8. Solve Problem 7 if the medium offers a resistance (lb) equal to (a) $v/64$ and (b) $64v$, where v is expressed in ft/sec.

a) Here $\dfrac{16}{g}\dfrac{d^2x}{dt^2} = -48x - \dfrac{1}{64}\dfrac{dx}{dt}$ or $\dfrac{d^2x}{dt^2} + \dfrac{1}{32}\dfrac{dx}{dt} + 96x = 0$. Using the D notation,

$$(D^2 + \frac{1}{32}D + 96)x = [D - (-0.0156 + 9.8i)][D - (-0.0156 - 9.8i)]x = 0,$$

and

$$x = e^{-0.0156t}(C_1 \cos 9.8\,t + C_2 \sin 9.8\,t).$$

Differentiating once with respect to t,

$$\frac{dx}{dt} = v = e^{-0.0156t}[(9.8C_2 - 0.0156\,C_1)\cos 9.8t - (9.8C_1 + 0.0156C_2)\sin 9.8t].$$

When $t = 0$, $v = 0$ and $x = 1/6$. Then $C_1 = 1/6$, $0 = 9.8C_2 - 0.0156C_1$, and $C_2 = 0.000265$. Thus,

$$x = e^{-0.0156t}(\frac{1}{6}\cos 9.8\,t + 0.000265 \sin 9.8\,t).$$

This represents a damped oscillatory motion. Note that the frequency $= \dfrac{9.8}{2\pi} = 1.56$ cycles/sec remains constant throughout the motion, while the amplitude of each oscillation is smaller than the preceding one due to the damping factor $e^{-0.0156t}$. At $t = 0$ the magnitude of the damping factor is 1. It will be 2/3 when $e^{-0.0156t} = 2/3$ or after $t = 26$ sec. It will be 1/3 when $e^{-0.0156t} = 1/3$ or after $t = 70$ sec.

b) Here $\dfrac{16}{32}\dfrac{d^2x}{dt^2} = -48x - 64\dfrac{dx}{dt}$ or $(D^2 + 128D + 96)x = 0$.

Integrating, $x = C_1e^{-0.76t} + C_2e^{-127.24t}$.

Differentiating once with respect to t,

$$v = -0.76\,C_1e^{-0.76t} - 127.24\,C_2e^{-127.24t}.$$

When $t = 0$, $x = 1/6$ and $v = 0$. Then $C_1 + C_2 = 1/6$, $-0.76\,C_1 - 127.24\,C_2 = 0$, $C_1 = 0.166$, $C_2 = -0.001$, and

$$x = 0.166\,e^{-0.76t} - 0.001\,e^{-127.24t}.$$

The motion is not vibratory. After the initial displacement, the mass moves slowly toward the position of equilibrium as t increases.

9. Solve Problem 8*a* if, in addition, the support of the spring is given a motion $y = \cos 4t$ ft.

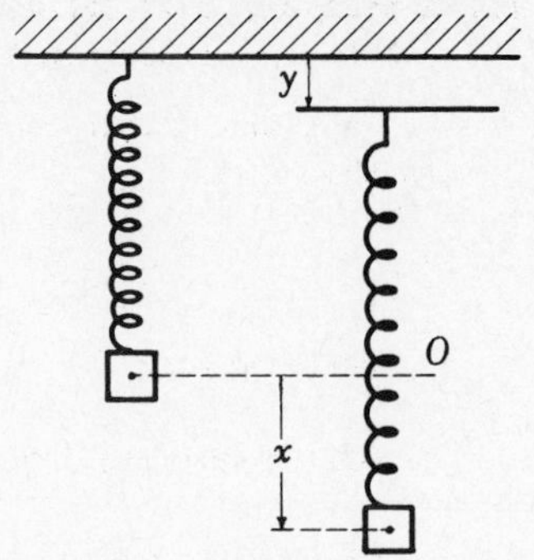

Take the origin as in Problem 8 and let x represent the change in position of the mass after t sec. From the figure, it is seen that the stretch in the spring is $(x - y)$ and the spring force is $-48(x - y) = -48(x - \cos 4t)$ lb. Hence,

$$\frac{16}{g}\frac{d^2x}{dt^2} = -48(x - \cos 4t) - \frac{1}{64}\frac{dx}{dt}$$

or
$$(D^2 + \frac{1}{32}D + 96)x = 96 \cos 4t.$$

Integrating, $$x = e^{-0.0156t}(C_1 \cos 9.8t + C_2 \sin 9.8t) + \frac{96}{D^2 + D/32 + 96}\cos 4t$$

$$= e^{-0.0156t}(C_1 \cos 9.8t + C_2 \sin 9.8t) + 0.0019 \sin 4t + 1.2 \cos 4t.$$

Differentiating once with respect to t,

$$v = e^{-0.0156t}\left[(9.8C_2 - 0.0156C_1)\cos 9.8t - (9.8C_1 + 0.0156C_2)\sin 9.8t\right] + 0.0076 \cos 4t - 4.8 \sin 4t.$$

When $t = 0$, $v = 0$ and $x = 1 + 1/6 = 7/6$. Then $C_1 = -1/30$, $C_2 = -0.0008$, and

$$x = e^{-0.0156t}(-0.0333 \cos 9.8t - 0.0008 \sin 9.8t) + 0.0019 \sin 4t + 1.2 \cos 4t.$$

The motion consists of a damped harmonic motion which gradually dies away (transient phenomenon) and a harmonic motion which remains (steady-state phenomenon). After a time the only effective motion is that of the steady-state. These steady-state oscillations will have a period and a frequency equal to those of the forcing function $y = \cos 4t$, namely, a period of $2\pi/4 = 1.57$ sec and a frequency of $4/2\pi = 0.637$ cycle/sec.

The amplitude is $\sqrt{(0.0019)^2 + (1.2)^2} = 1.2$ ft.

10. A mass of 20 lb is suspended from a spring which is thereby stretched 3 inches. The upper end of the spring is then given a motion $y = 4(\sin 2t + \cos 2t)$ft. Find the equation of the motion, neglecting air resistance.

Take the origin at the center of gravity of the mass when at rest. Let x represent the change in position of the mass at time t. The change in the length of the spring is $(x - y)$, the spring constant is $20/\frac{1}{4} = 80$ lb/ft, and the net spring force is $-80(x - y)$. Then

$$\frac{20}{32}\frac{d^2x}{dt^2} = -80(x - 4\sin 2t - 4\cos 2t) \quad \text{or} \quad \frac{d^2x}{dt^2} + 128x = 512(\sin 2t + \cos 2t).$$

Integrating, $$x = C_1 \cos\sqrt{128}\, t + C_2 \sin\sqrt{128}\, t + \frac{128}{31}(\sin 2t + \cos 2t).$$

Differentiating once with respect to t,

$$v = -\sqrt{128}\, C_1 \sin\sqrt{128}\, t + \sqrt{128}\, C_2 \cos\sqrt{128}\, t + \frac{256}{31}(-\sin 2t + \cos 2t).$$

When $t = 0$, $x = 4$ and $v = 0$.

Then $4 = C_1 + \frac{128}{31}$, $C_1 = -0.129$; and $\sqrt{128}\, C_2 + \frac{256}{31} = 0$, $C_2 = -0.730$.

Hence, $$x = -0.13 \cos\sqrt{128}\, t - 0.73 \sin\sqrt{128}\, t + 4.13(\sin 2t + \cos 2t).$$

11. A mass of 64 lb is attached to a spring for which $k = 50$ lb/ft and brought to rest. Find the position of the mass at time t if a force equal to $4\sin 2t$ is applied to it.

Take the origin at the center of gravity of the mass when at rest. The equation of motion is then

$$\frac{64}{32}\frac{d^2x}{dt^2} + 50x = 4\sin 2t \qquad \text{or} \qquad \frac{d^2x}{dt^2} + 25x = 2\sin 2t.$$

Integrating, $\quad x = C_1\cos 5t + C_2\sin 5t + \frac{2}{21}\sin 2t.$

Differentiating once with respect to t, $\quad v = -5C_1\sin 5t + 5C_2\cos 5t + \frac{4}{21}\cos 2t.$

Using the initial conditions $x = 0,\ v = 0$ when $t = 0$, $\quad C_1 = 0,\ C_2 = -\frac{4}{105}$, and

$$x = -0.038\sin 5t + 0.095\sin 2t.$$

The displacement here is the algebraic sum of two harmonic displacements of *different* periods.

12. A mass of 16 lb is attached to a spring for which $k = 48$ lb/ft and brought to rest. Find the motion of the mass if the support of the spring is given a motion $y = \sin\sqrt{3g}\,t$ ft.

Take the origin at the center of gravity of the mass when at rest and let x represent the change in position of the mass at time t.

The stretch in the spring is $(x - y)$ and the spring force is $-48(x - y)$. Thus,

$$\frac{16}{g}\frac{d^2x}{dt^2} = -48(x - \sin\sqrt{3g}\,t) \qquad \text{or} \qquad \frac{d^2x}{dt^2} + 3gx = 3g\sin\sqrt{3g}\,t.$$

Integrating, $\quad x = C_1\cos\sqrt{3g}\,t + C_2\sin\sqrt{3g}\,t - \frac{1}{2}\sqrt{3g}\,t\cos\sqrt{3g}\,t,$

and $\quad v = -C_1\sqrt{3g}\sin\sqrt{3g}\,t + C_2\sqrt{3g}\cos\sqrt{3g}\,t - \frac{1}{2}\sqrt{3g}\cos\sqrt{3g}\,t + \frac{3g}{2}t\sin\sqrt{3g}\,t.$

Using the initial conditions $x = 0,\ v = 0$ when $t = 0$, $\quad C_1 = 0,\ C_2 = \frac{1}{2}$, and

$$x = \frac{1}{2}\sin\sqrt{3g}\,t - \frac{\sqrt{3g}}{2}t\cos\sqrt{3g}\,t.$$

The first term represents a simple harmonic motion while the second represents a vibratory motion with increasing amplitude (because of the factor t). As t increases, the amplitude of the oscillation increases until there is a mechanical breakdown.

13. A cylindrical buoy 2 ft in diameter stands in water (density 62.4 lb/ft^3) with its axis vertical. When depressed slightly and released, it is found that the period of vibration is 2 seconds. Find the weight of the cylinder.

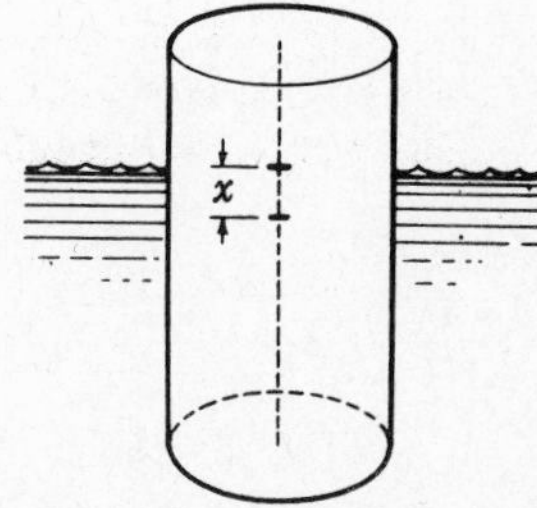

Take the origin at the intersection of the axis of the cylinder and the surface of the water when the buoy is in equilibrium, and take the downward direction as positive.

Let x (ft) denote the change in the position of the buoy at time t. By Archimedes' Principle, a body partly or totally submerged in a fluid is buoyed up by a force equal to the weight of the fluid it displaces. Thus, the corresponding change in the buoying force is $62.4\pi(1)^2x$ and

$$\frac{W}{g}\frac{d^2x}{dt^2} = -62.4\pi x \qquad \text{or} \qquad \frac{d^2x}{dt^2} + \frac{2009}{W}\pi x = 0,$$

where W (lb) is the weight of the buoy and $g = 32.2$ ft/sec^2.

Integrating, $x = C_1 \sin\sqrt{2009\pi/W}\,t + C_2 \cos\sqrt{2009\pi/W}\,t$.

Since the period is $\dfrac{2\pi}{\sqrt{2009\pi/W}} = 2\sqrt{\pi W/2009} = 2$, $W = \dfrac{2009}{\pi} = 640$ lb.

HANGING CABLE.

14. Determine the shape of a uniform cable which hangs under its own weight, w lb/ft of length.

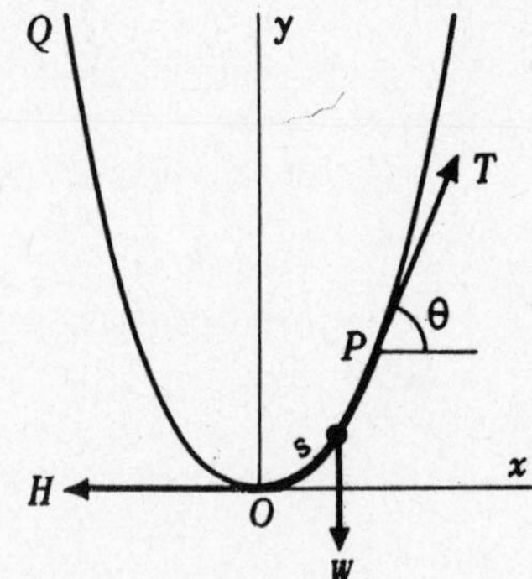

Choose the coordinate axes as in the figure, the origin being at the lowest point of the cable. Consider the part between O and a variable point $P(x,y)$. This part is in equilibrium under the action of (1) a horizontal force of magnitude H at O, (2) the tension T along the tangent at P, and (3) the weight W of OP.

Since OP is in equilibrium, all force acting horizontally toward the right and all force acting horizontally toward the left must be equal in magnitude, and, also, all force acting vertically upward and all force acting vertically downward.

Hence, $T\cos\theta = H$, $T\sin\theta = W$, and $\tan\theta = \dfrac{dy}{dx} = \dfrac{W}{H}$.

Now H is constant, being due to the part OQ of the cable, while $W = ws$, where s is the length of OP. Thus,

$$\frac{d^2y}{dx^2} = \frac{1}{H}\frac{dW}{dx} = \frac{w}{H}\frac{ds}{dx} = \frac{w}{H}\sqrt{1 + (dy/dx)^2}.$$

To solve the above equation, write $\dfrac{dy}{dx} = p$ and obtain

$$\frac{dp}{dx} = \frac{w}{H}\sqrt{1+p^2} \qquad \text{or} \qquad \frac{dp}{\sqrt{1+p^2}} = \frac{w}{H}\,dx.$$

Integrating between the limits $x = 0, p = 0$ and $x = x, p = p$,

$$\sinh^{-1} p = \frac{w}{H}x \qquad \text{and} \qquad p = \frac{dy}{dx} = \sinh\frac{w}{H}x.$$

Integrating $dy = \sinh\dfrac{w}{H}x\,dx$ between the limits $x = 0, y = 0$ and $x = x, y = y$,

$$y = \frac{H}{w}\left(\cosh\frac{w}{H}x - 1\right), \quad \text{a catenary.}$$

If the origin had been taken at a distance H/w under the lowest point of the cable (thus making H/w the y-intercept of the curve) the equation of the curve would have been

$$y = \frac{H}{w}\cosh\frac{w}{H}x.$$

HORIZONTAL BEAMS.

15. A horizontal beam of length $2l$ feet is freely supported at both ends. Find the equation of its elastic curve and its maximum deflection when the load is w lb/ft of length.

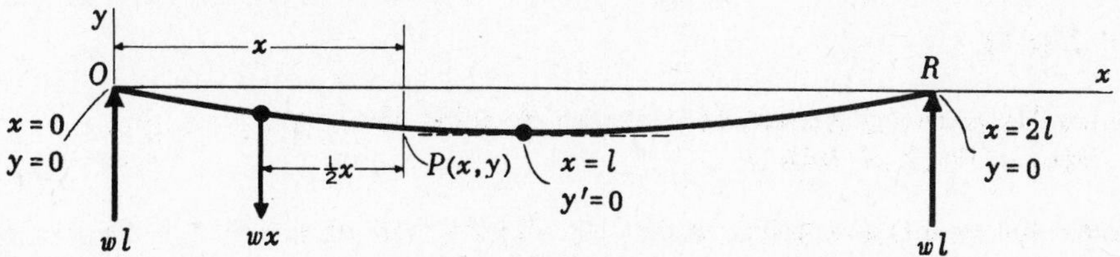

Take the origin at the left end of the beam with the x-axis horizontal as in the figure. Let P, any point on the elastic curve, have coordinates (x,y).

Consider the segment OP of the beam. There is an upward thrust wl lb at O, x ft from P, and the load wx lb at the midpoint of OP, $\frac{1}{2}x$ ft from P. Then, since $EI\, d^2y/dx^2 = M$,

1) $$EI\frac{d^2y}{dx^2} = wlx - wx(\tfrac{1}{2}x) = wlx - \tfrac{1}{2}wx^2.$$

Solution 1. Integrating 1) once, $EI\dfrac{dy}{dx} = \dfrac{1}{2}wlx^2 - \dfrac{1}{6}wx^3 + C_1.$

At the middle of the beam $x = l$ and $dy/dx = 0$. Then $C_1 = -\dfrac{1}{3}wl^3$ and

2) $$EI\frac{dy}{dx} = \frac{1}{2}wlx^2 - \frac{1}{6}wx^3 - \frac{1}{3}wl^3.$$

Integrating 2), $EIy = \dfrac{1}{6}wlx^3 - \dfrac{1}{24}wx^4 - \dfrac{1}{3}wl^3x + C_2.$ At O, $x = y = 0$. Then $C_2 = 0$ and

3) $$y = \frac{w}{24\,EI}(4lx^3 - x^4 - 8l^3x).$$

Solution 2. Integrating 1) twice, $EIy = \dfrac{1}{6}wlx^3 - \dfrac{1}{24}wx^4 + C_1x + C_2.$

At O, $x = y = 0$, while at R, $x = 2l$, $y = 0$. Using these boundary conditions in turn, we find $C_2 = 0$ and $C_1 = -\dfrac{1}{3}wl^3$, as before.

The deflection of the beam at any distance x from O is given by $-y$. The maximum deflection occurs at the middle of the beam ($x = l$) and is, from 3),

$$-y_{max} = -\frac{w}{24\,EI}(4l^4 - l^4 - 8l^4) = \frac{5wl^4}{24\,EI}.$$

16. Solve Problem 15 if there is in addition a load of W lb at the middle of the beam.

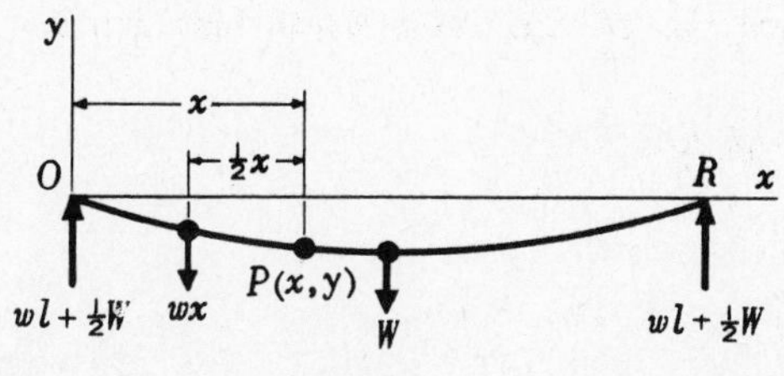
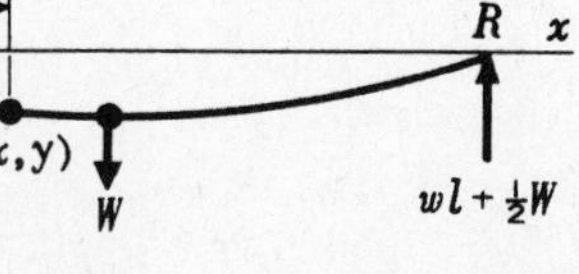

$0 < x < l$

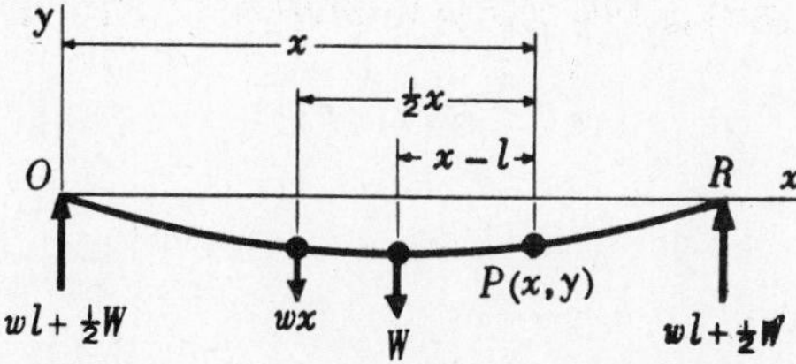

$l < x < 2l$

Choose the coordinate system as in Problem 15. Since the forces acting on a segment OP of the beam differ according as P lies to the left or right of the midpoint, two cases must be considered.

When $0 < x < l$, the forces acting on OP are an upward thrust of $(wl + \frac{1}{2}W)$ lb at O, x ft from P, and the load wx acting downward at the midpoint of OP, $\frac{1}{2}x$ ft from P. The bending moment is then

$$1)\quad M = (wl + \tfrac{1}{2}W)x - wx(\tfrac{1}{2}x) = wlx + \tfrac{1}{2}Wx - \tfrac{1}{2}wx^2.$$

When $l < x < 2l$, there is an additional force — the load W lb at the midpoint of the beam, $(x - l)$ ft from P. The bending moment is then

$$2)\quad M = (wl + \tfrac{1}{2}W)x - wx(\tfrac{1}{2}x) - W(x - l) = wlx + \tfrac{1}{2}Wx - \tfrac{1}{2}wx^2 - W(x - l).$$

Both 1) and 2) yield the bending moment $M = \frac{1}{2}wl^2 + \frac{1}{2}Wl$ when $x = l$. The two cases may be treated at the same time by noting that for 1)

$$wlx + \tfrac{1}{2}Wx - \tfrac{1}{2}wx^2 = wlx - \tfrac{1}{2}wx^2 - \tfrac{1}{2}W(l - x) + \tfrac{1}{2}Wl$$

and for 2) $\quad wlx + \frac{1}{2}Wx - \frac{1}{2}wx^2 - W(x - l) = wlx - \frac{1}{2}wx^2 + \frac{1}{2}W(l - x) + \frac{1}{2}Wl.$ Then

$$3)\quad EI\frac{d^2y}{dx^2} = wlx - \tfrac{1}{2}wx^2 \mp \tfrac{1}{2}W(l - x) + \tfrac{1}{2}Wl$$

with the understanding that the upper sign holds for $0 < x < l$ and the lower for $l < x < 2l$.

Integrating 3) twice, $\quad EIy = \frac{1}{6}wlx^3 - \frac{1}{24}wx^4 \mp \frac{1}{12}W(l - x)^3 + \frac{1}{4}Wlx^2 + C_1x + C_2.$

Using the boundary conditions $x = y = 0$ at O and $x = 2l$, $y = 0$ at R,

$C_2 = \frac{1}{12}Wl^3$, $\quad 2lC_1 + C_2 = -\frac{4}{3}wl^4 + \frac{2}{3}wl^4 + \frac{1}{12}Wl^3 - Wl^3$, and $C_1 = -\frac{1}{3}wl^3 - \frac{1}{2}Wl^2$. Then

$$\begin{aligned} EIy &= \frac{1}{6}wlx^3 - \frac{1}{24}wx^4 - \frac{1}{3}wl^3x \mp \frac{1}{12}W(l - x)^3 + \frac{1}{4}Wlx^2 - \frac{1}{2}Wl^2x + \frac{1}{12}Wl^3 \\ &= \frac{1}{6}wlx^3 - \frac{1}{24}wx^4 - \frac{1}{3}wl^3x - \frac{1}{12}W|l - x|^3 + \frac{1}{4}Wlx^2 - \frac{1}{2}Wl^2x + \frac{1}{12}Wl^3, \end{aligned}$$

and

$$y = \frac{w}{24EI}(4lx^3 - x^4 - 8l^3x) + \frac{W}{12EI}(3lx^2 - |l - x|^3 - 6l^2x + l^3).$$

The maximum deflection, occurring at the middle of the beam, is $-y_{max} = \dfrac{5wl^4}{24EI} + \dfrac{Wl^3}{6EI}$.

17. A horizontal beam of length l feet is fixed at one end but otherwise unsupported. Find the equation of its elastic curve and the maximum deflection when the uniform load is w lb/ft of length.

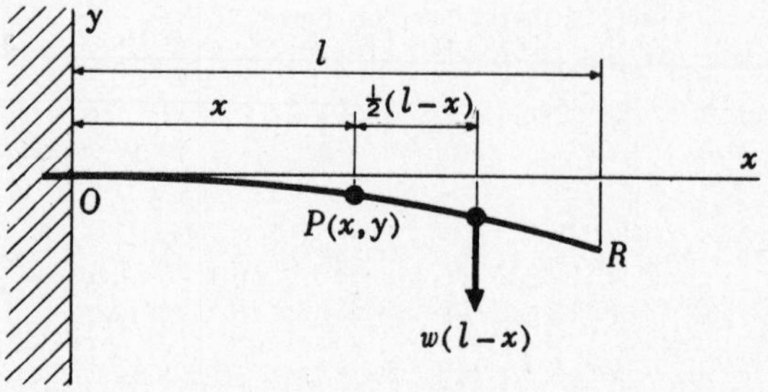

Take the origin at the fixed end and let P have coordinates (x,y). Consider the segment PR. The only force is the weight $w(l - x)$ lb at the midpoint of PR, $\frac{1}{2}(l - x)$ ft from P. Then

$$EI\frac{d^2y}{dx^2} = -w(l - x)\cdot\tfrac{1}{2}(l - x) = -\tfrac{1}{2}w(l - x)^2. \quad \text{Integrating once,} \quad EI\frac{dy}{dx} = \frac{1}{6}w(l - x)^3 + C_1.$$

At O: $x = 0$, $\frac{dy}{dx} = 0$; $C_1 = -\frac{1}{6}wl^3$ and $EI\frac{dy}{dx} = \frac{1}{6}w(l - x)^3 - \frac{1}{6}wl^3$.

Integrating again, $\quad EIy = -\frac{1}{24}w(l - x)^4 - \frac{1}{6}wl^3x + C_2.$

At O: $x = y = 0$; then $C_2 = \frac{1}{24}wl^4$, $EIy = -\frac{1}{24}w(l-x)^4 - \frac{1}{6}wl^3x + \frac{1}{24}wl^4$, and

$$y = \frac{w}{24EI}(4lx^3 - 6l^2x^2 - x^4).$$

The maximum deflection, occurring at $R\ (x = l)$, is $-y_{max} = \frac{1}{8}\frac{wl^4}{EI}$. Note that this is not a relative minimum as in Problem 16 but an absolute minimum occurring at an end of the interval $0 \leqq x \leqq l$.

18. A horizontal beam of length $3l$ feet is fixed at one end but otherwise unsupported. There is a uniform load of w lb/ft of length and two loads of W lb each at points l and $2l$ ft from the fixed end. Find the equation of the elastic curve and the maximum deflection.

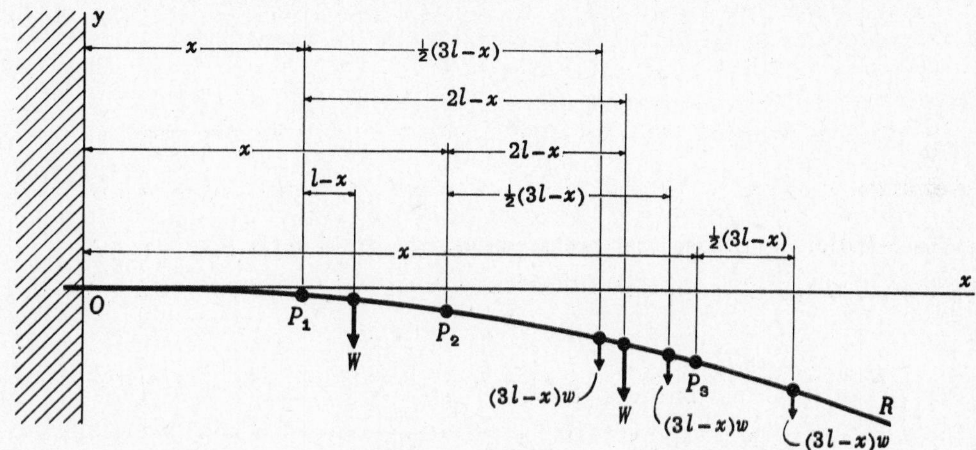

Take the origin at the fixed end and let P have coordinates (x,y). There are three cases to be considered according as P is on the interval $(0 < x < l)$, $(l < x < 2l)$, or $(2l < x < 3l)$. In each case, use will be made of the right hand segment of the beam in computing the three bending moments.

When $0 < x < l$, $(P = P_1$ in the figure), there are three forces acting on P_1R: the weight $(3l-x)w$ lb taken at the midpoint of P_1R, $\frac{1}{2}(3l-x)$ ft from P_1; the load W lb, $(l-x)$ ft from P_1; and the load W lb, $(2l-x)$ ft from P_1. The bending moment about P_1 is

$$M_1 = -(3l-x)w\cdot\tfrac{1}{2}(3l-x) - W(l-x) - W(2l-x) = -\tfrac{1}{2}w(3l-x)^2 - W(l-x) - W(2l-x),$$

and

$$EI\frac{d^2y}{dx^2} = -\tfrac{1}{2}w(3l-x)^2 - W(l-x) - W(2l-x).$$

Integrating, $EI\frac{dy}{dx} = \frac{1}{6}w(3l-x)^3 + \frac{1}{2}W(l-x)^2 + \frac{1}{2}W(2l-x)^2 + C_1$.

At O: $x = 0$ and $dy/dx = 0$; then $C_1 = -\frac{9}{2}wl^3 - \frac{5}{2}Wl^2$,

$$EI\frac{dy}{dx} = \frac{1}{6}w(3l-x)^3 + \frac{1}{2}W(l-x)^2 + \frac{1}{2}W(2l-x)^2 - \frac{9}{2}wl^3 - \frac{5}{2}Wl^2,$$

and

$$EIy = -\frac{1}{24}w(3l-x)^4 - \frac{1}{6}W(l-x)^3 - \frac{1}{6}W(2l-x)^3 - \frac{9}{2}wl^3x - \frac{5}{2}Wl^2x + C_2.$$

At O: $x = y = 0$; then $C_2 = \frac{27}{8}wl^4 + \frac{3}{2}Wl^3$ and

$$A)\quad EIy = -\frac{1}{24}w(3l-x)^4 - \frac{1}{6}W(l-x)^3 - \frac{1}{6}W(2l-x)^3 - \frac{9}{2}wl^3x - \frac{5}{2}Wl^2x + \frac{27}{8}wl^4 + \frac{3}{2}Wl^3.$$

When $l < x < 2l$, ($P = P_2$ in the figure), the bending moment about P_2 is

$$M_2 = -\tfrac{1}{2}w(3l-x)^2 - W(2l-x),$$

and

$$EI\frac{d^2y}{dx^2} = -\tfrac{1}{2}w(3l-x)^2 - W(2l-x).$$ Integrating twice, we obtain

$$B')\qquad EIy = -\frac{1}{24}w(3l-x)^4 - \frac{1}{6}W(2l-x)^3 + C_3x + C_4.$$

When $x = l$, B') and A) must agree in deflection and slope $\frac{dy}{dx}$ so that $C_3 = C_1$ and $C_4 = C_2$. Thus,

$$B)\qquad EIy = -\frac{1}{24}w(3l-x)^4 - \frac{1}{6}W(2l-x)^3 - \frac{9}{2}wl^3x - \frac{5}{2}Wl^2x + \frac{27}{8}wl^4 + \frac{3}{2}Wl^3.$$

When $2l < x < 3l$, ($P = P_3$ in the figure), the bending moment about P_3 is $M_3 = -\tfrac{1}{2}w(3l-x)^2$,

and

$$EI\frac{d^2y}{dx^2} = -\tfrac{1}{2}w(3l-x)^2.$$ Then

$$C)\quad EIy = -\frac{1}{24}w(3l-x)^4 + C_5x + C_6 = -\frac{1}{24}w(3l-x)^4 - \frac{9}{2}wl^3x - \frac{5}{2}Wl^2x + \frac{27}{8}wl^4 + \frac{3}{2}Wl^3,$$

since, when $x = 2l$, there must be agreement with B) in deflection and slope.

A), B), C) may be written in the form

$$y = \frac{w}{24EI}(12lx^3 - 54l^2x^2 - x^4) + \frac{W}{6EI}(2x^3 - 9lx^2), \qquad 0 \leqq x \leqq l,$$

$$y = \frac{w}{24EI}(12lx^3 - 54l^2x^2 - x^4) + \frac{W}{6EI}(x^3 - 6lx^2 - 3l^2x + l^3), \qquad l \leqq x \leqq 2l,$$

$$y = \frac{w}{24EI}(12lx^3 - 54l^2x^2 - x^4) + \frac{W}{2EI}(3l^3 - 5l^2x), \qquad 2l \leqq x \leqq 3l.$$

The maximum deflection, occurring at R ($x = 3l$), is $-y_{max} = \frac{1}{8EI}(81wl^4 + 48Wl^3)$.

Note that the elastic curve consists of arcs of three distinct curves, the slopes of each pair of arcs at a junction point being equal.

19. A horizontal beam of length l ft is fixed at both ends. Find the equation of the elastic curve and the maximum deflection if it carries a uniform load of w lb/ft of length.

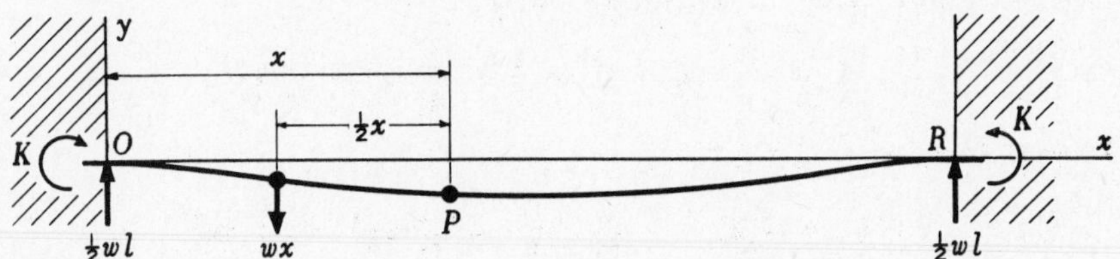

Take the origin at the left end of the beam and let P have coordinates (x,y).

The external forces acting on the segment OP are: a couple of unknown moment K exerted by the wall to keep the beam horizontal at O; an upward thrust of $\tfrac{1}{2}wl$ lb at O, x ft from P; and the load wx lb acting downward at the midpoint of OP, $\tfrac{1}{2}x$ ft from P. Thus,

$$EI\,\frac{d^2y}{dx^2} \;=\; K + \tfrac{1}{2}wlx - \tfrac{1}{2}wx^2.$$

Integrating once and using $x=0,\ dy/dx=0$ at O, $\quad EI\,\dfrac{dy}{dx} \;=\; Kx + \dfrac{1}{4}wlx^2 - \dfrac{1}{6}wx^3.$

At R: $x=l,\ dy/dx=0$ since the beam is fixed there. Then

$$Kl + \frac{1}{4}wl^3 - \frac{1}{6}wl^3 = 0, \quad K = -\frac{1}{12}wl^2, \quad \text{and} \quad EI\,\frac{dy}{dx} = -\frac{1}{12}wl^2x + \frac{1}{4}wlx^2 - \frac{1}{6}wx^3.$$

Integrating and using $x=y=0$ at O,

$$EIy \;=\; -\frac{1}{24}wl^2x^2 + \frac{1}{12}wlx^3 - \frac{1}{24}wx^4 \quad \text{and} \quad y \;=\; \frac{wx^2}{24\,EI}(2lx - l^2 - x^2).$$

The maximum deflection, occurring at the middle of the beam ($x=\frac{1}{2}l$), is $-y_{max} \;=\; \dfrac{wl^4}{384\,EI}$.

20. Solve Problem 19 when in addition there is a weight W lb at the middle of the beam.

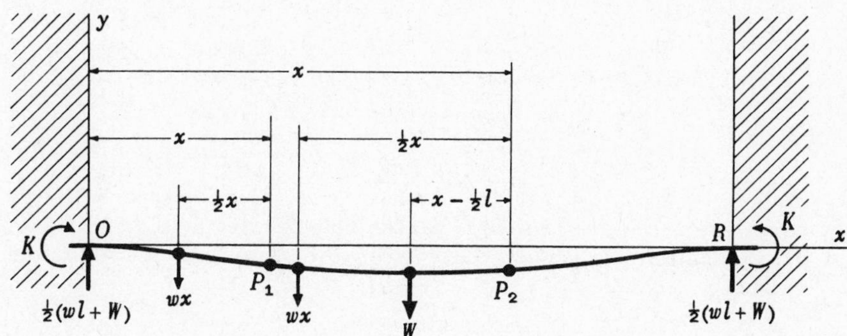

Using the coordinate system of Problem 19, there are two cases to be considered: from $x=0$ to $x=\frac{1}{2}l$ and from $x=\frac{1}{2}l$ to $x=l$.

When $0<x<\frac{1}{2}l$, the external forces on the segment to the left of $P_1(x,y)$ are: a couple of unknown moment K at O; an upward thrust of $\frac{1}{2}(wl+W)$ lb at O, x ft from P_1; and the load wx lb, $\frac{1}{2}x$ ft from P_1. Thus,

$$EI\,\frac{d^2y}{dx^2} \;=\; K + \frac{1}{2}(wl+W)x - \frac{1}{2}wx^2 \;=\; K + \frac{1}{2}wlx - \frac{1}{2}wx^2 + \frac{1}{2}Wx.$$

Integrating once and using $x=0,\ dy/dx=0$ at O,

A') $$EI\,\frac{dy}{dx} \;=\; Kx + \frac{1}{4}wlx^2 - \frac{1}{6}wx^3 + \frac{1}{4}Wx^2.$$

Integrating and using $x=y=0$ at O,

A) $$EIy \;=\; \frac{1}{2}Kx^2 + \frac{1}{12}wlx^3 - \frac{1}{24}wx^4 + \frac{1}{12}Wx^3.$$

When $\frac{1}{2}l<x<l$, there is in addition the weight W lb at the middle of the beam, $(x-\frac{1}{2}l)$ ft from P_2. Thus,

$$EI\,\frac{d^2y}{dx^2} \;=\; K + \frac{1}{2}wlx - \frac{1}{2}wx^2 + \frac{1}{2}Wx - W(x-\tfrac{1}{2}l).$$ Integrating twice,

$B')\qquad EIy = \frac{1}{2}Kx^2 + \frac{1}{12}wlx^3 - \frac{1}{24}wx^4 + \frac{1}{12}Wx^3 - \frac{1}{6}W(x-\frac{1}{2}l)^3 + C_1x + C_2.$

When $x = \frac{1}{2}l$, the values of y and dy/dx for $B')$ must agree respectively with those for $A)$. Thus, $C_1 = C_2 = 0$ and

$B)\qquad EIy = \frac{1}{2}Kx^2 + \frac{1}{12}wlx^3 - \frac{1}{24}wx^4 + \frac{1}{12}Wx^3 - \frac{1}{6}W(x-\frac{1}{2}l)^3.$

To determine K, use $x = \frac{1}{2}l$, $dy/dx = 0$ in $A')$. Then

$\frac{1}{2}lK + \frac{1}{16}wl^3 - \frac{1}{48}wl^3 + \frac{1}{16}Wl^2 = 0$ and $K = -\frac{1}{12}wl^2 - \frac{1}{8}Wl.$ Substituting in $A)$ and $B)$,

$$EIy = -\frac{1}{24}wl^2x^2 + \frac{1}{12}wlx^3 - \frac{1}{24}wx^4 + \frac{1}{12}Wx^3 - \frac{1}{16}Wlx^2 \quad\text{and}$$

$$y = \frac{w}{24EI}(2lx^3 - l^2x^2 - x^4) + \frac{W}{48EI}(4x^3 - 3lx^2), \qquad 0 \leqq x \leqq \tfrac{1}{2}l,$$

$$EIy = -\frac{1}{24}wl^2x^2 + \frac{1}{12}wlx^3 - \frac{1}{24}wx^4 + \frac{1}{12}Wx^3 - \frac{1}{6}W(x-\tfrac{1}{2}l)^3 - \frac{1}{16}Wlx^2 \quad\text{and}$$

$$y = \frac{w}{24EI}(2lx^3 - l^2x^2 - x^4) + \frac{W}{48EI}(l^3 - 6l^2x + 9lx^2 - 4x^3), \qquad \tfrac{1}{2}l \leqq x \leqq l.$$

The maximum deflection, occurring at the middle of the beam, is $-y_{max} = \frac{1}{384EI}(wl^4 + 2Wl^3)$.

21. A horizontal beam of length l ft is fixed at one end and freely supported at the other end. (a) Find the equation of the elastic curve if the beam carries a uniform load w lb/ft of length and a weight W lb at the middle. (b) Locate the point of maximum deflection when $l = 10$ and $W = 10w$.

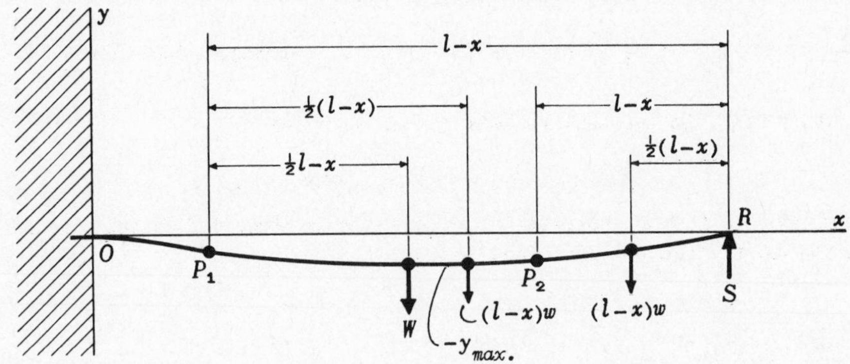

Take the origin at the fixed end and let P have coordinates (x,y). There are two cases to be considered.

When $0 < x < \frac{1}{2}l$, the external forces acting on the segment P_1R are: an unknown upward thrust S lb at R, $(l-x)$ ft from P_1; the load $w(l-x)$ lb at the midpoint of P_1R, $\frac{1}{2}(l-x)$ ft from P_1; and W lb, $(\frac{1}{2}l-x)$ ft from P_1. Thus,

$$EI\frac{d^2y}{dx^2} = S(l-x) - w(l-x)\cdot\tfrac{1}{2}(l-x) - W(\tfrac{1}{2}l-x) = S(l-x) - \tfrac{1}{2}w(l-x)^2 - W(\tfrac{1}{2}l-x).$$

Integrating once and using $x = 0$, $dy/dx = 0$ at O,

$$EI\frac{dy}{dx} = -\frac{1}{2}S(l-x)^2 + \frac{1}{6}w(l-x)^3 + \frac{1}{2}W(\tfrac{1}{2}l-x)^2 + \frac{1}{2}Sl^2 - \frac{1}{6}wl^3 - \frac{1}{8}Wl^2.$$

Integrating again and using $x = y = 0$ at O,

$$A)\quad EIy = \frac{1}{6}S(l-x)^3 - \frac{1}{24}w(l-x)^4 - \frac{1}{6}W(\tfrac{1}{2}l-x)^3 + (\frac{1}{2}Sl^2 - \frac{1}{6}wl^3 - \frac{1}{8}Wl^2)x - \frac{1}{6}Sl^3 + \frac{1}{24}wl^4 + \frac{1}{48}Wl^3.$$

When $\frac{1}{2}l < x < l$, the forces acting on P_2R are the unknown upward thrust S at R, $(l-x)$ ft from P_2 and the load $w(l-x)$ lb, $\frac{1}{2}(l-x)$ ft from P_2. Thus,

$$EI\frac{d^2y}{dx^2} = S(l-x) - \frac{1}{2}w(l-x)^2 \qquad \text{and}$$

$$B')\qquad EIy = \frac{1}{6}S(l-x)^3 - \frac{1}{24}w(l-x)^4 + C_1x + C_2.$$

When $x = \frac{1}{2}l$, the values of EIy and $EI\frac{dy}{dx}$ as given by A) and B') must agree. Hence, C_1 and C_2 in B') have the values of the corresponding constants of integration found in determining A), and B') becomes

$$B)\quad EIy = \frac{1}{6}S(l-x)^3 - \frac{1}{24}w(l-x)^4 + (\frac{1}{2}Sl^2 - \frac{1}{6}wl^3 - \frac{1}{8}Wl^2)x - \frac{1}{6}Sl^3 + \frac{1}{24}wl^4 + \frac{1}{48}Wl^3.$$

To determine S, use $x = l$, $y = 0$ at R in B); then $S = \frac{3}{8}wl + \frac{5}{16}W$. Making this replacement in A) and B),

$$y = \frac{w}{48EI}(5lx^3 - 3l^2x^2 - 2x^4) + \frac{W}{96EI}(11x^3 - 9lx^2), \qquad 0 \leqq x \leqq \tfrac{1}{2}l, \qquad \text{and}$$

$$y = \frac{w}{48EI}(5lx^3 - 3l^2x^2 - 2x^4) + \frac{W}{96EI}(2l^3 - 12l^2x + 15lx^2 - 5x^3), \qquad \tfrac{1}{2}l \leqq x \leqq l.$$

It is clear that the maximum deflection occurs to the right of the midpoint of the beam. When $l = 10$, $W = 10w$, the equation immediately above becomes

$$y = \frac{w}{48EI}(-2x^4 + 25x^3 + 450x^2 - 6000x + 10000).$$

Since $\frac{dy}{dx} = 0$ at the point of maximum deflection, we solve

$$8x^3 - 75x^2 - 900x + 6000 = 0$$

for the real root $x = 5.6$, approximately. Thus, the maximum deflection occurs at the point approximately 5.6 ft from the fixed end.

ELECTRIC CIRCUITS.

22. An electric circuit consists of an inductance of 0.1 henry, a resistance of 20 ohms and a condenser of capacitance 25 microfarads (1 microfarad = 10^{-6} farad). Find the charge q and the current i at time t, given the initial conditions (*a*) $q = 0.05$ coulomb, $i = dq/dt = 0$ when $t = 0$, (*b*) $q = 0.05$ coulomb, $i = -0.2$ ampere when $t = 0$.

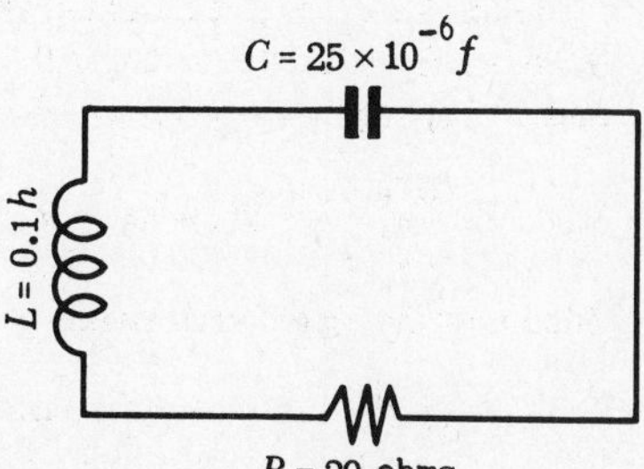

Since $L = 0.1$, $R = 20$, $C = 25\cdot10^{-6}$, $E(t) = 0$,

$$L\frac{d^2q}{dt^2} + R\frac{dq}{dt} + \frac{q}{C} = E(t)$$

reduces to $$\frac{d^2q}{dt^2} + 200\frac{dq}{dt} + 400{,}000q = 0.$$

Integrating, $q = e^{-100t}(A \cos 100\sqrt{39}\,t + B \sin 100\sqrt{39}\,t)$.

Differentiating once with respect to t,

$$i = \frac{dq}{dt} = 100\,e^{-100t}\,[(\sqrt{39}\,B - A)\cos 100\sqrt{39}\,t - (\sqrt{39}\,A + B)\sin 100\sqrt{39}\,t].$$

a) Using the initial conditions $q = 0.05$, $i = 0$ when $t = 0$, $A = 0.05$ and $B = \dfrac{0.05}{\sqrt{39}} = 0.008$.

Hence, $q = e^{-100t}(0.05 \cos 624.5t + 0.008 \sin 624.5t)$

and $i = -0.32\,e^{-100t} \sin 624.5\,t$.

b) Using the initial conditions $q = 0.05$, $i = -0.2$ when $t = 0$, $A = 0.05$ and $B = 0.0077$.

Hence, $q = e^{-100t}(0.05 \cos 624.5t + 0.0077 \sin 624.5t)$

and $i = e^{-100t}(-0.2 \cos 624.5t - 32.0 \sin 624.5t)$

Note that q and i are transients, each becoming negligible very quickly.

23. A circuit consists of an inductance of 0.05 henry, a resistance of 20 ohms, a condenser of capacitance 100 microfarads, and an emf of $E = 100$ volts. Find i and q, given the initial conditions $q = 0$, $i = 0$ when $t = 0$.

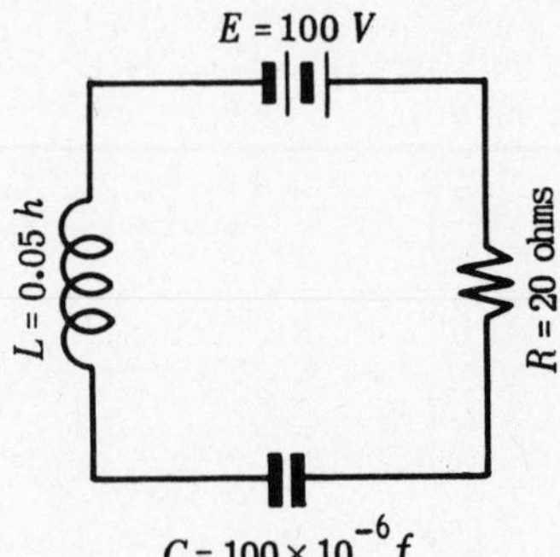

Here $0.05\dfrac{d^2q}{dt^2} + 20\dfrac{dq}{dt} + \dfrac{q}{100\cdot 10^{-6}} = 100$

or $\dfrac{d^2q}{dt^2} + 400\dfrac{dq}{dt} + 200{,}000q = 2000.$

Integrating, $q = e^{-200t}(A \cos 400t + B \sin 400t) + 0.01$.

Differentiating once with respect to t,

$$i = \frac{dq}{dt} = 200e^{-200t}[(-A + 2B)\cos 400t + (-B - 2A)\sin 400t].$$

Using the initial conditions: $A = -0.01$, $-A + 2B = 0$, and $B = -0.005$.

Then $q = e^{-200t}(-0.01 \cos 400t - 0.005 \sin 400t) + 0.01$

and $i = 5e^{-200t} \sin 400t$.

Here i becomes negligible very soon while q, for all purposes, becomes $q = 0.01$.

24. Solve Problem 23 assuming that there is a variable emf of $E(t) = 100 \cos 200t$.

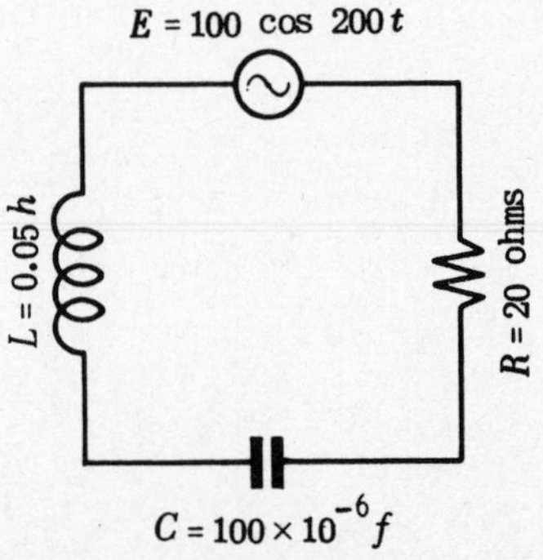

The differential equation is now

$\dfrac{d^2q}{dt^2} + 400\dfrac{dq}{dt} + 200{,}000q = 2000 \cos 200t.$ Then

$$q = e^{-200t}(A \cos 400t + B \sin 400t) + 0.01 \cos 200t + 0.005 \sin 200t$$ and

$$i = e^{-200t}[(-200A + 400B)\cos 400t + (-200B - 400A)\sin 400t] - 2 \sin 200t + \cos 200t.$$

Using the initial conditions: $A = -0.01$, $-200A + 400B + 1 = 0$ and $B = -0.0075$. Then

$$q = e^{-200t}(-0.01\cos 400t - 0.0075\sin 400t) + 0.01\cos 200t + 0.005\sin 200t$$

and

$$i = e^{-200t}(-\cos 400t + 5.5\sin 400t) - 2\sin 200t + \cos 200t.$$

Here the transient parts of q and i very quickly become negligible. For this reason, when the transients may be neglected, one needs find only the steady-state solutions

$$q = 0.01\cos 200t + 0.005\sin 200t \quad \text{and} \quad i = \cos 200t - 2\sin 200t.$$

The frequency $200/2\pi$ cycles/sec of the steady-state solutions is equal to the frequency of the applied emf. (See also Problem 25.)

25. For a circuit consisting of an inductance L, a resistance R, a capacitance C, and an emf $E(t) = E_0 \sin \omega t$, derive the formula for the steady-state current

$$i = \frac{E_0}{Z}(\frac{R}{Z}\sin\omega t - \frac{X}{Z}\cos\omega t) = \frac{E_0}{Z}\sin(\omega t - \theta),$$

where $X = L\omega - \dfrac{1}{C\omega}$, $Z = \sqrt{X^2 + R^2}$, and θ is determined from $\sin\theta = \dfrac{X}{Z}$ and $\cos\theta = \dfrac{R}{Z}$.

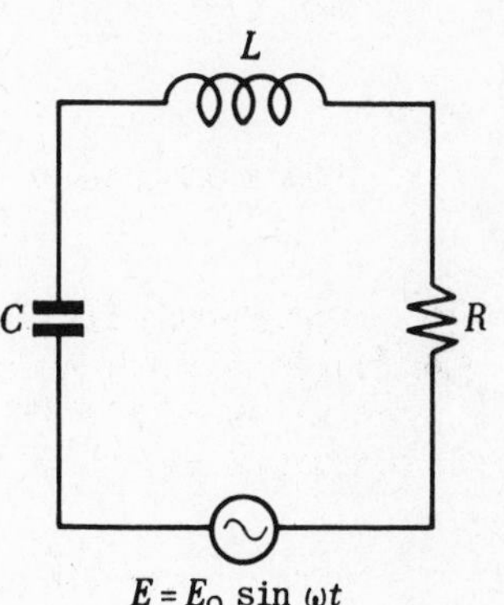

By differentiating $L\dfrac{d^2q}{dt^2} + R\dfrac{dq}{dt} + \dfrac{q}{C} = E_0\sin\omega t$

and using $i = \dfrac{dq}{dt}$, we obtain

1) $$L\frac{d^2i}{dt^2} + R\frac{di}{dt} + \frac{i}{C} = (LD^2 + RD + 1/C)i = \omega E_0\cos\omega t.$$

The required steady-state solution is the particular integral of 1):

$$i = \frac{\omega E_0}{LD^2 + RD + 1/C}\cos\omega t = \frac{\omega E_0}{RD - (L\omega - \frac{1}{C\omega})\omega}\cos\omega t$$

$$= \frac{\omega E_0(RD + X\omega)}{R^2D^2 - X^2\omega^2}\cos\omega t = \frac{E_0}{R^2 + X^2}(R\sin\omega t - X\cos\omega t)$$

$$= \frac{E_0}{Z}(\frac{R}{Z}\sin\omega t - \frac{X}{Z}\cos\omega t) = \frac{E_0}{Z}\sin(\omega t - \theta).$$

X is called the *reactance* of the circuit; when $X = 0$, the amplitude of i is greatest (the circuit is in resonance). Z, called the *impedance* of the circuit, is also the ratio of the amplitudes of the emf and the current. θ is called the phase angle.

At times $t = \pi/2\omega, 3\pi/2\omega, \cdots\cdots$ the emf attains maximum amplitude, while at times given by $\omega t - \theta = \pi/2, 3\pi/2, \cdots\cdots$, that is, when $t = \dfrac{\pi/2 + \theta}{\omega}, \dfrac{3\pi/2 + \theta}{\omega}, \cdots\cdots$ the current attains maximum amplitude. Thus the voltage *leads* the current by a time θ/ω or the current and voltage are out of phase by the phase angle θ.

Note that $\theta = 0$ when $X = 0$, that is, $\theta = 0$ if there is resonance.

26. The circuit consisting of an inductance L, a condenser of capacitance C, and an emf E is known as an harmonic oscillator. Find q and i when $E = E_0 \cos \omega t$ and the initial conditions are $q = q_0$, $i = i_0$ when $t = 0$.

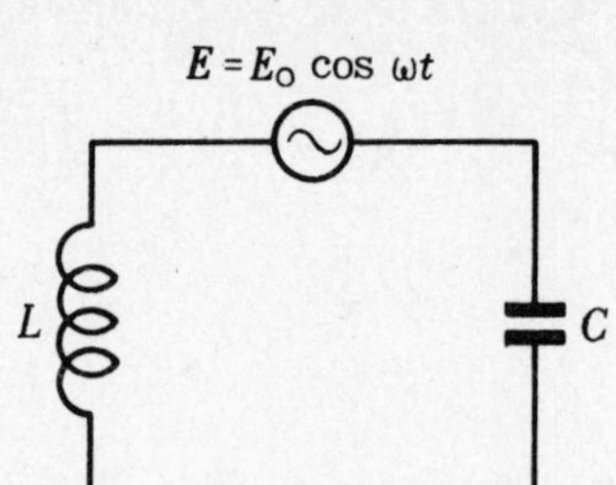

Since $R = 0$, the differential equation is

$$\frac{d^2q}{dt^2} + \frac{q}{CL} = \frac{E_0}{L}\cos \omega t.$$

There are two cases to be considered:

$$(a)\ \omega \neq \frac{1}{\sqrt{CL}} \quad \text{and} \quad (b)\ \omega = \frac{1}{\sqrt{CL}}.$$

a)
$$q = A\cos\frac{1}{\sqrt{CL}}t + B\sin\frac{1}{\sqrt{CL}}t + \frac{E_0}{L}\frac{1}{D^2+1/CL}\cos\omega t$$

$$= A\cos\frac{1}{\sqrt{CL}}t + B\sin\frac{1}{\sqrt{CL}}t + \frac{E_0 C}{1-\omega^2 CL}\cos\omega t$$

and
$$i = \frac{1}{\sqrt{CL}}(-A\sin\frac{1}{\sqrt{CL}}t + B\cos\frac{1}{\sqrt{CL}}t) - \frac{E_0 C\omega}{1-\omega^2 CL}\sin\omega t.$$

Using the initial conditions: $A = q_0 - \dfrac{E_0 C}{1-\omega^2 CL}$ and $B = \sqrt{CL}\, i_0$. Then

$$q = (q_0 - \frac{E_0 C}{1-\omega^2 CL})\cos\frac{1}{\sqrt{CL}}t + \sqrt{CL}\, i_0 \sin\frac{1}{\sqrt{CL}}t + \frac{E_0 C}{1-\omega^2 CL}\cos\omega t$$

and
$$i = i_0\cos\frac{1}{\sqrt{CL}}t - \frac{1}{\sqrt{CL}}(q_0 - \frac{E_0 C}{1-\omega^2 CL})\sin\frac{1}{\sqrt{CL}}t - \frac{E_0 C\omega}{1-\omega^2 CL}\sin\omega t.$$

b) Here
$$\frac{d^2q}{dt^2} + \omega^2 q = \frac{E_0}{L}\cos\omega t.$$

Then
$$q = A\cos\omega t + B\sin\omega t + \frac{E_0}{2L\omega}t\sin\omega t$$

and
$$i = \omega(-A\sin\omega t + B\cos\omega t) + \frac{E_0}{2L}(\frac{1}{\omega}\sin\omega t + t\cos\omega t).$$

Using the initial conditions: $A = q_0$ and $B = i_0/\omega$.

Then
$$q = q_0\cos\omega t + \frac{i_0}{\omega}\sin\omega t + \frac{E_0}{2L\omega}t\sin\omega t$$

and
$$i = i_0\cos\omega t - q_0\omega\sin\omega t + \frac{E_0}{2L}(\frac{1}{\omega}\sin\omega t + t\cos\omega t).$$

Note that in (b) the frequency of the emf is the natural frequency of the oscillator, that is, the frequency when there is no emf. The circuit is in resonance since the reactance $X = L\omega - \dfrac{1}{C\omega} = 0$ when $\omega = \dfrac{1}{\sqrt{CL}}$. The presence of the term $\dfrac{E_0 t}{2L}\cos\omega t$, whose amplitude increases with t, indicates that eventually such a circuit will be destroyed.

SUPPLEMENTARY PROBLEMS

27. Determine the curve for which the radius of curvature is proportional to the slope of the tangent.

$$Ans. \quad y = \pm\left(\sqrt{k^2-(x+C_1)^2} + k\ln\frac{\sqrt{k^2-(x+C_1)^2}-k}{x+C_1}\right) + C_2$$

28. A 6 inch pendulum is released with a velocity of 1/2 rad/sec, toward the vertical, from a position 1/5 rad from the vertical. Find the equation of motion.

$$Ans. \quad \theta = \frac{1}{5}\cos 8t - \frac{1}{16}\sin 8t$$

29. A particle of mass m is repelled from O with a force equal to $k > 0$ times the distance from O. If the particle starts from rest at a distance a from O, find its position t sec later.

$$Ans. \quad x = \tfrac{1}{2}a\left(e^{\sqrt{k/m}\,t} + e^{-\sqrt{k/m}\,t}\right)$$

30. If, in Problem 29, $k = m$ and $a = 12$ ft, determine a) the distance from O and the velocity when $t = 2$ sec, b) when it will be 18 ft from O and its velocity then.

Ans. a) $x = 45.1$ ft, $v = 43.5$ ft/sec; b) $t = 0.96$ sec, $v = 13.4$ ft/sec

31. A chain hangs over a smooth peg, 8 ft on one side and 10 ft on the other. If the force of friction is equal to the weight of 1 ft of chain, find the time required for it to slide off.

$$Ans. \quad \frac{3}{\sqrt{g}}\ln(17 + 12\sqrt{2}) \text{ sec}$$

32. When the inner of two concentric spheres of radii r_1 and r_2, $r_1 < r_2$, carries an electric charge, the differential equation for the potential V at any point between the two spheres at a distance r from their common center is

$$\frac{d^2V}{dr^2} + \frac{2}{r}\frac{dV}{dr} = 0.$$

Solve for V given $V = V_1$ when $r = r_1$ and $V = V_2$ when $r = r_2$.

$$Ans. \quad V = \frac{V_2 r_2(r-r_1) - V_1 r_1(r-r_2)}{r(r_2-r_1)}$$

33. A spring is such that it would be stretched 3 in. by a 9 lb weight. A 24 lb weight is attached and brought to rest. Find the equation of the motion if the weight is then

a) pulled down 4 in. and released.
b) pulled down 2 in. and given an upward velocity of 2 ft/sec.
c) pulled down 3 in. and given a downward velocity of 4 ft/sec.
d) pushed up 3 in. and released.
e) pushed up 4 in. and given an upward velocity of 5 ft/sec.

Ans. a) $x = \frac{1}{3}\cos 4\sqrt{3}\,t$, b) $x = \frac{1}{6}\cos 4\sqrt{3}\,t - \frac{\sqrt{3}}{6}\sin 4\sqrt{3}\,t$, c) $x = \frac{1}{4}\cos 4\sqrt{3}\,t + \frac{\sqrt{3}}{3}\sin 4\sqrt{3}\,t$,

d) $x = -\frac{1}{4}\cos 4\sqrt{3}\,t$, e) $x = -\frac{1}{3}\cos 4\sqrt{3}\,t - \frac{5\sqrt{3}}{12}\sin 4\sqrt{3}\,t$

34. A spring is such that it would be stretched 3 in. by a 30 lb weight. A 64 lb weight is attached and brought to rest. The resistance of the medium is numerically equal to 8 dx/dt lb. Find the equation of the motion of the weight if
a) it is started downward with velocity 10 ft/sec.
b) it is pulled down 6 in. and given an upward velocity of 10 ft/sec.

Ans. *a*) $x = \frac{5\sqrt{14}}{14}e^{-2t}\sin 2\sqrt{14}\,t$, *b*) $x = e^{-2t}(\frac{1}{2}\cos 2\sqrt{14}\,t - \frac{9\sqrt{14}}{28}\sin 2\sqrt{14}\,t)$

35. A spring is such that it would be stretched 6 in. by a 3 lb weight. A 3 lb weight is attached and brought to rest. The weight is then pulled down 3 in. and released. Determine the equation of motion if
a) an impressed force $\frac{3}{2}\sin 6t$ acts on the spring. *Ans.* $x = \frac{1}{4}\cos 8t - \frac{3}{7}\sin 8t + \frac{4}{7}\sin 6t$
b) an impressed force $\frac{3}{2}\sin 8t$ acts on the spring. *Ans.* $x = \frac{1}{4}(1-4t)\cos 8t + \frac{1}{8}\sin 8t$

36. A beam of length l ft is fixed at one end and otherwise unsupported. Find the equation of the elastic curve and the maximum deflection if there is a uniform load of w lb/ft of length and a load W lb at the free end.

Ans. $y = \frac{w}{24EI}(4lx^3 - 6l^2x^2 - x^4) + \frac{W}{6EI}(x^3 - 3lx^2)$, $-y_{max} = \frac{1}{24EI}(3wl^4 + 8Wl^3)$

37. A beam of length $2l$ ft is freely supported at both ends and carries a uniform load of w lb/ft of length. Taking the origin at the midpoint (low point) of the beam, find the equation of the elastic curve and the maximum deflection. Compare with Problem 15.

Hint: $EIy'' = wl(l-x) - \frac{1}{2}w(l-x)^2 = \frac{1}{2}w(l^2 - x^2)$ and $y = y' = 0$ when $x = 0$.

Ans. $y = \frac{w}{24EI}(6l^2x^2 - x^4)$, $y_{max} = \frac{5wl^4}{24EI}$

38. A beam of length $3l$ ft is freely supported at both ends. There is a uniform load of w lb/ft of length and loads of W lb at a distance l ft from each end. Taking the origin as in Problem 37, find the maximum deflection.

Hint: $M = \frac{w}{2}(\frac{9l^2}{4} - x^2) + W(\frac{3l}{2} - x)$, $\frac{l}{2} < x < \frac{3l}{2}$; and $M = \frac{w}{2}(\frac{9l^2}{4} - x^2) + Wl$, $0 < x < \frac{l}{2}$.

Ans. $y_{max} = \frac{1}{384EI}(405wl^4 + 368Wl^3)$

39. A circuit consists of an inductance of 0.05 henry, a resistance of 5 ohms, and a condenser of capacitance $4(10)^{-4}$ farad. If $q = i = 0$ when $t = 0$, find q and i in terms of t when *a*) there is a constant emf = 110 volts, *b*) there is an alternating emf = 200 cos $100t$. Find the steady state solutions in *b*).

Ans. *a*) $q = e^{-50t}(-\frac{11}{250}\cos 50\sqrt{19}\,t - \frac{11\sqrt{19}}{4750}\sin 50\sqrt{19}\,t) + \frac{11}{250}$, $i = \frac{44\sqrt{19}}{19}e^{-50t}\sin 50\sqrt{19}\,t$

b) $q = e^{-50t}(-\frac{16}{170}\cos 50\sqrt{19}\,t - \frac{12\sqrt{19}}{1615}\sin 50\sqrt{19}\,t) + \frac{4}{170}(4\cos 100t + \sin 100t)$,

$i = e^{-50t}(-\frac{40}{17}\cos 50\sqrt{19}\,t + \frac{1640\sqrt{19}}{323}\sin 50\sqrt{19}\,t) + \frac{40}{17}(\cos 100t - 4\sin 100t)$

40. Solve Problem 39 after replacing the 5 ohm resistance with a 50 ohm resistance.

Ans. *a*) $q = -0.047e^{-53t} + 0.0026e^{-947t} + 0.044$, $i = 2.46(e^{-53t} - e^{-947t})$

b) $q = -0.018e^{-53t} + 0.005e^{-947t} + 0.034\sin 100t + 0.014\cos 100t$,

$i = 0.98e^{-53t} - 4.43e^{-947t} + 3.45\cos 100t - 1.38\sin 100t$

CHAPTER 21

Systems of Simultaneous Linear Equations

IN PREVIOUS CHAPTERS, differential equations involving only two variables have been treated. In the next several chapters, equations involving more than two variables will be considered. If but one of the variables is independent, the equations are ordinary differential equations; if more of the variables are independent, the equations are called partial differential equations. In this chapter we shall be concerned with systems of ordinary linear differential equations with *constant* coefficients such as

$$A)\ \begin{cases} 2\dfrac{dx}{dt} + \dfrac{dy}{dt} - 4x - y = e^t \\[2ex] \dfrac{dx}{dt} \qquad\quad + 3x + y = 0 \end{cases} \qquad \text{or } A')\ \begin{cases} 2(D-2)x + (D-1)y = e^t \\ (D+3)x + \quad y \quad = 0, \end{cases} \quad \text{where } D = \frac{d}{dt}$$

and

$$B)\ \begin{cases} \dfrac{dx}{dt} + \dfrac{dy}{dt} + y = 1 \\[2ex] \dfrac{dx}{dt} - \dfrac{dz}{dt} + 2x + z = 1 \\[2ex] \dfrac{dy}{dt} + \dfrac{dz}{dt} + y + 2z = 0 \end{cases} \qquad \text{or } B')\ \begin{cases} Dx \quad + (D+1)y = 1 \\ (D+2)x - (D-1)z = 1 \\ (D+1)y + (D+2)z = 0 \end{cases}$$

in which the number of simultaneous equations is equal to the number of dependent variables.

THE BASIC PROCEDURE for solving a system of n ordinary differential equations in n dependent variables consists in obtaining, by differentiating the given equations, a set from which all but one of the dependent variables, say x, can be eliminated. The equation resulting from the elimination is then solved for this variable x. Each of the dependent variables is obtained in a similar manner.

EXAMPLE. Consider system A): 1) $2\dfrac{dx}{dt} + \dfrac{dy}{dt} - 4x - y = e^t$, 2) $\dfrac{dx}{dt} + 3x + y = 0$.

Solution 1.

First, we note that the general solution $x = x(t)$, $y = y(t)$ of this system will also satisfy

$$3)\qquad \frac{d^2x}{dt^2} + 3\frac{dx}{dt} + \frac{dy}{dt} = 0$$

obtained by differentiating 2). Moreover, multiplying 1) by −1, 2) by −1, 3) by 1, and adding, we obtain

$$4)\qquad \frac{d^2x}{dt^2} + x = -e^t$$

which is also satisfied by $x = x(t)$, $y = y(t)$. This latter differential equation, being free of y and its derivatives, may be solved readily; thus,

$$x = C_1 \cos t + C_2 \sin t - \frac{1}{D^2+1} e^t = C_1 \cos t + C_2 \sin t - \tfrac{1}{2}e^t.$$

To find y in a similar manner, we differentiate 1) to obtain

5) $$2\frac{d^2x}{dt^2} + \frac{d^2y}{dt^2} - 4\frac{dx}{dt} - \frac{dy}{dt} = e^t$$

and between this and equations 1),2),3) eliminate x and its derivatives. However, it is simpler here to proceed as follows. From 2), we have

$$y = -\frac{dx}{dt} - 3x = -(-C_1 \sin t + C_2 \cos t - \tfrac{1}{2}e^t) - 3(C_1 \cos t + C_2 \sin t - \tfrac{1}{2}e^t)$$
$$= (C_1 - 3C_2)\sin t - (3C_1 + C_2)\cos t + 2e^t.$$

Thus, $x = C_1 \cos t + C_2 \sin t - \tfrac{1}{2}e^t$, $y = (C_1 - 3C_2)\sin t - (3C_1 + C_2)\cos t + 2e^t$ is the general solution.

When the equations are written in the D notation, there is a striking similarity between the procedures used here and the method of solving a system of n equations in n unknowns. This is due to the fact, noted in previous chapters, that the operator D may at times be treated as a variable (letter).

Solution 2. Consider the system A'): 1) $2(D-2)x + (D-1)y = e^t$

2) $(D+3)x + y = 0.$

Proceeding as in the case of two equations in two unknowns x and y, we multiply 2) by $D-1$. Actually, we operate on 2) with $D-1 = (\frac{d}{dt} - 1)$, to get

$$(D-1)(D+3)x + (D-1)y = 0$$

and subtract 1) from it to obtain

$$[(D-1)(D+3) - 2(D-2)]x = -e^t \quad \text{or} \quad (D^2+1)x = -e^t.$$

Now this is 4) above as might have been anticipated, since operating on 2) with $D-1$ is equivalent to differentiating 2) and adding -1 times 2) as in the previous solution. The general solution is obtained as in Solution 1.

Solution 3. We may also effect a solution using determinants. From system A') we obtain

$$\begin{vmatrix} 2(D-2) & D-1 \\ D+3 & 1 \end{vmatrix} x = \begin{vmatrix} e^t & D-1 \\ 0 & 1 \end{vmatrix} \quad \text{and} \quad \begin{vmatrix} 2(D-2) & D-1 \\ D+3 & 1 \end{vmatrix} y = \begin{vmatrix} 2(D-2) & e^t \\ D+3 & 0 \end{vmatrix}$$

or $$(D^2+1)x = -e^t \quad \text{and} \quad (D^2+1)y = 4e^t.$$

The first of these equations is 4) above, and the second would have been obtained by the procedure rejected in Solution 1. We shall now show why it was rejected. When the two equations are solved, we have

6) $x = C_1 \cos t + C_2 \sin t - \tfrac{1}{2}e^t$ and 7) $y = C_3 \cos t + C_4 \sin t + 2e^t$.

We know from Solution 1 that 6) and 7) contain extraneous solutions. To eliminate them (that is, to reduce the number of arbitrary constants), we substitute in 2) and see that

$$(C_2 + 3C_1 + C_3)\cos t + (3C_2 - C_1 + C_4)\sin t = 0$$

for every value of t. Thus,

$$C_3 = -(3C_1 + C_2) \quad \text{and} \quad C_4 = C_1 - 3C_2.$$

When these values are substituted in 6) and 7), we obtain the general solution found above.

THE NUMBER OF INDEPENDENT ARBITRARY CONSTANTS appearing in the general solution of the system

$$f_1(D)x + g_1(D)y = h_1(t)$$
$$f_2(D)x + g_2(D)y = h_2(t)$$

is equal to the degree in D of the determinant $\Delta = \begin{vmatrix} f_1(D) & g_1(D) \\ f_2(D) & g_2(D) \end{vmatrix}$,

provided Δ does not vanish identically. If $\Delta \equiv 0$, the system is dependent; such systems will not be considered here.

For the system A'), $\Delta = \begin{vmatrix} 2(D-2) & D-1 \\ D+3 & 1 \end{vmatrix} = -(D^2+1)$.

The degree (2) in D agrees with the number of arbitrary constants appearing in the general solution.

The theorem may be extended readily to the case of n equations in n dependent variables.

SOLVED PROBLEMS

1. Solve the system: 1) $(D-1)x + Dy = 2t + 1$

2) $(2D+1)x + 2Dy = t$.

Subtracting twice 1) from 2), we have $3x = -3t - 2$. Substituting $x = -t - 2/3$ in 1), we obtain $Dy = 2t + 1 - (D-1)x = t + \frac{4}{3}$ and $y = \frac{1}{2}t^2 + \frac{4}{3}t + C_1$.

The complete solution is $x = -t - \frac{2}{3}$, $y = \frac{1}{2}t^2 + \frac{4}{3}t + C_1$.

Note that $\begin{vmatrix} D-1 & D \\ 2D+1 & 2D \end{vmatrix}$ is of degree 1 in D and there is but one arbitrary constant.

2. Solve the system: 1) $(D+2)x + 3y = 0$

2) $3x + (D+2)y = 2e^{2t}$.

Operating on 1) with $D+2$, multiplying 2) by -3, and adding: $(D^2+4D-5)x = -6e^{2t}$.

Then $x = C_1e^t + C_2e^{-5t} - \frac{6}{7}e^{2t}$. From 1), $y = -\frac{1}{3}(D+2)x = -C_1e^t + C_2e^{-5t} + \frac{8}{7}e^{2t}$.

3. Solve the system: 1) $(D-3)x + 2(D+2)y = 2\sin t$

2) $2(D+1)x + (D-1)y = \cos t$.

Operating on 1) with $D-1$ and on 2) with $2(D+2)$, we have

3) $(D-1)(D-3)x + 2(D-1)(D+2)y = (D-1)[2\sin t] = 2\cos t - 2\sin t$

4) $4(D+2)(D+1)x + 2(D+2)(D-1)y = 2(D+2)\cos t = 4\cos t - 2\sin t$.

Subtracting 3) from 4) and noting that $(D-1)(D+2) = (D+2)(D-1)$, since the operators have constant coefficients,

$$[4(D^2+3D+2) - (D^2-4D+3)]x = (3D^2+16D+5)x = 2\cos t$$

and $x = C_1e^{-5t} + C_2e^{-t/3} + \dfrac{2}{3D^2+16D+5}\cos t = C_1e^{-5t} + C_2e^{-t/3} + \dfrac{1}{8D+1}\cos t$

$= C_1e^{-5t} + C_2e^{-t/3} + (8\sin t + \cos t)/65.$

From 2), $(D-1)y = \cos t - 2(D+1)x$

$= \cos t + 8C_1e^{-5t} - \frac{4}{3}C_2e^{-t/3} - (18\cos t + 14\sin t)/65$

$= 8C_1e^{-5t} - \frac{4}{3}C_2e^{-t/3} + (47\cos t - 14\sin t)/65.$

Then $ye^{-t} = \displaystyle\int\left(8C_1e^{-6t} - \frac{4}{3}C_2e^{-4t/3} + \frac{47\cos t - 14\sin t}{65}e^{-t}\right)dt$

$= -\frac{4}{3}C_1e^{-6t} + C_2e^{-4t/3} + \dfrac{61\sin t - 33\cos t}{130}e^{-t} + C_3$

and $y = -\frac{4}{3}C_1e^{-5t} + C_2e^{-t/3} + \dfrac{61\sin t - 33\cos t}{130} + C_3e^{t}.$

Since the degree of Δ is 2, the general solution has but two arbitrary constants. Hence, when these expressions are substituted for x and y in 1) it is found that $C_3 = 0$. Then

$$x = C_1e^{-5t} + C_2e^{-t/3} + \frac{8\sin t + \cos t}{65}, \qquad y = -\frac{4}{3}C_1e^{-5t} + C_2e^{-t/3} + \frac{61\sin t - 33\cos t}{130}$$

is the general solution.

4. Solve the system: 1) $(D^2-2)x - 3y = e^{2t}$
2) $(D^2+2)y + x = 0.$

Find the particular solution satisfying the conditions $x = y = 1$, $Dx = Dy = 0$ when $t = 0$.

Operating on 1) with D^2 to obtain $D^4x - 2D^2x - 3D^2y = 4e^{2t}$ and making the replacements $D^2x = 2x + 3y + e^{2t}$ from 1) and $D^2y = -x - 2y$ from 2), we have $(D^4 - 1)x = 6e^{2t}$.

Then $x = C_1e^{t} + C_2e^{-t} + C_3\cos t + C_4\sin t + \frac{2}{5}e^{2t}$ and, using 1),

$$y = \frac{1}{3}[(D^2-2)x - e^{2t}] = -\frac{1}{3}(C_1e^{t} + C_2e^{-t}) - (C_3\cos t + C_4\sin t) - \frac{1}{15}e^{2t}.$$

Note that x could also be obtained by the use of determinants. Thus,

$$\begin{vmatrix} D^2-2 & -3 \\ 1 & D^2+2 \end{vmatrix} x = \begin{vmatrix} e^{2t} & 3 \\ 0 & D^2+2 \end{vmatrix} \quad\text{or}\quad (D^4-1)x = 6e^{2t}, \text{ etc.}$$

When $t = 0$, $x = C_1 + C_2 + C_3 + \frac{2}{5} = 1$ and $Dx = C_1 - C_2 + C_4 + \frac{4}{5} = 0,$

$y = -\frac{1}{3}(C_1 + C_2) - C_3 - \frac{1}{15} = 1$ and $Dy = -\frac{1}{3}(C_1 - C_2) - C_4 - \frac{2}{15} = 0.$

Then $C_1 = 3/4$, $C_2 = 7/4$, $C_3 = -19/10$, $C_4 = 1/5$, and the required particular solution is

$$x = \frac{1}{4}(3e^{t} + 7e^{-t}) - \frac{1}{10}(19\cos t - 2\sin t) + \frac{2}{5}e^{2t},$$

$$y = -\frac{1}{12}(3e^{t} + 7e^{-t}) + \frac{1}{10}(19\cos t - 2\sin t) - \frac{1}{15}e^{2t}.$$

5. Solve the system: 1) $(D+1)x + (D-1)y = e^t$, 2) $(D^2+D+1)x + (D^2-D+1)y = t^2$.

Operating on 1) with D^2+D+1 and on 2) with $D+1$, and subtracting, we have

$$2y = t^2 + 2t - 3e^t \qquad \text{and} \qquad y = \frac{1}{2}t^2 + t - \frac{3}{2}e^t.$$

Operating on 1) with D^2-D+1 and on 2) with $D-1$, and subtracting, we have

$$2x = t^2 - 2t + e^t \qquad \text{and} \qquad x = \frac{1}{2}t^2 - t + \frac{1}{2}e^t.$$

Note that $\begin{vmatrix} D+1 & D-1 \\ D^2+D+1 & D^2-D+1 \end{vmatrix} = 2$ is of degree 0 in D; hence, there are no arbitrary constants in the solution.

6. Solve the system: 1) $D^2x - m^2y = 0$, 2) $D^2y + m^2x = 0$.

Operating on 1) with D^2 and substituting $D^2y = -m^2x$ from 2), we obtain

$$D^4x - m^2(-m^2x) = D^4x + m^4x = (D^4 + m^4)x = 0. \quad \text{Then } D = \pm\frac{m}{\sqrt{2}}(1 \pm i)$$

and $x = e^{mt/\sqrt{2}}(C_1 \cos mt/\sqrt{2} + C_2 \sin mt/\sqrt{2}) + e^{-mt/\sqrt{2}}(C_3 \cos mt/\sqrt{2} + C_4 \sin mt/\sqrt{2})$.

Substituting for x in 1) and solving,

$$y = \frac{1}{m^2}D^2x = e^{mt/\sqrt{2}}(C_2 \cos mt/\sqrt{2} - C_1 \sin mt/\sqrt{2}) + e^{-mt/\sqrt{2}}(C_3 \sin mt/\sqrt{2} - C_4 \cos mt/\sqrt{2}).$$

7. Solve the system: 1) $(D^2+4)x - 3Dy = 0$, 2) $3Dx + (D^2+4)y = 0$.

Operating on 1) with D^2+4 and on 2) with $3D$, and adding, we have

$[(D^2+4)^2 + 9D^2]x = (D^2+16)(D^2+1)x = 0$ and $x = C_1 \cos 4t + C_2 \sin 4t + C_3 \cos t + C_4 \sin t$.

Operating on 1) with $-3D$ and on 2) with D^2+4, and adding, we have

$(D^2+16)(D^2+1)y = 0$ and $y = K_1 \cos 4t + K_2 \sin 4t + K_3 \cos t + K_4 \sin t$.

To eliminate the extraneous solutions, substitute for x and y in 1). We have

$$-12C_1 \cos 4t - 12C_2 \sin 4t + 3C_3 \cos t + 3C_4 \sin t + 12K_1 \sin 4t - 12K_2 \cos 4t + 3K_3 \sin t - 3K_4 \cos t = 0$$

for all values of t; thus, $K_1 = C_2$, $K_2 = -C_1$, $K_3 = -C_4$, $K_4 = C_3$.

The complete solution is:

$$x = C_1 \cos 4t + C_2 \sin 4t + C_3 \cos t + C_4 \sin t,$$
$$y = C_2 \cos 4t - C_1 \sin 4t - C_4 \cos t + C_3 \sin t.$$

8. Solve the system:

1) $Dx + (D+1)y = 1$
2) $(D+2)x - (D-1)z = 1$
3) $(D+1)y + (D+2)z = 0$.

Subtracting 3) from 1), we have 4) $Dx - (D+2)z = 1$ which is free of y.

Operating on 2) with D and on 4) with $D+2$, and subtracting, we have $(5D+4)z = -2$; then $z = -\frac{1}{2} + C_1e^{-4t/5}$. Substituting for z in 3), $(D+1)y = -(D+2)z = 1 - \frac{6}{5}C_1e^{-4t/5}$; then

$$y = e^{-t}\int(e^t - \frac{6}{5}C_1e^{t/5})dt = e^{-t}(e^t - 6C_1e^{t/5} + C_2) = 1 - 6C_1e^{-4t/5} + C_2e^{-t}.$$

Substituting for y in 1), $Dx = 1-(D+1)y = \frac{6}{5}C_1e^{-4t/5}$; then $x = -\frac{3}{2}C_1e^{-4t/5} + C_3$.

Since $\begin{vmatrix} D & D+1 & 0 \\ D+2 & 0 & -(D-1) \\ 0 & D+1 & D+2 \end{vmatrix} = -(5D^2+9D+4)$ is of degree 2 in D, there are but two arbitrary constants in the general solution. Substituting for x and z in 2), we obtain

$(\frac{6}{5}C_1e^{-4t/5} - 3C_1e^{-4t/5} + 2C_3) - (-\frac{4}{5}C_1e^{-4t/5} + \frac{1}{2} - C_1e^{-4t/5}) = 1$ and, hence, $C_3 = \frac{3}{4}$. Thus,

$$x = \frac{3}{4} - \frac{3}{2}C_1e^{-4t/5}, \quad y = 1 - 6C_1e^{-4t/5} + C_2e^{-t}, \quad z = -\frac{1}{2} + C_1e^{-4t/5}$$

is the general solution.

9. Solve the system:

$$\begin{aligned} &1)\quad (D+1)^2x + 2Dy + 3Dz = 1 \\ &2)\quad Dx + z = 0 \\ &3)\quad x - Dy - Dz = 0. \end{aligned}$$

Find the particular solution for which $x = z = 1$, $y = 0$ when $t = 0$.

First, operate on 2) with D to obtain 4) $D^2x + Dz = 0$.

Next, add twice 3) to 1) and subtract 4) to get $(2D+3)x = 1$; then

$$xe^{3t/2} = \frac{1}{2}\int e^{3t/2}\,dt = \frac{1}{3}e^{3t/2} + C_1 \quad \text{and} \quad x = \frac{1}{3} + C_1e^{-3t/2}.$$

From 2),
$$z = -Dx = \frac{3}{2}C_1e^{-3t/2}.$$

From 3), $Dy = x - Dz = \frac{1}{3} + C_1e^{-3t/2} + \frac{9}{4}C_1e^{-3t/2} = \frac{1}{3} + \frac{13}{4}C_1e^{-3t/2}$; then

$$y = \frac{1}{3}t - \frac{13}{6}C_1e^{-3t/2} + C_2.$$

Since $\begin{vmatrix} (D+1)^2 & 2D & 3D \\ D & 0 & 1 \\ 1 & -D & -D \end{vmatrix} = 2D^2 + 3D$ is of degree 2 in D, there are 2 arbitrary constants and the general solution is $x = \frac{1}{3} + C_1e^{-3t/2}$, $y = \frac{1}{3}t - \frac{13}{6}C_1e^{-3t/2} + C_2$, $z = \frac{3}{2}C_1e^{-3t/2}$.

When $t = 0$: $x = \frac{1}{3} + C_1 = 1$ and $C_1 = \frac{2}{3}$; $y = (-\frac{13}{6})(\frac{2}{3}) + C_2 = 0$ and $C_2 = \frac{13}{9}$.

Thus, the required particular solution is

$$x = \frac{1}{3} + \frac{2}{3}e^{-3t/2}, \quad y = \frac{1}{3}t - \frac{13}{9}e^{-3t/2} + \frac{13}{9}, \quad z = e^{-3t/2}.$$

Note that a particular solution satisfying a given set of initial conditions cannot always be found. For example, there is no solution satisfying the conditions $x = 1$, $y = z = 0$ when $t = 0$ since $x = 1$, $y = 0$ contradicts $x = 1/3 + 2z/3$. Similarly, $y = 0$, $z = 1$, $dx/dt = 1$ when $t = 0$ contradicts $dx/dt = -z$.

SUPPLEMENTARY PROBLEMS

Solve the following simultaneous equations.

10. $Dx - (D+1)y = -e^t$, $x + (D-1)y = e^{2t}$

Ans. $x = (C_1 - C_2)\cos t + (C_1 + C_2)\sin t + 3e^{2t}/5$
$y = C_1 \cos t + C_2 \sin t + 2e^{2t}/5 + e^t/2$

11. $(D+2)x + (D+1)y = t$, $5x + (D+3)y = t^2$

$x = \dfrac{C_1 - 3C_2}{5}\sin t - \dfrac{3C_1 + C_2}{5}\cos t - t^2 + t + 3$
$y = C_1 \cos t + C_2 \sin t + 2t^2 - 3t - 4$

12. $(D+1)x + (2D+7)y = e^t + 2$, $-2x + (D+3)y = e^t - 1$

$x = \frac{1}{2} C_1 e^{-4t}[\cos(t + C_2) - \sin(t + C_2)] - \dfrac{5e^t}{26} + \dfrac{13}{17}$
$y = C_1 e^{-4t} \sin(t + C_2) + \dfrac{2e^t}{13} + \dfrac{3}{17}$

13. $(D-1)x + (D+3)y = e^{-t} - 1$, $(D+2)x + (D+1)y = e^{2t} + t$.

$x = 2C_1 e^{-7t/5} + \dfrac{5}{17}e^{2t} + \dfrac{3}{7}t - \dfrac{1}{49}$
$y = 3C_1 e^{-7t/5} - \dfrac{1}{17}e^{2t} + \dfrac{1}{2}e^{-t} + \dfrac{1}{7}t - \dfrac{26}{49}$

14. $(D^2 + 16)x - 6Dy = 0$, $6Dx + (D^2 + 16)y = 0$

$x = C_1 \cos 2t - C_2 \sin 2t + C_3 \cos 8t + C_4 \sin 8t$
$y = C_2 \cos 2t + C_1 \sin 2t + C_4 \cos 8t - C_3 \sin 8t$

15. $(D^2 + 4)x + y = \sin^2 z$, $(D^2 + 1)y - 2x = \cos^2 z$

$x = C_1 \cos(\sqrt{2}z + C_2) + C_3 \cos(\sqrt{3}z + C_4) + \frac{1}{2}\cos 2z$
$y = -2C_1 \cos(\sqrt{2}z + C_2) - C_3 \cos(\sqrt{3}z + C_4) + \frac{1}{2} - \frac{1}{2}\cos 2z$

16. $(D^2 + D + 1)x + (D^2 + 1)y = e^t$, $(D^2 + D)x + D^2 y = e^{-t}$

$x = -e^t - 2e^{-t} - C_1$
$y = 2e^t + e^{-t} + C_1$

17. $(D-1)x + (D+2)y = 1 + e^t$, $(D+2)y + (D+1)z = 2 + e^t$, $(D-1)x + (D+1)z = 3 + e^t$

$x = -1 + te^t/2 + C_2 e^t$
$y = e^t/6 + C_1 e^{-2t}$
$z = 2 + e^t/4 + C_3 e^{-t}$

CHAPTER 22

Total Differential Equations

THE DIFFERENTIAL EQUATIONS

$A)\quad (3x^2y^2 - e^x z)dx + (2x^3y + \sin z)dy + (y\cos z - e^x)dz = 0,$

$B)\quad (3xz + 2y)dx + x\,dy + x^2\,dz = 0,$

$C)\quad y\,dx + dy + dz = 0,$

being of the general form

$$P(x,y,z,\cdots,t)\,dx + Q(x,y,z,\cdots,t)\,dy + \cdots\cdots + S(x,y,z,\cdots,t)\,dt = 0,$$

are called *total differential equations*.

It may be verified readily that $A)$ is the exact differential of

$$f(x,y,z) = x^3y^2 - e^x z + y\sin z = C,$$

C being an arbitrary constant. Such an equation is called *exact*.

Equation $B)$ is not exact, but the use of x as an integrating factor yields

$$(3x^2z + 2xy)\,dx + x^2\,dy + x^3\,dz = 0$$

which is the exact differential of $x^3z + x^2y = C$. Equations $A)$ and $B)$ are called *integrable*.

Equation $C)$ is not integrable; that is, no primitive

$1)\quad f(x,y,z) = C$

can be found for it. It will be shown later (Problem 32) that for such equations a solution 1) can be obtained consistent with any prescribed relation $g(x,y,z) = 0$ of the variables.

THE CONDITION OF INTEGRABILITY of the total differential equation

$2)\quad P(x,y,z)\,dx + Q(x,y,z)\,dy + R(x,y,z)\,dz = 0$

is

$3)\quad P(\frac{\partial Q}{\partial z} - \frac{\partial R}{\partial y}) + Q(\frac{\partial R}{\partial x} - \frac{\partial P}{\partial z}) + R(\frac{\partial P}{\partial y} - \frac{\partial Q}{\partial x}) = 0,$ identically. See Problem 1.

EXAMPLE 1. For equation $B)$,

$P = 3xz + 2y,\ \frac{\partial P}{\partial y} = 2,\ \frac{\partial P}{\partial z} = 3x;\quad Q = x,\ \frac{\partial Q}{\partial x} = 1,\ \frac{\partial Q}{\partial z} = 0;\quad R = x^2,\ \frac{\partial R}{\partial x} = 2x,\ \frac{\partial R}{\partial y} = 0,$ and

3) becomes $(3xz + 2y)(0-0) + x(2x - 3x) + x^2(2-1) = 0 - x^2 + x^2 = 0$. The equation is integrable.

EXAMPLE 2. For equation $C)$,

$P = y,\ \frac{\partial P}{\partial y} = 1,\ \frac{\partial P}{\partial z} = 0;\quad Q = 1,\ \frac{\partial Q}{\partial x} = \frac{\partial Q}{\partial z} = 0;\quad R = 1,\ \frac{\partial R}{\partial x} = \frac{\partial R}{\partial y} = 0,$ and 3) becomes

$y(0-0) + 1(0-0) + 1(1-0) \neq 0$. The equation is not integrable.

THE CONDITIONS FOR EXACTNESS of 2) are

4) $$\frac{\partial P}{\partial y} = \frac{\partial Q}{\partial x}, \quad \frac{\partial Q}{\partial z} = \frac{\partial R}{\partial y}, \quad \frac{\partial R}{\partial x} = \frac{\partial P}{\partial z}.$$

EXAMPLE 3. For equation *A*),

$$P = 3x^2y^2 - e^x z, \quad \frac{\partial P}{\partial y} = 6x^2y, \quad \frac{\partial P}{\partial z} = -e^x;$$

$$Q = 2x^3y + \sin z, \quad \frac{\partial Q}{\partial x} = 6x^2y, \quad \frac{\partial Q}{\partial z} = \cos z;$$

$$R = y\cos z - e^x, \quad \frac{\partial R}{\partial x} = -e^x, \quad \frac{\partial R}{\partial y} = \cos z,$$

and the conditions 4) are satisfied. The equation is exact.

EXAMPLE 4. From Example 1 it is readily seen that 4) is not satisfied; hence, equation *B*) is not exact.

TO SOLVE AN INTEGRABLE TOTAL DIFFERENTIAL EQUATION in three variables:

a) If 2) is exact, the solution is evident after, at most, a regrouping of terms. See Problem 3.

b) If 2) is not exact, it may be possible to find an integrating factor. See Problems 4-6.

c) If 2) is homogeneous, one variable, say z, can be separated from the others by the transformation $x = uz$, $y = vz$. See Problems 7-10.

d) If no integrating factor can be found, consider one of the variables, say z, as a constant. Integrate the resulting equation, denoting the arbitrary constant of integration by $\phi(z)$. Take the total differential of the integral just obtained and compare the coefficients of its differentials with those of the given differential equation, thus determining $\phi(z)$. This procedure is illustrated in Problem 13. See also Problems 14-16.

PAIRS OF TOTAL DIFFERENTIAL EQUATIONS IN THREE VARIABLES. The solution of the simultaneous total differential equations

5) $$P_1\,dx + Q_1\,dy + R_1\,dz = 0$$

6) $$P_2\,dx + Q_2\,dy + R_2\,dz = 0$$

consists of a pair of relations

7) $$f(x,y,z) = C_1$$

8) $$g(x,y,z) = C_2.$$

To solve a given pair of equations:

e) If 5) and 6) are both integrable, each may be solved by one or more of the procedures *a*)-*d*). Then, 7), say, is the complete solution (primitive) of 5), and 8) is the complete solution of 6). See Problem 18.

f) If 5) is integrable but 6) is not, then 7), say, is the complete solution of 5). To obtain 8), we use 5),6),7) to eliminate one variable and its differential, and integrate the resulting equation. See Problem 19.

g) If neither equation is integrable, we may use the method of Chapter 21, treating two of the variables, say x and y, as functions of the third variable z.

At times it may be simpler to proceed as follows: Eliminate in turn dy and dz (or any other pair) between 5) and 6) to obtain

$$\begin{vmatrix} P_1 & Q_1 \\ P_2 & Q_2 \end{vmatrix} dx - \begin{vmatrix} Q_1 & R_1 \\ Q_2 & R_2 \end{vmatrix} dz = 0, \qquad \begin{vmatrix} R_1 & P_1 \\ R_2 & P_2 \end{vmatrix} dx - \begin{vmatrix} Q_1 & R_1 \\ Q_2 & R_2 \end{vmatrix} dy = 0$$

and express them in the symmetric form

9) $$\frac{dx}{X} = \frac{dy}{Y} = \frac{dz}{Z},$$

where $X = \lambda\begin{vmatrix} Q_1 & R_1 \\ Q_2 & R_2 \end{vmatrix}$, $Y = \lambda\begin{vmatrix} R_1 & P_1 \\ R_2 & P_2 \end{vmatrix}$, $Z = \lambda\begin{vmatrix} P_1 & Q_1 \\ P_2 & Q_2 \end{vmatrix}$, $\lambda \neq 0$.

(Note that this is the procedure for obtaining the symmetric form of the equations of a straight line when the two-plane form is given.)

Of the three equations

9') $$Y\,dx = X\,dy, \qquad X\,dz = Z\,dx, \qquad Z\,dy = Y\,dz$$

given by 9), any one may be obtained from the other two. Hence, in obtaining 9), we merely replace the original pair of differential equations by an equivalent pair, that is, any two of 9').

If two of 9') are integrable, we proceed as in e). See Problem 20.
If but one of 9') is integrable, we proceed as in f). See Problem 21.
If no one of 9') is integrable, we increase the number of possible equations. By a well known principle,

$$\frac{dx}{X} = \frac{dy}{Y} = \frac{dz}{Z} = \frac{l_1\,dx + m_1\,dy + n_1\,dz}{l_1 X + m_1 Y + n_1 Z} = \frac{l_2\,dx + m_2\,dy + n_2\,dz}{l_2 X + m_2 Y + n_2 Z}$$

where the l, m, n are arbitrary functions of the variables such that

$$lX + mY + nZ \neq 0.$$

By a proper choice of the multipliers, it may be possible to obtain an integrable equation, say

$$\frac{dy}{Y} = \frac{l\,dx + m\,dy + n\,dz}{lX + mY + nZ} \quad \text{or} \quad \frac{a\,dx + b\,dy + c\,dz}{aX + bY + cZ} = \frac{p\,dx + q\,dy + r\,dz}{pX + qY + rZ}.$$

If so, we proceed as in f). See Problem 22.

In actual practice, it may be simpler at times to find by means of multipliers a second integrable equation, rather than to proceed as in f). See Problems 23-24.

If $lX + mY + nZ = 0$, then also $l\,dx + m\,dy + n\,dz = 0$.
If now $l\,dx + m\,dy + n\,dz = 0$ is integrable, we integrate and have one of the required relations. See Problems 25-29.

SOLVED PROBLEMS

1. Obtain the condition of integrability of $P\,dx + Q\,dy + R\,dz = 0$.

Suppose that the given equation is obtained by differentiating

1) $$f(x,y,z) = C$$

and, perhaps, removing a common factor $\mu(x,y,z)$. Since from 1) $\frac{\partial f}{\partial x}dx + \frac{\partial f}{\partial y}dy + \frac{\partial f}{\partial z}dz = 0$,

it follows that $\frac{\partial f}{\partial x} = \mu P$, $\frac{\partial f}{\partial y} = \mu Q$, and $\frac{\partial f}{\partial z} = \mu R$.

Now assuming the existence and continuity conditions,

$$A)\quad \frac{\partial^2 f}{\partial y\,\partial x} = \mu\frac{\partial P}{\partial y} + P\frac{\partial \mu}{\partial y} = \mu\frac{\partial Q}{\partial x} + Q\frac{\partial \mu}{\partial x} = \frac{\partial^2 f}{\partial x\,\partial y}$$

$$B)\quad \frac{\partial^2 f}{\partial z\,\partial y} = \mu\frac{\partial Q}{\partial z} + Q\frac{\partial \mu}{\partial z} = \mu\frac{\partial R}{\partial y} + R\frac{\partial \mu}{\partial y} = \frac{\partial^2 f}{\partial y\,\partial z}$$

$$C)\quad \frac{\partial^2 f}{\partial x\,\partial z} = \mu\frac{\partial R}{\partial x} + R\frac{\partial \mu}{\partial x} = \mu\frac{\partial P}{\partial z} + P\frac{\partial \mu}{\partial z} = \frac{\partial^2 f}{\partial z\,\partial x}.$$

Upon multiplying these relations by R,P,Q, respectively and adding,

$$\mu\left(R\frac{\partial P}{\partial y} + P\frac{\partial Q}{\partial z} + Q\frac{\partial R}{\partial x}\right) = \mu\left(R\frac{\partial Q}{\partial x} + P\frac{\partial R}{\partial y} + Q\frac{\partial P}{\partial z}\right)$$

and the condition $P\left(\frac{\partial Q}{\partial z} - \frac{\partial R}{\partial y}\right) + Q\left(\frac{\partial R}{\partial x} - \frac{\partial P}{\partial z}\right) + R\left(\frac{\partial P}{\partial y} - \frac{\partial Q}{\partial x}\right) = 0$ follows.

2. If $\mu(x,y,z) = 1$ in Problem 1, the differential equation is exact. Show that this implies

$$\frac{\partial P}{\partial y} = \frac{\partial Q}{\partial x},\quad \frac{\partial Q}{\partial z} = \frac{\partial R}{\partial y},\quad \frac{\partial R}{\partial x} = \frac{\partial P}{\partial z}.$$

These relations follow from $A), B), C)$ in Prob. 1. For example, if $\mu = 1$, $A)$ yields $\frac{\partial P}{\partial y} = \frac{\partial Q}{\partial x}$.

3. Solve $(x - y)dx - x\,dy + z\,dz = 0$.

Since $\frac{\partial P}{\partial y} = -1 = \frac{\partial Q}{\partial x}$, $\frac{\partial Q}{\partial z} = 0 = \frac{\partial R}{\partial y}$, $\frac{\partial R}{\partial x} = 0 = \frac{\partial P}{\partial z}$, the equation is exact.

Upon regrouping thus $x\,dx - (x\,dy + y\,dx) + z\,dz = 0$ and integrating, we have

$$\tfrac{1}{2}x^2 - xy + \tfrac{1}{2}z^2 = K \quad\text{or}\quad x^2 - 2xy + z^2 = C.$$

4. Solve $y^2\,dx - z\,dy + y\,dz = 0$.

Here $P = y^2$, $\frac{\partial P}{\partial y} = 2y$, $\frac{\partial P}{\partial z} = 0$; $Q = -z$, $\frac{\partial Q}{\partial x} = 0$, $\frac{\partial Q}{\partial} = -1$; $R = y$, $\frac{\partial R}{\partial x} = 0$, $\frac{\partial R}{\partial y} = 1$;

then $P\left(\frac{\partial Q}{\partial z} - \frac{\partial R}{\partial y}\right) + Q\left(\frac{\partial R}{\partial x} - \frac{\partial P}{\partial z}\right) + R\left(\frac{\partial P}{\partial y} - \frac{\partial Q}{\partial x}\right) = y^2(-1-1) - z(0-0) + y(2y-0) = 0$ and the equation is integrable. The integrating factor $1/y^2$ reduces the equation to $dx + \frac{y\,dz - z\,dy}{y^2} = 0$ whose solution is $x + z/y = C$.

5. Solve $(2x^3y+1)dx + x^4\,dy + x^2\tan z\,dz = 0.$

The condition of integrability is satisfied since

$$(2x^3y+1)(0-0) + x^4(2x\tan z - 0) + x^2\tan z(2x^3 - 4x^3) = 0.$$

The integrating factor $1/x^2$ reduces the equation to

$$(2xy + \frac{1}{x^2})dx + x^2 dy + \tan z\,dz = 0 \qquad \text{or} \qquad (2xy\,dx + x^2 dy) + \frac{1}{x^2}dx + \tan z\,dz = 0$$

whose solution is $x^2y - \dfrac{1}{x} + \ln\sec z = C.$

6. Solve $(2x^3 - z)z\,dx + 2x^2yz\,dy + x(z+x)dz = 0.$

The normal procedure here would be to show that the equation is integrable and then to seek an integrating factor. By examining the preceding problems it will be found that, upon using the integrating factor, one variable appears only in an exact differential, for example, the variable z in $\tan z\,dz$ in Problem 5.

When the equation of this problem is divided by x^2z, the variable y appears only in the term $2y\,dy$ which is an exact differential. Thus, we shall use $1/x^2z$ as a *possible* integrating factor. The result is $2x\,dx + 2y\,dy + \dfrac{1}{z}dz + \dfrac{x\,dz - z\,dx}{x^2} = 0$ whose solution is $x^2 + y^2 + \ln z + \dfrac{z}{x} = C.$

Of course, the separation of the variable here does not indicate that the equation is integrable; for example, $x\,dx + z\,dy + dz = 0$ is not integrable although x appears only in an exact differential.

7. Show that if $P\,dx + Q\,dy + R\,dz = 0$ is homogeneous (*i.e.*, if P, Q, R are homogeneous and of the same degree) then the substitution $x = uz,\ y = vz$ will separate the variable z from the variables u and v.

Let the coefficients P, Q, R be of degree n in the variables.
Substituting $x = uz,\ y = vz$, the given equation becomes

$$P(uz,vz,z)[u\,dz + z\,du] + Q(uz,vz,z)[v\,dz + z\,dv] + R(uz,vz,z)dz = 0.$$

Dividing out the common factor z^n and rearranging, we have

$$z[P(u,v,1)du + Q(u,v,1)dv] + [uP(u,v,1) + vQ(u,v,1) + R(u,v,1)]dz = 0$$

or $$z(P_1 du + Q_1 dv) + (uP_1 + vQ_1 + R_1)dz = 0, \quad \text{where } P_1 = P(u,v,1), \text{ etc.}$$

This may be written as $A)\quad \dfrac{P_1}{uP_1 + vQ_1 + R_1}du + \dfrac{Q_1}{uP_1 + vQ_1 + R_1}dv + \dfrac{1}{z}dz = 0$ in which the variable z occurs only in the last term.

Now the condition of integrability for $A)$, $\dfrac{1}{z}\left(\dfrac{\partial}{\partial u}\dfrac{Q_1}{uP_1 + vQ_1 + R_1} - \dfrac{\partial}{\partial v}\dfrac{P_1}{uP_1 + vQ_1 + R_1}\right) = 0,$ is satisfied provided the original equation is integrable and, when this occurs, the sum of the first two terms of $A)$ is an exact differential. Moreover, since the third term is an exact differential, $A)$ is an exact differential equation provided only that $P\,dx + Q\,dy + R\,dz = 0$ is integrable.

8. Solve the homogeneous equation $2(y+z)dx - (x+z)dy + (2y - x + z)dz = 0.$

The equation is integrable since $2(y+z)(-1-2) - (x+z)(-1-2) + (2y-x+z)(2+1) = 0.$

The transformation $x = uz,\ y = vz$ reduces the given equation to

$$2z(v+1)(u\,dz + z\,du) - z(u+1)(v\,dz + z\,dv) + z(2v - u + 1)dz = 0.$$

Dividing by z and rearranging, we have $2z(v+1)du - z(u+1)dv + (uv+u+v+1)dz = 0$ or, dividing by $z(uv+u+v+1) = z(u+1)(v+1)$, $\dfrac{2\,du}{u+1} - \dfrac{dv}{v+1} + \dfrac{dz}{z} = 0$.

Then $2\ln(u+1) - \ln(v+1) + \ln z = \ln K$, $z(u+1)^2 = K(v+1)$,

$$(x+z)^2 = K(y+z) \quad \text{or} \quad y+z = C(x+z)^2.$$

9. Solve the homogeneous equation $yz\,dx - z^2\,dy - xy\,dz = 0$.

The equation is integrable since $yz(-2z+x) - z^2(-y-y) - xy(z-0) = 0$.
The transformation $x = uz$, $y = vz$ reduces it to

$$vz^2(u\,dz + z\,du) - z^2(v\,dz + z\,dv) - uvz^2\,dz = 0.$$

Dividing by z^2 and rearranging, $vz\,du - z\,dv - v\,dz = 0$ or $du - \dfrac{dv}{v} - \dfrac{dz}{z} = 0$.

Then $u - \ln v - \ln z = \ln K$, $vz = Ce^{u}$ or $y = Ce^{x/z}$.

10. Solve $(2y-z)dx + 2(x-z)dy - (x+2y)dz = 0$.

The equation is homogeneous and, by inspection, is seen to be exact since it may be written as

$$2(y\,dx + x\,dy) - (z\,dx + x\,dz) - 2(z\,dy + y\,dz) = 0.$$

The solution is $2xy - xz - 2yz = C$.

11. Show that $xP + yQ + zR = C$ is the solution of $P\,dx + Q\,dy + R\,dz = 0$ when the equation is exact and homogeneous of degree $n \neq -1$.

First, we check the theorem using the equation of Problem 10. Here

$$xP + yQ + zR = x(2y-z) + 2y(x-z) - z(x+2y) = 2(2xy - xz - 2yz)$$

and we obtain the solution above.

From $xP + yQ + zR = C$, we obtain by differentiation

$$A)\quad (P + x\frac{\partial P}{\partial x} + y\frac{\partial Q}{\partial x} + z\frac{\partial R}{\partial x})dx + (Q + x\frac{\partial P}{\partial y} + y\frac{\partial Q}{\partial y} + z\frac{\partial R}{\partial y})dy + (R + x\frac{\partial P}{\partial z} + y\frac{\partial Q}{\partial z} + z\frac{\partial R}{\partial z})dz = 0.$$

Since the given equation is exact, $\dfrac{\partial Q}{\partial x} = \dfrac{\partial P}{\partial y}$, $\dfrac{\partial R}{\partial x} = \dfrac{\partial P}{\partial z}$, $\dfrac{\partial R}{\partial y} = \dfrac{\partial Q}{\partial z}$.

Making these replacements, A) becomes

$$B)\quad (P + x\frac{\partial P}{\partial x} + y\frac{\partial P}{\partial y} + z\frac{\partial P}{\partial z})dx + (Q + x\frac{\partial Q}{\partial x} + y\frac{\partial Q}{\partial y} + z\frac{\partial Q}{\partial z})dy + (R + x\frac{\partial R}{\partial x} + y\frac{\partial R}{\partial y} + z\frac{\partial R}{\partial z})dz = 0.$$

Since the given equation is homogeneous, $x\dfrac{\partial P}{\partial x} + y\dfrac{\partial P}{\partial y} + z\dfrac{\partial P}{\partial z} = nP$, etc., according to Euler's Formula on homogeneous functions.

Making these replacements, B) becomes

$$(n+1)P\,dx + (n+1)Q\,dy + (n+1)R\,dz = 0$$

or, since $n \neq -1$, $\quad P\,dx + Q\,dy + R\,dz = 0.$

12. Solve $(y^2+z^2+2xy+2xz)\,dx + (x^2+z^2+2xy+2yz)\,dy + (x^2+y^2+2xz+2yz)\,dz = 0.$

The equation is homogeneous of degree 2 and is also exact since

$$\frac{\partial P}{\partial y} = 2(y+x) = \frac{\partial Q}{\partial x}, \quad \frac{\partial Q}{\partial z} = 2(z+y) = \frac{\partial R}{\partial y}, \quad \frac{\partial R}{\partial x} = 2(x+z) = \frac{\partial P}{\partial z}.$$

The solution is $x(y^2+z^2+2xy+2xz) + y(x^2+z^2+2xy+2yz) + z(x^2+y^2+2xz+2yz) = K$

or

$$x(y^2+z^2) + y(x^2+z^2) + z(x^2+y^2) = C.$$

13. Solve the differential equation $P\,dx + Q\,dy + R\,dz = 0$ given only that the condition of integrability is satisfied.

Consider one of the variables, say z, as a constant for the moment and let the solution of the resulting equation

1) $$P\,dx + Q\,dy = 0$$

be

2) $$u(x,y,z) = \phi(z).$$

Differentiating 2) with respect to all the variables,

3) $$\frac{\partial u}{\partial x}dx + \frac{\partial u}{\partial y}dy + \frac{\partial u}{\partial z}dz = \phi'(z)\,dz = d\phi.$$

Now $\frac{\partial u}{\partial x} = \mu P$ and $\frac{\partial u}{\partial y} = \mu Q$, where $\mu = \mu(x,y,z)$ is an integrating factor of 1). Substituting in 3), we have $\mu P\,dx + \mu Q\,dy + \frac{\partial u}{\partial z}dz = d\phi.$

But from the given equation $\mu P\,dx + \mu Q\,dy + \mu R\,dz = 0$ so that

$$d\phi = \frac{\partial u}{\partial z}dz - \mu R\,dz = \left(\frac{\partial u}{\partial z} - \mu R\right)dz.$$

This relation is free of dx and dy and, using 2) if necessary, can be written as a differential equation in z and ϕ. Solving the integral for ϕ and substituting in 2), we have the required solution.

14. Solve $2(y+z)\,dx - (x+z)\,dy + (2y-x+z)\,dz = 0.$ (See Problem 8.)

We treat z as a constant and solve $2(y+z)\,dx - (x+z)\,dy = 0$ or $\dfrac{dy}{dx} - \dfrac{2}{x+z}y = \dfrac{2z}{x+z}$, using the integrating factor $e^{-2\int dx/(x+z)} = \dfrac{1}{(x+z)^2}$, to obtain

A) $$\frac{y}{(x+z)^2} = \int \frac{2z}{(x+z)^3}dx = -\frac{z}{(x+z)^2} + \phi(z).$$

Differentiating A) with respect to all variables,

$$\frac{dy}{(x+z)^2} - \frac{2y}{(x+z)^3}(dx+dz) = -\frac{dz}{(x+z)^2} + \frac{2z}{(x+z)^3}(dx+dz) + d\phi$$

or

$$2(y+z)\,dx - (x+z)\,dy + (2y-x+z)\,dz + (x+z)^3\,d\phi = 0.$$

Comparing this with the given equation, it is seen that $(x+z)^3\,d\phi = 0$ and $\phi = C$. Since, from A), $y+z = \phi(x+z)^2$, the solution is $y+z = C(x+z)^2$.

15. Solve $(e^x y + e^z)dx + (e^y z + e^x)dy + (e^y - e^x y - e^y z)dz = 0.$

The equation is integrable since

$$(e^x y + e^z)(e^y - e^y + e^x + e^y z) + (e^y z + e^x)(-e^x y - e^z) + (e^y - e^x y - e^y z)(e^x - e^x) = 0.$$

Considering z as a constant and solving the resulting equation

$$(e^x y\,dx + e^x dy) + e^y z\,dy + e^z dx = 0,$$

we have $$e^x y + e^y z + e^z x = \phi(z).$$

Differentiation with respect to all variables yields

$$(e^x y + e^z)dx + (e^y z + e^x)dy + (e^y + e^z x)dz = d\phi.$$

From the given equation, $(e^x y + e^z)dx + (e^y z + e^x)dy + (e^y + e^z x)dz = (e^x y + e^y z + e^z x)dz.$

Thus, $d\phi = (e^x y + e^y z + e^z x)dz = \phi dz$ and $\phi = Ce^z$. The required solution is

$$e^x y + e^y z + e^z x = Ce^z.$$

16. Solve $yz\,dx + (xz - yz^3)dy - 2xy\,dz = 0.$

The equation is integrable since $yz(x - 3yz^2 + 2x) + (xz - yz^3)(-2y - y) - 2xy(z - z) = 0.$

Considering y as a constant and solving the resulting equation

$$yz\,dx - 2xy\,dz = 0 \qquad \text{or} \qquad z\,dx - 2x\,dz = 0,$$

we obtain $$\ln x - 2\ln z = \ln\phi(y) \qquad \text{or} \qquad x = \phi z^2.$$

Differentiating this result and making the replacement $\phi = x/z^2$, we have

$$dx - 2\phi z\,dz - z^2 d\phi = 0, \qquad dx - 2\frac{x}{z}\,dz - z^2 d\phi = 0, \qquad \text{or} \qquad yz\,dx - 2xy\,dz - yz^3 d\phi = 0.$$

Comparing this with the given differential equation, we have

$$(xz - yz^3)dy + yz^3 d\phi = (\phi z^3 - yz^3)dy + yz^3 d\phi = 0 \qquad \text{or} \qquad \phi dy + y\,d\phi - y\,dy = 0.$$

Then $\phi y - \frac{1}{2}y^2 = K$ or $\phi = \frac{1}{2}y + K/y$, so that the solution is

$$x = \phi z^2 = z^2(\tfrac{1}{2}y + K/y) \qquad \text{or} \qquad 2xy = y^2 z^2 + Cz^2.$$

17. Discuss geometrically the solution of the integrable total differential equation

$$P\,dx + Q\,dy + R\,dz = 0.$$

Let (x_0, y_0, z_0) be a general point in space for which not all of $P_0 = P(x_0, y_0, z_0)$, $Q_0 = Q(x_0, y_0, z_0)$, $R_0 = R(x_0, y_0, z_0)$ are zero.

Assuming that P, Q, R are single-valued, the set (P_0, Q_0, R_0) may be considered as direction numbers of a unique line through the point. Hence, the given differential equation may be thought of as defining at each point (x_0, y_0, z_0)

a line $$\frac{x - x_0}{P_0} = \frac{y - y_0}{Q_0} = \frac{z - z_0}{R_0}$$

and a plane $$P_0(x - x_0) + Q_0(y - y_0) + R_0(z - z_0) = 0$$ normal to the line.

The solution $f(x,y,z) = C$ of the given differential equation represents a family of surfaces such that through a general point (x_0, y_0, z_0) of space there passes a single surface S_0 of the family. The equation of the tangent plane π_0 to this surface at the point is

$$(x - x_0)\frac{\partial f}{\partial x_0} + (y - y_0)\frac{\partial f}{\partial y_0} + (z - z_0)\frac{\partial f}{\partial z_0} = 0$$

and the equations of the normal line L_0 are $\dfrac{x-x_0}{\frac{\partial f}{\partial x_0}} = \dfrac{y-y_0}{\frac{\partial f}{\partial y_0}} = \dfrac{z-z_0}{\frac{\partial f}{\partial z_0}}$.

From Problem 1, $\frac{\partial f}{\partial x} = \lambda P$, $\frac{\partial f}{\partial y} = \lambda Q$, $\frac{\partial f}{\partial z} = \lambda R$. Hence, the solution of an integrable total differential equation in three variables is a family of surfaces whose tangent plane and normal at each point are respectively the plane and line associated with the point by the differential equation.

PAIRS OF TOTAL DIFFERENTIAL EQUATIONS IN THREE VARIABLES.

18. Solve the system: $(y+z)dx + (z+x)dy + (x+y)dz = 0$
$(x+z)dx + y\,dy + x\,dz = 0$.

Both equations are integrable. The first may be written as

$$(y\,dx + x\,dy) + (z\,dy + y\,dz) + (x\,dz + z\,dx) = 0$$

and the solution is $xy + yz + zx = C_1$.

The second may be written as $x\,dx + y\,dy + (z\,dx + x\,dz) = 0$ and the solution is

$$x^2 + y^2 + 2xz = C_2.$$

Thus, $xy+yz+zx = C_1$, $x^2+y^2+2xz = C_2$ constitute the general solution.

Through each point in space there passes a single surface of each of the two families. Since the two surfaces on a point have a curve in common, the solution of the pair of differential equations is a family of curves. This family of curves may be given by the equations of any two families of surfaces passing through the family of curves. For example,

$$xy+yz+zx = C_1, \quad x^2+y^2+2(C_1-xy-yz) = C_2$$

also constitute the general solution.

19. Solve the system: 1) $yz\,dx + xz\,dy + xy\,dz = 0$
2) $z^2(dx+dy) + (xz+yz-xy)dz = 0$.

The first equation is integrable, with solution 3) $xyz = C_1$, but the second is not.

Multiply 1) by z, multiply 2) by y, and subtract to obtain $z^2(y-x)dy + y^2(z-x)dz = 0$. Multiply this by yz, and substitute $xyz = C_1$ from 3). The result is

$$z^2(y^2z - C_1)dy + y^2(yz^2 - C_1)dz = 0 \qquad \text{or} \qquad z\,dy + y\,dz - C_1\left(\frac{dy}{y^2} + \frac{dz}{z^2}\right) = 0$$

whose solution is 4) $yz + C_1\left(\dfrac{y+z}{yz}\right) = C_2$.

Equations 3) and 4) constitute a general solution. However, 4) may be replaced by the simpler form 4') $xy+yz+xz = C_2$, obtained from 4) by substituting for C_1.

20. Solve the system: $dx + 2dy - (x+2y)dz = 0$
$2dx + dy + (x-y)dz = 0$.

Here $X = \lambda\begin{vmatrix} 2 & -(x+2y) \\ 1 & x-y \end{vmatrix} = 3\lambda x$, $Y = \lambda\begin{vmatrix} -(x+2y) & 1 \\ x-y & 2 \end{vmatrix} = -3\lambda(x+y)$, $Z = \lambda\begin{vmatrix} 1 & 2 \\ 2 & 1 \end{vmatrix} = -3\lambda$.

For the choice $\lambda = -1/3$, $X = -x$, $Y = x+y$, $Z = 1$, and we write the system in the symmetric form

$$\frac{dx}{-x} = \frac{dy}{x+y} = \frac{dz}{1}.$$

From the integrable equation $\dfrac{dx}{-x} = \dfrac{dz}{1}$, we obtain $z + \ln x = C_1$.

From the integrable equation $\dfrac{dx}{-x} = \dfrac{dy}{x+y}$, we obtain $x^2 + 2xy = C_2$.

Thus, $z + \ln x = C_1$, $x^2 + 2xy = C_2$ constitute the general solution.

21. Solve the system $\dfrac{dx}{x} = \dfrac{dy}{x+z} = \dfrac{dz}{-z}$. Find the equations of the integral curves through the points *a*) (1,1,1) and *b*) (2,1,1).

Consider the equations $\dfrac{dx}{x} = \dfrac{dz}{-z}$ and $\dfrac{dx}{x} = \dfrac{dy}{x+z}$. The first is integrable and yields $xz = C_1$. The second is not integrable but is reduced to $dy = (1 + C_1/x^2)dx$ by the substitution $z = C_1/x$. Integrating, we have $y = x - C_1/x + C_2$ or, substituting $C_1 = xz$, $y - x + z = C_2$. Thus, $xz = C_1$, $y - x + z = C_2$ constitute the general solution.

The integral curve through the point (1,1,1) is the intersection of the hyperbolic cylinder $xz = 1$ and the plane $y - x + z = 1$. The integral curve through (2,1,1) is the intersection of the cylinder $xz = 2$ and the plane $y - x + z = 0$.

22. Solve $\dfrac{dx}{y-z} = \dfrac{dy}{z-x} = \dfrac{dz}{y-x}$.

No equation is integrable. By means of the multipliers $l = m = 1$, $n = 0$, we obtain

$$\frac{dz}{y-x} = \frac{l\,dx + m\,dy + n\,dz}{l(y-z) + m(z-x) + n(y-x)} = \frac{dx+dy}{y-x} \quad \text{or} \quad dx + dy - dz = 0. \quad \text{Then}$$

A) $$x + y - z = C_1.$$

Using *A*) to eliminate z in $\dfrac{dx}{y-z} = \dfrac{dy}{z-x}$, we obtain $\dfrac{dx}{C_1 - x} = \dfrac{dy}{y - C_1}$. Then $\ln(x - C_1) + \ln(y - C_1) = \ln C_2$, or $(x - C_1)(y - C_1) = C_2$, or, eliminating C_1 by means of *A*),

B) $$(z-y)(z-x) = C_2.$$

A) and *B*) constitute the general solution.

23. Solve $\dfrac{x^2 dx}{y^3} = \dfrac{y^2 dy}{x^3} = \dfrac{dz}{z}$.

From the integrable equation $\dfrac{x^2 dx}{y^3} = \dfrac{y^2 dy}{x^3}$ or $x^5 dx - y^5 dy = 0$, we obtain

A) $$x^6 - y^6 = C_1.$$

We may then use *A*) to eliminate x in the non-integrable equation $\dfrac{dz}{z} = \dfrac{y^2 dy}{x^3}$. However, it is simpler to use the multipliers $l = m = 1$, $n = 0$ to obtain $\dfrac{dz}{z} = \dfrac{x^2 dx + y^2 dy}{x^3 + y^3}$. Then $x^3 + y^3 = C_2 z^3$.

24. Solve the system $\dfrac{dx}{x^2 + y^2} = \dfrac{dy}{2xy} = \dfrac{dz}{(x+y)^3 z}$.

Using $l = m = 1$, $n = 0$, we obtain $\dfrac{dz}{(x+y)^3 z} = \dfrac{dx + dy}{(x+y)^2}$ or $\dfrac{dz}{z} = (x+y)(dx+dy)$. Then

$$(x+y)^2 - 2\ln z = C_1.$$

Using $l_1 = m_1 = 1,\ n_1 = 0$ and $l_2 = 1,\ m_2 = -1,\ n_2 = 0$, we obtain $\dfrac{dx+dy}{(x+y)^2} = \dfrac{dx-dy}{(x-y)^2}$. Then

$$\frac{1}{x+y} = \frac{1}{x-y} + K \qquad \text{or} \qquad y = C_2(x^2-y^2).$$

25. Solve the system $\dfrac{dx}{y} = \dfrac{dy}{-x} = \dfrac{dz}{2x-3y}$.

The equation $\dfrac{dx}{y} = \dfrac{dy}{-x}$ or $x\,dx + y\,dy = 0$ is integrable and we obtain $x^2+y^2 = C_1$.

Using $l=3,\ m=2,\ n=1$, we find $3(y)+2(-x)+(2x-3y) = 0$. Hence, $3\,dx + 2\,dy + dz = 0$ and $3x+2y+z = C_2$.

26. Solve the system $\dfrac{dx}{4y-3z} = \dfrac{dy}{4x-2z} = \dfrac{dz}{2y-3x}$.

We seek multipliers l,m,n such that $A)$ $l(4y-3z) + m(4x-2z) + n(2y-3x) = 0$.

Rearranging $A)$ in the form $(4m-3n)x + (4l+2n)y + (-3l-2m)z = 0$, we see that it will be satisfied when $4m-3n = 0,\ 4l+2n = 0,\ -3l-2m = 0$ or $l:m:n = 2:-3:-4$. Then

$$2\,dx - 3\,dy - 4\,dz = 0 \qquad \text{and} \qquad 2x-3y-4z = C_1.$$

Using the arrangement $4(ly+mx) + 3(-lz-nx) + 2(ny-mz) = 0$ and setting $ly+mx = 0$, $-lz-nx = 0,\ ny-mz = 0$, we obtain $l:m:n = x:-y:-z$. Then

$$x\,dx - y\,dy - z\,dz = 0 \qquad \text{and} \qquad x^2-y^2-z^2 = C_2.$$

27. Solve the system $\dfrac{p\,dx}{(q-r)yz} = \dfrac{q\,dy}{(r-p)xz} = \dfrac{r\,dz}{(p-q)xy}$.

Consider $l(q-r)yz + m(r-p)xz + n(p-q)xy = 0$.

From $q(lyz-nxy) + r(mxz-lyz) + p(nxy-mxz) = 0$ we obtain $l:m:n = x:y:z$. Then

$$px\,dx + qy\,dy + rz\,dz = 0 \qquad \text{and} \qquad px^2+qy^2+rz^2 = C_1.$$

From $z(lqy-mpx) + y(npx-lrz) + x(mrz-nqy) = 0$ we obtain $l:m:n = px:qy:rz$. Then

$$p^2x\,dx + q^2y\,dy + r^2z\,dz = 0 \qquad \text{and} \qquad p^2x^2+q^2y^2+r^2z^2 = C_2.$$

28. Solve the system $\dfrac{dx}{x^2+y^2-yz} = \dfrac{dy}{-x^2-y^2+xz} = \dfrac{dz}{(x-y)z}$.

Using $l=m=1,\ n=-1$, we obtain $(x^2+y^2-yz) + (-x^2-y^2+xz) - (x-y)z = 0$. Then

$$dx + dy - dz = 0 \qquad \text{and} \qquad x+y-z = C_1.$$

Using $l = xz,\ m = yz,\ n = -(x^2+y^2)$, we find

$$xz(x^2+y^2-yz) + yz(-x^2-y^2+xz) - (x^2+y^2)(x-y)z = 0.$$

Then $$xz\,dx + yz\,dy - (x^2+y^2)dz = 0 \qquad \text{or} \qquad \frac{x\,dx+y\,dy}{x^2+y^2} - \frac{dz}{z} = 0$$

and $$\ln(x^2+y^2) - 2\ln z = \ln C_2$$

or $$x^2+y^2 = C_2z^2.$$

29. Solve the system $\dfrac{dx}{2x} = \dfrac{dy}{-y} = \dfrac{dz}{4xy^2 - 2z}$.

From $\dfrac{dx}{2x} = \dfrac{dy}{-y}$ we have $xy^2 = C_1$. By inspection, $2y^4(2x) + 2yz(-y) - y^2(4xy^2 - 2z) = 0$; then $2y^4\,dx + 2yz\,dy - y^2\,dz = 0$ or $2\,dx - \dfrac{y^2\,dz - 2yz\,dy}{y^4} = 0$, and $2x - \dfrac{z}{y^2} = C_2$.

30. Discuss geometrically the general solution of $\dfrac{dx}{P} = \dfrac{dy}{Q} = \dfrac{dz}{R}$.

For convenience, let us assume that in solving the given system we have obtained a pair of integrable equations

$$P_1 dx + Q_1 dy + R_1 dz = 0 \qquad \text{and} \qquad P_2 dx + Q_2 dy + R_2 dz = 0$$

whose integrals are respectively

$$g(x,y,z) = C_1 \qquad \text{and} \qquad h(x,y,z) = C_2.$$

Through a general point (x_0,y_0,z_0) of space there pass two surfaces (one of each of the above families) whose curve of intersection C_0 is the integral curve of the given system through the point. The tangent planes to the two surfaces at (x_0,y_0,z_0) are normal to the directions (P_1,Q_1,R_1) and (P_2,Q_2,R_2) evaluated at the point, and the line of intersection L_0 of these planes is normal to the two directions. Let (X,Y,Z) be a set of direction numbers for L_0; then

$$X = \lambda\begin{vmatrix} Q_1 & R_1 \\ Q_2 & R_2 \end{vmatrix}, \qquad Y = \lambda\begin{vmatrix} R_1 & P_1 \\ R_2 & P_2 \end{vmatrix}, \qquad Z = \lambda\begin{vmatrix} P_1 & Q_1 \\ P_2 & Q_2 \end{vmatrix}$$

are proportional to P,Q,R (all evaluated at the point).

Now L_0 is the tangent to C_0 at (x_0,y_0,z_0), since the tangent to a space curve at one of its points lies in the tangent plane at the point of any surface containing the curve. Hence, the integral curves of the system $\dfrac{dx}{P} = \dfrac{dy}{Q} = \dfrac{dz}{R}$ consist of a doubly infinite system of curves characterized by the fact that at any point (x_0,y_0,z_0) the tangent to the curve through the point has (P_0,Q_0,R_0) as direction numbers.

31. Show that the family of integral surfaces of 1) $P\,dx + Q\,dy + R\,dz = 0$ and the family of integral curves of 2) $\dfrac{dx}{P} = \dfrac{dy}{Q} = \dfrac{dz}{R}$ are orthogonal.

This follows from the fact that at any general point (x_0,y_0,z_0) the direction (P_0,Q_0,R_0) is: *a*) normal to the integral surface of 1) through the point (see Problem 17) and *b*) the direction of the integral curve of 2) through the point (see Problem 30).

32. Solve 1) $y\,dx + x\,dy - (x+y+2z)dz = 0$ consistent with *a*) $z = a$, *b*) $x + y + 2z = 0$, *c*) $x + y = 0$, *d*) $xy = a$.

Equation 1) is not integrable. From each given surface we may obtain an integrable total differential equation. Our problem then is to solve this differential equation simultaneously with 1) using the particular solution of the former rather than the general solution as in *f*) of the introduction of this chapter.

a) Here $z = a$, $dz = 0$. Substituting in 1), we obtain $y\,dx + x\,dy = 0$; then $xy = C$. Equations $z = a$, $xy = C$ are said to constitute a solution of 1).

b) Substituting $x+y+2z = 0$ in 1), we obtain $y\,dx + x\,dy = 0$ and $xy = C$.
The solution is $xy = C$, $x+y+2z = 0$.

c) Here $y = -x$, $dy = -dx$. Substituting in 1), we obtain $x\,dx + z\,dz = 0$ and $x^2+z^2 = C$.
The solution is $x^2+z^2 = C$, $x+y = 0$.

d) Here $xy = a$, $x\,dy + y\,dx = 0$. Equation 1) reduces to $(x+y+2z)dz = 0$.
Then, either $x+2y+2z = 0$ or $dz = 0$ and $z = C$.
$xy = a$, $x+y+2z = 0$ and $z = C$, $xy = a$ constitute the solution.

33. Discuss geometrically the problem of solving $P\,dx + Q\,dy + R\,dz = 0$ consistent with the given relation $g(x,y,z) = 0$.

From the relation $g(x,y,z) = 0$, we obtain $\dfrac{\partial g}{\partial x}dx + \dfrac{\partial g}{\partial y}dy + \dfrac{\partial g}{\partial z}dz = 0$.

We solve the system $P\,dx + Q\,dy + R\,dz = 0$, $\dfrac{\partial g}{\partial x}dx + \dfrac{\partial g}{\partial y}dy + \dfrac{\partial g}{\partial z}dz = 0$ using the particular solution $g(x,y,z) = 0$ of the latter. Let

$$f(x,y,z) = C, \quad g(x,y,z) = 0$$

constitute the solution. The integral curves are those cut out on the surface $g(x,y,z) = 0$ by the system of surfaces $f(x,y,z) = C$. Thus, Problem 32*c* may be stated as: Find all curves lying on the surface (plane) $x+y = 0$ which satisfy the differential equation

$$y\,dx + x\,dy - (x+y+2z)dz = 0.$$

At a general point (x_0,y_0,z_0) on the surface $g(x,y,z) = 0$, the line of intersection L_0 of the tangent planes to $g(x,y,z) = 0$ and the surface of the system $f(x,y,z) = C$, through the point, is tangent to the curve of intersection of the two surfaces. Thus, we have found the *family of curves* on the given surface $g(x,y,z) = 0$ whose tangent at any point lies in the plane, through this point, determined by the differential equation. (See Problem 17.)

For example, consider Problem 32*c*. On the prescribed surface $x+y = 0$, choose any point $(a,-a,b)$. At this point, the tangent plane to $x+y = 0$ (here, the plane itself) is normal to the direction $(1,1,0)$ and the tangent plane to the surface (of the family) $x^2+z^2 = a^2+b^2$ is normal to the direction $(a,0,b)$. A set of direction numbers for the line of intersection L of these planes [the tangent to the curve through $(a,-a,b)$] is $(-b,b,a)$.

Now the plane through $(a,-a,b)$ determined by the given differential equation is normal to the direction $[y,x,-(x+y+2z)]_{(a,-a,b)} = (-a,a,-2b)$. Since $(-b,b,a)$ and $(-a,a,-2b)$ are normal directions, the line L lies in the plane determined by the differential equation.

34. Solve 1) $2z\,dx + dy + y\,dz = 0$ consistent with 2) $x+y+z = 0$.

From 2), $y = -x-z$ and $dy = -dx - dz$. Substituting for y and dy in 1), we obtain

3) $(2z-1)dx - (x+z+1)dz = 0$.

The transformation $z = z_1 + 1/2$, $x = x_1 - 3/2$ reduces 3) to

4) $2z_1 dx_1 - (x_1+z_1)dz_1 = 0$, a homogeneous equation.

The transformation $x_1 = uz_1$ reduces 4) to $(u-1)dz_1 + 2z_1 du = 0$ or $\dfrac{dz_1}{z_1} + \dfrac{2\,du}{u-1} = 0$.

Then
$$\ln z_1 + 2\ln(u-1) = \ln K$$
or
$$z_1(u-1)^2 = K.$$

Replacing u by x_1/z_1, x_1 by $x+3/2$ and z_1 by $z-1/2$, this becomes

$$(x-z+2)^2 = C(2z-1).$$

SUPPLEMENTARY PROBLEMS

Test for integrability and solve when possible.

35. $(y+3z)dx + (x+2z)dy + (3x+2y)dz = 0$ *Ans.* $xy + 2yz + 3xz = C$

36. $(\cos x + e^x y)dx + (e^x + e^y z)dy + e^y dz = 0$ $e^x y + e^y z + \sin x = C$

37. $dx + (x+z)dy + dz = 0$ $y + \ln(x+z) = C$

38. $z^3 dx + z\,dy - 2y\,dz = 0$ $xz^2 + y = Cz^2$

39. $x^2 dx - z^2 dy - xy\,dz = 0$ Not integrable.

40. $(x+z)^2 dy + y^2(dx+dz) = 0$ $y(x+z) = C(x+y+z)$

41. $2x(y+z)dx + (2yz - x^2 + y^2 - z^2)dy + (2yz - x^2 - y^2 + z^2)dz = 0$ $x^2 + y^2 + z^2 = C(y+z)$

42. $yz\,dx - 2xz\,dy + xy\,dz = 0$ $y^2 = Cxz$

43. $x\,dx + y\,dy + (x^2+y^2+z^2+1)z\,dz = 0$ $(x^2+y^2+z^2)e^{z^2} = C$

44. $z(x^2 - yz - z^2)dx + xz(x+z)dy + x(z^2 - x^2 - xy)dz = 0$ $(x+y)/z + (y+z)/x = C$

Solve the following pairs of equations.

45. $dx + dy + (x+y)dz = 0$
$z(dx+dy) + (x+y)dz = 0$ *Ans.* $x + y = C_1 e^{-z}, \quad x + y = C_2/z$

46. $2yz\,dx + x(z\,dy + y\,dz) = 0$
$y\,dx - x^2 z\,dy + y\,dz = 0$ $x^2 yz = C_1, \quad x^2 z + x + z = C_2$

47. $\dfrac{x\,dx}{y^3 z} = \dfrac{dy}{x^2 z} = \dfrac{dz}{y^3}$ $x^4 - y^4 = C_1, \quad x^2 - z^2 = C_2$

48. $\dfrac{3\,dx}{yz} = \dfrac{dy}{xz} = \dfrac{dz}{xy}$ $3x^2 - y^2 = C_1, \quad y^2 - z^2 = C_2$

49. $\dfrac{dx}{1} = \dfrac{dy}{1} = \dfrac{dz}{(x+y)(1+2xy+3x^2y^2)}$ $x - y = C_1, \quad z = xy + x^2y^2 + x^3y^3 + C_2$

50. $\dfrac{dx}{x^2 - y^2 - z^2} = \dfrac{dy}{2xy} = \dfrac{dz}{2xz}$ $y = C_1 z, \quad x^2 + y^2 + z^2 = C_2 z$

51. $\dfrac{dx}{3y - 2z} = \dfrac{dy}{z - 3x} = \dfrac{dz}{2x - y}$ $x + 2y + 3z = C_1, \quad x^2 + y^2 + z^2 = C_2$

52. $\dfrac{dx}{x(2y^4 - z^4)} = \dfrac{dy}{y(z^4 - 2x^4)} = \dfrac{dz}{z(x^4 - y^4)}$ $xyz^2 = C_1, \quad x^4 + y^4 + z^4 = C_2$

53. $\dfrac{dx}{x(z^2 - y^2)} = \dfrac{dy}{y(x^2 - z^2)} = \dfrac{dz}{z(y^2 - x^2)}$ $xyz = C_1, \quad x^2 + y^2 + z^2 = C_2$

CHAPTER 23

Applications of Total and Simultaneous Equations

WHEN A MASS m moves in a plane subject to a force F, its acceleration continues to satisfy Newton's Second Law of Motion: mass × acceleration = force.

To obtain the equations of motion, when rectangular coordinates are used, consider the components of the vectors force and acceleration along the axes. The components of acceleration a_x and a_y are given by

$$a_x = \frac{d^2x}{dt^2}, \qquad a_y = \frac{d^2y}{dt^2}$$

and, denoting the components of the force by F_x and F_y, the equations of motion are

$$m\frac{d^2x}{dt^2} = F_x, \qquad m\frac{d^2y}{dt^2} = F_y.$$

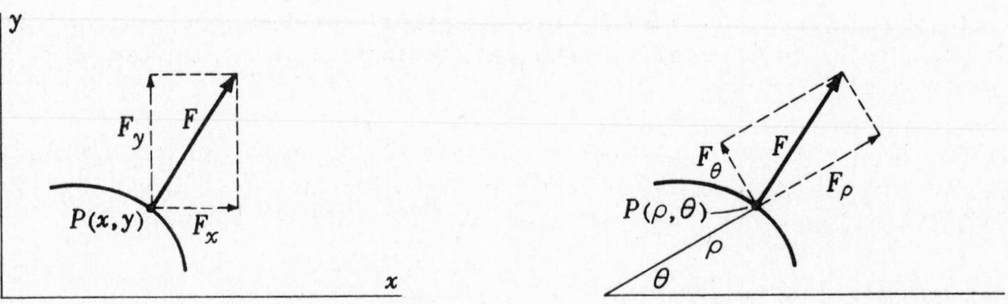

COMPONENTS OF F IN RECTANGULAR AND POLAR COORDINATES.

In polar coordinates, the corresponding equations are

$$m\{\frac{d^2\rho}{dt^2} - \rho(\frac{d\theta}{dt})^2\} = F_\rho, \qquad m\{2\frac{d\rho}{dt}\frac{d\theta}{dt} + \rho\frac{d^2\theta}{dt^2}\} = F_\theta,$$

where F_ρ and F_θ are the radial and transverse components of force, i.e., the components along the radius vector at P and a line perpendicular to it.

SOLVED PROBLEMS

1. Find the family of curves orthogonal to the surfaces $x^2 + 2y^2 + 4z^2 = C$.

Since $x^2 + 2y^2 + 4z^2 = C$ is the primitive of the total differential equation

$$x\,dx + 2y\,dy + 4z\,dz = 0,$$

the differential equation of the family of orthogonal curves is

$$\frac{dx}{x} = \frac{dy}{2y} = \frac{dz}{4z}.$$ (See Chapter 22, Problem 31.)

Solving $\dfrac{dx}{x} = \dfrac{dy}{2y}$, we have $y = Ax^2$. Solving $\dfrac{dy}{2y} = \dfrac{dz}{4z}$, we have $z = By^2$.

The required family of curves has equations $y = Ax^2,\ z = By^2$.

2. Show that there is no family of surfaces orthogonal to the system of curves

$$x^2 - y^2 = ay, \quad x + y = bz.$$

Differentiating the given equations and eliminating the constants, we have

$$2x\,dx - 2y\,dy = \frac{x^2-y^2}{y}\,dy, \qquad dx + dy = \frac{x+y}{z}\,dz.$$

The first can be written as $\dfrac{dx}{x^2+y^2} = \dfrac{dy}{2xy}$. Solving it for dx, $dx = \dfrac{x^2+y^2}{2xy}\,dy$, and substituting in the second, we have $\left(\dfrac{x^2+y^2}{2xy} + 1\right)dy = \dfrac{x+y}{z}\,dz$ or $\dfrac{dy}{2xy} = \dfrac{dz}{(x+y)z}$.

Thus, the differential equations in symmetric form of the given family of curves are

$$\frac{dx}{x^2+y^2} = \frac{dy}{2xy} = \frac{dz}{(x+y)z}.$$

Since the equation $(x^2+y^2)dx + 2xy\,dy + (x+y)z\,dz = 0$ does not satisfy the condition of integrability, there is no family of surfaces cutting the curves orthogonally.

3. The x-component of the acceleration of a particle of unit mass, moving in a plane, is equal to its ordinate and the y-component is equal to twice its abscissa. Find the equation of its path, given the initial conditions $x = y = 0$, $dx/dt = 2$, $dy/dt = 4$ when $t = 0$.

The equations of motion are $\dfrac{d^2x}{dt^2} = y$, $\dfrac{d^2y}{dt^2} = 2x$.

Differentiating the first twice and substituting from the second, $\dfrac{d^4x}{dt^4} = \dfrac{d^2y}{dt^2} = 2x$ and

$$x = C_1e^{at} + C_2e^{-at} + C_3\cos at + C_4\sin at, \quad \text{where } a^4 = 2.$$

Then, $$y = \frac{d^2x}{dt^2} = a^2(C_1e^{at} + C_2e^{-at} - C_3\cos at - C_4\sin at),$$

$$\frac{dx}{dt} = a(C_1e^{at} - C_2e^{-at} - C_3\sin at + C_4\cos at),$$

and $$\frac{dy}{dt} = a^3(C_1e^{at} - C_2e^{-at} + C_3\sin at - C_4\cos at).$$

Using the initial conditions: $C_1 + C_2 + C_3 = 0$, $C_1 + C_2 - C_3 = 0$, $C_1 - C_2 + C_4 = \dfrac{2}{a}$, $C_1 - C_2 - C_4 = \dfrac{4}{a^3}$.

Then $C_1 = -C_2 = \dfrac{a^2+2}{2a^3}$, $C_3 = 0$, and $C_4 = \dfrac{a^2-2}{a^3}$.

The parametric equations of the path are:

$$x = \tfrac{1}{4}(2+\sqrt{2})\sqrt[4]{2}\,(e^{\sqrt[4]{2}\,t} - e^{-\sqrt[4]{2}\,t}) - \tfrac{1}{2}(2-\sqrt{2})\sqrt[4]{2}\sin\sqrt[4]{2}\,t,$$

$$y = \tfrac{1}{4}(2+\sqrt{2})\sqrt[4]{8}\,(e^{\sqrt[4]{2}\,t} - e^{-\sqrt[4]{2}\,t}) + \tfrac{1}{2}(2-\sqrt{2})\sqrt[4]{8}\sin\sqrt[4]{2}\,t.$$

4. A particle of mass m is repelled from the origin O by a force varying inversely as the cube of the distance ρ from O. If it starts at $\rho = a$, $\theta = 0$ with velocity v_0, perpendicular to the initial line, find the equation of the path.

The radial and transverse components of the repelling force are: $F_\rho = \dfrac{K}{\rho^3} = \dfrac{mk^2}{\rho^3}$, $F_\theta = 0$.

Hence, $$m\left(\frac{d^2\rho}{dt^2} - \rho\left(\frac{d\theta}{dt}\right)^2\right) = \frac{mk^2}{\rho^3}, \qquad m\left(2\frac{d\rho}{dt}\frac{d\theta}{dt} + \rho\frac{d^2\theta}{dt^2}\right) = 0$$

or $$1)\quad \frac{d^2\rho}{dt^2} - \rho\left(\frac{d\theta}{dt}\right)^2 = \frac{k^2}{\rho^3}, \qquad 2)\quad \rho\frac{d^2\theta}{dt^2} + 2\frac{d\rho}{dt}\frac{d\theta}{dt} = 0.$$

Integrating 2), $\rho^2\dfrac{d\theta}{dt} = C_1$. When $t = 0$, $\rho = a$ and $\rho\dfrac{d\theta}{dt} = v_0$; then $C_1 = av_0$ and $\dfrac{d\theta}{dt} = \dfrac{av_0}{\rho^2}$.

Substituting for $\dfrac{d\theta}{dt}$ in 1), $\dfrac{d^2\rho}{dt^2} = \dfrac{a^2v_0^2}{\rho^3} + \dfrac{k^2}{\rho^3}$. Multiplying by $2\dfrac{d\rho}{dt}$,

$$2\frac{d\rho}{dt}\frac{d^2\rho}{dt^2} = 2\frac{a^2v_0^2 + k^2}{\rho^3}\frac{d\rho}{dt} \qquad \text{and} \qquad \left(\frac{d\rho}{dt}\right)^2 = -\frac{a^2v_0^2 + k^2}{\rho^2} + C_2.$$

When $t = 0$, $\rho = a$ and $\dfrac{d\rho}{dt} = 0$; then $C_2 = \dfrac{a^2v^2 + k^2}{a^2}$ and

$$\left(\frac{d\rho}{dt}\right)^2 = (a^2v_0^2 + k^2)\left(\frac{1}{a^2} - \frac{1}{\rho^2}\right) = (a^2v_0^2 + k^2)\frac{\rho^2 - a^2}{a^2\rho^2}.$$

Dividing by $\left(\dfrac{d\theta}{dt}\right)^2 = \dfrac{a^2v_0^2}{\rho^4}$, $\left(\dfrac{d\rho}{d\theta}\right)^2 = \dfrac{(a^2v_0^2 + k^2)\rho^2(\rho^2 - a^2)}{a^4v_0^2}$ and $\dfrac{d\rho}{\rho\sqrt{\rho^2 - a^2}} = \dfrac{\sqrt{a^2v_0^2 + k^2}}{a^2v_0}\,d\theta$.

Integrating, $$\frac{1}{a}\operatorname{arc\,sec}\frac{\rho}{a} = \frac{\sqrt{a^2v_0^2 + k^2}}{a^2v_0}\,\theta + C_3.$$

When $t = 0$, $\rho = a$ and $\theta = 0$; then $C_3 = 0$ and $\rho = a\sec\dfrac{\sqrt{a^2v_0^2 + k^2}}{av_0}\,\theta$.

5. A projectile of mass m is fired into the air with initial velocity v_0 at an angle θ with the ground. Neglecting all forces except gravity and the resistance of the air, assumed proportional to the velocity, find the position of the projectile at time t.

In its horizontal motion, the projectile is affected only by the x-component of the resistance. Hence,

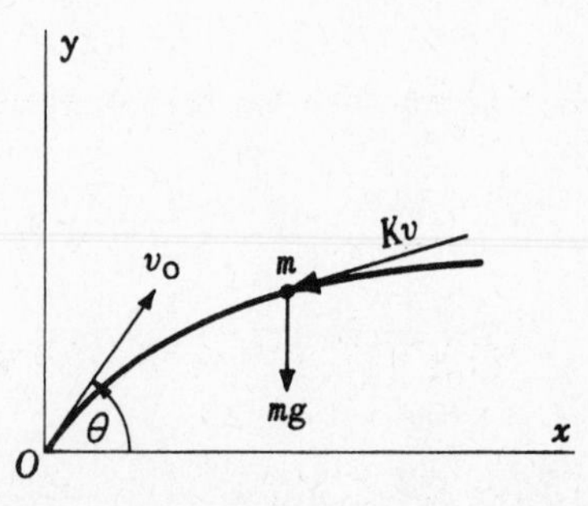

$$1)\quad m\frac{d^2x}{dt^2} = -K\frac{dx}{dt} = -mk\frac{dx}{dt} \qquad \text{or} \qquad \frac{d^2x}{dt^2} = -k\frac{dx}{dt}.$$

In its vertical motion, the projectile is affected by gravity and by the y-component of the resistance. Hence,

$$2)\quad m\frac{d^2y}{dt^2} = -mg - mk\frac{dy}{dt} \qquad \text{or} \qquad \frac{d^2y}{dt^2} = -g - k\frac{dy}{dt}.$$

Integrating 1), $\dfrac{dx}{dt} = -kx + C_1$ and $x = \dfrac{1}{k}C_1 + C_2e^{-kt}$.

Integrating 2), $\dfrac{dy}{dt} = -gt - ky + K_1$ and $y = \dfrac{1}{k}K_1 + K_2e^{-kt} - g(\dfrac{1}{k}t - \dfrac{1}{k^2})$.

Using the initial conditions $x = y = 0$, $\dfrac{dx}{dt} = v_0 \cos\theta$, $\dfrac{dy}{dt} = v_0 \sin\theta$ when $t = 0$:

$$C_1 = v_0\cos\theta,\quad C_2 = -\frac{1}{k}v_0\cos\theta;\quad K_1 = v_0\sin\theta,\quad K_2 = -\frac{1}{k}v_0\sin\theta - \frac{1}{k^2}g.$$

Thus, $x = \dfrac{1}{k}(v_0\cos\theta)(1 - e^{-kt})$, $y = \dfrac{1}{k}\{(\dfrac{g}{k} + v_0\sin\theta)(1 - e^{-kt}) - gt\}$.

6. Two masses, m_1 and m_2, are separated by a spring for which $k = k_2$ lb/ft and m_1 is attached to a support by a spring for which $k = k_1$ lb/ft as in the figure. After the system is brought to rest, the masses are displaced a feet downward and released. Discuss their motion.

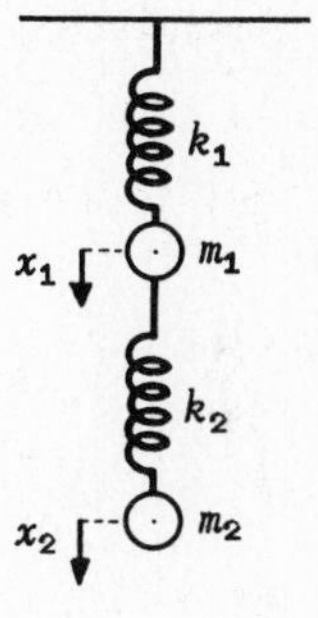

Let positive direction be downward and let x_1 and x_2 denote the displacement of the masses at time t from their respective positions at rest. The elongation of the upper spring is then x_1 and that of the lower spring is $x_2 - x_1$. The corresponding restoring forces in the springs are

$-k_1x_1 + k_2(x_2 - x_1)$ acting on m_1

and $-k_2(x_2 - x_1)$ acting on m_2.

The equations of motion are

$$m_1\frac{d^2x_1}{dt^2} = -k_1x_1 + k_2(x_2 - x_1) \quad\text{and}\quad m_2\frac{d^2x_2}{dt^2} = -k_2(x_2 - x_1)$$

or 1) $[m_1D^2 + (k_1 + k_2)]x_1 - k_2x_2 = 0$ and 2) $(m_2D^2 + k_2)x_2 - k_2x_1 = 0$.

Operating on 1) with $(m_2D^2 + k_2)$ and substituting from 2),

$$(m_2D^2 + k_2)(m_1D^2 + k_1 + k_2)x_1 - k_2(m_2D^2 + k_2)x_2 = (m_2D^2 + k_2)(m_1D^2 + k_1 + k_2)x_1 - k_2^2x_1 = 0$$

or

$$(D^4 + (\frac{k_1 + k_2}{m_1} + \frac{k_2}{m_2})D^2 + \frac{k_1k_2}{m_1m_2})x_1 = 0.$$

Denoting the roots of the characteristic equation by $\pm i\alpha$, $\pm i\beta$, where

$$\alpha^2,\ \beta^2 = \frac{1}{2}\left|-(\frac{k_1 + k_2}{m_1} + \frac{k_2}{m_2}) \pm \sqrt{(\frac{k_1 + k_2}{m_1} + \frac{k_2}{m_2})^2 - 4\frac{k_1k_2}{m_1m_2}}\right|,$$

$$x_1 = C_1e^{i\alpha t} + C_2e^{-i\alpha t} + C_3e^{i\beta t} + C_4e^{-i\beta t} \quad\text{and}$$

$$x_2 = \frac{1}{k_2}(m_1D^2 + k_1 + k_2)x_1 = \frac{k_1 + k_2 - m_1\alpha^2}{k_2}(C_1e^{i\alpha t} + C_2e^{-i\alpha t}) + \frac{k_1 + k_2 - m_1\beta^2}{k_2}(C_3e^{i\beta t} + C_4e^{-i\beta t})$$

$$= \mu(C_1e^{i\alpha t} + C_2e^{-i\alpha t}) + \nu(C_3e^{i\beta t} + C_4e^{-i\beta t}).$$

Using the initial conditions $x_1 = x_2 = a$, $\dfrac{dx_1}{dt} = \dfrac{dx_2}{dt} = 0$ when $t = 0$,

$$C_1 = C_2 = \frac{a}{2}(\frac{\nu - 1}{\nu - \mu}) = \frac{a}{2m_1}(\frac{k_1 - m_1\beta^2}{\alpha^2 - \beta^2}) \quad\text{and}\quad C_3 = C_4 = -\frac{a}{2m_1}(\frac{k_1 - m_1\alpha^2}{\alpha^2 - \beta^2}).$$

7. A uniform shaft carries three disks as in the adjoining figure. The polar moment of inertia of the disk at either end is I, and that of the disk at the middle is $4I$. The torsional stiffness constant of the shaft between two disks (the torque required to produce an angular displacement difference of one radian between successive disks) is k. Find the motion of the disks if a torque $2T_0 \sin \omega t$ is applied to the middle disk, assuming that at $t = 0$ the disks are at rest and there is no twist in the shaft.

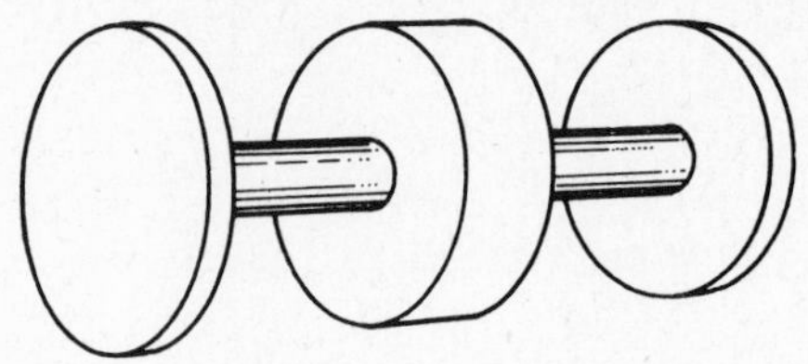

At time t, let the angular displacement of the disk at either end be θ_1 and that of the disk at the middle be θ_2. The differences of the angular twists of the ends of the two pieces of shaft, from left to right, are $\theta_2 - \theta_1$ and $\theta_1 - \theta_2$. The restoring torques acting on the disks are $k(\theta_2 - \theta_1)$, $k(\theta_1 - \theta_2) - k(\theta_2 - \theta_1)$ and $-k(\theta_1 - \theta_2)$ respectively. The net torque acting on a mass when rotating is equal to the product of the polar moment of inertia of the mass about the axis of rotation and its angular acceleration; hence the equation of motion of the middle disk is

1) $$4I \frac{d^2\theta_2}{dt^2} = k(\theta_1 - \theta_2) - k(\theta_2 - \theta_1) + 2T_0 \sin \omega t \qquad \text{or} \qquad (2ID^2 + k)\theta_2 = k\theta_1 + T_0 \sin \omega t$$

and that of either end disk is

2) $$I \frac{d^2\theta_1}{dt^2} = k(\theta_2 - \theta_1) \qquad \text{or} \qquad (ID^2 + k)\theta_1 = k\theta_2.$$

Operating on 2) with $(2ID^2 + k)$ and substituting from 1),

$$(2ID^2 + k)(ID^2 + k)\theta_1 = k(2ID^2 + k)\theta_2 = k^2\theta_1 + T_0 k \sin \omega t, \qquad \text{or}$$

3) $$D^2(2I^2D^2 + 3kI)\theta_1 = T_0 k \sin \omega t.$$

The characteristic roots are $0, 0, ai, -ai$, where $a^2 = 3k/2I$, and

4) $$\theta_1 = C_1 + C_2 t + C_3 \cos at + C_4 \sin at + \frac{T_0 k \sin \omega t}{I\omega^2(2I\omega^2 - 3k)}$$

$$= C_1 + C_2 t + C_3 \cos at + C_4 \sin at + \frac{T_0 k}{2I^2\omega^2(\omega^2 - a^2)} \sin \omega t.$$

From 2), $\theta_2 = (\frac{I}{k} D^2 + 1)\theta_1$ and

5) $$\theta_2 = C_1 + C_2 t + C_3(1 - \frac{I}{k} a^2)\cos at + C_4(1 - \frac{I}{k} a^2)\sin at + \frac{T_0 k - T_0\omega^2 I}{2I^2\omega^2(\omega^2 - a^2)} \sin \omega t.$$

From 4) and 5), we obtain by differentiation,

4′) $$\frac{d\theta_1}{dt} = C_2 - C_3 a \sin at + C_4 a \cos at + \frac{T_0 k}{2I^2\omega(\omega^2 - a^2)} \cos \omega t, \qquad \text{and}$$

5′) $$\frac{d\theta_2}{dt} = C_2 - C_3 a(1 - \frac{I}{k} a^2)\sin at + C_4 a(1 - \frac{I}{k} a^2)\cos at + \frac{T_0 k - T_0\omega^2 I}{2I^2\omega(\omega^2 - a^2)} \cos \omega t.$$

Using the initial conditions $\theta_1 = \theta_2 = 0$, $\frac{d\theta_1}{dt} = \frac{d\theta_2}{dt} = 0$ when $t = 0$, we have $C_1 + C_3 = 0$,

$C_1 + C_3(1 - \frac{I}{k}a^2) = 0$, $C_2 + C_4 a + \frac{T_0 k}{2I^2\omega(\omega^2 - a^2)} = 0$, and $C_2 + C_4 a(1 - \frac{I}{k}a^2) + \frac{T_0 k - T_0\omega^2 I}{2I^2\omega(\omega^2 - a^2)} = 0$.

Then $C_1 = C_3 = 0$, $C_4 = -\frac{T_0\omega}{3Ia(\omega^2 - a^2)}$, $C_2 = \frac{T_0}{3I\omega}$,

$$\theta_1 = \frac{T_0}{3I}\left(\frac{t}{\omega} + \frac{a^2 \sin\omega t}{\omega^2(\omega^2 - a^2)} - \frac{\omega \sin at}{a(\omega^2 - a^2)}\right) = \frac{T_0}{3I}\left(\frac{t}{\omega} + \frac{a^3 \sin\omega t - \omega^3 \sin at}{a\omega^2(\omega^2 - a^2)}\right), \quad \text{and}$$

$$\theta_2 = \theta_1 - \frac{T_0(a \sin\omega t - \omega \sin at)}{2Ia(\omega^2 - a^2)}.$$

8. The fundamental equations of a transformer are

$$1)\ M\frac{di_1}{dt} + L_2\frac{di_2}{dt} + R_2 i_2 = 0, \qquad 2)\ M\frac{di_2}{dt} + L_1\frac{di_1}{dt} + R_1 i_1 = E(t),$$

where $i_1(t)$ and $i_2(t)$ denote the currents, while M, L_1, L_2, R_1, R_2 are constants.

Assuming $M^2 < L_1L_2$, show that

$$A)\ (L_1L_2 - M^2)\frac{d^2 i_1}{dt^2} + (R_1L_2 + R_2L_1)\frac{di_1}{dt} + R_1R_2 i_1 = R_2E(t) + L_2E'(t),$$

$$B)\ (L_1L_2 - M^2)\frac{d^2 i_2}{dt^2} + (R_1L_2 + R_2L_1)\frac{di_2}{dt} + R_1R_2 i_2 = -ME'(t).$$

Solve the system when $E(t) = E_0$, a constant.

Differentiating 1) and 2) with respect to t,

$$3)\ M\frac{d^2 i_1}{dt^2} + L_2\frac{d^2 i_2}{dt^2} + R_2\frac{di_2}{dt} = 0, \qquad 4)\ M\frac{d^2 i_2}{dt^2} + L_1\frac{d^2 i_1}{dt^2} + R_1\frac{di_1}{dt} = E'(t).$$

Multiplying 3) by M and 4) by L_2, and subtracting,

$$(L_1L_2 - M^2)\frac{d^2 i_1}{dt^2} + R_1L_2\frac{di_1}{dt} - MR_2\frac{di_2}{dt} = L_2E'(t).$$

Substituting for $\frac{di_2}{dt}$ from 2), we obtain A).

Multiplying 3) by L_1 and 4) by M, and subtracting,

$$(L_1L_2 - M^2)\frac{d^2 i_2}{dt^2} + R_2L_1\frac{di_2}{dt} - R_1M\frac{di_1}{dt} = -ME'(t).$$

Substituting for $\frac{di_1}{dt}$ from 1), we obtain B).

When $E(t) = E_0$, equation A) is $(L_1L_2 - M^2)\frac{d^2 i_1}{dt^2} + (R_1L_2 + R_2L_1)\frac{di_1}{dt} + R_1R_2 i_1 = R_2E_0$.

Let $\alpha, \beta = \frac{1}{2}\,\frac{-(R_1L_2 + R_2L_1) \pm \sqrt{(R_1L_2 - R_2L_1)^2 + 4M^2R_1R_2}}{L_1L_2 - M^2}$ denote the characteristic roots.

Then $$i_1 = C_1 e^{\alpha t} + C_2 e^{\beta t} + \frac{E_0}{R_1}.$$

To find i_2, multiply 1) by M and 2) by L_2, and subtract to obtain

$$MR_2 i_2 = (L_1 L_2 - M^2)\frac{di_1}{dt} + L_2 R_1 i_1 - L_2 E_0.$$

Then $$i_2 = \frac{1}{MR_2}[(L_1L_2 - M^2)(\alpha C_1 e^{\alpha t} + \beta C_2 e^{\beta t}) + L_2 R_1 (C_1 e^{\alpha t} + C_2 e^{\beta t})].$$

Note that since $M^2 < L_1 L_2$, both α and β are negative. Then after a time, the primary current becomes approximately constant $= E_0/R_1$ and the secondary current i_2 becomes negligible.

9. A moving particle of mass m is attracted to a fixed point O by a central force which varies inversely as the square of the distance of the particle from O. Show that the equation of its path is a conic having the fixed point as focus.

Using polar coordinates with O as pole, the equations of motion are

1) $$m\left[\frac{d^2\rho}{dt^2} - \rho\left(\frac{d\theta}{dt}\right)^2\right] = -\frac{K}{\rho^2} = -\frac{mk^2}{\rho^2} \quad \text{or} \quad \frac{d^2\rho}{dt^2} - \rho\left(\frac{d\theta}{dt}\right)^2 = -\frac{k^2}{\rho^2},$$

2) $$m\left[2\frac{d\rho}{dt}\frac{d\theta}{dt} + \rho\frac{d^2\theta}{dt^2}\right] = 0 \quad \text{or} \quad 2\frac{d\rho}{dt}\frac{d\theta}{dt} + \rho\frac{d^2\theta}{dt^2} = 0.$$

From 2), $\frac{d}{dt}(\rho^2 \frac{d\theta}{dt}) = 0$ and $\rho^2 \frac{d\theta}{dt} = C_1$.

Let $\sigma = \frac{1}{\rho}$. Then $\frac{d\theta}{dt} = \frac{C_1}{\rho^2} = C_1\sigma^2$, $\frac{d\rho}{dt} = \frac{d\rho}{d\sigma}\frac{d\sigma}{dt} = -\frac{1}{\sigma^2}\frac{d\sigma}{d\theta}\frac{d\theta}{dt} = -C_1\frac{d\sigma}{d\theta}$, and

$\frac{d^2\rho}{dt^2} = \frac{d}{dt}(-C_1\frac{d\sigma}{d\theta}) = -C_1\frac{d^2\sigma}{d\theta^2}\frac{d\theta}{dt} = -C_1^2\sigma^2\frac{d^2\sigma}{d\theta^2}$. Substituting in 1) and simplifying, we have

1') $\frac{d^2\sigma}{d\theta^2} + \sigma = \frac{k^2}{C_1^2}$, a linear equation with constant coefficients. Solving,

$$\sigma = C_2\cos(\theta + C_3) + \frac{k^2}{C_1^2} \quad \text{or} \quad \rho = \frac{1}{\frac{k^2}{C_1^2} + C_2\cos(\theta + C_3)} = \frac{C_1^2/k^2}{1 + \frac{C_2C_1^2}{k^2}\cos(\theta + C_3)}.$$

Writing $C_1^2/k^2 = l$, $|C_2C_1^2/k^2| = e$, $C_3 = \alpha$, this becomes $\rho = \frac{l}{1 \pm e\cos(\theta + \alpha)}$, the equation of a conic having O as focus.

SUPPLEMENTARY PROBLEMS

10. Find the family of curves orthogonal to the family of surfaces $x^2 + y^2 + 2z^2 = C$.

Ans. $y = Ax,\ z = By^2$

11. Find the family of surfaces orthogonal to the family of curves $y = C_1 x,\ x^2 + y^2 + 2z^2 = C_2$.

Ans. $z = C(x^2 + y^2)$

12. A particle of mass m is attracted to the origin O by a force varying directly as its distance from O. If it starts at $(a, 0)$ with velocity v_0 in a direction making an angle θ with the horizontal, find the position at time t.

Ans. $x = a\cos kt + \dfrac{v_0 \cos\theta}{k}\sin kt,\quad y = \dfrac{v_0 \sin\theta}{k}\sin kt$

13. The currents $i_1,\ i_2,\ i = i_1 + i_2$ in a certain network satisfy the equations

$$20i + 0.1\frac{di_2}{dt} = 5, \qquad 4i + i_1 + 1000q_1 = 1.$$

Determine the currents subject to the initial conditions $i = i_1 = i_2 = 0$ when $t = 0$.

Hint: Use $i_1 = \dfrac{dq_1}{dt}$ to obtain $\dfrac{d^2 q_1}{dt^2} + 240\dfrac{dq_1}{dt} + 40{,}000 q_1 = 0$.

Ans. $i_1 = -\dfrac{1}{4}e^{-120t}\sin 160t,\quad i_2 = \dfrac{1}{4}(1 - e^{-120t}\cos 160t) + \dfrac{1}{8}e^{-120t}\sin 160t$

14. Initially tank I contains 100 gal of brine with 200 lb of salt, and tank II contains 50 gal of fresh water. Brine from tank I runs into tank II at 3 gal/min, and from tank II into tank I at 2 gal/min. If each tank is kept well stirred, how much salt will tank I contain after 50 minutes?

Hint: $q_1 + q_2 = 200,\quad \dfrac{dq_1}{dt} = \dfrac{2q_2}{50 + t} - \dfrac{3q_1}{100 - t}$. *Ans.* 68.75 lb

CHAPTER 24

Numerical Approximations to Solutions

IN MANY APPLICATIONS it is required to find the value $\bar{y}$ of y corresponding to $x = x_0+h$ from the particular solution of a given differential equation

1) $$y' = f(x,y)$$

satisfying the initial conditions $y=y_0$ when $x=x_0$. Such problems have been solved by first finding the primitive

2) $$y = F(x) + C$$

of 1), then selecting the particular solution

3) $$y = g(x)$$

through (x_0, y_0), and finally computing the required value $\bar{y} = g(x_0+h)$.

When no method is available for finding the primitive, it is necessary to use some procedure for approximating the desired value. Integrating 1) between the limits $x=x_0, y=y_0$ and $x=x, y=y$, we obtain

4) $$y = y_0 + \int_{x_0}^{x} f(x,y)\,dx.$$

The value of y when $x=x_0+h$ is then

5) $$\bar{y} = y_0 + \int_{x_0}^{x_0+h} f(x,y)\,dx.$$

The methods of this chapter will consist of procedures for approximating 4) or 5).

PICARD'S METHOD. For values of x near $x=x_0$, the corresponding value of $y = g(x)$ is near $y_0 = g(x_0)$. Thus, a first approximation y_1 of $y=g(x)$ is obtained by replacing y by y_0 in the right member of 4), that is,

$$y_1 = y_0 + \int_{x_0}^{x} f(x,y_0)\,dx.$$

A second approximation, y_2, is then obtained by replacing y by y_1 in the right member of 4), that is,

$$y_2 = y_0 + \int_{x_0}^{x} f(x,y_1)\,dx.$$

Continuing this procedure, a succession of functions of x

$$y_0,\ y_1,\ y_2,\ y_3,\ \cdots\cdots$$

is obtained, each giving a better approximation of the required solution than the preceding one. See Problems 1-2.

Picard's method is of considerable theoretical value. In general, it is unsatisfactory as a practical means of approximation because of difficulties which arise in performing the necessary integrations.

TAYLOR SERIES. The Taylor expansion of $y = g(x)$ near (x_0, y_0) is

$$6)\quad y = g(x_0) + (x-x_0)\,g'(x_0) + \frac{1}{2}(x-x_0)^2\,g''(x_0) + \frac{1}{6}(x-x_0)^3\,g'''(x_0) + \cdots\cdots$$

From 1), $y' = g'(x) = f(x,y)$; hence, by repeated differentiation,

$$y'' = g''(x) = \frac{\partial f}{\partial x} + \frac{\partial f}{\partial y}\frac{dy}{dx} = \frac{\partial f}{\partial x} + f\frac{\partial f}{\partial y},$$

$$7)\quad y''' = g'''(x) = \frac{d}{dx}\left(\frac{\partial f}{\partial x} + f\frac{\partial f}{\partial y}\right) = \left(\frac{\partial}{\partial x} + f\frac{\partial}{\partial y}\right)\left(\frac{\partial f}{\partial x} + f\frac{\partial f}{\partial y}\right)$$

$$= \frac{\partial^2 f}{\partial x^2} + \frac{\partial f}{\partial x}\frac{\partial f}{\partial y} + 2f\frac{\partial^2 f}{\partial x\,\partial y} + f\left(\frac{\partial f}{\partial y}\right)^2 + f^2\frac{\partial^2 f}{\partial y^2}, \qquad \text{etc.}$$

For convenience, write $p = \frac{\partial f}{\partial x}$, $q = \frac{\partial f}{\partial y}$, $r = \frac{\partial^2 f}{\partial x^2}$, $s = \frac{\partial^2 f}{\partial x\,\partial y}$, $t = \frac{\partial^2 f}{\partial y^2}$ and let $f_0, p_0, q_0, \cdots$ denote the values of $f, p, q, \cdots$ at (x_0, y_0). Substituting in 6) the results of 7) and evaluating for $x = x_0 + h$, we obtain

$$8)\quad \bar{y} = y_0 + h\cdot f_0 + \frac{1}{2}h^2(p_0 + f_0\cdot q_0) + \frac{1}{6}h^3(r_0 + p_0\cdot q_0 + 2f_0\cdot s_0 + f_0\cdot q_0^2 + f_0^2\cdot t_0)$$
$$+ \cdots\cdots\cdots$$

This series may be used to compute $\bar{y}$; it is evident, however, that additional terms will be increasingly complex. See Problems 3-4.

FIRST DERIVATIVE METHOD. A procedure involving only first derivatives, that is, using only the first two terms of Taylor series, follows.

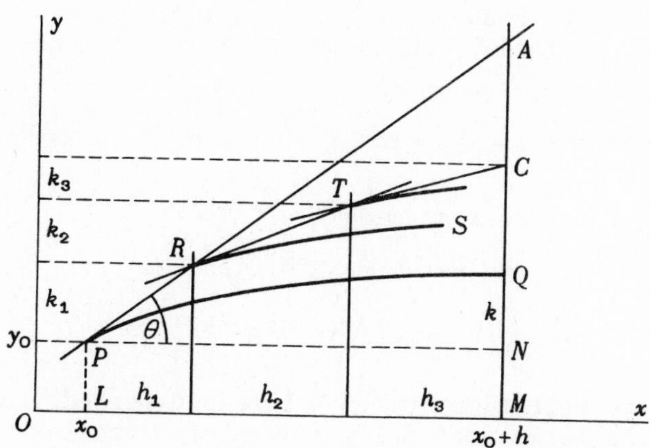

As a first approximation of $\bar{y}$, take the first two terms of 8)

$$\bar{y} \approx y_0 + h\,f(x_0, y_0).$$

To interpret this approximation geometrically, let PQ be the integral curve of 1) through $P(x_0, y_0)$ and let Q be the point on the curve corresponding to $x = x_0 + h$. Then $\bar{y} = MQ = y_0 + k$. If θ is the angle of inclination of the tangent at P, then from 1) $\tan\theta = f(x_0, y_0)$ and the approximation

$$y_0 + h\,f(x_0, y_0) = LP + h\tan\theta = MN + NA = MA.$$

To obtain a better approximation, let the interval LM of width h be divided into n subintervals of widths $h_1, h_2, \cdots h_n$. (In the figure, $n = 3$.) Let the line $x = x_0 + h_1$ meet PA in $R(x_0+h_1, y_0+k_1) = (x_1, y_1)$. Then

$$y_1 = y_0 + k_1 = y_0 + h_1 f(x_0, y_0).$$

Let RS be the integral curve of 1) through R, and on its tangent at R take T having coordinates $(x_1+h_2, y_1+k_2) = (x_2, y_2)$. Then

$$y_2 = y_1 + k_2 = y_1 + h_2 f(x_1, y_1) = y_1 + h_2 f(x_0+h_1,\ y_0+h_1 f_0).$$

After a sufficient number of repetitions, we reach finally an approximation MC of MQ. It is clear from the figure that the accuracy will increase as the number of subintervals is increased in such a manner that the widths of the subintervals decrease. See Problems 5-6.

RUNGE'S METHOD. From 5) and 8) we obtain

$$9)\quad k = \bar{y} - y_0 = \int_{x_0}^{x_0+h} f(x,y)\,dx$$

$$= h f_0 + \frac{1}{2}h^2(p_0 + f_0 q_0) + \frac{1}{6}h^3(r_0 + p_0 q_0 + 2f_0 s_0 + f_0 q_0^2 + f_0^2 t_0) + \cdots\cdots$$

Assume for the moment that the values y_0, y_1, y_2 of $y = g(x)$ corresponding to $x_0,\ x_1 = x_0 + \frac{1}{2}h,\ x_2 = x_0 + h$ are known. Then by Simpson's Rule,

$$10)\quad k = \int_{x_0}^{x_0+h} f(x,y)\,dx \approx \frac{h}{6}[f(x_0\ y_0) + 4f(x_0+\tfrac{1}{2}h,\ y_1) + f(x_0+h,\ y_2)].$$

Actually, only y_0 is known. Runge's Method is based on certain approximations of y_1 and y_2,

$$y_1 \approx y_0 + \tfrac{1}{2}hf(x_0, y_0) = y_0 + \tfrac{1}{2}hf_0,$$

$$y_2 \approx y_0 + hf(x_0+h,\ y_0+hf_0),$$

chosen so that when k, found by 10), is expanded as a power series in h the first three terms coincide with those of the right member of 9). Thus 10) becomes

$$11)\quad k \approx \frac{h}{6}\{f_0 + 4f(x_0+\tfrac{1}{2}h,\ y_0+\tfrac{1}{2}hf_0) + f[x_0+h,\ y_0+hf(x_0+h,\ y_0+hf_0)]\}.$$

These calculations are best made as follows:

$$k_1 = hf_0,\quad k_2 = hf(x_0+h, y_0+k_1),\quad k_3 = hf(x_0+h, y_0+k_2),\quad k_4 = hf(x_0+\tfrac{1}{2}h, y_0+\tfrac{1}{2}k_1),$$

$$k \approx \frac{1}{6}(k_1 + 4k_4 + k_3).$$

Note. Since the approximation of k obtained here differs from the value as given by 8) in the terms containing powers of h greater than 3, the approximation may be poor if $f_0 > 1$. See Problems 7-11.

KUTTA-SIMPSON METHOD. Various modifications of the Runge Method have been made by Kutta. One of these, known as Kutta's Simpson's Rule uses the following calculations:

$$k_1 = hf_0,\quad k_2 = hf(x_0+\tfrac{1}{2}h, y_0+\tfrac{1}{2}k_1),\quad k_3 = hf(x_0+\tfrac{1}{2}h, y_0+\tfrac{1}{2}k_2),\quad k_4 = hf(x_0+h, y_0+k_3),$$

$$k \approx \frac{1}{6}(k_1 + 2k_2 + 2k_3 + k_4).$$

See Problem 12.

SIMULTANEOUS FIRST ORDER DIFFERENTIAL EQUATIONS. Approximations to that solution of the simultaneous differential equations

$$\frac{dy}{dx} = f(x,y,z), \qquad \frac{dz}{dx} = g(x,y,z)$$

for which $y = y_0$ and $z = z_0$ when $x = x_0$, may be obtained by the use of Picard's Method, Taylor Series, Runge's Method, or Kutta-Simpson Method. The necessary modifications of the formulas given above are made in Solved Problems 13-14. Further extensions to three or more simultaneous first order equations may be readily made.

DIFFERENTIAL EQUATIONS OF ORDER n. The differential equation

$$\frac{d^n y}{dx^n} = f(x,y,y',y'',\cdots\cdots,y^{n-1})$$

where $y' = \frac{dy}{dx}$, $y'' = \frac{d^2y}{dx^2}$, ..., may be reduced to the system of simultaneous first order equations

$$\frac{dy}{dx} = y_1, \quad \frac{dy_1}{dx} = y_2, \quad \cdots\cdots\cdots, \quad \frac{dy_{n-2}}{dx} = y_{n-1}, \quad \frac{dy_{n-1}}{dx} = f(x,y,y_1,y_2,\cdots\cdots,y_{n-1}).$$

When initial conditions $x = x_0$, $y = y_0$, $y' = (y_1)_0$, $y'' = (y_2)_0$, ..., $y^{n-1} = (y_{n-1})_0$ are given, the methods of the preceding paragraph apply.

EXAMPLE. The second order differential equation $\frac{d^2y}{dx^2} + 2x\frac{dy}{dx} - 4y = 0$ is equivalent to the system of simultaneous first order differential equations

$$\frac{dy}{dx} = z, \qquad \frac{dz}{dx} = 4y - 2xz.$$

See Problems 15-16.

SOLVED PROBLEMS

1. Use Picard's Method to approximate y when $x = 0.2$, given that $y = 1$ when $x = 0$, and $dy/dx = x - y$.

Here $f(x,y) = x - y$, $x_0 = 0$, $y_0 = 1$. Then

$$y_1 = y_0 + \int_0^x f(x,y_0)\,dx = 1 + \int_0^x (x-1)\,dx = \frac{1}{2}x^2 - x + 1,$$

$$y_2 = y_0 + \int_0^x f(x,y_1)\,dx = 1 + \int_0^x (-\frac{1}{2}x^2 + 2x - 1)\,dx = -\frac{1}{6}x^3 + x^2 - x + 1,$$

$$y_3 = y_0 + \int_0^x f(x,y_2)\,dx = 1 + \int_0^x (\frac{1}{6}x^3 - x^2 + 2x - 1)\,dx = \frac{1}{24}x^4 - \frac{1}{3}x^3 + x^2 - x + 1,$$

$$y_4 = y_0 + \int_0^x f(x,y_3)\,dx = 1 + \int_0^x (-\frac{1}{24}x^4 + \frac{1}{3}x^3 - x^2 + 2x - 1)\,dx = -\frac{x^5}{120} + \frac{x^4}{12} - \frac{x^3}{3} + x^2 - x + 1,$$

$$y_5 = \frac{1}{720}x^6 - \frac{1}{60}x^5 + \frac{1}{12}x^4 - \frac{1}{3}x^3 + x^2 - x + 1, \quad \cdots\cdots\cdots$$

When $x = 0.2$, $y_0 = 1$, $y_1 = 0.82$, $y_2 = 0.83867$, $y_3 = 0.83740$, $y_4 = 0.83746$, $y_5 = 0.83746$. Thus, to five decimal places, $\bar{y} = 0.83746$.

Note. The primitive of the given differential equation is $y = x - 1 + Ce^{-x}$. The particular solution satisfying the initial conditions $x = 0, y = 1$ is $y = x - 1 + 2e^{-x}$. Replacing e^{-x} by its MacLaurin series, we have $y = 1 - x + x^2 - \frac{1}{3}x^3 + \frac{1}{12}x^4 - \frac{1}{60}x^5 + \frac{1}{360}x^6 + \cdots\cdots\cdots$. Upon comparing this with the successive approximations obtained above, it seems reasonable to suppose that the sequence of approximations given by Picard's Method tends to the exact solution as a limit.

2. Use Picard's Method to approximate the value of y when $x = 0.1$, given that $y = 1$ when $x = 0$, and $dy/dx = 3x + y^2$.

Here $f(x,y) = 3x + y^2$, $x_0 = 0$, $y_0 = 1$. Then

$$y_1 = y_0 + \int_0^x (3x + y_0^2)\,dx = 1 + \int_0^x (3x+1)\,dx = \frac{3}{2}x^2 + x + 1,$$

$$y_2 = y_0 + \int_0^x (3x + y_1^2)\,dx = 1 + \int_0^x (\frac{9}{4}x^4 + 3x^3 + 4x^2 + 5x + 1)\,dx = \frac{9}{20}x^5 + \frac{3}{4}x^4 + \frac{4}{3}x^3 + \frac{5}{2}x^2 + x + 1,$$

$$y_3 = 1 + \int_0^x (\frac{81}{400}x^{10} + \frac{27}{40}x^9 + \frac{141}{80}x^8 + \frac{17}{4}x^7 + \frac{1157}{180}x^6 + \frac{136}{15}x^5 + \frac{125}{12}x^4 + \frac{23}{3}x^3 + 6x^2 + 5x + 1)\,dx$$

$$= \frac{81}{4400}x^{11} + \frac{27}{400}x^{10} + \frac{47}{240}x^9 + \frac{17}{32}x^8 + \frac{1157}{1260}x^7 + \frac{68}{45}x^6 + \frac{25}{12}x^5 + \frac{23}{12}x^4 + 2x^3 + \frac{5}{2}x^2 + x + 1.$$

When $x = 0.1$, $y_0 = 1$, $y_1 = 1.115$, $y_2 = 1.1264$, $y_3 = 1.12721$.

3. If $\frac{dy}{dx} = x - y$, use the Taylor Series Method to approximate y when:

a) $x = 0.2$, given that $y = 1$ when $x = 0$.
b) $x = 1.6$, given that $y = 0.4$ when $x = 1$.

a) Here
$y = g(x)$, $g(x_0) = 1$, $y''' = g'''(x) = -y''$, $g'''(x_0) = -2$,
$y' = g'(x) = x - y$, $g'(x_0) = -1$, $y^{iv} = g^{iv}(x) = -y'''$, $g^{iv}(x_0) = 2$,
$y'' = g''(x) = 1 - y'$, $g''(x_0) = 2$, $y^{v} = g^{v}(x) = -y^{iv}$, $g^{v}(x_0) = -2$, etc.,

and equation 6) becomes $y = 1 - x + x^2 - \frac{1}{3}x^3 + \frac{1}{12}x^4 - \frac{1}{60}x^5 + \cdots\cdots$. Then

$$\bar{y} = 1 - 0.2 + 0.04 - \frac{1}{3}(0.008) + \frac{1}{12}(0.0016) - \frac{1}{60}(0.00032) + \cdots \approx 0.83746.$$ (See Problem 1.)

b) Here $g(x_0) = 0.4$, $g'(x_0) = 0.6$, $g''(x_0) = 0.4$, $g'''(x_0) = -0.4$, $g^{iv}(x_0) = 0.4$, etc., and equation 6) becomes

$$y = 0.4 + 0.6h + 0.4\frac{h^2}{2} - 0.4\frac{h^3}{6} + 0.4\frac{h^4}{24} - 0.4\frac{h^5}{120} + 0.4\frac{h^6}{720} + \cdots\cdots,$$ where $h = x - x_0$.

When $x = 1.6$, $h = 0.6$ and

$$\bar{y} = 0.4 + 0.6(0.6) + 0.4(0.18) - 0.4(0.036) + 0.4(0.0054) - 0.4(0.000648) + 0.4(0.0000648) + \cdots\cdots$$

$\approx 0.81953.$

4. If $\frac{dy}{dx} = 3x + y^2$, use the Taylor Series Method to approximate y when:

a) $x = 0.1$, given that $y = 1$ when $x = 0$.
b) $x = 1.1$, given that $y = 1.2$ when $x = 1$.

a) Here $(x_0, y_0) = (0, 1)$, $g(x_0) = 1$,

$$y' = g'(x) = 3x + y^2, \qquad g'(x_0) = 1,$$
$$y'' = g''(x) = 3 + 2yy', \qquad g''(x_0) = 5,$$
$$y''' = g'''(x) = 2(y')^2 + 2yy'', \qquad g'''(x_0) = 12,$$
$$y^{iv} = g^{iv}(x) = 6y'y'' + 2yy''', \qquad g^{iv}(x_0) = 54,$$
$$y^{v} = g^{v}(x) = 6(y'')^2 + 8y'y''' + 2yy^{iv}, \qquad g^{v}(x_0) = 354,$$ and 6) becomes

$$y = 1 + x + \frac{5}{2}x^2 + 2x^3 + \frac{9}{4}x^4 + \frac{177}{60}x^5 + \cdots\cdots\cdots\cdots.$$ When $x = 0.1$,

$$\bar{y} = 1 + 0.1 + 0.025 + 0.002 + 0.00022 + 0.00003 + \cdots\cdots \approx 1.12725.$$ (See Problem 2.)

b) Here $(x_0, y_0) = (1, 1.2)$, $g(x_0) = 1.2$, $g'(x_0) = 4.44$, $g''(x_0) = 13.656$, $g'''(x_0) = 72.202$, $g^{iv}(x_0) = 537.078$, $g^{v}(x_0) = 4973$, $\cdots\cdots\cdots\cdots$, and 6) becomes

$$y = 1.2 + 4.44h + 13.656\frac{h^2}{2} + 72.202\frac{h^3}{6} + 537.078\frac{h^4}{24} + 4973\frac{h^5}{120} + \cdots\cdots\cdots\cdots,$$

where $h = x - x_0$. When $x = 1.1$, $h = 0.1$ and

$$\bar{y} = 1.2 + 0.1(4.44) + 0.01(6.828) + 0.001(12.03) + 0.0001(22.4) + 0.00001(41) + \cdots \approx 1.7270.$$

5. Use the First Derivative Method, with $n = 4$, to approximate y when $x = 1.1$, given that $y = 1.2$ when $x = 1$ and $dy/dx = 3x + y^2$. See Problem 4b.

Here $h = 0.1$ and we take $h_1 = h_2 = h_3 = h_4 = 0.025$. We seek $y_0 + k_1 + k_2 + k_3 + k_4 = y_3 + k_4$.

a) $(x_0, y_0) = (1, 1.2)$, $h_1 = 0.025$, $f(x_0, y_0) = 4.44$, $k_1 = h_1 f(x_0, y_0) = 0.111$;
$y_1 = y_0 + k_1 = 1.311$.

b) $(x_1, y_1) = (1.025, 1.311)$, $h_2 = 0.025$, $f(x_1, y_1) = 4.7937$, $k_2 = h_2 f(x_1, y_1) = 0.1198$;
$y_2 = y_1 + k_2 = 1.4308$.

c) $(x_2, y_2) = (1.05, 1.4308)$, $h_3 = 0.025$, $f(x_2, y_2) = 5.1972$, $k_3 = h_3 f(x_2, y_2) = 0.1299$;
$y_3 = y_2 + k_3 = 1.5607$.

d) $(x_3, y_3) = (1.075, 1.5607)$, $h_4 = 0.025$, $f(x_3, y_3) = 5.6608$, $k_4 = h_4 f(x_3, y_3) = 0.1415$;
$\bar{y} \approx y_3 + k_4 = 1.7022$.

6. Use the First Derivative Method, with $n = 4$, to approximate y when $x = 1.4$, given that $y = 0.2$ when $x = 1$ and $\frac{dy}{dx} = (x^2 + 2y)^{1/2}$.

Here $h = 0.4$ and we take $h_1 = h_2 = h_3 = h_4 = 0.1$.

a) $(x_0, y_0) = (1, 0.2)$, $h_1 = 0.1$, $f(x_0, y_0) = \sqrt{1.4} = 1.183$, $k_1 = h_1 f(x_0, y_0) = 0.1183$;
$y_1 = y_0 + k_1 = 0.3183$.

b) $(x_1, y_1) = (1.1, 0.3183)$, $h_2 = 0.1$, $f(x_1, y_1) = 1.359$, $k_2 = h_2 f(x_1, y_1) = 0.1359$;
$y_2 = y_1 + k_2 = 0.4542$.

c) $(x_2, y_2) = (1.2, 0.4542)$, $h_3 = 0.1$, $f(x_2, y_2) = 1.532$, $k_3 = h_3 f(x_2, y_2) = 0.1532$;
$y_3 = y_2 + k_3 = 0.6074$.

d) $(x_3, y_3) = (1.3, 0.6074)$, $h_4 = 0.1$, $f(x_3, y_3) = 1.704$, $k_4 = h_4 f(x_3, y_3) = 0.1704$;

$$\bar{y} \approx y_3 + k_4 = 0.7778.$$

7. Use Runge's Method to approximate y when $x = 1.6$, given that $y = 0.4$ when $x = 1$ and $dy/dx = x - y$. (See Problem 3*b*.)

Here $(x_0, y_0) = (1, 0.4)$, $h = 0.6$, $f_0 = 1 - 0.4 = 0.6$. Then

$$k_1 = hf_0 = 0.36,$$

$$k_2 = hf(x_0+h,\ y_0+k_1) = 0.6[(1+0.6) - (0.4+0.36)] = 0.504,$$

$$k_3 = hf(x_0+h,\ y_0+k_2) = 0.6[(1+0.6) - (0.4+0.504)] = 0.4176,$$

$$k_4 = hf(x_0+\tfrac{1}{2}h,\ y_0+\tfrac{1}{2}k_1) = 0.6[(1+0.3) - (0.4+0.18)] = 0.432,$$

$$k \approx \frac{1}{6}(k_1 + 4k_4 + k_3) = \frac{1}{6}[0.36 + 4(0.432) + 0.4176] = 0.4176, \quad \text{and} \quad \bar{y} = y_0 + k \approx 0.8176.$$

The difference between this approximation and that found in Problem 3*b* arises from the fact that $h = 0.6$. In finding the value of y when $x = 1.1$, (that is, $h = 0.1$), the Taylor series gives $\bar{y} = 0.4 + 0.6(0.1) + 0.4(0.005) - 0.4(0.00017) + 0.4(0.000004) - \cdots\cdots \approx 0.46193$, while by Runge's Method

$$k_1 = 0.1(0.6) = 0.06, \quad k_2 = 0.1(1.1 - 0.46) = 0.064, \quad k_3 = 0.1(1.1 - 0.464) = 0.0636,$$

$$k_4 = 0.1(1.05 - 0.43) = 0.062, \quad k \approx \frac{1}{6}(k_1 + 4k_4 + k_3) = 0.06193, \quad \text{and} \quad \bar{y} \approx 0.46193.$$

8. Use Runge's Method to approximate y when $x = 0.1$, given that $y = 1$ when $x = 0$ and $dy/dx = 3x + y^2$.

Here $(x_0, y_0) = (0, 1)$, $h = 0.1$, $f_0 = 1$. Then

$$k_1 = hf_0 = 0.1,$$

$$k_2 = hf(x_0+h,\ y_0+k_1) = 0.1[3(0+0.1) + (1+0.1)^2] = 0.151,$$

$$k_3 = hf(x_0+h,\ y_0+k_2) = 0.1[3(0+0.1) + (1+0.151)^2] = 0.16248,$$

$$k_4 = hf(x_0+\tfrac{1}{2}h,\ y_0+\tfrac{1}{2}k_1) = 0.1[3(0+0.05) + (1+0.05)^2] = 0.12525,$$

$$k \approx \frac{1}{6}(k_1 + 4k_4 + k_3) = \frac{1}{6}[0.1 + 4(0.12525) + 0.16248] = 0.12725, \quad \text{and} \quad \bar{y} = y_0 + k \approx 1.12725.$$

(See Problems 2 and 4*a*.)

9. Use Runge's Method to approximate y when $x = 1.1$, given that $y = 1.2$ when $x = 1$ and $dy/dx = 3x + y^2$.

Here $(x_0, y_0) = (1, 1.2)$, $h = 0.1$, $f_0 = 4.44$. Then

$$k_1 = hf_0 = 0.444,$$

$$k_2 = hf(x_0+h,\ y_0+k_1) = 0.1[3(1+0.1) + (1.2+0.444)^2] = 0.600274,$$

$$k_3 = hf(x_0+h,\ y_0+k_2) = 0.1[3(1+0.1) + (1.2+0.60027)^2] = 0.654097,$$

$$k_4 = hf(x_0+\tfrac{1}{2}h,\ y_0+\tfrac{1}{2}k_1) = 0.1[3(1+0.05) + (1.2+0.222)^2] = 0.517208,$$

$$k \approx \frac{1}{6}(k_1 + 4k_4 + k_3) = \frac{1}{6}[0.444 + 4(0.517208) + 0.654097] = 0.527822, \quad \text{and}$$

$$\bar{y} = y_0 + k \approx 1.727822.$$

Comparing this result with that obtained in Problem 4*b*, it is to be noted that the approximation is better than might have been expected in view of the value $f_0 = 4.44$.

10. Use Runge's Method to approximate y when $x = 0.8$ for that particular solution of $dy/dx = \sqrt{x+y}$ satisfying $y = 0.41$ when $x = 0.4$.

Here $(x_0, y_0) = (0.4,\ 0.41)$, $h = 0.4$, $f_0 = \sqrt{0.81} = 0.9$. Then

$$k_1 = hf_0 = 0.36,$$
$$k_2 = hf(x_0+h,\ y_0+k_1) = 0.4\sqrt{1.57} = 0.50120,$$
$$k_3 = hf(x_0+h,\ y_0+k_2) = 0.4\sqrt{1.7112} = 0.52325,$$
$$k_4 = hf(x_0+\tfrac{1}{2}h,\ y_0+\tfrac{1}{2}k_1) = 0.4\sqrt{1.19} = 0.43635,$$
$$k \approx \frac{1}{6}(k_1 + 4k_4 + k_3) = 0.43811, \quad \text{and} \quad \bar{y} = y_0 + k \approx 0.84811.$$

11. Solve Problem 10, first approximating y when $x = 0.6$ and then, using this pair of values as (x_0, y_0), approximate the required value of y.

First, $(x_0, y_0) = (0.4,\ 0.41)$, $h = 0.2$, $f_0 = \sqrt{0.81} = 0.9$. Then

$$k_1 = hf_0 = 0.18,$$
$$k_2 = hf(x_0+h,\ y_0+k_1) = 0.2\sqrt{1.19} = 0.21817,$$
$$k_3 = hf(x_0+h,\ y_0+k_2) = 0.2\sqrt{1.22817} = 0.22165,$$
$$k_4 = hf(x_0+\tfrac{1}{2}h,\ y_0+\tfrac{1}{2}k_1) = 0.2,$$
$$k \approx \frac{1}{6}(k_1 + 4k_4 + k_3) = 0.20028, \quad \text{and} \quad \bar{y} = y_0 + k \approx 0.61028.$$

Next, take $(x_0, y_0) = (0.6,\ 0.61028)$, $h = 0.2$. Then $f_0 = \sqrt{1.21028} = 1.1001$,

$$k_1 = hf_0 = 0.22002,$$
$$k_2 = hf(x_0+h,\ y_0+k_1) = 0.2\sqrt{1.63030} = 0.25537,$$
$$k_3 = hf(x_0+h,\ y_0+k_2) = 0.2\sqrt{1.66565} = 0.25812,$$
$$k_4 = hf(x_0+\tfrac{1}{2}h,\ y_0+\tfrac{1}{2}k_1) = 0.2\sqrt{1.42029} = 0.23836,$$
$$k \approx \frac{1}{6}(k_1 + 4k_4 + k_3) = 0.23860, \quad \text{and} \quad \bar{y} = y_0 + k \approx 0.84888.$$

12. Solve Problem 10, using the Kutta-Simpson Method.

Here $(x_0, y_0) = (0.4,\ 0.41)$, $h = 0.4$, $f_0 = \sqrt{0.81} = 0.9$. Then

$$k_1 = hf_0 = 0.36,$$
$$k_2 = hf(x_0+\tfrac{1}{2}h,\ y_0+\tfrac{1}{2}k_1) = 0.4\sqrt{1.19} = 0.43635,$$
$$k_3 = hf(x_0+\tfrac{1}{2}h,\ y_0+\tfrac{1}{2}k_2) = 0.4\sqrt{1.22817} = 0.44329,$$
$$k_4 = hf(x_0+h,\ y_0+k_3) = 0.4\sqrt{1.65329} = 0.51432,$$
$$k \approx \frac{1}{6}(k_1 + 2k_2 + 2k_3 + k_4) = 0.43893, \quad \text{and} \quad \bar{y} = y_0 + k \approx 0.84893.$$

13. Use Picard's Method to approximate y and z corresponding to $x = 0.1$ for that particular solution of

$$\frac{dy}{dx} = f(x,y,z) = x + z, \qquad \frac{dz}{dx} = g(x,y,z) = x - y^2$$

satisfying $y = 2$, $z = 1$ when $x = 0$.

For the first approximations,

$$y_1 = y_0 + \int_0^x f(x,y_0,z_0)\,dx = 2 + \int_0^x (1+x)\,dx = 2 + x + \tfrac{1}{2}x^2,$$

$$z_1 = z_0 + \int_0^x g(x,y_0,z_0)\,dx = 1 + \int_0^x (-4+x)\,dx = 1 - 4x + \tfrac{1}{2}x^2.$$

For the second approximations,

$$y_2 = y_0 + \int_0^x f(x,y_1,z_1)\,dx = 2 + \int_0^x (1-3x+\tfrac{1}{2}x^2)\,dx = 2 + x - \frac{3}{2}x^2 + \frac{1}{6}x^3,$$

$$z_2 = z_0 + \int_0^x g(x,y_1,z_1)\,dx = 1 + \int_0^x (-4-3x-3x^2-x^3-\tfrac{1}{4}x^4)\,dx$$

$$= 1 - 4x - \frac{3}{2}x^2 - x^3 - \frac{1}{4}x^4 - \frac{1}{20}x^5.$$

For the third approximations,

$$y_3 = y_0 + \int_0^x f(x,y_2,z_2)\,dx = 2 + \int_0^x (1 - 3x - \frac{3}{2}x^2 - x^3 - \frac{1}{4}x^4 - \frac{1}{20}x^5)\,dx$$

$$= 2 + x - \frac{3}{2}x^2 - \frac{1}{2}x^3 - \frac{1}{4}x^4 - \frac{1}{20}x^5 - \frac{1}{120}x^6,$$

$$z_3 = z_0 + \int_0^x g(x,y_2,z_2)\,dx = 1 + \int_0^x (-4 - 3x + 5x^2 + \frac{7}{3}x^3 - \frac{31}{12}x^4 + \frac{1}{2}x^5 - \frac{1}{36}x^6)\,dx$$

$$= 1 - 4x - \frac{3}{2}x^2 + \frac{5}{3}x^3 + \frac{7}{12}x^4 - \frac{31}{60}x^5 + \frac{1}{12}x^6 - \frac{1}{252}x^7,$$

and so on.

When $x = 0.1$:

$y_1 = 2.105$ $\quad z_1 = 0.605$

$y_2 = 2.08517$ $\quad z_2 = 0.58397$

$y_3 = 2.08447$ $\quad z_3 = 0.58672$.

14. Use Runge's Method to approximate y and z when $x = 0.3$ for that particular solution of the system $\dfrac{dy}{dx} = x + \sqrt{z} = f(x,y,z)$, $\dfrac{dz}{dx} = y - \sqrt{z} = g(x,y,z)$ satisfying $y = 0.5$, $z = 0$ when $x = 0.2$.

Here $(x_0,y_0,z_0) = (0.2, 0.5, 0)$, $h = 0.1$, $f_0 = 0.2$, $g_0 = 0.5$. Then

$$k_1 = hf_0 = 0.02,$$

$$l_1 = hg_0 = 0.05,$$

$$k_2 = hf(x_0+h,\ y_0+k_1,\ z_0+l_1) = 0.1(0.3 + \sqrt{0.05}) = 0.05236,$$

$$l_2 = hg(x_0+h,\ y_0+k_1,\ z_0+l_1) = 0.1(0.52 - \sqrt{0.05}) = 0.02964,$$

$$k_3 = hf(x_0+h,\ y_0+k_2,\ z_0+l_2) = 0.1(0.3 + \sqrt{0.02964}) = 0.047216,$$

$$l_3 = hg(x_0+h,\ y_0+k_2,\ z_0+l_2) = 0.1(0.52 - \sqrt{0.02964}) = 0.034784,$$

$$k_4 = hf(x_0+\tfrac{1}{2}h,\ y_0+\tfrac{1}{2}k_1,\ z_0+\tfrac{1}{2}l_1) = 0.1(0.25 + \sqrt{0.025}) = 0.040811,$$

$$l_4 = hg(x_0+\tfrac{1}{2}h,\ y_0+\tfrac{1}{2}k_1,\ z_0+\tfrac{1}{2}l_1) = 0.1(0.51 - \sqrt{0.025}) = 0.035189,$$

$$k \approx \frac{1}{6}(k_1 + 4k_4 + k_3) = 0.03841, \qquad l \approx \frac{1}{6}(l_1 + 4l_4 + l_3) = 0.03759,$$

and $\quad \bar{y} = y_0 + k \approx 0.53841, \qquad \bar{z} = z_0 + l \approx 0.03759.$

15. Use the Taylor Series Method to approximate the value of θ corresponding to $t = 0.05$ for that particular solution of $\dfrac{d^2\theta}{dt^2} = -8\sin\theta$ satisfying $\theta = \pi/4$, $\dfrac{d\theta}{dt} = 1$ when $t = 0$.

The given differential equation is equivalent to the system

$$\frac{d\theta}{dt} = \phi = f(t,\theta,\phi), \qquad \frac{d\phi}{dt} = -8\sin\theta = g(t,\theta,\phi)$$

with initial conditions $t=0$, $\theta=\pi/4$, $\phi=1$. Then

$$\frac{d\theta}{dt} = \theta' = \phi \qquad \theta_0' = 1 \qquad \phi' = -8\sin\theta \qquad \phi_0' = -4\sqrt{2}$$

$$\theta'' = \phi' \qquad \theta_0'' = -4\sqrt{2} \qquad \phi'' = -8\theta'\cos\theta \qquad \phi_0'' = -4\sqrt{2}$$

$$\theta''' = \phi'' \qquad \theta_0''' = -4\sqrt{2} \qquad \phi''' = 8(\theta')^2\sin\theta - 8\theta''\cos\theta$$

$$\theta^{iv} = \phi''' \qquad \theta_0^{iv} = 4\sqrt{2} + 32 \qquad \phi_0''' = 4\sqrt{2}(1+4\sqrt{2})$$

and $\theta = \pi/4 + t - 4\sqrt{2}\,\dfrac{t^2}{2} - 4\sqrt{2}\,\dfrac{t^3}{6} + 4(8+\sqrt{2})\,\dfrac{t^4}{24} + \cdots\cdots\cdots = 0.82821.$

16. Use the Kutta-Simpson Method to approximate y corresponding to $x=0.1$ for that particular solution of $\dfrac{d^2y}{dx^2} + 2x\dfrac{dy}{dx} - 4y = 0$ satisfying $y=0.2$, $\dfrac{dy}{dx} = 0.5$ when $x=0$.

The given equation with initial conditions is equivalent to the system

$$\frac{dy}{dx} = z = f(x,y,z), \qquad \frac{dz}{dx} = 4y - 2xz = g(x,y,z)$$

with initial conditions $x=0$, $y=0.2$, $z=0.5$.

Here $(x_0,y_0,z_0) = (0,\ 0.2,\ 0.5)$, $h=0.1$, $f_0 = 0.5$, $g_0 = 0.8$. Then

$$k_1 = hf_0 = 0.05,$$

$$l_1 = hg_0 = 0.08,$$

$$k_2 = hf(x_0+\tfrac{1}{2}h,\ y_0+\tfrac{1}{2}k_1,\ z_0+\tfrac{1}{2}l_1) = 0.1(0.54) = 0.054,$$

$$l_2 = hg(x_0+\tfrac{1}{2}h,\ y_0+\tfrac{1}{2}k_1,\ z_0+\tfrac{1}{2}l_1) = 0.1(0.846) = 0.0846,$$

$$k_3 = hf(x_0+\tfrac{1}{2}h,\ y_0+\tfrac{1}{2}k_2,\ z_0+\tfrac{1}{2}l_2) = 0.1(0.5423) = 0.05423,$$

$$l_3 = hg(x_0+\tfrac{1}{2}h,\ y_0+\tfrac{1}{2}k_2,\ z_0+\tfrac{1}{2}l_2) = 0.1(0.85377) = 0.085377,$$

$$k_4 = hf(x_0+h,\ y_0+k_3,\ z_0+l_3) = 0.1(0.585377) = 0.0585377,$$

$$k \approx \frac{1}{6}(k_1 + 2k_2 + 2k_3 + k_4) = 0.0541663, \quad \text{and} \quad \bar{y} = y_0 + k \approx 0.25417.$$

SUPPLEMENTARY PROBLEMS

17. Approximate y when $x = 0.2$ if $dy/dx = x + y^2$ and $y = 1$ when $x = 0$, using *a*) Picard's method, *b*) Taylor series, and *c*) the First Derivative method with $n = 4$.

    Ans. *a*) $y_1 = 1.22$, $y_2 = 1.2657$, $y_3 = 1.2727$; *b*) 1.2735; *c*) 1.2503

18. Approximate y when $x = 0.1$ if $dy/dx = x - y^2$ and $y = 1$ when $x = 0$, using *a*) Picard's method, *b*) Taylor series, and *c*) the First Derivative method with $n = 4$.

    Ans. *a*) $y_1 = 0.905$, $y_2 = 0.9143$, $y_3 = 0.9138$; *b*) 0.9138; *c*) 0.9107

19. Use Runge's method to approximate y when $x = 0.025$ if $dy/dx = x + y$ and $y = 1$ when $x = 0$.
    Ans. 1.0256

20. Use Runge's method to approximate y when $x = 2.2$ if $dy/dx = 1 + y/x$ and $y = 2$ when $x = 2$.
    Ans. 2.4096

21. Use Runge's method to approximate y when $x = 0.5$ if $dy/dx = \sqrt{x + 2y}$ and $y = 0.17$ when $x = 0.3$.
    Ans. 0.3607

22. Solve Problem 21 using the Kutta-Simpson method. *Ans.* 0.3611

23. Use Runge's method to approximate y and z when $x = 0.2$ for the particular solution of the system $dy/dx = y + z$, $dz/dx = x^2 + y$ satisfying $y = 0.4$, $z = 0.1$ when $x = 0.1$.
    Ans. $y \approx 0.4548$, $z \approx 0.1450$

24. Use the Kutta-Simpson method to approximate y when $x = 0.2$ for that particular solution of $\frac{d^2y}{dx^2} + 3x\frac{dy}{dx} + y = 0$ satisfying $y = 0.1$, $\frac{dy}{dx} = 0.2$ when $x = 0.1$. *Ans.* 0.1191

CHAPTER 25

Integration in Series

EQUATIONS OF ORDER ONE. The existence theorem of Chapter 2 for a differential equation of the form

$$1) \qquad \frac{dy}{dx} = f(x,y)$$

gives a sufficient condition for a solution. In the proof using power series, y is found in the form of a Taylor series

$$2) \qquad y = A_0 + A_1(x-x_0) + A_2(x-x_0)^2 + \cdots\cdots + A_n(x-x_0)^n + \cdots\cdots,$$

where for convenience y_0 has been replaced by A_0. This series *i*) satisfies the differential equation 1), *ii*) has the value $y=y_0$ when $x=x_0$, and *iii*) is convergent for all values of x sufficiently near $x=x_0$.

A. To obtain the solution of 1) satisfying the condition $y=y_0$ when $x=0$:

a) Assume the solution to be of the form

$$y = A_0 + A_1x + A_2x^2 + A_3x^3 + \cdots\cdots + A_nx^n + \cdots\cdots$$

in which $A_0=y_0$ and the remaining A's are constants to be determined.

b) Substitute the assumed series in the differential equation and proceed as in the Method of Undetermined Coefficients of Chapter 15.

EXAMPLE 1. Solve $y' = x^2+y$ in series satisfying the condition $y=y_0$ when $x=0$.

Since $f(x,y) = x^2+y$ is single valued and continuous while $\partial f/\partial y = 1$ is continuous over any rectangle of values (x,y) enclosing $(0,y_0)$, the conditions of the Existence Theorem are satisfied and we assume the solution

$$y = A_0 + A_1x + A_2x^2 + A_3x^3 + A_4x^4 + \cdots\cdots + A_nx^n + \cdots\cdots$$

Now, within the region of convergence, this series may be differentiated term by term yielding a series which converges to the derivative y'. Hence,

$$y' = A_1 + 2A_2x + 3A_3x^2 + 4A_4x^3 + \cdots\cdots + nA_nx^{n-1} + \cdots\cdots$$

and

$$y' - x^2 - y = (A_1 - A_0) + (2A_2 - A_1)x + (3A_3 - A_2 - 1)x^2 + (4A_4 - A_3)x^3 + \cdots\cdots\cdots + (nA_n - A_{n-1})x^{n-1} + \cdots\cdots\cdots = 0.$$

In order that this series vanish for all values of x in some region surrounding $x=0$, it is necessary and sufficient that the coefficients of each power of x vanish. Thus,

$A_1 - A_0 = 0$ and $A_1 = A_0 = y_0$, $\qquad 3A_3 - A_2 - 1 = 0$ and $A_3 = \frac{1}{3} + \frac{1}{6}y_0$,

$2A_2 - A_1 = 0$ and $A_2 = \frac{1}{2}A_1 = \frac{1}{2}A_0 = \frac{1}{2}y_0$, $\qquad 4A_4 - A_3 = 0$ and $A_4 = \frac{1}{12} + \frac{1}{24}y_0$,

....

$$nA_n - A_{n-1} = 0 \quad \text{and} \quad A_n = \frac{1}{n}A_{n-1}, \quad n \geqq 4.$$

This latter relation, called a *recursion formula*, may be used to compute additional coefficients; thus,

$$A_5 = \frac{1}{5}A_4 = \frac{1}{60} + \frac{1}{120}y_0, \qquad A_6 = \frac{1}{6}A_5 = \frac{1}{360} + \frac{1}{720}y_0, \qquad \cdots\cdots\cdots$$

It is also possible to obtain the coefficients as follows:

Since $A_n = \frac{1}{n}A_{n-1}$ and $A_{n-1} = \frac{1}{n-1}A_{n-2}$, $A_n = \frac{1}{n(n-1)}A_{n-2}$. But $A_{n-2} = \frac{1}{n-2}A_{n-3}$, $\cdots\cdots$;

hence, $A_n = \frac{1}{n(n-1)(n-2)\cdots\cdots 4}A_3 = \frac{1}{n(n-1)(n-2)\cdots\cdots 4\cdot 3}(1+\frac{1}{2}A_0) = \frac{1}{n!}(2+y_0)$, $n \geqq 3$.

When the values of the A's are substituted in the assumed series, we have

$$\begin{aligned} y &= y_0 + y_0x + \frac{1}{2}y_0x^2 + (\frac{1}{3}+\frac{1}{6}y_0)x^3 + (\frac{1}{12}+\frac{1}{24}y_0)x^4 + \cdots\cdots + \frac{1}{n!}(2+y_0)x^n + \cdots\cdots \\ &= (y_0+2)(1 + x + \frac{1}{2!}x^2 + \frac{1}{3!}x^3 + \cdots\cdots + \frac{1}{n!}x^n + \cdots\cdots) - x^2 - 2x - 2 \\ &= (y_0+2)e^x - x^2 - 2x - 2. \end{aligned}$$

The given differential equation may be solved using the integrating factor e^{-x}; thus,

$$ye^{-x} = \int x^2e^{-x}\,dx = (-x^2-2x-2)e^{-x} + C \qquad \text{and} \qquad y = Ce^x - x^2 - 2x - 2.$$

Using the initial condition, $y = y_0$ when $x = 0$, $C = y_0 + 2$, and $y = (y_0+2)e^x - x^2 - 2x - 2$, as before.

B. To obtain the solution of 1) satisfying the condition $y = y_0$ when $x = x_0$:

a) Make the substitution $x - x_0 = v$, that is,

$$x = v + x_0, \qquad \frac{dy}{dx} = \frac{dy}{dv}$$

resulting in $dy/dv = F(y,y)$.

b) Use the procedure of *A* to obtain the solution of this equation satisfying the condition $y = y_0$ when $v = 0$.

c) Make the substitution $v = x - x_0$ in the solution.

EXAMPLE 2. Solve $y' = x^2 - 4x + y + 1$ satisfying the condition $y = 3$ when $x = 2$.

First make the substitution $x = v + 2$ and obtain $\frac{dy}{dv} = v^2 + y - 3$. We seek the solution satisfying $y = 3$ when $v = 0$; hence, we assume the series solution

$$y = 3 + A_1v + A_2v^2 + A_3v^3 + A_4v^4 + \cdots\cdots\cdots + A_nv^n + \cdots\cdots\cdots$$

Then $$\frac{dy}{dv} = A_1 + 2A_2v + 3A_3v^2 + 4A_4v^3 + \cdots\cdots\cdots + nA_nv^{n-1} + \cdots\cdots\cdots$$

and

$$\begin{aligned}\frac{dy}{dv} - v^2 - y + 3 &= A_1 + (2A_2 - A_1)v + (3A_3 - A_2 - 1)v^2 + (4A_4 - A_3)v^3 + \cdots\cdots\cdots \\ &\quad + (nA_n - A_{n-1})v^{n-1} + \cdots\cdots\cdots = 0.\end{aligned}$$

Equating the coefficients to zero, we have: $A_1 = 0$, $2A_2 - A_1 = 0$ and $A_2 = 0$, $3A_3 - A_2 - 1 = 0$ and $A_3 = 1/3$, $4A_4 - A_3 = 0$ and $A_4 = 1/12$, $\cdots\cdots\cdots$

The recursion formula $A_n = \frac{1}{n}A_{n-1}$ yields

$$A_n = \frac{1}{n}A_{n-1} = \frac{1}{n(n-1)}A_{n-2} = \cdots\cdots = \frac{1}{n(n-1)(n-2)\cdots\cdot 4}A_3 = \frac{2}{n!}, \quad n \geqq 3.$$

Thus,
$$y = 3 + \frac{1}{3}v^3 + \frac{1}{12}v^4 + \cdots\cdots + \frac{2}{n!}v^n + \cdots\cdots$$

$$= 3 + \frac{2}{3!}(x-2)^3 + \frac{2}{4!}(x-2)^4 + \cdots\cdots + \frac{2}{n!}(x-2)^n + \cdots\cdots$$

See also Problems 1-4.

LINEAR EQUATIONS OF ORDER TWO. Consider the equation

3) $$P_0(x)\,y'' + P_1(x)\,y' + P_2(x)\,y = 0$$

where the P's are polynomials in x. We shall call $x=a$ an *ordinary point* of 3) if $P_0(a) \neq 0$; otherwise, a *singular point*.

If $x=0$ is an ordinary point, 3) may be solved in series about $x=0$ as

4) $$y = A\{\text{series in } x\} + B\{\text{series in } x\},$$

in which A and B are arbitrary constants. The two series are linearly independent and both are convergent in a region surrounding $x=0$. The procedure for equations of order one in the section above may be used to obtain 4).

See Problems 5-7.

SOLVED PROBLEMS

EQUATIONS OF ORDER ONE.

1. Solve $\dfrac{dy}{dx} = \dfrac{2x-y}{1-x}$ in series satisfying the condition $y = y_0$ when $x = 0$.

Assume the series to be $y = A_0 + A_1x + A_2x^2 + A_3x^3 + A_4x^4 + \cdots\cdots + A_nx^n + \cdots\cdots$,

where $A_0 = y_0$. Then $y' = A_1 + 2A_2x + 3A_3x^2 + 4A_4x^3 + \cdots\cdots + nA_nx^{n-1} + \cdots\cdots$.

Substituting in the given differential equation $(1-x)y' - 2x + y = 0$, we have

$$(1-x)(A_1 + 2A_2x + 3A_3x^2 + 4A_4x^3 + \cdots\cdots + nA_nx^{n-1} + \cdots\cdots)$$
$$- 2x + (A_0 + A_1x + A_2x^2 + A_3x^3 + \cdots\cdots + A_nx^n + \cdots\cdots) = 0,$$

or

$$(A_1 + A_0) + (2A_2 - 2)x + (3A_3 - A_2)x^2 + (4A_4 - 2A_3)x^3 + \cdots\cdot + [(n+1)A_{n+1} - (n-1)A_n]x^n + \cdots\cdot = 0.$$

(Note. In finding the general term in the line immediately above, we may write a number of terms on either side of the general term of the assumed series for y, differentiate each in getting y', carry out the required multiplications, and pick out the terms in x^n *OR* learn to write the required term using the general term of the assumed series and its derivative. In the present problem we wish the term in x^n when the substitutions are made in $y' - xy' - 2x + y = 0$. First, we need the term in x^n of y' when we have the term in x^{n-1}. We simply replace n

by $(n+1)$ in nA_nx^{n-1} and obtain $(n+1)A_{n+1}x^n$. The remaining terms $-nA_nx^n + A_nx^n$ are obvious.)

Equating the coefficients of distinct powers of x to zero yields

$$A_1 + A_0 = 0 \quad\text{and}\quad A_1 = -A_0, \qquad 3A_3 - A_2 = 0 \quad\text{and}\quad A_3 = \frac{1}{3}A_2 = \frac{1}{3},$$

$$2A_2 - 2 = 0 \quad\text{and}\quad A_2 = 1, \qquad 4A_4 - 2A_3 = 0 \quad\text{and}\quad A_4 = \frac{1}{2}A_3 = \frac{1}{6},$$

....

$$(n+1)A_{n+1} - (n-1)A_n = 0 \quad\text{and}\quad A_{n+1} = \frac{n-1}{n+1}A_n, \quad (n \geqq 2).$$

Now $$A_n = \frac{n-2}{n}A_{n-1} = \frac{(n-2)(n-3)}{n(n-1)}A_{n-2} = \frac{(n-2)(n-3)(n-4)}{n(n-1)(n-2)}A_{n-3} = \cdots\cdots$$

$$= \frac{(n-2)(n-3)(n-4)\cdots\cdots 2\cdot 1}{n(n-1)(n-2)\cdots\cdots 4\cdot 3}A_2 = \frac{2}{n(n-1)}, \quad n \geqq 2.$$

Thus, $$y = y_0(1-x) + x^2 + \frac{1}{3}x^3 + \frac{1}{6}x^4 + \frac{1}{10}x^5 + \cdots\cdots + \frac{2}{n(n-1)}x^n + \cdots\cdots$$

$$= y_0(1-x) + \sum_{n=2}^{\infty}\frac{2}{n(n-1)}x^n.$$

Using the ratio test, $$\lim_{n\to\infty}\left|\frac{A_{n+1}x^{n+1}}{A_nx^n}\right| = |x| \lim_{n\to\infty}\frac{n-1}{n+1} = |x|.$$

The series converges for $|x| < 1$.

Note. By means of the integrating factor $1/(1-x)$ the solution of the differential equation is $y = 2(1-x)\ln(1-x) + 2x + C(1-x)$. The particular integral required is

$$y = y_0(1-x) + 2(1-x)\ln(1-x) + 2x.$$

2. Solve $(1-xy)y' - y = 0$ in powers of x.

Assume the series to be $y = A_0 + A_1x + A_2x^2 + A_3x^3 + A_4x^4 + \cdots\cdots + A_nx^n + \cdots\cdots$. Then

$$y' = A_1 + 2A_2x + 3A_3x^2 + 4A_4x^3 + \cdots\cdots + nA_nx^{n-1} + \cdots\cdots \quad\text{and}$$

$$(1-xy)y' - y$$

$$= (1 - A_0x - A_1x^2 - A_2x^3 - A_3x^4 - \cdots\cdots - A_nx^{n+1} - \cdots\cdots)(A_1 + 2A_2x + 3A_3x^2 + 4A_4x^3 + \cdots\cdots + nA_nx^{n-1} + \cdots\cdots) - (A_0 + A_1x + A_2x^2 + A_3x^3 + \cdots\cdots + A_nx^n + \cdots\cdots)$$

$$= (A_1-A_0) + (2A_2-A_0A_1-A_1)x + (3A_3-2A_0A_2-A_1^2-A_2)x^2 + (4A_4-3A_0A_3-3A_1A_2-A_3)x^3 + \cdots\cdots = 0.$$

Equating to zero the coefficients of distinct powers of x,

$$A_1 - A_0 = 0 \quad\text{and}\quad A_1 = A_0,$$

$$2A_2 - A_0A_1 - A_1 = 0 \quad\text{and}\quad A_2 = \tfrac{1}{2}A_1(1+A_0) = \tfrac{1}{2}A_0(1+A_0),$$

$$3A_3 - 2A_0A_2 - A_1^2 - A_2 = 0 \quad\text{and}\quad A_3 = \frac{1}{3}(2A_0A_2 + A_1^2 + A_2) = \frac{1}{6}A_0(1+5A_0+2A_0^2),$$

$$4A_4 - 3A_0A_3 - 3A_1A_2 - A_3 = 0 \quad\text{and}\quad A_4 = \frac{1}{24}A_0(1 + 17A_0 + 26A_0^2 + 6A_0^3),$$

....

Thus, $y = A_0[1 + x + \frac{1}{2!}(1+A_0)x^2 + \frac{1}{3!}(1+5A_0+2A_0^2)x^3 + \frac{1}{4!}(1+17A_0+26A_0^2+6A_0^3)x^4 + \cdots\cdots]$.

We shall not attempt to obtain a recursion formula here nor to test for convergence.

3. Solve $xy' - y - x - 1 = 0$ in powers of $(x-1)$.

Setting $x = z+1$, the equation becomes $(z+1)\frac{dy}{dz} - y - z - 2 = 0$. Since we seek its solution in powers of z, assume the series to be

$$y = A_0 + A_1 z + A_2 z^2 + A_3 z^3 + A_4 z^4 + \cdots\cdots + A_n z^n + \cdots\cdots. \quad \text{Then}$$

$$\frac{dy}{dz} = A_1 + 2A_2 z + 3A_3 z^2 + 4A_4 z^3 + \cdots\cdots + nA_n z^{n-1} + \cdots\cdots \quad \text{and}$$

$$(z+1)\frac{dy}{dz} - y - z - 2$$

$$= (z+1)(A_1 + 2A_2 z + 3A_3 z^2 + 4A_4 z^3 + \cdots\cdots + nA_n z^{n-1} + \cdots\cdots) - z - 2 - (A_0 + A_1 z + A_2 z^2 + A_3 z^3 + \cdots\cdots + A_n z^n + \cdots\cdots)$$

$$= (A_1 - 2 - A_0) + (2A_2 - 1)z + (3A_3 + A_2)z^2 + (4A_4 + 2A_3)z^3 + \cdots\cdots + [(n+1)A_{n+1} + (n-1)A_n]z^n + \cdots\cdots = 0.$$

Equating to zero the coefficients of the distinct powers of z,

$A_1 - 2 - A_0 = 0$ and $A_1 = 2 + A_0$, $\quad 3A_3 + A_2 = 0$ and $A_3 = -\frac{1}{3}A_2 = -\frac{1}{6}$,

$2A_2 - 1 = 0$ and $A_2 = \frac{1}{2}$, $\quad 4A_4 + 2A_3 = 0$ and $A_4 = -\frac{1}{2}A_3 = \frac{1}{12}$,

$\cdots\cdots\cdots\cdots\cdots\cdots\cdots\cdots$

$$(n+1)A_{n+1} + (n-1)A_n = 0 \quad \text{and} \quad A_{n+1} = -\frac{n-1}{n+1}A_n, \quad n \geqq 2.$$

From Problem 1, $A_n = (-1)^n \frac{(n-2)(n-3)\cdots\cdots 2\cdot 1}{n(n-1)\cdots\cdots\cdots 4\cdot 3} A_2 = (-1)^n \frac{1}{n(n-1)}$, $n \geqq 2$,

and $$y = A_0 + (2+A_0)z + \frac{1}{2}z^2 - \frac{1}{6}z^3 + \frac{1}{12}z^4 - \cdots\cdots + (-1)^n \frac{1}{n(n-1)} z^n + \cdots\cdots.$$

Replacing z by $(x-1)$, we have

$$y = A_0 x + 2(x-1) + \frac{1}{2}(x-1)^2 - \frac{1}{6}(x-1)^3 + \frac{1}{12}(x-1)^4 - \cdots\cdots$$

$$= A_0 x + 2(x-1) + \sum_{n=2}^{\infty} (-1)^n \frac{1}{n(n-1)}(x-1)^n.$$

Using the ratio test, $\lim_{n\to\infty} \left| \frac{A_{n+1}z^{n+1}}{A_n z^n} \right| = |z| \lim_{n\to\infty} \frac{n-1}{n+1} = |z| = |x-1|$.

The series converges for $|x-1| < 1$.

4. Solve $y' - x^2 - e^y = 0$ satisfying the condition $y = 0$ when $x = 0$.

In view of the initial condition, assume the series to be

$$y = A_1x + A_2x^2 + A_3x^3 + A_4x^4 + A_5x^5 + \cdots\cdots\cdots.$$

Then $$y' = A_1 + 2A_2x + 3A_3x^2 + 4A_4x^3 + 5A_5x^4 + \cdots\cdots\cdots.$$

Also, $$e^y = 1 + y + \frac{1}{2!}y^2 + \frac{1}{3!}y^3 + \frac{1}{4!}y^4 + \cdots\cdots\cdots$$

$$= 1 + (A_1x + A_2x^2 + A_3x^3 + A_4x^4 + \cdots) + \frac{1}{2!}[A_1^2x^2 + 2A_1A_2x^3 + (A_2^2 + 2A_1A_3)x^4 + \cdots\cdots]$$

$$+ \frac{1}{3!}(A_1^3x^3 + 3A_1^2A_2x^4 + \cdots\cdots) + \frac{1}{4!}(A_1^4x^4 + \cdots\cdots) + \cdots\cdots\cdots$$

$$= 1 + A_1x + (A_2 + \frac{1}{2}A_1^2)x^2 + (A_3 + A_1A_2 + \frac{1}{6}A_1^3)x^3$$

$$+ (A_4 + \frac{1}{2}A_2^2 + A_1A_3 + \frac{1}{2}A_1^2A_2 + \frac{1}{24}A_1^4)x^4 + \cdots\cdots\cdots.$$

Substituting in the differential equation,

$$(A_1 - 1) + (2A_2 - A_1)x + (3A_3 - 1 - A_2 - \frac{1}{2}A_1^2)x^2 + (4A_4 - A_3 - A_1A_2 - \frac{1}{6}A_1^3)x^3$$

$$+ (5A_5 - A_4 - \frac{1}{2}A_2^2 - A_1A_3 - \frac{1}{2}A_1^2A_2 - \frac{1}{24}A_1^4)x^4 + \cdots\cdots\cdots = 0.$$

Equating coefficients of distinct powers of x to zero,

$$A_1 - 1 = 0 \text{ and } A_1 = 1, \quad 2A_2 - A_1 = 0 \text{ and } A_2 = \frac{1}{2}A_1 = \frac{1}{2},$$

$$3A_3 - 1 - A_2 - \frac{1}{2}A_1^2 = 0 \text{ and } A_3 = \frac{1}{3}(1 + A_2 + \frac{1}{2}A_1^2) = \frac{2}{3},$$

$$4A_4 - A_3 - A_1A_2 - \frac{1}{6}A_1^3 = 0 \text{ and } A_4 = \frac{1}{4}(A_3 + A_1A_2 + \frac{1}{6}A_1^3) = \frac{1}{3},$$

$$5A_5 - A_4 - \frac{1}{2}A_2^2 - A_1A_3 - \frac{1}{2}A_1^2A_2 - \frac{1}{24}A_1^4 = 0 \text{ and } A_5 = \frac{17}{60}, \cdots\cdots\cdots$$

and $$y = x + \frac{1}{2}x^2 + \frac{2}{3}x^3 + \frac{1}{3}x^4 + \frac{17}{60}x^5 + \cdots\cdots\cdots.$$

LINEAR EQUATIONS OF ORDER TWO.

5. Solve $(1+x^2)y'' + xy' - y = 0$ in powers of x.

Here $P_0(x) = 1 + x^2$, $P_0(0) \neq 0$ and $x = 0$ is an ordinary point.

We assume the series

$$y = A_0 + A_1x + A_2x^2 + A_3x^3 + A_4x^4 + \cdots\cdots + A_nx^n + \cdots\cdots.$$

Then $$y' = A_1 + 2A_2x + 3A_3x^2 + 4A_4x^3 + \cdots\cdots + nA_nx^{n-1} + \cdots\cdots,$$

and $$y'' = 2A_2 + 6A_3x + 12A_4x^2 + \cdots\cdots + n(n-1)A_nx^{n-2} + \cdots\cdots.$$

Substituting in the given differential equation,

$$(1+x^2)[2A_2 + 6A_3x + 12A_4x^2 + \cdots + n(n-1)A_nx^{n-2} + \cdots] + x(A_1 + 2A_2x + 3A_3x^2 + 4A_4x^3 + \cdots + nA_nx^{n-1} + \cdots) - (A_0 + A_1x + A_2x^2 + A_3x^3 + A_4x^4 + \cdots + A_nx^n + \cdots) = 0,$$

or $$(2A_2 - A_0) + 6A_3x + (12A_4 + 3A_2)x^2 + \cdots + [(n+2)(n+1)A_{n+2} + (n^2-1)A_n]x^n + \cdots = 0.$$

Equating to zero the coefficients of the distinct powers of x,

$2A_2 - A_0 = 0$ and $A_2 = \frac{1}{2}A_0$, $6A_3 = 0$ and $A_3 = 0$, $12A_4 + 3A_2 = 0$ and $A_4 = -\frac{1}{8}A_0$, $\cdots\cdots$

$$(n+2)(n+1)A_{n+2} + (n^2-1)A_n = 0 \quad\text{and}\quad A_{n+2} = -\frac{n-1}{n+2}A_n.$$

From the latter relation it is clear that $A_3 = A_5 = A_7 = \cdots = 0$, that is, $A_{n+2} = 0$ if n is odd. If n is even, $(n = 2k)$, then

$$A_{2k} = -\frac{2k-3}{2k}A_{2k-2} = \frac{(2k-3)(2k-5)}{2k(2k-2)}A_{2k-4} = \cdots = (-1)^{k+1}\frac{1\cdot3\cdot5\cdots\cdots(2k-3)}{2^k\,k!}A_0.$$

Thus, the complete solution is

$$y = A_0(1 + \frac{1}{2}x^2 - \frac{1}{8}x^4 + \frac{1}{16}x^6 - \frac{5}{128}x^8 + \cdots\cdots) + A_1x$$

$$= A_0[1 + \frac{1}{2}x^2 + \sum_{k=2}^{\infty}(-1)^{k+1}\frac{1\cdot3\cdot5\cdots\cdots(2k-3)}{2^k\,k!}x^{2k}] + A_1x$$

$$= A_0[1 + \frac{1}{2}x^2 - \sum_{k=2}^{\infty}(-1)^{k}\frac{1\cdot3\cdot5\cdots\cdots(2k-3)}{2^k\,k!}x^{2k}] + A_1x.$$

Here $\lim_{n\to\infty}\left|\dfrac{A_{n+2}x^{n+2}}{A_nx^n}\right| = x^2\lim_{n\to\infty}\dfrac{n-1}{n+2} = x^2$, and the series converges for $|x| < 1$.

6. Solve $y'' - x^2y' - y = 0$ in powers of x.

Here $P_0(x) = 1$ and $x = 0$ is an ordinary point. We assume the series

$$y = A_0 + A_1x + A_2x^2 + A_3x^3 + \cdots\cdots\cdots + A_nx^n + \cdots\cdots\cdots. \quad\text{Then}$$

$$y' = A_1 + 2A_2x + 3A_3x^2 + \cdots\cdots\cdots\cdot + nA_nx^{n-1} + \cdots\cdots\cdots,$$

$$y'' = 2A_2 + 6A_3x + 12A_4x^2 + 20A_5x^3 + \cdots\cdots\cdots + n(n-1)A_nx^{n-2} + \cdots\cdots\cdots, \quad\text{and}$$

$$y'' - x^2y' - y = (2A_2 - A_0) + (6A_3 - A_1)x + (12A_4 - A_1 - A_2)x^2 + (20A_5 - 2A_2 - A_3)x^3 + \cdots + [(n+2)(n+1)A_{n+2} - (n-1)A_{n-1} - A_n]x^n + \cdots = 0.$$

Equating to zero the coefficients of the distinct powers of x,

$2A_2 - A_0 = 0$ and $A_2 = \frac{1}{2}A_0$, $6A_3 - A_1 = 0$ and $A_3 = \frac{1}{6}A_1$, $12A_4 - A_1 - A_2 = 0$ and $A_4 = \frac{1}{24}A_0 + \frac{1}{12}A_1$,

$\cdots\cdots\cdots\cdots\cdots\cdots\cdots\cdots\cdots\cdots$

$$(n+2)(n+1)A_{n+2} - (n-1)A_{n-1} - A_n = 0 \quad \text{and} \quad A_{n+2} = \frac{(n-1)A_{n-1} + A_n}{(n+1)(n+2)}, \quad n \geqq 1.$$

The complete solution is
$$y = A_0\left(1 + \frac{1}{2}x^2 + \frac{1}{24}x^4 + \frac{1}{20}x^5 + \frac{1}{720}x^6 + \frac{13}{2520}x^7 + \cdots\cdots\right)$$
$$+ A_1\left(x + \frac{1}{6}x^3 + \frac{1}{12}x^4 + \frac{1}{120}x^5 + \frac{7}{360}x^6 + \frac{41}{5040}x^7 + \cdots\cdots\right).$$

7. Solve $y'' - 2x^2y' + 4xy = x^2 + 2x + 2$ in powers of x.

Assume the series to be

$$y = A_0 + A_1x + A_2x^2 + A_3x^3 + A_4x^4 + A_5x^5 + \cdots\cdots + A_nx^n + \cdots\cdots. \quad \text{Then}$$
$$y' = A_1 + 2A_2x + 3A_3x^2 + 4A_4x^3 + 5A_5x^4 + \cdots\cdots + nA_nx^{n-1} + \cdots\cdots,$$
$$y'' = 2A_2 + 6A_3x + 12A_4x^2 + 20A_5x^3 + \cdots\cdots + n(n-1)A_nx^{n-2} + \cdots\cdots, \quad \text{and}$$

$$y'' - 2x^2y' + 4xy - x^2 - 2x - 2 = (2A_2 - 2) + (6A_3 + 4A_0 - 2)x + (12A_4 + 2A_1 - 1)x^2 + 20A_5x^3 + \cdots\cdots$$
$$+ [(n+2)(n+1)A_{n+2} - 2(n-1)A_{n-1} + 4A_{n-1}]x^n + \cdots\cdots = 0.$$

Equating the coefficients to zero, we obtain

$$2A_2 - 2 = 0 \text{ and } A_2 = 1, \quad 6A_3 + 4A_0 - 2 = 0 \text{ and } A_3 = \frac{1}{3} - \frac{2}{3}A_0, \quad A_4 = \frac{1}{12} - \frac{1}{6}A_1, \quad A_5 = 0,$$
$$\cdots\cdots\cdots\cdots$$

$$(n+2)(n+1)A_{n+2} - 2(n-3)A_{n-1} = 0 \quad \text{and} \quad A_{n+2} = \frac{2(n-3)}{(n+1)(n+2)}A_{n-1}, \quad n \geqq 3.$$

The complete solution is

$$y = A_0\left(1 - \frac{2}{3}x^3 - \frac{2}{45}x^6 - \frac{2}{405}x^9 - \cdots\cdots\right) + A_1\left(x - \frac{1}{6}x^4 - \frac{1}{63}x^7 - \frac{1}{567}x^{10} - \cdots\cdots\right)$$
$$+ x^2 + \frac{1}{3}x^3 + \frac{1}{12}x^4 + \frac{1}{45}x^6 + \frac{1}{126}x^7 + \frac{1}{405}x^9 + \frac{1}{1134}x^{10} + \cdots\cdots$$

8. Solve $y'' + (x-1)y' + y = 0$ in powers of $x - 2$.

Put $x = v + 2$ in the given equation and obtain $\dfrac{d^2y}{dv^2} + (v+1)\dfrac{dy}{dv} + y = 0$ which is to be integrated in powers of v. Assume the series

$$y = A_0 + A_1v + A_2v^2 + A_3v^3 + A_4v^4 + \cdots\cdots + A_nv^n + \cdots\cdots. \quad \text{Then}$$
$$\frac{dy}{dv} = A_1 + 2A_2v + 3A_3v^2 + 4A_4v^3 + \cdots\cdots + nA_nv^{n-1} + \cdots\cdots,$$
$$\frac{d^2y}{dv^2} = 2A_2 + 6A_3v + 12A_4v^2 + \cdots\cdots + n(n-1)A_nv^{n-2} + \cdots\cdots, \quad \text{and}$$

$$\frac{d^2y}{dv^2} + (v+1)\frac{dy}{dv} + y = (2A_2 + A_1 + A_0) + (6A_3 + 2A_1 + 2A_2)v + (12A_4 + 3A_2 + 3A_3)v^2 + \cdots\cdots$$
$$+ [(n+2)(n+1)A_{n+2} + (n+1)A_n + (n+1)A_{n+1}]v^n + \cdots\cdots = 0.$$

Equating the coefficients of powers of v to zero, we obtain

$$A_2 = -\frac{1}{2}(A_0 + A_1), \quad A_3 = -\frac{1}{3}(A_1 + A_2) = \frac{1}{6}(A_0 - A_1), \quad A_4 = -\frac{1}{4}(A_2 + A_3) = \frac{1}{12}(A_0 + 2A_1), \quad \cdots\cdots$$

$$(n+2)(n+1)A_{n+2} + (n+1)A_n + (n+1)A_{n+1} = 0 \quad \text{and} \quad A_{n+2} = -\frac{1}{n+2}(A_n + A_{n+1}).$$

Thus, noting that $v = x-2$, the complete solution is

$$y = A_0[1 - \frac{1}{2}(x-2)^2 + \frac{1}{6}(x-2)^3 + \frac{1}{12}(x-2)^4 - \frac{1}{20}(x-2)^5 - \frac{1}{180}(x-2)^6 + \cdots\cdots]$$

$$+ A_1[(x-2) - \frac{1}{2}(x-2)^2 - \frac{1}{6}(x-2)^3 + \frac{1}{6}(x-2)^4 - \frac{1}{36}(x-2)^6 + \cdots\cdots].$$

SUPPLEMENTARY PROBLEMS

9. Solve $(1-x)y' = x^2 - y$ in powers of x.

Ans. $y = A_0(1-x) + x^3(\frac{1}{3} + \frac{1}{6}x + \frac{1}{10}x^2 + \cdots\cdots + \frac{1\cdot 2}{(n+2)(n+3)}x^n + \cdots\cdots)$

10. Solve $xy' = 1 - x + 2y$ in powers of $x-1$. Also integrate directly.

Hint: Let $x-1 = z$ and solve $(z+1)\frac{dy}{dz} = -z + 2y$ in powers of z.

Ans. $y = A_0[1 + 2(x-1) + (x-1)^2] + \frac{1}{2} + (x-1)$

11. Solve $y' = 2x^2 + 3y$ in powers of x.

Ans. $y = A_0[1 + 3x + 9x^2/2 + 9x^3/2 + 27x^4/8 + \cdots\cdots] + (2x^3/3 + x^4/2 + \cdots\cdots)$

12. Solve $(x+1)y' = x^2 - 2x + y$ in powers of x.

Ans. $y = A_0(1+x) - x^2 + 2x^3/3 - x^4/3 + x^5/5 - 2x^6/15 + \cdots\cdots$

13. Solve $y'' + xy = 0$ in powers of x.

R.F. $A_n = -\frac{1}{n(n-1)}A_{n-3}$, $n \geqq 3$; convergent for all x.

Ans. $y = A_0(1 - x^3/6 + x^6/180 - \cdots\cdots) + A_1(x - x^4/12 + x^7/504 - \cdots\cdots)$

14. Solve $y'' + 2x^2y = 0$ in powers of x.

R.F. $A_n = -\frac{2}{n(n-1)}A_{n-4}$; convergent for all x.

Ans. $y = A_0(1 - x^4/6 + x^8/168 - \cdots\cdots) + A_1(x - x^5/10 + x^9/360 - \cdots\cdots)$

15. Solve $y'' - xy' + x^2y = 0$ in powers of x.

R.F. $n(n-1)A_n - (n-2)A_{n-2} + A_{n-4} = 0$, $n \geqq 4$.

Ans. $y = A_0(1 - x^4/12 - x^6/90 + x^8/3360 + \cdots\cdot) + A_1(x + x^3/6 - x^5/40 - x^7/144 - \cdots\cdot)$

16. Solve $(1-x^2)y'' - 2xy' + p(p+1)y = 0$, where p is a constant, in powers of x. (Legendre Equation)

R.F. $A_n = \frac{(n-2-p)(n+p-1)}{n(n-1)}A_{n-2}$; convergent for $|x| < 1$.

$$Ans.\quad y = A_0(1 - \frac{p(p+1)}{2!}x^2 + \frac{(p-2)p(p+1)(p+3)}{4!}x^4 - \cdots\cdots\cdots\cdots)$$

$$+ A_1(x - \frac{(p-1)(p+2)}{3!}x^3 + \frac{(p-3)(p-1)(p+2)(p+4)}{5!}x^5 - \cdots\cdots\cdots\cdots)$$

17. Solve $y'' + x^2y = 1 + x + x^2$ in powers of x. R.F. $A_n = -\frac{1}{n(n-1)}A_{n-4}$; convergent for all x.

$$Ans.\quad y = A_0(1 - x^4/12 + x^8/672 - \cdots\cdots) + A_1(x - x^5/20 + x^9/1440 - \cdots\cdots)$$

$$+ x^2/2 + x^3/6 + x^4/12 - x^6/60 - x^7/252 - x^8/672 + \cdots\cdots$$

CHAPTER 26

Integration in Series

WHEN $x = a$ IS A SINGULAR POINT OF THE DIFFERENTIAL EQUATION

1) $$P_0(x)\,y'' + P_1(x)\,y' + P_2(x)\,y = 0,$$

in which $P_i(x)$ are polynomials, the procedure of the preceding chapter will not yield a complete solution in series about $x = a$.

EXAMPLE 1. For the equation $x^2y'' + (x^2 - x)y' + 2y = 0$, $x = 0$ is a singular point since $P_0(0) = 0$. If we assume a solution of the form

(i) $$y = A_0 + A_1x + A_2x^2 + A_3x^3 + \cdots\cdots\cdots$$

and substitute in the given equation, we obtain

$$2A_0 + A_1x + (2A_2 + A_1)x^2 + (5A_3 + 2A_2)x^3 + \cdots\cdots\cdots = 0.$$

In order that this relation be satisfied identically, it is necessary that $A_0 = 0$, $A_1 = 0$, $A_2 = 0$, $A_3 = 0$, $\cdots\cdots$; hence, there is no series of the form (i) satisfying the given equation.

A SINGULAR POINT $x = a$ OF 1) IS CALLED ***REGULAR*** IF, when 1) is put in the form

1') $$y'' + \frac{R_1(x)}{x-a}\,y' + \frac{R_2(x)}{(x-a)^2}\,y = 0,$$

$R_1(x)$ and $R_2(x)$ can be expanded in Taylor series about $x = a$.

EXAMPLE 2. For the equation $(1+x)y'' + 2xy' - 3y = 0$, $x = -1$ is a singular point since $P_0(-1) = 1 + (-1) = 0$. When the equation is put in the form

$$y'' + \frac{R_1(x)}{x+1}\,y' + \frac{R_2(x)}{(x+1)^2}\,y = y'' + \frac{2x}{x+1}\,y' + \frac{-3(x+1)}{(x+1)^2}\,y = 0,$$

the Taylor expansions about $x = -1$ of $R_1(x)$ and $R_2(x)$ are

$$R_1(x) = 2x = 2(x+1) - 2 \qquad \text{and} \qquad R_2(x) = -3(x+1).$$

Thus, $x = -1$ is a regular singular point.

EXAMPLE 3. For the equation $x^3y'' + x^2y' + y = 0$, $x = 0$ is a singular point. Writing the equation in the form

$$y'' + \frac{1}{x}\,y' + \frac{1/x}{x^2}\,y = 0,$$

it is seen that $R_2(x) = 1/x$ cannot be expanded in a Taylor series about $x = 0$. Thus, $x = 0$ is not a regular singular point.

WHEN $x=0$ IS A REGULAR SINGULAR POINT OF 1), there always exists a series solution of the form

$$2)\quad y = x^m \sum_{n=0}^{\infty} A_n x^n = A_0 x^m + A_1 x^{m+1} + A_2 x^{m+2} + \cdots + A_n x^{m+n} + \cdots\cdots,$$

with $A_0 \neq 0$, and we shall proceed to determine m and the A's so that 2) satisfies 1).

EXAMPLE 4. Solve in series $2xy'' + (x+1)y' + 3y = 0$.

Here, $x=0$ is a regular singular point. Substituting

$$y = A_0x^m + A_1x^{m+1} + A_2x^{m+2} + \cdots\cdots + A_nx^{m+n} + \cdots\cdots,$$

$$y' = mA_0x^{m-1} + (m+1)A_1x^m + (m+2)A_2x^{m+1} + \cdots\cdots + (m+n)A_nx^{m+n-1} + \cdots\cdots,$$

$$y'' = (m-1)mA_0x^{m-2} + m(m+1)A_1x^{m-1} + (m+1)(m+2)A_2x^m + \cdots + (m+n-1)(m+n)A_nx^{m+n-2} + \cdots$$

in the given differential equation, we have

$$\text{(i)}\quad m(2m-1)A_0x^{m-1} + [(m+1)(2m+1)A_1 + (m+3)A_0]x^m + [(m+2)(2m+3)A_2 + (m+4)A_1]x^{m+1}$$
$$+ \cdots\cdots + [(m+n)(2m+2n-1)A_n + (m+n+2)A_{n-1}]x^{m+n-1} + \cdots\cdots\cdots\cdots = 0.$$

Since $A_0 \neq 0$, the coefficient of the first term will vanish provided $m(2m-1) = 0$, that is, provided $m = 0$ or $m = \frac{1}{2}$. However, without regard to m, all terms after the first will vanish provided the A's satisfy the recursion formula

$$A_n = -\frac{m+n+2}{(m+n)(2m+2n-1)}A_{n-1}, \quad n \geqq 1.$$

Thus, the series

$$2')\quad \bar{y} = A_0x^m\Big[1 - \frac{m+3}{(m+1)(2m+1)}x + \frac{(m+3)(m+4)}{(m+1)(m+2)(2m+1)(2m+3)}x^2$$
$$- \frac{(m+4)(m+5)}{(m+1)(m+2)(2m+1)(2m+3)(2m+5)}x^3 + \cdots\cdots\cdots\cdots\Big]$$

satisfies the equation

$$\text{(ii)}\quad 2x\bar{y}'' + (x+1)\bar{y}' + 3\bar{y} = m(2m-1)A_0x^{m-1}.$$

The right hand member of (ii) will be zero when $m=0$ or $m=\frac{1}{2}$. When $m = 0$, we have from 2') with $A_0 = 1$, the particular solution

$$y_1 = 1 - 3x + 2x^2 - 2x^3/3 + \cdots\cdots\cdots,$$

and when $m = \frac{1}{2}$ with $A_0 = 1$, the particular solution

$$y_2 = \sqrt{x}(1 - 7x/6 + 21x^2/40 - 11x^3/80 + \cdots\cdots).$$

The complete solution is then

$$y = Ay_1 + By_2$$
$$= A(1 - 3x + 2x^2 - 2x^3/3 + \cdots\cdots) + B\sqrt{x}(1 - 7x/6 + 21x^2/40 - 11x^3/80 + \cdots\cdots).$$

The coefficient of the lowest power of x in (i), (also, the coefficient in the right hand member of (ii)), has the form $f(m)A_0$. The equation $f(m)=0$ is called the *indicial equation*. The linearly independent solutions y_1 and y_2 above correspond to the distinct roots $m = 0$ and $m = \frac{1}{2}$ of this equation.

In the Solved Problems below, the roots of the indicial equation will be:
a) distinct and do not differ by an integer,
b) equal, or
c) distinct and differ by an integer.

The first case is illustrated in the example above and also in Problems 1-2.

When the roots m_1 and m_2 of the indicial equation are equal, the solutions corresponding will be identical. The complete solution is then obtained as

$$y = A\bar{y}\Big|_{m=m_1} + B\frac{\partial\bar{y}}{\partial m}\Big|_{m=m_1}.$$ See Problems 3-4.

When the two roots $m_1 < m_2$ of the indicial equation differ by an integer, the greater of the roots m_2 will always yield a solution while the smaller root m_1 may or may not. In the latter case, we set $A_0 = B_0(m-m_1)$ and obtain the complete solution as

$$y = A\bar{y}\Big|_{m=m_1} + B\frac{\partial\bar{y}}{\partial m}\Big|_{m=m_1}$$ See Problems 5-7.

The series, expanded about $x=0$, which appear in these complete solutions converge *always* in the region of the complex plane bounded by two circles centered at $x=0$. The radius of one of the circles is arbitrarily small while that of the other extends to the finite singular point of the differential equation nearest $x=0$. It is clear that the series obtained in Example 4 converge also at $x=0$; moreover, since the differential equation has but one singular point $x=0$, these series converge for all finite values of x.

THE COMPLETE SOLUTION OF

3) $$P_0(x)\,y'' + P_1(x)\,y' + P_2(x)\,y = Q$$

consists of the sum of the complementary function (complete solution of 1)), and any particular integral of 3). A procedure for obtaining a particular integral when Q is a sum of positive and negative powers of x is illustrated in Problem 8.

LARGE VALUES OF x. It is at times necessary to solve a differential equation 1) for large values of x. In such instances the series thus far obtained, even when valid for all finite values of x, are impractical.

To solve an equation in series convergent for large values of x or "about the point at infinity", we transform the given equation by means of the substitution

$$x = 1/z$$

and solve, if possible, the resulting equation in series near $z=0$.

See Problems 9-10.

SOLVED PROBLEMS

1. Solve in series $2x^2y'' - xy' + (x^2+1)y = 0$.

Substituting

$$y = A_0x^m + A_1x^{m+1} + A_2x^{m+2} + \cdots\cdots + A_nx^{m+n} + \cdots$$

$$y' = mA_0x^{m-1} + (m+1)A_1x^m + (m+2)A_2x^{m+1} + \cdots\cdots + (m+n)A_nx^{m+n-1} + \cdots$$

$$y'' = (m-1)mA_0x^{m-2} + (m+1)mA_1x^{m-1} + (m+1)(m+2)A_2x^m + \cdots + (m+n-1)(m+n)A_nx^{m+n-2} + \cdots$$

in the given differential equation, we obtain

$$(m-1)(2m-1)A_0x^m + m(2m+1)A_1x^{m+1} + \{[(m+2)(2m+1)+1]A_2 + A_0\}x^{m+2} + \cdots$$
$$+ \{[(m+n)(2m+2n-3)+1]A_n + A_{n-2}\}x^{m+n} + \cdots = 0.$$

Now all terms except the first two will vanish if $A_2, A_3, \cdots$ satisfy the recursion formula

1) $$A_n = -\frac{1}{(m+n)(2m+2n-3)+1}A_{n-2}, \quad n \geqq 2.$$

The roots of the indicial equation, $(m-1)(2m-1) = 0$, are $m = \frac{1}{2}, 1$, and for either value the first term will vanish. Since, however, neither of these values of m will cause the second term to vanish, we take $A_1 = 0$. Using 1), it follows that $A_1 = A_3 = A_5 = \cdots = 0$. Thus,

$$\bar{y} = A_0x^m\left(1 - \frac{1}{(m+2)(2m+1)+1}x^2 + \frac{1}{[(m+2)(2m+1)+1][(m+4)(2m+5)+1]}x^4 - \cdots\right)$$

satisfies $$2x^2\bar{y}'' - x\bar{y}' + (x^2+1)\bar{y} = (m-1)(2m-1)A_0x^m$$
and the right hand member will be 0 when $m = \frac{1}{2}$ or $m = 1$.

When $m = \frac{1}{2}$ and $A_0 = 1$, we have $\quad y_1 = \sqrt{x}(1 - x^2/6 + x^4/168 - x^6/11088 + \cdots)$

and when $m = 1$, with $A_0 = 1$, we have $\quad y_2 = x(1 - x^2/10 + x^4/360 - x^6/28080 + \cdots)$.

The complete solution is then

$$y = Ay_1 + By_2$$
$$= A\sqrt{x}(1 - x^2/6 + x^4/168 - x^6/11088 + \cdots) + Bx(1 - x^2/10 + x^4/360 - x^6/28080 + \cdots).$$

Since $x=0$ is the only finite singular point, the series converge for all finite values of x.

2. Solve in series $3xy'' + 2y' + x^2y = 0$.

Substituting for y, y', and y'' as in the problem above, we have

$$m(3m-1)A_0x^{m-1} + (m+1)(3m+2)A_1x^m + (m+2)(3m+5)A_2x^{m+1} + [(m+3)(3m+8)A_3 + A_0]x^{m+2}$$
$$+ \cdots + [(m+n)(3m+3n-1)A_n + A_{n-3}]x^{m+n-1} + \cdots = 0.$$

All terms after the third will vanish if $A_3, A_4, \cdots$ satisfy the recursion formula

$$A_n = -\frac{1}{(m+n)(3m+3n-1)}A_{n-3}, \quad n \geqq 3.$$

The roots of the indicial equation $m(3m-1)=0$ are $m = 0,\ 1/3$. Since neither will cause the second and third terms to vanish, we take $A_1 = A_2 = 0$. Then, using the recursion formula, $A_1 = A_4 = A_7 = \cdots = 0$ and $A_2 = A_5 = A_8 = \cdots = 0$. Thus the series

$$1)\qquad \bar{y} = A_0 x^m \left(1 - \frac{1}{(m+3)(3m+8)}x^3 + \frac{1}{(m+3)(m+6)(3m+8)(3m+17)}x^6 - \cdots\cdots\right)$$

satisfies $$3x\bar{y}'' + 2\bar{y}' + x^2\bar{y} = m(3m-1)A_0 x^{m-1}.$$

For $m = 0$, with $A_0 = 1$, we obtain from 1) $\quad y_1 = 1 - x^3/24 + x^6/2448 - \cdots\cdots\cdot$

and for $m = 1/3$, with $A_0 = 1$, we obtain $\quad y_2 = x^{1/3}(1 - x^3/30 + x^6/3420 - \cdots\cdots\cdot)$.

The complete solution is

$$y = Ay_1 + By_2 = A(1 - x^3/24 + x^6/2448 - \cdots\cdots) + Bx^{1/3}(1 - x^3/30 + x^6/3420 - \cdots\cdots).$$

The series converge for all finite values of x.

ROOTS OF INDICIAL EQUATION EQUAL.

3. Solve in series $\quad xy'' + y' - y = 0$.

Substituting for y, y', and y'' as in Problems 1 and 2 above, we obtain

$$m^2 A_0 x^{m-1} + [(m+1)^2 A_1 - A_0]x^m + [(m+2)^2 A_2 - A_1]x^{m+1} + \cdots\cdots\cdot + [(m+n)^2 A_n - A_{n-1}]x^{m+n-1} + \cdots\cdots\cdot = 0.$$

All terms except the first will vanish if $A_1, A_2, \cdots\cdots$ satisfy the recursion formula

$$1)\qquad A_n = \frac{1}{(m+n)^2}A_{n-1}, \quad n \geqq 1.$$

Thus,

$$\bar{y} = A_0 x^m \left(1 + \frac{1}{(m+1)^2}x + \frac{1}{(m+1)^2(m+2)^2}x^2 + \frac{1}{(m+1)^2(m+2)^2(m+3)^2}x^3 + \cdots\cdots\cdot\right)$$

satisfies

$$2)\qquad x\bar{y}'' + \bar{y}' - \bar{y} = m^2 A_0 x^{m-1}.$$

The roots of the indicial equation are $m = 0,0$. Hence, there corresponds but one series solution satisfying 2) with $m = 0$. However, regarding $\bar{y}$ as a function of the independent variables x and m,

$$\frac{\partial \bar{y}'}{\partial m} = \frac{\partial}{\partial m}\left(\frac{\partial \bar{y}}{\partial x}\right) = \frac{\partial}{\partial x}\left(\frac{\partial \bar{y}}{\partial m}\right) = \left(\frac{\partial \bar{y}}{\partial m}\right)'$$

and $$\frac{\partial \bar{y}''}{\partial m} = \frac{\partial}{\partial m}\frac{\partial}{\partial x}\left(\frac{\partial \bar{y}}{\partial x}\right) = \frac{\partial}{\partial x}\frac{\partial}{\partial m}\left(\frac{\partial \bar{y}}{\partial x}\right) = \frac{\partial}{\partial x}\frac{\partial}{\partial x}\left(\frac{\partial \bar{y}}{\partial m}\right) = \left(\frac{\partial \bar{y}}{\partial m}\right)'',$$

and we have by differentiating 2) partially with respect to m,

$$3)\qquad x\left(\frac{\partial \bar{y}}{\partial m}\right)'' + \left(\frac{\partial \bar{y}}{\partial m}\right)' - \left(\frac{\partial \bar{y}}{\partial m}\right) = 2mA_0 x^{m-1} + m^2 A_0 x^{m-1}\ln x.$$

From 2) and 3) it follows that $y_1 = \bar{y}\big|_{m=0}$ and $y_2 = \dfrac{\partial \bar{y}}{\partial m}\Big|_{m=0}$ are solutions of the given differential equation. Taking $A_0 = 1$, we find

$$\frac{\partial \bar{y}}{\partial m} = x^m \ln x \, [1 + \frac{1}{(m+1)^2} x + \frac{1}{(m+1)^2 (m+2)^2} x^2 + \frac{1}{(m+1)^2 (m+2)^2 (m+3)^2} x^3 + \cdots\cdots\cdots\cdots]$$

$$+ x^m [- \frac{2}{(m+1)^3} x - (\frac{2}{(m+1)^3 (m+2)^2} + \frac{2}{(m+1)^2 (m+2)^3}) x^2 - (\frac{2}{(m+1)^3 (m+2)^2 (m+3)^2}$$

$$+ \frac{2}{(m+1)^2 (m+2)^3 (m+3)^2} + \frac{2}{(m+1)^2 (m+2)^2 (m+3)^3}) x^3 - \cdots\cdots\cdots\cdots]$$

$$= \bar{y} \ln x - 2x^m [\frac{1}{(m+1)^3} x + (\frac{1}{(m+1)^3 (m+2)^2} + \frac{1}{(m+1)^2 (m+2)^3}) x^2$$

$$+ (\frac{1}{(m+1)^3 (m+2)^2 (m+3)^2} + \frac{1}{(m+1)^2 (m+2)^3 (m+3)^2} + \frac{1}{(m+1)^2 (m+2)^2 (m+3)^3}) x^3 + \cdots\cdots].$$

Then $y_1 = \bar{y}\Big|_{m=0} = 1 + x + \frac{x^2}{(2!)^2} + \frac{x^3}{(3!)^2} + \cdots\cdots\cdots\cdots,$

$$y_2 = \frac{\partial \bar{y}}{\partial m}\Big|_{m=0} = y_1 \ln x - 2[x + \frac{1}{(2!)^2}(1 + \frac{1}{2})x^2 + \frac{1}{(3!)^2}(1 + \frac{1}{2} + \frac{1}{3})x^3 + \cdots\cdots\cdots\cdots],$$

and the complete solution is

$$y = Ay_1 + By_2 = (A + B \ln x)[1 + x + \frac{1}{(2!)^2} x^2 + \frac{1}{(3!)^2} x^3 + \cdots\cdots\cdots]$$

$$- 2B[x + \frac{1}{(2!)^2}(1 + \frac{1}{2})x^2 + \frac{1}{(3!)^2}(1 + \frac{1}{2} + \frac{1}{3})x^3 + \cdots\cdots\cdots].$$

The series converge for all finite values of $x \neq 0$.

4. Solve in series $xy'' + y' + x^2 y = 0$.

Substituting for y, y', and y'', we obtain

$$m^2 A_0 x^{m-1} + (m+1)^2 A_1 x^m + (m+2)^2 A_2 x^{m+1} + [(m+3)^2 A_3 + A_0] x^{m+2} + \cdots\cdots\cdots\cdots\cdots\cdots$$

$$+ [(m+n)^2 A_n + A_{n-3}] x^{m+n-1} + \cdots\cdots\cdots\cdots = 0.$$

The two roots of the indicial equation are equal. We take $A_0 = 1$, $A_1 = A_2 = 0$, and the remaining A's satisfying the recursion formula $A_n = -\frac{1}{(m+n)^2} A_{n-3}$.

Then $A_1 = A_4 = A_7 = \cdots = 0$, $A_2 = A_5 = A_8 = \cdots = 0$,

$$\bar{y} = x^m (1 - \frac{1}{(m+3)^2} x^3 + \frac{1}{(m+3)^2 (m+6)^2} x^6 - \frac{1}{(m+3)^2 (m+6)^2 (m+9)^2} x^9 + \cdots\cdots\cdots)$$

and, following the procedure of Problem 3 above,

$$\frac{\partial \bar{y}}{\partial m} = \bar{y} \ln x + 2x^m [\frac{1}{(m+3)^3} x^3 - (\frac{1}{(m+3)^3 (m+6)^2} + \frac{1}{(m+3)^2 (m+6)^3}) x^6 +$$

$$(\frac{1}{(m+3)^3 (m+6)^2 (m+9)^2} + \frac{1}{(m+3)^2 (m+6)^3 (m+9)^2} + \frac{1}{(m+3)^2 (m+6)^2 (m+9)^3}) x^9 - \cdots\cdots].$$

Using the root $m = 0$ of the indicial equation,

$$y_1 = \bar{y}\Big|_{m=0} = 1 - \frac{1}{3^2}x^3 + \frac{1}{3^4(2!)^2}x^6 - \frac{1}{3^6(3!)^2}x^9 + \cdots\cdots\cdots$$

and $$y_2 = \frac{\partial\bar{y}}{\partial m}\Big|_{m=0} = y_1 \ln x + 2\left[\frac{1}{3^3}x^3 - \frac{1}{3^5(2!)^2}(1+\frac{1}{2})x^6 + \frac{1}{3^7(3!)^2}(1+\frac{1}{2}+\frac{1}{3})x^9 - \cdots\cdots\right].$$

The complete solution is

$$y = Ay_1 + By_2 = (A + B\ln x)\left[1 - \frac{1}{3^2}x^3 + \frac{1}{3^4(2!)^2}x^6 - \frac{1}{3^6(3!)^2}x^9 + \cdots\cdots\right]$$

$$+ 2B\left[\frac{1}{3^3}x^3 - \frac{1}{3^5(2!)^2}(1+\frac{1}{2})x^6 + \frac{1}{3^7(3!)^2}(1+\frac{1}{2}+\frac{1}{3})x^9 - \cdots\cdots\cdots\right].$$

The series converge for all finite values of $x \neq 0$.

ROOTS OF INDICIAL EQUATION DIFFERING BY AN INTEGER.

5. Solve in series $xy'' - 3y' + xy = 0$.

Substituting for y, y', and y'', we obtain

$$(m-4)mA_0x^{m-1} + (m-3)(m+1)A_1x^m + [(m-2)(m+2)A_2 + A_0]x^{m+1} + \cdots\cdots$$
$$+ [(m+n-4)(m+n)A_n + A_{n-2}]x^{m+n-1} + \cdots\cdots = 0.$$

The roots of the indicial equation are $m = 0,4$, and we have the second special case mentioned above since the difference of the two roots is an integer. We take $A_1 = 0$ and choose the remaining A's to satisfy the recursion formula

$$A_n = -\frac{1}{(m+n-4)(m+n)}A_{n-2}, \qquad n \geqq 2.$$

It is clear that this relation yields finite values when $m = 4$, the larger of the roots, but when $m = 0$, $A_4 \to \infty$. Since the root $m = 0$ gives difficulty, we replace A_0 by $B_0(m - 0) = B_0m$ and note that the series

$$\bar{y} = A_0x^m\left[1 - \frac{1}{(m-2)(m+2)}x^2 + \frac{1}{m(m-2)(m+2)(m+4)}x^4 - \frac{1}{m(m-2)(m+2)^2(m+4)(m+6)}x^6\right.$$
$$\left. + \frac{1}{m(m-2)(m+2)^2(m+4)^2(m+6)(m+8)}x^8 - \cdots\cdots\cdots\right]$$

$$= B_0x^m\left[m - \frac{m}{(m-2)(m+2)}x^2 + \frac{1}{(m-2)(m+2)(m+4)}x^4 - \frac{1}{(m-2)(m+2)^2(m+4)(m+6)}x^6\right.$$
$$\left. + \frac{1}{(m-2)(m+2)^2(m+4)^2(m+6)(m+8)}x^8 - \cdots\cdots\cdots\right]$$

satisfies the equation
$$x\bar{y}'' - 3\bar{y}' + x\bar{y} = (m-4)mA_0x^{m-1} = (m-4)m^2B_0x^{m-1}.$$

Since the right hand member contains the factor m^2, it follows by the argument made in Problem 3 that $\bar{y}$ and $\dfrac{\partial \bar{y}}{\partial m}$, with $m = 0$, are solutions of the given differential equation. We find

$$\frac{\partial \bar{y}}{\partial m} = \bar{y}\ln x + B_0 x^m \Big[1 + \frac{m^2+4}{[(m-2)(m+2)]^2}x^2 - \frac{1}{(m-2)(m+2)(m+4)}\Big(\frac{1}{m-2} + \frac{1}{m+2} + \frac{1}{m+4}\Big)x^4$$

$$+ \frac{1}{(m-2)(m+2)^2(m+4)(m+6)}\Big(\frac{1}{m-2} + \frac{2}{m+2} + \frac{1}{m+4} + \frac{1}{m+6}\Big)x^6$$

$$- \frac{1}{(m-2)(m+2)^2(m+4)^2(m+6)(m+8)}\Big(\frac{1}{m-2} + \frac{2}{m+2} + \frac{2}{m+4} + \frac{1}{m+6} + \frac{1}{m+8}\Big)x^8 + \cdots\cdots\Big].$$

Using the root $m = 0$, with $B_0 = 1$, we obtain

$$y_1 = \bar{y}\Big|_{m=0} = -\frac{1}{2\cdot2\cdot4}x^4 + \frac{1}{2\cdot2^2\cdot4\cdot6}x^6 - \frac{1}{2\cdot2^2\cdot4^2\cdot6\cdot8}x^8 + \cdots\cdots\cdots\cdots$$

and

$$y_2 = \frac{\partial \bar{y}}{\partial m}\Big|_{m=0} = y_1\ln x + 1 + \frac{1}{2^2}x^2 + \frac{1}{2^5\,2!}x^4 - \frac{1}{2^6\,3!\,1!}\Big(1 + \frac{1}{2} + \frac{1}{3}\Big)x^6$$

$$+ \frac{1}{2^8\,4!\,2!}\Big[\Big(1+\frac{1}{2}+\frac{1}{3}+\frac{1}{4}\Big) + \frac{1}{2}\Big]x^8 - \frac{1}{2^{10}\,5!\,3!}\Big[\Big(1+\frac{1}{2}+\frac{1}{3}+\frac{1}{4}+\frac{1}{5}\Big) + \Big(\frac{1}{2}+\frac{1}{3}\Big)\Big]x^{10} + \cdots.$$

The complete solution is

$$y = Ay_1 + By_2$$

$$= (A + B\ln x)\Big\{-\frac{1}{2^3\,2!}x^4 + \frac{1}{2^5\,3!\,1!}x^6 - \frac{1}{2^7\,4!\,2!}x^8 + \cdots\cdots\cdots\Big\}$$

$$+ B\Big\{1 + \frac{1}{2^2}x^2 + \frac{1}{2^5\,2!}x^4 - \frac{1}{2^6\,3!\,1!}\Big(1+\frac{1}{2}+\frac{1}{3}\Big)x^6 + \frac{1}{2^8\,4!\,2!}\Big[\Big(1+\frac{1}{2}+\frac{1}{3}+\frac{1}{4}\Big) + \frac{1}{2}\Big]x^8$$

$$- \frac{1}{2^{10}\,5!\,3!}\Big[\Big(1+\frac{1}{2}+\frac{1}{3}+\frac{1}{4}+\frac{1}{5}\Big) + \Big(\frac{1}{2}+\frac{1}{3}\Big)\Big]x^{10} + \cdots\cdots\cdots\Big].$$

Tre series converge for all finite values of $x \neq 0$.

6. Solve in series $(x - x^2)y'' - 3y' + 2y = 0$.

Substituting for y, y', and y'', we obtain

$$(m-4)mA_0x^{m-1} + [(m-3)(m+1)A_1 - (m-2)(m+1)A_0]x^m + [(m-2)(m+2)A_2 - (m-1)(m+2)A_1]x^{m+1}$$

$$+ \cdots\cdots\cdots + [(m+n-4)(m+n)A_n - (m+n-3)(m+n)A_{n-1}]x^{m+n-1} + \cdots\cdots\cdots = 0.$$

The recursion formula is $A_n = \dfrac{m+n-3}{m+n-4}A_{n-1}$ so that

$$1)\quad \bar{y} = A_0x^m\Big[1 + \frac{m-2}{m-3}x + \frac{m-1}{m-3}x^2 + \frac{m}{m-3}x^3 + \frac{m+1}{m-3}x^4 + \frac{m+2}{m-3}x^5 + \frac{m+3}{m-3}x^6 + \cdots\cdots\cdots\Big]$$

satisfies the differential equation

$$(x - x^2)\bar{y}'' - 3\bar{y}' + 2\bar{y} = (m-4)mA_0x^{m-1}.$$

The roots $m = 0,4$ of the indicial equation differ by an integer. However, when $m=0$ the expected vanishing of the denominator in the coefficient of x^4 does not occur since the factor m appears in both numerator and denominator and thus cancels out. Note that the coefficient of x^3 is zero when $m = 0$.

Thus, with $A_0 = 1$,

$$y_1 = \bar{y}\Big|_{m=0} = 1 + 2x/3 + x^2/3 + 0 - x^4/3 - 2x^5/3 - 3x^6/3 - 4x^7/3 - \cdots\cdots$$

and

$$y_2 = \bar{y}\Big|_{m=4} = x^4(1 + 2x + 3x^2 + 4x^3 + \cdots\cdots)$$

so that $y_1 = (1 + 2x/3 + x^2/3) - y_2/3$.

The complete solution is $y = C_1y_1 + C_2y_2 = C_1(1 + 2x/3 + x^2/3) + (C_2 - C_1/3)y_2$

$$= A(x^2 + 2x + 3) + Bx^4(1 + 2x + 3x^2 + 4x^3 + \cdots\cdots)$$

$$= A(x^2 + 2x + 3) + B\frac{x^4}{(1-x)^2}.$$

There are finite singular points at $x=0$ and $x=1$. The series converge for $|x|<1$.

7. Solve in series $xy'' + (x-1)y' - y = 0$.

Substituting for y, y', and y'', we obtain

$$(m-2)mA_0x^{m-1} + [(m-1)(m+1)A_1 + (m-1)A_0]x^m + [m(m+2)A_2 + mA_1]x^{m+1} + \cdots\cdots$$
$$+ [(m+n-2)(m+n)A_n + (m+n-2)A_{n-1}]x^{m+n-1} + \cdots\cdots = 0.$$

The roots of the indicial equation are $m = 0,2$ which differ by an integer. We choose the A's to satisfy the recursion formula

$$A_n = -\frac{m+n-2}{(m+n-2)(m+n)}A_{n-1} = -\frac{1}{m+n}A_{n-1}.$$

At this point we see that no $A_i \to \infty$ for $m = 0$, the smaller root, as in Problem 5. This is due, of course, to the fact that the factor $m+n-2$ cancels out. Thus, since

$$\bar{y} = A_0x^m[1 - \frac{1}{m+1}x + \frac{1}{(m+1)(m+2)}x^2 - \frac{1}{(m+1)(m+2)(m+3)}x^3 + \cdots\cdots]$$

satisfies

$$x\bar{y}'' + (x-1)\bar{y}' - \bar{y} = (m-2)mA_0x^{m-1},$$

we obtain, with $A_0 = 1$ and $m=0$, $m=2$ respectively,

$$y_1 = \bar{y}\Big|_{m=0} = 1 - x + x^2/2! - x^3/3! + \cdots\cdots = e^{-x}$$

and

$$y_2 = \bar{y}\Big|_{m=2} = x^2 - 2x^3/3! + 2x^4/4! - 2x^5/5! + \cdots\cdots = 2(e^{-x} + x - 1).$$

The complete solution is $y = C_1e^{-x} + C_2[2(e^{-x} + x - 1)] = Ae^{-x} + B(1-x)$, convergent for all finite values of x.

PARTICULAR INTEGRAL.

8. Solve $(x^2 - x)y'' + 3y' - 2y = x + 3/x^2$ near $x = 0$.

Substituting for y, y', and y'' as in Problem 6, we obtain the condition

1) $$m(4-m)A_0x^{m-1} + [(m+1)(3-m)A_1 + (m+1)(m-2)A_0]x^m + \cdots\cdots\cdots$$
$$+ [(m+n)(4-m-n)A_n + (m+n)(m+n-3)A_{n-1}]x^{m+n-1} + \cdots\cdots\cdots = x + 3/x^2.$$

To find the complementary function, we set the left member of 1) equal to zero and proceed as before.

The recursion formula is $A_n = \dfrac{m+n-3}{m+n-4}A_{n-1}$, and thus

$$\bar{y} = A_0x^m\left(1 + \frac{m-2}{m-3}x + \frac{m-1}{m-3}x^2 + \frac{m}{m-3}x^3 + \frac{m+1}{m-3}x^4 + \cdots\cdots\cdots\right)$$

satisfies

2) $$(x^2-x)\bar{y}'' + 3\bar{y}' - 2\bar{y} = m(4-m)A_0x^{m-1}.$$

The right hand member of 2) will be 0 when $m = 0,4$. For $m = 0$ with $A_0 = 1$, we have

$$y_1 = 1 + 2x/3 + x^2/3 - x^4/3 - 2x^5/3 - 3x^6/3 - 4x^7/3 - \cdots\cdots\cdots\cdots$$

and for $m = 4$ with $A_0 = 1$, we have

$$y_2 = x^4(1 + 2x + 3x^2 + 4x^3 + 5x^4 + \cdots\cdots\cdots\cdots).$$

Then $y_1 = (1 + 2x/3 + x^2/3) - y_2/3$ and (See Problem 6) the complementary function is

$$y = A(x^2 + 2x + 3) + Bx^4/(1-x)^2.$$

In finding a particular integral, we consider each of the terms of the right member of the given differential equation separately. Setting the right member of 2) equal to x, that is,

$$m(4-m)A_0x^{m-1} = x, \quad \text{identically},$$

we have $m = 2$ and $A_0 = \frac{1}{4}$. For $m = 2$, the recursion formula is $A_n = \dfrac{n-1}{n-2}A_{n-1}$; thus, $A_1 = A_2 = A_3 = \cdots\cdots = 0$. The particular integral corresponding to the term x is $x^2/4$.

Again, setting the right member of 2) equal to $3/x^2$, that is,

$$m(4-m)A_0x^{m-1} = 3/x^2, \quad \text{identically},$$

we have $m = -1$ and $A_0 = -3/5$. For $m = -1$, $A_n = \dfrac{n-4}{n-5}A_{n-1}$; thus, $A_1 = \dfrac{3}{4}A_0$, $A_2 = \dfrac{1}{2}A_0$, $A_3 = \dfrac{1}{4}A_0$, $A_4 = A_5 = A_6 = \cdots\cdots = 0$. The particular integral corresponding to the term $3/x^2$ is $-\dfrac{3}{5}x^{-1}\left(1 + \dfrac{3}{4}x + \dfrac{1}{2}x^2 + \dfrac{1}{4}x^3\right)$. The required complete solution is

$$y = A(x^2+2x+3) + \frac{Bx^4}{(1-x)^2} - \frac{3}{5x} - \frac{9}{20} - \frac{3}{10}x + \frac{1}{10}x^2$$

$$= C(x^2+2x+3) + \frac{Bx^4}{(1-x)^2} + \frac{1}{4}x^2 - \frac{3}{5x}.$$

Note. A partial check of the work is obtained by showing that the particular integral $y = x^2/4 - 3/5x$ satisfies the differential equation.

Since $x = 1$ is the only other finite singular point, the series converge in the annular region bounded by a circle of arbitrarily small radius and a circle of radius one, both centered at $x = 0$.

EXPANSION FOR LARGE VALUES OF x.

9. Solve $2x^2(x-1)y'' + x(3x+1)y' - 2y = 0$ in series convergent near $x = \infty$.

The substitution

$$x = \frac{1}{z},\qquad y' = \frac{dy}{dz}\frac{dz}{dx} = -\frac{1}{x^2}\frac{dy}{dz} = -z^2\frac{dy}{dz},\qquad y'' = \frac{2}{x^3}\frac{dy}{dz} + \frac{1}{x^4}\frac{d^2y}{dz^2} = z^4\frac{d^2y}{dz^2} + 2z^3\frac{dy}{dz}$$

transforms the given equation into

$$2(z-z^2)\frac{d^2y}{dz^2} + (1-5z)\frac{dy}{dz} - 2y = 0$$

for which $z = 0$, the transform of $x = \infty$, is a regular singular point. We next assume the series solution

$$y = A_0z^m + A_1z^{m+1} + A_2z^{m+2} + \cdots\cdots + A_nz^{m+n} + \cdots\cdots$$

and obtain the condition

$$m(2m-1)A_0z^{m-1} + \{(m+1)(2m+1)A_1 - (2m^2+3m+2)A_0\}z^m + \cdots\cdots\cdots$$
$$+ \{(m+n)(2m+2n-1)A_n - [2(m+n)^2 - (m+n) + 1]A_{n-1}\}z^{m+n-1} + \cdots\cdots = 0.$$

The recursion formula is $A_n = \dfrac{2(m+n)^2 - (m+n) + 1}{(m+n)(2m+2n-1)}A_{n-1}$, and thus the series

$$\bar{y} = A_0z^m\left(1 + \frac{2m^2+3m+2}{(m+1)(2m+1)}z + \frac{2m^2+3m+2}{(m+1)(2m+1)}\cdot\frac{2m^2+7m+7}{(m+2)(2m+3)}z^2 + \cdots\cdots\right)$$

satisfies

$$2(z-z^2)\frac{d^2\bar{y}}{dz^2} + (1-5z)\frac{d\bar{y}}{dz} - 2\bar{y} = m(2m-1)A_0z^{m-1}.$$

For $m = 0$, with $A_0 = 1$, we have

$$y_1 = 1 + 2z + 7z^2/3 + 112z^3/45 + \cdots\cdots = 1 + \frac{2}{x} + \frac{7}{3x^2} + \frac{112}{45x^3} + \cdots\cdots\cdots,$$

and for $m = \frac{1}{2}$, with $A_0 = 1$, we have

$$y_2 = z^{\frac{1}{2}}(1 + 4z/3 + 22z^2/15 + 484z^3/315 + \cdots\cdots) = x^{-\frac{1}{2}}\left(1 + \frac{4}{3x} + \frac{22}{15x^2} + \frac{484}{315x^3} + \cdots\cdots\right).$$

The complete solution is

$$y = Ay_1 + By_2 = A\left(1 + \frac{2}{x} + \frac{7}{3x^2} + \frac{112}{45x^3} + \cdots\cdots\right) + Bx^{-\frac{1}{2}}\left(1 + \frac{4}{3x} + \frac{22}{15x^2} + \frac{484}{315x^3} + \cdots\cdots\right).$$

The series in z converge for $|z| < 1$, that is, for all z inside a circle of radius 1, centered at $z = 0$. The series in x converge for $|x| > 1$, that is, for all x outside a circle of radius 1, centered at $x = 0$.

10. Solve $x^3y'' + x(1-x)y' + y = 0$ in series convergent near $x = \infty$.

Making the substitution $x = 1/z$ as in Problem 9, we obtain

1) $$z\frac{d^2y}{dz^2} + (3-z)\frac{dy}{dz} + y = 0$$

for which $z = 0$ is a regular singular point. We next assume the series solution

$$y = A_0 z^m + A_1 z^{m+1} + A_2 z^{m+2} + \cdots\cdots + A_n z^{m+n} + \cdots\cdots,$$

substitute in 1), and obtain

$$m(m+2)A_0 z^{m-1} + [(m+1)(m+3)A_1 - (m-1)A_0]z^m + [(m+2)(m+4)A_2 - mA_1]z^{m+1} + \cdots\cdots$$
$$+ [(m+n)(m+n+2)A_n - (m+n-2)A_{n-1}]z^{m+n-1} + \cdots\cdots = 0.$$

The roots of the indicial equation are $m = 0,-2$ and differ by an integer. From the recursion formula $A_n = \dfrac{m+n-2}{(m+n)(m+n+2)}A_{n-1}$ it is seen that $A_2 \to \infty$ when $m = -2$. We replace A_0 by $B_0(m+2)$ and note that the series

$$\bar{y} = B_0 z^m[(m+2) + \frac{(m-1)(m+2)}{(m+1)(m+3)}z + \frac{(m-1)m}{(m+1)(m+3)(m+4)}z^2 + \frac{(m-1)m}{(m+3)^2(m+4)(m+5)}z^3$$
$$+ \frac{(m-1)m(m+2)}{(m+3)^2(m+4)^2(m+5)(m+6)}z^4 + \cdots\cdots]$$

satisfies the equation

$$z\frac{d^2\bar{y}}{dz^2} + (3-z)\frac{d\bar{y}}{dz} + \bar{y} = B_0 m(m+2)^2 z^{m-1}.$$

Hence,

$$\frac{\partial\bar{y}}{\partial m} = \bar{y}\ln z + B_0 z^m\{1 + [\frac{2m+1}{(m+1)(m+3)} - \frac{(m-1)(m+2)}{(m+1)(m+3)}(\frac{1}{m+1} + \frac{1}{m+3})]z +$$
$$[\frac{2m-1}{(m+1)(m+3)(m+4)} - \frac{(m-1)m}{(m+1)(m+3)(m+4)}(\frac{1}{m+1} + \frac{1}{m+3} + \frac{1}{m+4})]z^2 +$$
$$[\frac{2m-1}{(m+3)^2(m+4)(m+5)} - \frac{(m-1)m}{(m+3)^2(m+4)(m+5)}(\frac{2}{m+3} + \frac{1}{m+4} + \frac{1}{m+5})]z^3 +$$
$$[\frac{3m^2+2m-2}{(m+3)^2(m+4)^2(m+5)(m+6)} - \frac{(m-1)m(m+2)}{(m+3)^2(m+4)^2(m+5)(m+6)}(\frac{2}{m+3} + \frac{2}{m+4} + \frac{1}{m+5} + \frac{1}{m+6})]z^4$$
$$+ \cdots\cdots\}$$ also satisfies this equation.

Using $m = -2$ with $B_0 = 1$, we find

$$y_1 = \bar{y}\Big|_{m=-2} = z^{-2}(-3z^2 + z^3) = \frac{1}{x} - 3 \quad \text{and}$$

$$y_2 = \frac{\partial\bar{y}}{\partial m}\Big|_{m=-2} = y_1 \ln z + z^{-2}(1 + 3z + 4z^2 - 11z^3/3 + z^4/8 + \cdots\cdots)$$
$$= y_1 \ln\frac{1}{x} + x^2 + 3x + 4 - 11/3x + 1/8x^2 + \cdots\cdots.$$ The complete solution

is $y = Ay_1 + By_2 = (A + B\ln\frac{1}{x})(1/x - 3) + B(x^2 + 3x + 4 - 11/3x + 1/8x^2 + \cdots\cdots).$

The series converge for all values of $x \neq 0$.

SUPPLEMENTARY PROBLEMS

Solve in series near $x=0$.

11. $2(x^2+x^3)y'' - (x-3x^2)y' + y = 0.$

R.F. $A_n = -A_{n-1}$

Ans. $y = (A\sqrt{x}+Bx)(1-x+x^2-x^3+\cdots\cdots).$ Converges for $|x|<1$.

12. $4xy'' + 2(1-x)y' - y = 0.$

R.F. $A_n = \dfrac{1}{2(m+n)}A_{n-1}$

Ans. $y = A(1+\dfrac{x}{2\cdot 1!}+\dfrac{x^2}{2^2\cdot 2!}+\dfrac{x^3}{2^3\cdot 3!}+\cdots\cdot) + B\sqrt{x}(1+\dfrac{x}{1\cdot 3}+\dfrac{x^2}{1\cdot 3\cdot 5}+\dfrac{x^3}{1\cdot 3\cdot 5\cdot 7}+\cdots\cdot).$

Converges for all finite values of x.

13. $2x^2y'' - xy' + (1-x^2)y = 0.$

R.F. $A_n = \dfrac{1}{(m+n-1)(2m+2n-1)}A_{n-2}$, n even; $A_n = 0$, n odd.

Ans. $y = Ax(1+\dfrac{x^2}{2\cdot 5}+\dfrac{x^4}{2\cdot 4\cdot 5\cdot 9}+\dfrac{x^6}{2\cdot 4\cdot 6\cdot 5\cdot 9\cdot 13}+\cdots\cdots\cdots\cdots\cdots)$

$+ B\sqrt{x}(1+\dfrac{x^2}{2\cdot 3}+\dfrac{x^4}{2\cdot 4\cdot 3\cdot 7}+\dfrac{x^6}{2\cdot 4\cdot 6\cdot 3\cdot 7\cdot 11}+\cdots\cdots\cdots\cdots\cdots\cdot).$

Converges for all finite values of x.

14. $xy'' + y' + xy = 0.$

R.F. $A_n = -\dfrac{1}{(m+n)^2}A_{n-2}$, n even; $A_n = 0$, n odd.

Ans. $y = (A+B\ln x)(1-\dfrac{x^2}{2^2}+\dfrac{x^4}{2^2\cdot 4^2}-\dfrac{x^6}{2^2\cdot 4^2\cdot 6^2}+\cdots\cdots\cdots\cdot)$

$+ B[\dfrac{x^2}{2^2}-\dfrac{x^4}{2^2\cdot 4^2}(1+\dfrac{1}{2})+\dfrac{x^6}{2^2\cdot 4^2\cdot 6^2}(1+\dfrac{1}{2}+\dfrac{1}{3})-\cdots\cdots\cdots\cdot].$

Converges for all finite values of $x \neq 0$.

15. $x^2y'' - xy' + (x^2+1)y = 0.$ R.F. $A_n = -\dfrac{1}{(m+n-1)^2}A_{n-2}$, n even; $A_n = 0$, n odd.

Ans. $y = (A+B\ln x)x(1-\dfrac{x^2}{2^2}+\dfrac{x^4}{2^4(2!)^2}-\dfrac{x^6}{2^6(3!)^2}+\cdots\cdots\cdots\cdot)$

$+ Bx[\dfrac{x^2}{2^2}-\dfrac{x^4}{2^4(2!)^2}(1+\dfrac{1}{2})+\dfrac{x^6}{2^6(3!)^2}(1+\dfrac{1}{2}+\dfrac{1}{3})+\cdots\cdots\cdots\cdot].$

Converges for all finite values of $x \neq 0$.

16. $xy'' - 2y' + y = 0.$ R.F. $A_n = -\dfrac{1}{(m+n-3)(m+n)} A_{n-1}$

Ans. $y = (A + B\ln x)(-\dfrac{x^3}{12} + \dfrac{x^4}{48} - \dfrac{x^5}{480} + \cdots) + B(1 + \dfrac{x}{2} + \dfrac{x^2}{4} + \dfrac{x^3}{36} - \dfrac{19x^4}{576} + \dfrac{137x^5}{28800} - \cdots).$

Converges for all finite values of $x \neq 0$.

17. $xy'' + 2y' + xy = 0.$ R.F. $A_n = -\dfrac{1}{(m+n)(m+n+1)} A_{n-2}$, n even; $A_n = 0$, n odd.

Ans. $y = Ax^{-1}(1 - \dfrac{x^2}{2!} + \dfrac{x^4}{4!} - \cdots\cdots) + B(1 - \dfrac{x^2}{3!} + \dfrac{x^4}{5!} - \cdots\cdots).$

Converges for all finite values of $x \neq 0$.

18. $x^2(x+1)y'' + x(x+1)y' - y = 0.$

Singular points: $x = 0, -1$. R.F. $A_n = -\dfrac{m+n-1}{m+n+1} A_{n-1}$.

Ans. $y = Ax(1 - x/3 + x^2/6 - x^3/10 + \cdots\cdots\cdots) + Bx^{-1}(1 + x).$

Converges in the annular region bounded by a circle of arbitrarily small radius and a circle of radius one, both centered at $x = 0$.

19. $2xy'' + y' - y = x + 1.$ R.F. $A_n = \dfrac{1}{(m+n)(2m+2n-1)} A_{n-1}$

Ans. $y = A(1 + x + x^2/6 + x^3/90 + \cdots\cdots) + B\sqrt{x}(1 + x/3 + x^2/30 + x^3/630 + \cdots\cdots\cdots)$

$+ \dfrac{1}{6}x^2(1 + x/15 + x^2/420 + x^3/18900 + \cdots\cdots\cdots) - 1.$

Converges for all finite values of x.

Solve in series near $x = \infty$.

20. $2x^3y'' + x^2y' + y = 0.$ R.F. $A_n = -\dfrac{1}{(m+n)(2m+2n+1)} A_{n-1}$

Ans. $y = A(1 - \dfrac{1}{3x} + \dfrac{1}{30x^2} - \dfrac{1}{630x^3} + \cdots\cdots) + B\sqrt{x}(1 - \dfrac{1}{x} + \dfrac{1}{6x^2} - \dfrac{1}{90x^3} + \cdots\cdots).$

Converges for all finite values of $x \neq 0$.

21. $x^3y'' + (x^2 + x)y' - y = 0.$ R.F. $A_n = \dfrac{1}{m+n} A_{n-1}$

Ans. $y = (A + B\ln\dfrac{1}{x})(1 + \dfrac{1}{x} + \dfrac{1}{2x^2} + \dfrac{1}{6x^3} + \cdots) + B[\dfrac{1}{x} + \dfrac{1}{2x^2}(1 + \dfrac{1}{2}) + \dfrac{1}{6x^3}(1 + \dfrac{1}{2} + \dfrac{1}{3}) + \cdots].$

Converges for all finite values of $x \neq 0$.

CHAPTER 27

The Legendre, Bessel, and Gauss Equations

THE THREE DIFFERENTIAL EQUATIONS to be considered here are solved by the methods of the preceding chapter. The first two have important applications in mathematical physics. The solutions of all three have many interesting properties.

THE LEGENDRE EQUATION

$$(1-x^2)y'' - 2xy' + p(p+1)y = 0.$$

A solution of this equation in series convergent near $x=0$, an ordinary point, was called for in Problem 16, Chapter 25. Under certain conditions on p which will be stated later, we shall obtain here the solution convergent near $x = \infty$. Using the substitution $x = 1/z$ (see Chapter 26) the equation becomes

$$(z^4 - z^2)\frac{d^2y}{dz^2} + 2z^3\frac{dy}{dz} + p(p+1)y = 0$$

for which $z=0$ is a regular singular point.

Putting $y = A_0z^m + A_1z^{m+1} + A_2z^{m+2} + \cdots\cdots + A_nz^{m+n} + \cdots\cdots$, we have

$$\{-m(m-1) + p(p+1)\}A_0z^m + \{-m(m+1) + p(p+1)\}A_1z^{m+1} + \{[-(m+1)(m+2) + p(p+1)]A_2 + m(m+1)A_0\}z^{m+2} + \cdots\cdots + \{[-(m+n)(m+n-1) + p(p+1)]A_n + (m+n-2)(m+n-1)A_{n-2}\}z^{m+n} + \cdots\cdots = 0.$$

We take $A_1 = 0$ and $A_n = \dfrac{(m+n-2)(m+n-1)}{(m+n)(m+n-1) - p(p+1)}A_{n-2}$, and see that

$$\bar{y} = A_0z^m\Big[1 + \frac{m(m+1)}{(m+1)(m+2)-p(p+1)}z^2 + \frac{m(m+1)(m+2)(m+3)}{[(m+1)(m+2)-p(p+1)][(m+3)(m+4)-p(p+1)]}z^4 + \frac{m(m+1)(m+2)(m+3)(m+4)(m+5)}{[(m+1)(m+2)-p(p+1)][(m+3)(m+4)-p(p+1)][(m+5)(m+6)-p(p+1)]}z^6 + \cdots\cdots\Big]$$

satisfies the equation

$$(z^4-z^2)\frac{d^2\bar{y}}{dz^2} + 2z^3\frac{d\bar{y}}{dz} + p(p+1)\bar{y} = [-m(m-1)+p(p+1)]A_0z^m = (m+p)(-m+p+1)A_0z^m.$$

For $m = -p$ with $A_0 = 1$, we obtain

1) $$y_1 = z^{-p}\Big[1 - \frac{p(p-1)}{2(2p-1)}z^2 + \frac{p(p-1)(p-2)(p-3)}{2\cdot4(2p-1)(2p-3)}z^4 - \frac{p(p-1)(p-2)(p-3)(p-4)(p-5)}{2\cdot4\cdot6(2p-1)(2p-3)(2p-5)}z^6 + \cdots\cdots\cdots\Big]$$

$$= x^p\Big[1 - \frac{p(p-1)}{2(2p-1)}x^{-2} + \frac{p(p-1)(p-2)(p-3)}{2\cdot4(2p-1)(2p-3)}x^{-4} - \frac{p(p-1)(p-2)(p-3)(p-4)(p-5)}{2\cdot4\cdot6(2p-1)(2p-3)(2p-5)}x^{-6} + \cdots\cdots\cdots\Big].$$

For $m = p+1$ with $A_0 = 1$, we obtain

2) $$y_2 = z^{p+1}\left[1 + \frac{(p+1)(p+2)}{2(2p+3)}z^2 + \frac{(p+1)(p+2)(p+3)(p+4)}{2\cdot 4(2p+3)(2p+5)}z^4 + \frac{(p+1)(p+2)(p+3)(p+4)(p+5)(p+6)}{2\cdot 4\cdot 6(2p+3)(2p+5)(2p+7)}z^6 + \cdots\cdots\cdots\right]$$

$$= x^{-p-1}\left[1 + \frac{(p+1)(p+2)}{2(2p+3)}x^{-2} + \frac{(p+1)(p+2)(p+3)(p+4)}{2\cdot 4(2p+3)(2p+5)}x^{-4} + \frac{(p+1)(p+2)(p+3)(p+4)(p+5)(p+6)}{2\cdot 4\cdot 6(2p+3)(2p+5)(2p+7)}x^{-6} + \cdots\cdots\cdots\right]$$

Thus, $$y = Ay_1 + By_2$$
is the complete solution, convergent for $|x| > 1$, provided that $p \neq 1/2,\ 3/2,\ 5/2, \cdots\cdots$ or $p \neq -3/2,\ -5/2, \cdots\cdots$.

Suppose p is a positive integer including 0 and consider the solution y_1 which is a polynomial, say $u_p(x)$. Putting $p = 0, 1, 2, 3, \cdots\cdots$ in 1), we have

$u_0(x) = 1,\quad u_1(x) = x,\quad u_2(x) = x^2 - 1/3,\quad u_3(x) = x^3 - 3x/5,\quad \cdots\cdots\cdots\cdots\cdots\cdots,$

$$u_k(x) = \sum_{n=0}^{[\frac{1}{2}k]} (-1)^n \frac{k(k-1)\cdots\cdots(k-2n+1)}{2^n\, n!\,(2k-1)\cdots\cdots(2k-2n+1)} x^{k-2n}, \quad \cdots\cdots,$$

where $[\frac{1}{2}k]$ denotes the greatest integer $\leqq \frac{1}{2}k$ (i.e., $[\frac{1}{2}k] = 3$ if $k = 7$, $[\frac{1}{2}k] = 4$ if $k = 8$).

The polynomials defined by

3) $$P_p(x) = \frac{(2p)!}{2^p(p!)^2}u_p(x) = \frac{1\cdot 3\cdot 5\cdots(2p-1)}{p!}u_p(x), \qquad p = 0, 1, 2, \cdots\cdots,$$

are called Legendre polynomials. The first few of these are:

$P_0(x) = u_0(x) = 1,$

$P_1(x) = u_1(x) = x,$

$$P_2(x) = \frac{1\cdot 3}{2!}u_2(x) = \frac{3}{2}x^2 - \frac{1}{2},$$

$$P_3(x) = \frac{1\cdot 3\cdot 5}{3!}u_3(x) = \frac{5}{2}x^3 - \frac{3}{2}x,$$

$$P_4(x) = \frac{1\cdot 3\cdot 5\cdot 7}{4!}u_4(x) = \frac{5\cdot 7}{2\cdot 4}x^4 - 2\frac{3\cdot 5}{2\cdot 4}x^2 + \frac{1\cdot 3}{2\cdot 4},$$

$$P_5(x) = \frac{1\cdot 3\cdot 5\cdot 7\cdot 9}{5!}u_5(x) = \frac{7\cdot 9}{2\cdot 4}x^5 - 2\frac{5\cdot 7}{2\cdot 4}x^3 + \frac{3\cdot 5}{2\cdot 4}x,$$

$$P_6(x) = \frac{1\cdot 3\cdots 11}{6!}u_6(x) = \frac{7\cdot 9\cdot 11}{2\cdot 4\cdot 6}x^6 - 3\frac{5\cdot 7\cdot 9}{2\cdot 4\cdot 6}x^4 + 3\frac{3\cdot 5\cdot 7}{2\cdot 4\cdot 6}x^2 - \frac{1\cdot 3\cdot 5}{2\cdot 4\cdot 6},$$

$$P_7(x) = \frac{1\cdot 3\cdots 13}{7!}u_7(x) = \frac{9\cdot 11\cdot 13}{2\cdot 4\cdot 6}x^7 - 3\frac{7\cdot 9\cdot 11}{2\cdot 4\cdot 6}x^5 + 3\frac{5\cdot 7\cdot 9}{2\cdot 4\cdot 6}x^3 - \frac{3\cdot 5\cdot 7}{2\cdot 4\cdot 6}x, \quad \text{etc.}$$

It is clear from 3) that $P_p(x)$ is a particular solution of the Legendre equation $(1-x^2)y'' - 2xy' + p(p+1) = 0$.

See Problems 1-6.

THE BESSEL EQUATION

$$x^2y'' + xy' + (x^2 - k^2)y = 0.$$

It is evident that $x = 0$ is a regular singular point. To obtain the solution in series, convergent near $x = 0$, we substitute

$$y = A_0x^m + A_1x^{m+1} + A_2x^{m+2} + \cdots\cdots + A_nx^{m+n} + \cdots\cdots$$

and obtain

$$(m^2 - k^2)A_0x^m + \{(m+1)^2 - k^2\}A_1x^{m+1} + \{[(m+2)^2 - k^2]A_2 + A_0\}x^{m+2} + \cdots\cdots\cdots$$
$$+ \{[(m+n)^2 - k^2]A_n + A_{n-2}\}x^{m+n} + \cdots\cdots = 0.$$

We take $A_1 = 0$ and $A_n = -\dfrac{1}{(m+n)^2 - k^2}A_{n-2}$ and see that

$$\bar{y} = A_0x^m\{1 - \frac{1}{(m+2)^2 - k^2}x^2 + \frac{1}{[(m+2)^2 - k^2][(m+4)^2 - k^2]}x^4$$
$$- \frac{1}{[(m+2)^2 - k^2][(m+4)^2 - k^2][(m+6)^2 - k^2]}x^6 + \cdots\cdots\cdots\}$$

satisfies the equation $x^2\bar{y}'' + x\bar{y}' + (x^2 - k^2)\bar{y} = (m^2 - k^2)A_0x^m.$

For $m = k$ with $A_0 = 1$, we obtain

$$y_1 = x^k\{1 - \frac{1}{4(k+1)}x^2 + \frac{1}{4^2\cdot 2!(k+1)(k+2)}x^4 - \frac{1}{4^3\cdot 3!(k+1)(k+2)(k+3)}x^6 + \cdots\cdots\}$$

and for $m = -k$ with $A_0 = 1$, we obtain

$$y_2 = x^{-k}\{1 - \frac{1}{4(1-k)}x^2 + \frac{1}{4^2\cdot 2!(1-k)(2-k)}x^4 - \frac{1}{4^3\cdot 3!(1-k)(2-k)(3-k)}x^6 + \cdots\cdots\}.$$

Note that $y_2 = y_1$ if $k = 0$, y_1 is meaningless if k is a negative integer, and y_2 is meaningless if k is a positive integer. Except for these cases, the complete solution of the Bessel equation is $y = Ay_1 + By_2$, convergent for all $x \neq 0$.

The Bessel functions of the *first kind* are defined by

$$J_k(x) = \frac{1}{2^k\cdot k!}y_1 = (\frac{x}{2})^k\{\frac{1}{k!} - \frac{1}{1!(k+1)!}(\frac{x}{2})^2 + \frac{1}{2!(k+2)!}(\frac{x}{2})^4 - \frac{1}{3!(k+3)!}(\frac{x}{2})^6 + \cdots\},$$

$J_{-k}(x) = (-1)^kJ_k(x)$, where k is a positive integer including 0.

Of these, $$J_0(x) = 1 - \frac{1}{(1!)^2}(\frac{x}{2})^2 + \frac{1}{(2!)^2}(\frac{x}{2})^4 - \frac{1}{(3!)^2}(\frac{x}{2})^6 + \cdots\cdots\cdots$$

and $$J_1(x) = (\frac{x}{2})\{1 - \frac{1}{1!\,2!}(\frac{x}{2})^2 + \frac{1}{2!\,3!}(\frac{x}{2})^4 - \frac{1}{3!\,4!}(\frac{x}{2})^6 + \cdots\cdots\cdots\}$$

are more frequently used.

See Problems 7-10.

THE GAUSS EQUATION

$$(x-x^2)y'' + [\gamma-(\alpha+\beta+1)x]y' - \alpha\beta y = 0.$$

To obtain the solution in series, convergent near $x=0$, substitute

$$y = A_0x^m + A_1x^{m+1} + A_2x^{m+2} + \cdots\cdots + A_nx^{m+n} + \cdots\cdots$$

and obtain

$$m(m+\gamma-1)A_0x^{m-1} + \{(m+1)(m+\gamma)A_1 - [m(m+\alpha+\beta)+\alpha\beta]A_0\}x^m + \cdots\cdots$$
$$+ \{(m+n)(m+n+\gamma-1)A_n - [(m+n-1)(m+n+\alpha+\beta-1)+\alpha\beta]A_{n-1}\}x^{m+n-1} + \cdots\cdots = 0.$$

We take $A_n = \dfrac{(m+n-1)(m+n+\alpha+\beta-1)+\alpha\beta}{(m+n)(m+n+\gamma-1)}A_{n-1}$ and see that

$$\bar{y} = A_0x^m\Big[1 + \frac{m(m+\alpha+\beta)+\alpha\beta}{(m+1)(m+\gamma)}x + \frac{m(m+\alpha+\beta)+\alpha\beta}{(m+1)(m+\gamma)}\cdot\frac{(m+1)(m+\alpha+\beta+1)+\alpha\beta}{(m+2)(m+\gamma+1)}x^2$$
$$+ \frac{m(m+\alpha+\beta)+\alpha\beta}{(m+1)(m+\gamma)}\cdot\frac{(m+1)(m+\alpha+\beta+1)+\alpha\beta}{(m+2)(m+\gamma+1)}\cdot\frac{(m+2)(m+\alpha+\beta+2)+\alpha\beta}{(m+3)(m+\gamma+2)}x^3 + \cdots\cdots\Big]$$

satisfies the equation

$$(x-x^2)\bar{y}'' + [\gamma-(\alpha+\beta+1)x]\bar{y}' - \alpha\beta\bar{y} = m(m+\gamma-1)A_0x^{m-1}.$$

For $m=0$, with $A_0=1$, we obtain

$$y_1 = 1 + \frac{\alpha\cdot\beta}{1\cdot\gamma}x + \frac{\alpha(\alpha+1)\beta(\beta+1)}{1\cdot2\cdot\gamma(\gamma+1)}x^2 + \frac{\alpha(\alpha+1)(\alpha+2)\beta(\beta+1)(\beta+2)}{1\cdot2\cdot3\cdot\gamma(\gamma+1)(\gamma+2)}x^3 + \cdots\cdots,$$

and for $m=1-\gamma$, $\gamma\neq1$, with $A_0=1$, we obtain

$$y_2 = x^{1-\gamma}\Big[1 + \frac{(\alpha-\gamma+1)(\beta-\gamma+1)}{1(2-\gamma)}x + \frac{(\alpha-\gamma+1)(\alpha-\gamma+2)(\beta-\gamma+1)(\beta-\gamma+2)}{1\cdot2(2-\gamma)(3-\gamma)}x^2$$
$$+ \frac{(\alpha-\gamma+1)(\alpha-\gamma+2)(\alpha-\gamma+3)(\beta-\gamma+1)(\beta-\gamma+2)(\beta-\gamma+3)}{1\cdot2\cdot3(2-\gamma)(3-\gamma)(4-\gamma)}x^3 + \cdots\cdots\Big].$$

The series y_1, known as the *hypergeometric series*, is convergent for $|x|<1$ and is represented by

$$y_1 = F(\alpha,\beta,\gamma,x).$$

Note that

$$y_2 = x^{1-\gamma}F(\alpha-\gamma+1,\ \beta-\gamma+1,\ 2-\gamma,\ x)$$

is of the same type. Thus, if γ is non-integral (including 0), the general solution is

$$y = Ay_1 + By_2 = AF(\alpha,\beta,\gamma,x) + Bx^{1-\gamma}F(\alpha-\gamma+1,\ \beta-\gamma+1,\ 2-\gamma,\ x).$$

There are numerous special cases, depending upon the values of α, β, and γ. Some of these will be treated in the Solved Problems.

SOLVED PROBLEMS

THE LEGENDRE EQUATION.

1. Verify that $2^p p!\,P_p(x) = \dfrac{d^p}{dx^p}(x^2-1)^p$. (Rodrigues' Formula)

By the binomial theorem, $(x^2-1)^p = \sum_{n=0}^{p} (-1)^n \dfrac{p!}{n!\,(p-n)!} x^{2p-2n}$. Then

$$\frac{d^p}{dx^p}(x^2-1)^p = \sum_{n=0}^{[\frac{1}{2}p]} (-1)^n \frac{p!}{n!\,(p-n)!}(2p-2n)(2p-2n-1)\cdots\cdots(p-2n+1)x^{p-2n}$$

$$= \sum_{n=0}^{[\frac{1}{2}p]} (-1)^n \frac{2p(2p-1)\cdots(2p-2n+1)}{2p(2p-1)\cdots(2p-2n+1)}(2p-2n)(2p-2n-1)\cdots(p-2n+1)\frac{(p-2n)(p-2n-1)\cdots1}{(p-2n)(p-2n-1)\cdots1}\cdot\frac{p!}{n!\,(p-n)!}x^{p-2n}.$$

Now (in the denominator) $2p(2p-1)\cdots(2p-2n+1) = 2^n[p(p-1)\cdots(p-n+1)][(2p-1)(2p-3)\cdots(2p-2n+1)]$ and when multiplied by $(p-n)!$ yields $2^n p!\,[(2p-1)(2p-3)\cdots\cdot(2p-2n+1)]$. Hence,

$$\frac{d^p}{dx^p}(x^2-1)^p = \sum_{n=0}^{[\frac{1}{2}p]} (-1)^n \frac{(2p)!\,p!}{2^n p!\,[(2p-1)(2p-3)\cdots(2p-2n+1)](p-2n)!\,n!} x^{p-2n}$$

$$= \sum_{n=0}^{[\frac{1}{2}p]} (-1)^n \frac{(2p)!}{2^n\, n!\, p!}\cdot\frac{p(p-1)\cdots(p-2n+1)}{(2p-1)(2p-3)\cdots(2p-2n+1)} x^{p-2n}$$

$$= \frac{(2p)!}{p!}u_p(x) = 2^p\cdot p!\,P_p(x).$$

2. Show that $P_p(x) = \sum_{n=0}^{[\frac{1}{2}p]} (-1)^n \dfrac{(2p-2n)!}{2^p\, n!\,(p-n)!\,(p-2n)!} x^{p-2n}$. From Problem 1 above,

$$\frac{d^p}{dx^p}(x^2-1)^p = \sum_{n=0}^{[\frac{1}{2}p]} (-1)^n \frac{p!}{n!\,(p-n)!}(2p-2n)(2p-2n-1)\cdots\cdot(p-2n+1)x^{p-2n}$$

$$= \sum_{n=0}^{[\frac{1}{2}p]} (-1)^n (2p-2n)(2p-2n-1)\cdots\cdot(p-2n+1)\frac{(p-2n)!}{(p-2n)!}\cdot\frac{p!}{n!\,(p-n)!} x^{p-2n}$$

$$= \sum_{n=0}^{[\frac{1}{2}p]} (-1)^n \frac{(2p-2n)!\,p!}{n!\,(p-n)!\,(p-2n)!} x^{p-2n}.$$

Hence, $P_p(x) = \dfrac{1}{2^p\, p!}\cdot\dfrac{d^p}{dx^p}(x^2-1)^p = \sum_{n=0}^{[\frac{1}{2}p]} (-1)^n \dfrac{(2p-2n)!}{2^p\, n!\,(p-n)!\,(p-2n)!} x^{p-2n}$.

3. Evaluate $\int_{-1}^{1} P_r(x)\, P_s(x)\, dx$.

Using Rodrigues' formula (Problem 1),

$$\int_{-1}^{1} P_r(x)\, P_s(x)\, dx \;=\; \frac{1}{2^{r+s}\, r!\, s!} \int_{-1}^{1} \frac{d^r}{dx^r}(x^2-1)^r \cdot \frac{d^s}{dx^s}(x^2-1)^s\, dx.$$

Let $u = \dfrac{d^r}{dx^r}(x^2-1)^r$ and $dv = \dfrac{d^s}{dx^s}(x^2-1)^s\, dx$. Then $du = \dfrac{d^{r+1}}{dx^{r+1}}(x^2-1)^r\, dx$, $v = \dfrac{d^{s-1}}{dx^{s-1}}(x^2-1)^s$,

and $$\int_{x=-1}^{x=1} u\, dv \;=\; uv\Big]_{x=-1}^{x=1} - \int_{x=-1}^{x=1} v\, du$$

$$= \frac{d^r}{dx^r}(x^2-1)^r \cdot \frac{d^{s-1}}{dx^{s-1}}(x^2-1)^s\Big]_{-1}^{1} - \int_{-1}^{1} \frac{d^{r+1}}{dx^{r+1}}(x^2-1)^r \cdot \frac{d^{s-1}}{dx^{s-1}}(x^2-1)^s\, dx.$$

Now $\dfrac{d^{s-j}}{dx^{s-j}}(x^2-1)^s\Big]_{-1}^{1} = 0$, for $j = 1, 2, \cdots, s-1$; hence, after one integration by parts,

$$\int_{-1}^{1} P_r(x)\, P_s(x)\, dx \;=\; -\frac{1}{2^{r+s}\, r!\, s!} \int_{-1}^{1} \frac{d^{r+1}}{dx^{r+1}}(x^2-1)^r \cdot \frac{d^{s-1}}{dx^{s-1}}(x^2-1)^s\, dx.$$

A second integration by parts yields

$$\int_{-1}^{1} P_r(x)\, P_s(x)\, dx \;=\; \frac{1}{2^{r+s}\, r!\, s!} \int_{-1}^{1} \frac{d^{r+2}}{dx^{r+2}}(x^2-1)^r \cdot \frac{d^{s-2}}{dx^{s-2}}(x^2-1)^s\, dx$$

and, after s integrations by parts, we have formally

$$A)\qquad \int_{-1}^{1} P_r(x)\, P_s(x)\, dx \;=\; \frac{(-1)^s}{2^{r+s}\, r!\, s!} \int_{-1}^{1} (x^2-1)^s \cdot \frac{d^{r+s}}{dx^{r+s}}(x^2-1)^r\, dx.$$

Suppose $s > r$. Then, since $(x^2-1)^r = x^{2r} - rx^{2r-2} + \cdots\cdots + (-1)^r$, $\dfrac{d^{r+s}}{dx^{r+s}}(x^2-1)^r = 0$

and $\int_{-1}^{1} P_r(x)\, P_s(x)\, dx = 0$. Since r and s enter symmetrically, this relation holds also when $r > s$. Thus, it holds when $r \neq s$.

Suppose $s = r$. Then $A)$ becomes

$$B)\qquad \int_{-1}^{1} P_r^2(x)\, dx \;=\; \frac{(-1)^r}{2^{2r}\,(r!)^2} \int_{-1}^{1} (x^2-1)^r \cdot \frac{d^{2r}}{dx^{2r}}(x^2-1)^r\, dx.$$

Now $\dfrac{d^{2r}}{dx^{2r}}(x^2-1)^r = (2r)!$. Hence, $\displaystyle\int_{-1}^{1} P_r^2(x)\, dx = \frac{(-1)^r (2r)!}{2^{2r}(r!)^2} \int_{-1}^{1} (x^2-1)^r\, dx$

$$= \frac{(-1)^r (2r)!}{2^{2r}(r!)^2} \cdot (-1)^r \cdot 2 \int_0^{\frac{1}{2}\pi} \sin^{2r+1}\theta\, d\theta \;=\; \frac{(2r)!}{2^{2r}(r!)^2} \cdot \frac{2^{r+1}\, r!}{1\cdot 3 \cdots (2r+1)} \;=\; \frac{2}{2r+1},$$ using the

substitution $x = \cos\theta$ and Wallis' Formula $\int_0^{\frac{1}{2}\pi} \sin^{2n+1}\theta\, d\theta = \frac{2^n n!}{1\cdot 3\cdots(2n+1)}$.

4. Express $f(x) = x^4 + 2x^3 + 2x^2 - x - 3$ in terms of Legendre polynomials.

Since $P_4(x) = \frac{35}{8}x^4 - \frac{15}{4}x^2 + \frac{3}{8}$, then $x^4 = \frac{8}{35}P_4(x) + \frac{6}{7}x^2 - \frac{3}{35}$ and

$$f(x) = (\frac{8}{35}P_4(x) + \frac{6}{7}x^2 - \frac{3}{35}) + 2x^3 + 2x^2 - x - 3 = \frac{8}{35}P_4(x) + 2x^3 + \frac{20}{7}x^2 - x - \frac{108}{35}.$$

Now $x^3 = \frac{2}{5}P_3(x) + \frac{3}{5}x$ and $f(x) = \frac{8}{35}P_4(x) + \frac{4}{5}P_3(x) + \frac{20}{7}x^2 + \frac{1}{5}x - \frac{108}{35}$;

$x^2 = \frac{2}{3}P_2(x) + \frac{1}{3}$ and

$$f(x) = \frac{8}{35}P_4(x) + \frac{4}{5}P_3(x) + \frac{40}{21}P_2(x) + \frac{1}{5}x - \frac{224}{105}$$
$$= \frac{8}{35}P_4(x) + \frac{4}{5}P_3(x) + \frac{40}{21}P_2(x) + \frac{1}{5}P_1(x) - \frac{224}{105}P_0(x).$$

5. Show that $(1-2xt+t^2)^{-\frac{1}{2}} = P_0(x) + P_1(x)\,t + P_2(x)\,t^2 + \cdots\cdots + P_k(x)\,t^k + \cdots\cdots$.

Now
$$(1-2xt+t^2)^{-\frac{1}{2}} = [1-(2xt-t^2)]^{-\frac{1}{2}} = 1 + \frac{1}{2}(2xt-t^2) + \frac{(1/2)(3/2)}{2}(2xt-t^2)^2 + \cdots\cdots$$
$$+ \frac{1\cdot 3\cdots(2k-5)}{2^{k-2}(k-2)!}(2xt-t^2)^{k-2} + \frac{1\cdot 3\cdots(2k-3)}{2^{k-1}(k-1)!}(2xt-t^2)^{k-1} + \frac{1\cdot 3\cdots(2k-1)}{2^k k!}(2xt-t^2)^k + \cdots\cdots.$$

But
$$(2xt-t^2)^k = (2x)^k t^k - \cdots\cdots,$$
$$(2xt-t^2)^{k-1} = (2x)^{k-1}t^{k-1} - (k-1)(2x)^{k-2}t^k + \cdots\cdots,$$
$$(2xt-t^2)^{k-2} = (2x)^{k-2}t^{k-2} - (k-2)(2x)^{k-3}t^{k-1} + \frac{(k-2)(k-3)}{2!}(2x)^{k-4}t^k - \cdots\cdots, \text{ etc.}$$

Hence,
$$(1-2xt+t^2)^{-\frac{1}{2}} = 1 + xt + (\frac{3}{2}x^2 - \frac{1}{2})t^2 + \cdots\cdots\cdots\cdots + [\frac{1\cdot 3\cdots(2k-1)}{2^k k!}2^k x^k$$
$$- \frac{1\cdot 3\cdots(2k-3)}{2^{k-1}(k-1)!}(k-1)2^{k-2}x^{k-2} + \frac{1\cdot 3\cdots(2k-5)}{2^{k-2}(k-2)!}\frac{(k-2)(k-3)}{2!}2^{k-4}x^{k-4} + \cdots\cdots]t^k + \cdots\cdots\cdots$$
$$= 1 + xt + (\frac{3}{2}x^2 - \frac{1}{2})t^2 + \cdots\cdots\cdots\cdots\cdots + \frac{1\cdot 3\cdots(2k-1)}{k!}[x^k$$
$$- \frac{k(k-1)}{2(2k-1)}x^{k-2} + \frac{k(k-1)(k-2)(k-3)}{2\cdot 4(2k-1)(2k-3)}x^{k-4} + \cdots\cdots]t^k + \cdots\cdots$$
$$= P_0(x) + P_1(x)t + P_2(x)t^2 + \cdots\cdots + P_k(x)t^k + \cdots\cdots.$$

6. Show that $P_p(1) = 1$, $p = 0,1,2,3,\cdots\cdots\cdots\cdots$.

Put $x=1$ in the identity established in Problem 5. Then
$$(1-2t+t^2)^{-\frac{1}{2}} = (1-t)^{-1} = 1 + t + t^2 + \cdots\cdots + t^p + \cdots\cdots$$
$$= P_0(1) + P_1(1)t + P_2(1)t^2 + \cdots\cdots + P_p(1)t^p + \cdots\cdots, \text{ identically.}$$

Hence, $P_0(1) = P_1(1) = \cdots\cdots = P_p(1) = \cdots\cdots = 1$.

THE BESSEL EQUATION.

7. Prove $\frac{d}{dx} J_0(x) = -J_1(x)$.

$$J_0(x) = \sum_{n=0}^{\infty} (-1)^n \frac{1}{(n!)^2} (\frac{x}{2})^{2n}$$

$$= 1 - (\frac{x}{2})^2 + \frac{1}{(2!)^2}(\frac{x}{2})^4 - \frac{1}{(3!)^2}(\frac{x}{2})^6 + \cdots + (-1)^{n+1} \frac{1}{[(n+1)!]^2}(\frac{x}{2})^{2n+2} + \cdots$$

and

$$\frac{d}{dx} J_0(x) = -(\frac{x}{2}) + \frac{1}{1!\,2!}(\frac{x}{2})^3 - \frac{1}{2!\,3!}(\frac{x}{2})^5 + \cdots + (-1)^{n+1} \frac{1}{n!\,(n+1)!}(\frac{x}{2})^{2n+1} + \cdots$$

$$= -[\frac{x}{2} - \frac{1}{1!\,2!}(\frac{x}{2})^3 + \cdots + (-1)^n \frac{1}{n!\,(n+1)!}(\frac{x}{2})^{2n+1} + \cdots]$$

$$= -\sum_{n=0}^{\infty} (-1)^n \frac{1}{n!\,(n+1)!}(\frac{x}{2})^{2n+1} = -J_1(x).$$

More briefly,

$$\frac{d}{dx} J_0(x) = \frac{d}{dx} \sum_{n=0}^{\infty} (-1)^n \frac{1}{(n!)^2}(\frac{x}{2})^{2n} = \frac{d}{dx}[1 + \sum_{n=1}^{\infty} (-1)^n \frac{1}{(n!)^2}(\frac{x}{2})^{2n}]$$

$$= \frac{d}{dx}[1 - \sum_{n=0}^{\infty} (-1)^n \frac{1}{[(n+1)!]^2}(\frac{x}{2})^{2n+2}] = -\sum_{n=0}^{\infty} (-1)^n \frac{1}{n!\,(n+1)!}(\frac{x}{2})^{2n+1} = -J_1(x).$$

8. Prove a) $\frac{d}{dx} x^k \cdot J_k(x) = x^k J_{k-1}(x)$, b) $\frac{d}{dx} x^{-k} \cdot J_k(x) = -x^{-k} J_{k+1}(x)$,

where k is a positive integer.

a) $$\frac{d}{dx} x^k \cdot J_k(x) = \frac{d}{dx} \sum_{n=0}^{\infty} (-1)^n \frac{1}{2^{k+2n}\, n!\,(k+n)!} x^{2k+2n}$$

$$= \sum_{n=0}^{\infty} (-1)^n \frac{2k+2n}{2^{k+2n}\, n!\,(k+n)!} x^{2k+2n-1}$$

$$= \sum_{n=0}^{\infty} (-1)^n \frac{1}{2^{k+2n-1}\, n!\,(k+n-1)!} x^{2k+2n-1}$$

$$= x^k \sum_{n=0}^{\infty} (-1)^n \frac{1}{n!\,(k+n-1)!} (\frac{x}{2})^{k+2n-1} = x^k J_{k-1}(x).$$

b) $$\frac{d}{dx} x^{-k} \cdot J_k(x) = \frac{d}{dx} \sum_{n=0}^{\infty} (-1)^n \frac{1}{2^{k+2n}\, n!\,(k+n)!} x^{2n}$$

$$= \frac{d}{dx}[\frac{1}{2^k\, k!} - \sum_{n=0}^{\infty} (-1)^n \frac{1}{2^{k+2n+2}\,(n+1)!\,(k+n+1)!} x^{2n+2}]$$

$$= -\sum_{n=0}^{\infty} (-1)^n \frac{1}{2^{k+2n+1} n!\,(k+n+1)!} x^{2n+1} = -x^{-k} \sum_{n=0}^{\infty} (-1)^n \frac{1}{n!\,(k+n+1)!}(\frac{x}{2})^{k+2n+1} = -x^{-k} J_{k+1}(x).$$

9. Prove $a)\ J_{k-1}(x) - J_{k+1}(x) = 2\frac{d}{dx}J_k(x)$, $b)\ J_{k-1}(x) + J_{k+1}(x) = \frac{2k}{x}J_k(x)$, where k is a positive integer.

From Problem 8,

$$A)\quad \frac{d}{dx}x^k\cdot J_k(x) = x^k\frac{d}{dx}J_k(x) + kx^{k-1}J_k(x) = x^k J_{k-1}(x) \quad \text{and}$$

$$B)\quad \frac{d}{dx}x^{-k}\cdot J_k(x) = x^{-k}\frac{d}{dx}J_k(x) - kx^{-k-1}J_k(x) = -x^{-k}J_{k+1}(x).$$

Then from A),

$$1)\quad \frac{d}{dx}J_k(x) + \frac{k}{x}J_k(x) = J_{k-1}(x);$$

and from B),

$$2)\quad \frac{d}{dx}J_k(x) - \frac{k}{x}J_k(x) = -J_{k+1}(x).$$

When 1) and 2) are added, we have a); when 2) is subtracted from 1), we have b).

Note that when b) is subtracted from a), we have

$$2\frac{d}{dx}J_k(x) - \frac{2k}{x}J_k(x) = -2J_{k+1}(x) \quad \text{or} \quad \frac{d}{dx}J_k(x) = \frac{k}{x}J_k(x) - J_{k+1}(x).$$

Note also that b) is a recursion formula for Bessel functions.

10. Show that $e^{\frac{1}{2}x(t-1/t)} = J_0(x) + tJ_1(x) + \cdots\cdots + t^k J_k(x) + \cdots\cdots + \frac{1}{t}J_{-1}(x) + \cdots\cdot$

$$+ \frac{1}{t^k}J_{-k}(x) + \cdots\cdots\cdots = \sum_{n=-\infty}^{+\infty} t^n J_n(x).$$

$$e^{\frac{1}{2}x(t-1/t)} = e^{\frac{1}{2}xt}\cdot e^{-x/2t}$$

$$= [1 + \frac{xt}{2} + \frac{x^2t^2}{2^2\,2!} + \frac{x^3t^3}{2^3\,3!} + \cdots\cdot + \frac{x^nt^n}{2^n\,n!} + \cdots\cdot][1 - \frac{x}{2t} + \frac{x^2}{2^2\,2!\,t^2}$$

$$- \frac{x^3}{2^3\,3!\,t^3} + \cdots\cdots + (-1)^n\frac{x^n}{2^n\,n!\,t^n} + \cdots\cdots].$$

In this product, the terms free of t are

$$1 - (\tfrac{x}{2})^2 + \frac{1}{(2!)^2}(\tfrac{x}{2})^4 - \frac{1}{(3!)^2}(\tfrac{x}{2})^6 + \cdots\cdot + (-1)^n\frac{1}{(n!)^2}(\tfrac{x}{2})^{2n} + \cdots\cdot = J_0(x),$$

the coefficient of t^k is

$$\frac{x^k}{2^k\,k!} - \frac{x^{k+1}}{2^{k+1}(k+1)!}\cdot\frac{x}{2} + \frac{x^{k+2}}{2^{k+2}(k+2)!}\cdot\frac{x^2}{2^2\,2!} - \frac{x^{k+3}}{2^{k+3}(k+3)!}\cdot\frac{x^3}{2^3\,3!} + \cdots\cdots\cdot$$

$$= \frac{1}{k!}(\tfrac{x}{2})^k - \frac{1}{1!(k+1)!}(\tfrac{x}{2})^{k+2} + \frac{1}{2!(k+2)!}(\tfrac{x}{2})^{k+4} - \frac{1}{3!(k+3)!}(\tfrac{x}{2})^{k+6} + \cdots\cdots\cdot$$

$$= \sum_{n=0}^{\infty}(-1)^n\frac{1}{n!(k+n)!}(\tfrac{x}{2})^{k+2n} = J_k(x),$$ and the coefficient of $\frac{1}{t^k}$ is

$$(-1)^k \left[\frac{x^k}{2^k\, k!} - \frac{x^{k+1}}{2^{k+1}\,(k+1)!}\cdot\frac{x}{2} + \frac{x^{k+2}}{2^{k+2}\,(k+2)!}\cdot\frac{x^2}{2^2\,2!} - \frac{x^{k+3}}{2^{k+3}\,(k+3)!}\cdot\frac{x^3}{2^3\,3!} + \cdots\right]$$

$$= (-1)^k \left[\frac{1}{k!}\left(\frac{x}{2}\right)^k - \frac{1}{1!\,(k+1)!}\left(\frac{x}{2}\right)^{k+2} + \frac{1}{2!\,(k+2)!}\left(\frac{x}{2}\right)^{k+4} - \frac{1}{3!\,(k+3)!}\left(\frac{x}{2}\right)^{k+6} + \cdots\right]$$

$$= (-1)^k J_k(x) = J_{-k}(x).$$

THE GAUSS EQUATION.

11. Solve in series $(x - x^2)y'' + (\frac{3}{2} - 2x)y' - \frac{1}{4}y = 0.$

Here $\alpha+\beta+1 = 2$, $\gamma = 3/2$, $\alpha\beta = 1/4$; thus $\alpha = \beta = 1/2$, and $\gamma = 3/2$.

Then $y_1 = F(\alpha,\beta,\gamma,x) = F(\frac{1}{2}, \frac{1}{2}, \frac{3}{2}, x) = 1 + \frac{x}{6} + \frac{3x^2}{40} + \frac{5x^3}{112} + \cdots\cdots$

and $y_2 = x^{1-\gamma}F(\alpha-\gamma+1,\ \beta-\gamma+1,\ 2-\gamma,\ x) = x^{-\frac{1}{2}}F(0,0,\frac{1}{2},x) = 1/\sqrt{x}$,

and the complete solution is $y = A\,F(\frac{1}{2}, \frac{1}{2}, \frac{3}{2}, x) + B/\sqrt{x}$.

12. Solve in series $(x - x^2)y'' + 4(1-x)y' - 2y = 0.$

Here $\alpha+\beta+1 = 4$, $\gamma = 4$, $\alpha\beta = 2$; then $\alpha = 1$, $\beta = 2$, $\gamma = 4$ or $\alpha = 2$, $\beta = 1$, $\gamma = 4$.

For either choice, $y_1 = F(1,2,4,x) = F(2,1,4,x)$

$$= 1 + \frac{x}{2} + \frac{3x^2}{10} + \frac{x^3}{5} + \frac{x^4}{7} + \frac{3x^5}{28} + \cdots\cdots$$

Since $\gamma = 4$, the fourth term in y_2 has zero for denominator. However, one of $\alpha-\gamma+2$ or $\beta-\gamma+2$ in the third term is zero so that

$$y_2 = x^{-3}\,F(-2,-1,-2,x) = x^{-3}\,F(-1,-2,-2,x) = x^{-3}(1-x)$$

and the complete solution is

$$y = A\,F(1,2,4,x) + B\,\frac{1-x}{x^3}.$$

13. Show that a) $F(\alpha,\beta,\beta,x) = (1-x)^{-\alpha}$, b) $xF(1,1,2,-x) = \ln(1+x)$.

a) $F(\alpha,\beta,\beta,x) = 1 + \dfrac{\alpha\cdot\beta}{1\cdot\beta}x + \dfrac{\alpha(\alpha+1)\beta(\beta+1)}{1\cdot 2\cdot\beta\,(\beta+1)}x^2 + \cdots\cdots$

$$= 1 + \alpha x + \frac{\alpha(\alpha+1)}{2!}x^2 + \frac{\alpha(\alpha+1)(\alpha+2)}{3!}x^3 + \cdots\cdots = (1-x)^{-\alpha}.$$

b) $xF(1,1,2,-x) = x\left[1 + \dfrac{1\cdot 1}{1\cdot 2}(-x) + \dfrac{1\cdot2\cdot1\cdot2}{1\cdot2\cdot2\cdot3}(-x)^2 + \dfrac{1\cdot2\cdot3\cdot1\cdot2\cdot3}{1\cdot2\cdot3\cdot2\cdot3\cdot4}(-x)^3 + \cdots\cdots\right]$

$$= x\left(1 - \frac{1}{2}x + \frac{1}{3}x^2 - \frac{1}{4}x^4 + \cdots\cdots\right) = \ln(1+x).$$

SUPPLEMENTARY PROBLEMS

14. Compute *a*) $P_4(2) = 55.3750$, *b*) $J_0(1) = 0.7652$, *c*) $J_1(1) = 0.4401$, *d*) $F(1,1,10,-1) = 0.9147$.

15. Verify each of the following by using the series expansion of $P_p(x)$.

a) $(x^2-1)P_p'(x) = (p+1)[P_{p+1}(x) - xP_p(x)] = p[x\,P_p(x) - P_{p-1}(x)]$.

b) $P_{p+1}'(x) = xP_p'(x) + (p+1)P_p(x)$.

c) $(2p+1)P_p(x) = P_{p+1}'(x) - P_{p-1}'(x) = \frac{1}{x}[(p+1)P_{p+1}(x) + pP_{p-1}(x)]$.

16. If $P_6(2) = a$ and $P_7(2) = b$, show that

a) $P_6'(2) = \frac{7}{3}(b-2a)$, *b*) $P_7'(2) = \frac{7}{3}(2b-a)$, *c*) $P_8(2) = \frac{1}{8}(30b-7a)$, *d*) $P_8'(2) = \frac{1}{3}(52b-14a)$.

17. If $J_0(2) = a$ and $J_1(2) = b$, show that *a*) $J_2(2) = b-a$, *b*) $J_1'(2) = a - \frac{1}{2}b$, *c*) $J_2'(2) = a$.

18. Show that the change of independent variable $x^2 = t$ reduces the Legendre equation to a Gauss equation.

19. *a*) Show that the change of dependent variable $y = x^{1/2}z$ transforms $y'' + y = 0$ into a Bessel equation.

b) Write the solution of the Bessel equation as $y = C_1x^{1/2}J_{1/2}(x) + C_2x^{1/2}J_{-1/2}(x)$ and show that $J_{1/2}(x)$ and $J_{-1/2}(x)$ may be defined as $ax^{-1/2}\sin x$ and $bx^{-1/2}\cos x$ respectively.

c) Show that if the relations of Problem 8 are to hold for $k = \pm\frac{1}{2}$, then $a = b$.

Note. These functions are defined with $a = \sqrt{2/\pi}$.

20. Use the substitution $y = x^{1/2}z$ and then $x = (3t/2)^{2/3}$ to show that $y'' + xy = 0$ is a special case of the Bessel equation, and solve.

Hint: $z'' + tz' + (t^2 - 1/9)z = 0$.

Ans.
$$y = Ax\left[1 - \frac{x^3}{2^2\,3} + \frac{x^6}{2!\,2^2\,3^2\,7} - \frac{x^9}{3!\,2^2\,3^3\,7\cdot 10} + \cdots\cdots\right] + B\left[1 - \frac{x^3}{3\cdot 2} + \frac{x^6}{2!\,3^2\,2\cdot 5} - \frac{x^9}{3!\,3^3\,2\cdot 5\cdot 8} + \cdots\cdots\cdots\right].$$

21. Solve $(x^2-3x+2)y'' + 4xy' + 2y = 0$ after reducing it to a Gauss equation by a substitution of the form $x = \xi z + \eta$.

Hint: $y = AF(1,2,-4,x-1) + B(x-1)^5F(6,7,6,x-1)$ is not a complete solution since the sixth term of $F(1,2,-4,x-1)$ becomes infinite.

Ans. $y = A\,F(1,2,8,2-x) + B(2-x)^{-7}F(-6,-5,-6,2-x)$

22. Express each of the following as Gauss functions.

a) $\frac{1}{1-x} = F(1,\beta,\beta,x)$

b) $\arcsin x = x\,F(\frac{1}{2},\frac{1}{2},\frac{3}{2},x^2)$

c) $\arctan x = x\,F(1,\frac{1}{2},\frac{3}{2},-x^2)$

d) $e^x = \lim_{\alpha\to\infty} F(\alpha,1,1,x/\alpha)$

e) $\sin x = \lim_{\substack{\alpha\to\infty\\ \beta\to\infty}} x\,F(\alpha,\beta,\frac{3}{2},-\frac{x^2}{4\alpha\beta})$.

CHAPTER 28

Partial Differential Equations

PARTIAL DIFFERENTIAL EQUATIONS are those which contain one or more partial derivatives. They must, therefore, involve at least two independent variables. The *order* of a partial differential equation is that of the derivative of highest order in the equation. For example, considering z as dependent variable and x, y as independent variables,

1) $$x\frac{\partial z}{\partial x} + y\frac{\partial z}{\partial y} = z \qquad \text{or} \qquad 1')\ xp + yq = z$$

is of order one and

2) $$\frac{\partial^2 z}{\partial x^2} + 3\frac{\partial^2 z}{\partial x\,\partial y} + \frac{\partial^2 z}{\partial y^2} = 0 \qquad \text{or} \qquad 2')\ r + 3s + t = 0$$

is of order two. In writing 1') and 2'), use has been made of the standard notation: $p = \dfrac{\partial z}{\partial x}$, $q = \dfrac{\partial z}{\partial y}$, $r = \dfrac{\partial^2 z}{\partial x^2}$, $s = \dfrac{\partial^2 z}{\partial x\,\partial y}$, $t = \dfrac{\partial^2 z}{\partial y^2}$.

Partial differential equations may be derived by the elimination of arbitrary constants from a given relation between the variables and by the elimination of arbitrary functions of the variables. They also may arise in connection with geometrical and physical problems.

ELIMINATION OF ARBITRARY CONSTANTS. Consider z to be a function of two independent variables x and y defined by

3) $$g(x, y, z, a, b) = 0,$$

in which a and b are two arbitrary constants. By differentiating 3) partially with respect to x and y, we obtain

4) $$\frac{\partial g}{\partial x} + \frac{\partial g}{\partial z}\frac{\partial z}{\partial x} = \frac{\partial g}{\partial x} + p\frac{\partial g}{\partial z} = 0$$

and

5) $$\frac{\partial g}{\partial y} + \frac{\partial g}{\partial z}\frac{\partial z}{\partial y} = \frac{\partial g}{\partial y} + q\frac{\partial g}{\partial z} = 0.$$

In general, the arbitrary constants may be eliminated from 3), 4), 5) yielding a partial differential equation of order one

6) $$f(x, y, z, p, q) = 0.$$

EXAMPLE 1. Eliminate the arbitrary constants a and b from $z = ax^2 + by^2 + ab$.

Differentiating partially with respect to x and y, we have

$$\frac{\partial z}{\partial x} = p = 2ax \qquad \text{and} \qquad \frac{\partial z}{\partial y} = q = 2by.$$

Solving for a and b from these equations and substituting in the given relation, we obtain

$$z = (\frac{1}{2}\frac{p}{x})x^2 + (\frac{1}{2}\frac{q}{y})y^2 + (\frac{1}{2}\frac{p}{x})(\frac{1}{2}\frac{q}{y}) \qquad \text{or} \qquad pq + 2px^2y + 2qxy^2 = 4xyz,$$

a partial differential equation of order one.

If z is a function of x and y defined by a relation involving but one arbitrary constant, it is usually possible to obtain two distinct partial differential equations of order one by eliminating the constant.

EXAMPLE 2. Eliminate a from $z = a(x+y)$.

Differentiation with respect to x gives $p = a$, so that the partial differential equation $z = p(x+y)$ is obtained. Similarly, differentiation with respect to y gives $q = a$ and the equation $z = q(x+y)$.

If the number of arbitrary constants to be eliminated exceeds the number of independent variables, the resulting partial differential equation (or equations) is usually of order higher than the first.

EXAMPLE 3. Eliminate a,b,c from $z = ax + by + cxy$.

Differentiating partially with respect to x and y, we have

$$\text{(i)}\quad p = a + cy \qquad \text{and} \qquad \text{(ii)}\quad q = b + cx.$$

These, together with the given relation, are not sufficient for the elimination of three constants. Differentiating (i) partially with respect to x, we have

$$\frac{\partial}{\partial x}p = \frac{\partial^2 z}{\partial x^2} = r = 0,$$

a partial differential equation of order two. Differentiating (ii) partially with respect to y, we have

$$\frac{\partial}{\partial y}q = \frac{\partial^2 z}{\partial y^2} = t = 0, \qquad \text{of order two.}$$

Differentiating (i) partially with respect to y or (ii) with respect to x, we obtain

$$\frac{\partial}{\partial y}p = \frac{\partial}{\partial x}q = \frac{\partial^2 z}{\partial x\,\partial y} = s = c.$$

From (i), $p = a + sy$ and $a = p - sy$; from (ii), $b = q - sx$.

Substituting for a,b,c in the given relation, we obtain

$$z = (p - sy)x + (q - sx)y + sxy = px + qy - sxy,$$

of order two.

Thus, we have three partial differential equations $r = 0$, $t = 0$, $z = px + qy - sxy$ of the same (minimum) order associated with the given relation.

See also Problems 1-4.

ELIMINATION OF ARBITRARY FUNCTIONS. Let $u = u(x,y,z)$ and $v = v(x,y,z)$ be independent functions of the variables x,y,z, and let

7) $$\phi(u,v) = 0$$

be an arbitrary relation between them. Regarding z as the dependent variable and differentiating partially with respect to x and y, we obtain

8) $$\frac{\partial\phi}{\partial u}(\frac{\partial u}{\partial x} + p\frac{\partial u}{\partial z}) + \frac{\partial\phi}{\partial v}(\frac{\partial v}{\partial x} + p\frac{\partial v}{\partial z}) = 0 \qquad \text{and}$$

9) $$\frac{\partial\phi}{\partial u}\left(\frac{\partial u}{\partial y} + q\frac{\partial u}{\partial z}\right) + \frac{\partial\phi}{\partial v}\left(\frac{\partial v}{\partial y} + q\frac{\partial v}{\partial z}\right) = 0.$$

Eliminating $\frac{\partial\phi}{\partial u}$ and $\frac{\partial\phi}{\partial v}$ from 8) and 9), we have

$$\begin{vmatrix} \frac{\partial u}{\partial x} + p\frac{\partial u}{\partial z} & \frac{\partial v}{\partial x} + p\frac{\partial v}{\partial z} \\ \frac{\partial u}{\partial y} + q\frac{\partial u}{\partial z} & \frac{\partial v}{\partial y} + q\frac{\partial v}{\partial z} \end{vmatrix} = \left(\frac{\partial u}{\partial x} + p\frac{\partial u}{\partial z}\right)\left(\frac{\partial v}{\partial y} + q\frac{\partial v}{\partial z}\right) - \left(\frac{\partial u}{\partial y} + q\frac{\partial u}{\partial z}\right)\left(\frac{\partial v}{\partial x} + p\frac{\partial v}{\partial z}\right)$$

$$= \frac{\partial u}{\partial x}\frac{\partial v}{\partial y} - \frac{\partial u}{\partial y}\frac{\partial v}{\partial x} + p\left(\frac{\partial u}{\partial z}\frac{\partial v}{\partial y} - \frac{\partial u}{\partial y}\frac{\partial v}{\partial z}\right) + q\left(\frac{\partial u}{\partial x}\frac{\partial v}{\partial z} - \frac{\partial u}{\partial z}\frac{\partial v}{\partial x}\right) = 0.$$

Writing $\lambda P = \frac{\partial u}{\partial y}\frac{\partial v}{\partial z} - \frac{\partial u}{\partial z}\frac{\partial v}{\partial y}$, $\lambda Q = \frac{\partial u}{\partial z}\frac{\partial v}{\partial x} - \frac{\partial u}{\partial x}\frac{\partial v}{\partial z}$, $\lambda R = \frac{\partial u}{\partial x}\frac{\partial v}{\partial y} - \frac{\partial u}{\partial y}\frac{\partial v}{\partial x}$,

this takes the form

$$Pp + Qq = R,$$

a partial differential equation linear in p and q and free of the arbitrary function $\phi(u,v)$.

EXAMPLE 4. Find the differential equation arising from $\phi(z/x^3,\ y/x) = 0$, where ϕ is an arbitrary function of the arguments.

We write the functional relation in the form $\phi(u,v) = 0$ with $u = z/x^3$ and $v = y/x$. Differentiating partially with respect to x and y, we have

$$\frac{\partial\phi}{\partial u}\left(\frac{p}{x^3} - \frac{3z}{x^4}\right) + \frac{\partial\phi}{\partial v}\left(-\frac{y}{x^2}\right) = 0, \qquad \frac{\partial\phi}{\partial u}\left(\frac{q}{x^3}\right) + \frac{\partial\phi}{\partial v}\left(\frac{1}{x}\right) = 0.$$

The elimination of $\frac{\partial\phi}{\partial u}$ and $\frac{\partial\phi}{\partial v}$ yields

$$\begin{vmatrix} p/x^3 - 3z/x^4 & -y/x^2 \\ q/x^3 & 1/x \end{vmatrix} = p/x^4 - 3z/x^5 + qy/x^5 = 0 \quad \text{or} \quad px + qy = 3z.$$

The arbitrary functional relation may also be given by $\frac{z}{x^3} = f\left(\frac{y}{x}\right)$ or $z = x^3 f\left(\frac{y}{x}\right)$, where f is an arbitrary function of its argument. Using $v = y/x$ and differentiating $z = x^3 f(v)$ with respect to x and y yields

$$p = 3x^2 f(v) + x^3\frac{df}{dv}\frac{\partial v}{\partial x} = 3x^2 f(v) + x^3\left(\frac{df}{dv}\right)\left(-\frac{y}{x^2}\right) = 3x^2 f(v) - xy\,f'(v),$$

$$q = x^3\frac{df}{dv}\frac{\partial v}{\partial y} = x^3\left(\frac{df}{dv}\right)\left(\frac{1}{x}\right) = x^2 f'(v).$$

When $f'(v)$ is eliminated from these, we have

$$px + qy = 3x^3 f(v) = 3z$$

as before.

See also Problems 5-8.

SOLVED PROBLEMS

1. Eliminate a and b from $z = (x^2 + a)(y^2 + b)$.

Differentiating partially with respect to x and y, $p = 2x(y^2 + b)$ and $q = 2y(x^2 + a)$. Then $y^2 + b = \dfrac{p}{2x}$, $x^2 + a = \dfrac{q}{2y}$, and $z = (x^2 + a)(y^2 + b) = (\dfrac{q}{2y})(\dfrac{p}{2x})$ or $pq = 4xyz$.

We could also eliminate a and b as follows: $pq = 4xy(y^2 + b)(x^2 + a) = 4xyz$.

2. Find the differential equation of the family of spheres of radius 5 with centers on the plane $x = y$.

The equation of the family of spheres is 1) $(x - a)^2 + (y - a)^2 + (z - b)^2 = 25$, a and b being arbitrary constants. Differentiating partially with respect to x and y, and dividing by 2, we have

$$(x - a) + (z - b)p = 0 \qquad \text{and} \qquad (y - a) + (z - b)q = 0.$$

Let $z - b = -m$; then $x - a = pm$ and $y - a = qm$. Making these replacements in 1), we get

$$m^2(p^2 + q^2 + 1) = 25.$$

Now $x - y = (p - q)m$. Then $m = \dfrac{x - y}{p - q}$, $m^2(p^2 + q^2 + 1) = \dfrac{(x - y)^2}{(p - q)^2}(p^2 + q^2 + 1) = 25$, and the required differential equation is $(x - y)^2(p^2 + q^2 + 1) = 25(p - q)^2$.

3. Show that the partial differential equation obtained by eliminating the arbitrary constants a, c from $z = ax + h(a)\,y + c$, where $h(a)$ is an arbitrary function of a, is free of the variables x, y, z.

Differentiating $z = ax + h(a)\,y + c$ partially with respect to x and y, we obtain $p = a$ and $q = h(a)$. The differential equation resulting from the elimination of a is $q = h(p)$ or $f(p,q) = 0$, where f is an arbitrary function of its arguments. This equation contains p and q but none of the variables x, y, z.

4. Show that the partial differential equation obtained by eliminating the arbitrary constants a and b from

$$z = ax + by + f(a,b),$$

the *extended Clairaut equation*, is

$$z = px + qy + f(p,q).$$

Differentiating $z = ax + by + f(a,b)$ with respect to x and y yields $p = a$ and $q = b$, and the required differential equation follows immediately.

5. Find the differential equation arising from $\phi(x+y+z,\ x^2+y^2-z^2) = 0$.

Let $u = x + y + z$, $v = x^2 + y^2 - z^2$ so that the given relation is $\phi(u,v) = 0$.

Differentiation with respect to x and y yields

$\dfrac{\partial\phi}{\partial u}(1 + p) + \dfrac{\partial\phi}{\partial v}(2x - 2zp) = 0$ and $\dfrac{\partial\phi}{\partial u}(1 + q) + \dfrac{\partial\phi}{\partial v}(2y - 2zq) = 0$. Eliminating $\dfrac{\partial\phi}{\partial u}$ and $\dfrac{\partial\phi}{\partial v}$, we have

$$\begin{vmatrix} 1 + p & 2x - 2zp \\ 1 + q & 2y - 2zq \end{vmatrix} = 2(y - x) + 2p(y + z) - 2q(z + x) = 0 \quad \text{or} \quad (y + z)p - (x + z)q = x - y.$$

6. Eliminate the arbitrary function $\phi(x+y)$ from $z = \phi(x+y)$.

Let $x+y=u$ so that the given relation is $z = \phi(u)$.

Differentiating with respect to x and y yields $p = \dfrac{d\phi}{du} = \phi'(u)$ and $q = \phi'(u)$.

Thus, $p = q$ is the resulting differential equation.

7. The equation of any cone with vertex at $P_0(x_0, y_0, z_0)$ is of the form $\phi(\dfrac{x-x_0}{z-z_0}, \dfrac{y-y_0}{z-z_0}) = 0$. Find the differential equation.

Let $\dfrac{x-x_0}{z-z_0} = u$, $\dfrac{y-y_0}{z-z_0} = v$ so that the given relation is $\phi(u,v) = 0$.

Differentiating with respect to x and y, we have

$$\frac{\partial\phi}{\partial u}\left(\frac{1}{z-z_0} - p\frac{x-x_0}{(z-z_0)^2}\right) + \frac{\partial\phi}{\partial v}\left(-p\frac{y-y_0}{(z-z_0)^2}\right) = 0$$

$$\frac{\partial\phi}{\partial u}\left(-q\frac{x-x_0}{(z-z_0)^2}\right) + \frac{\partial\phi}{\partial v}\left(\frac{1}{z-z_0} - q\frac{y-y_0}{(z-z_0)^2}\right) = 0.$$

Eliminating $\dfrac{\partial\phi}{\partial u}$ and $\dfrac{\partial\phi}{\partial v}$, we obtain $p(x-x_0) + q(y-y_0) = z - z_0$.

8. Eliminate the arbitrary functions $f(x)$ and $g(y)$ from $z = yf(x) + xg(y)$.

Differentiating partially with respect to x and y, we have

1) $p = y\,f'(x) + g(y)$ and 2) $q = f(x) + x\,g'(y)$.

Since it is not possible to eliminate f, g, f', g' from these relations and the given one, we find the second partial derivatives

3) $r = yf''(x)$, $s = f'(x) + g'(y)$, $t = xg''(y)$.

From 1) and 2) we find $f'(x) = \dfrac{1}{y}[p-g(y)]$ and $g'(y) = \dfrac{1}{x}[q-f(x)]$. Hence,

$$s = f'(x) + g'(y) = \frac{1}{y}[p-g(y)] + \frac{1}{x}[q-f(x)].$$

Thus, $xys = x[p-g(y)] + y[q-f(x)] = px + qy - [y\,f(x) + x\,g(y)] = px + qy - z$ is the resulting partial differential equation.

Note that the differential equation is of order two although, in general, a higher order is expected. However, since one of the relations 3) involves only the first derivatives of f and g, it is possible to eliminate f, g, f', g' between this relation, 1), 2), and the given relation.

9. Find the differential equation of all surfaces cutting the family of cones $x^2 + y^2 - a^2z^2 = 0$ orthogonally.

Let $z = f(x,y)$ be the equation of the required surfaces. At a point $P(x,y,z)$ on the surface, a set of direction numbers of the normal to the surface is $[p, q, -1]$. Likewise, at P a set of direction numbers of the normal to the cone through P is $[x, y, -a^2z]$. Since these directions are orthogonal,

$$px + qy + a^2z = 0.$$

The elimination of a^2 between this and the given equation yields the required differential equation

$$z(px + qy) + x^2 + y^2 = 0.$$

10. A surface which is the envelope of a one-parameter family of planes is called a developable surface. (Such a surface can be deformed (developed) into a plane without stretching or tearing.) Obtain the differential equation of developable surfaces.

Let $z = f(x,y)$ be the equation of a developable surface.

The tangent plane at a point (x_0, y_0, z_0) of the surface has equation

$$1)\quad F = (x - x_0)p + (y - y_0)q - (z - z_0) = 0.$$

Now when p and q satisfy a relation $\phi(p,q) = 0$, 1) is a one-parameter family of planes having $z = f(x,y)$ as envelope. Thus $\phi(p,q) = 0$ or $q = \lambda(p)$ is the required differential equation.

The cone of Problem 9 is a developable surface since $p = \dfrac{x}{a^2 z}$, $q = \dfrac{y}{a^2 z}$ satisfies $\phi(p,q) = a^2(p^2 + q^2) - 1 = 0$.

11. Eliminate the arbitrary functions ϕ_1 and ϕ_2 from

$$z = \phi_1(y + m_1 x) + \phi_2(y + m_2 x) = \phi_1(u) + \phi_2(v)$$

in which $m_1 \neq m_2$ are fixed constants.

Differentiating partially, we obtain

$$r = m_1^2 \frac{d^2\phi_1}{du^2} + m_2^2 \frac{d^2\phi_2}{dv^2}, \qquad s = m_1 \frac{d^2\phi_1}{du^2} + m_2 \frac{d^2\phi_2}{dv^2}, \qquad t = \frac{d^2\phi_1}{du^2} + \frac{d^2\phi_2}{dv^2}.$$

Eliminating $\dfrac{d^2\phi_1}{du^2}, \dfrac{d^2\phi_2}{dv^2}$ we have $\begin{vmatrix} m_1^2 & m_2^2 & r \\ m_1 & m_2 & s \\ 1 & 1 & t \end{vmatrix} = (m_1 - m_2)r - (m_1^2 - m_2^2)s + (m_1^2 m_2 - m_1 m_2^2)t = 0$

or, since $m_1 \neq m_2$, $\quad r - (m_1 + m_2)s + m_1 m_2 t = 0.$

12. Show that (a) $z = ax^3 + by^3$ and (b) $z = ax^3 + bx^2y + cxy^2 + dy^4/x$ give rise to the same differential equation.

$a)$ Differentiating $z = ax^3 + by^3$ partially with respect to x and y, we have

$$p = 3ax^2 \quad \text{and} \quad q = 3by^2.$$

Thus, $px + qy = 3(ax^3 + by^3) = 3z$ is the resulting differential equation.

$b)$ Differentiating $z = ax^3 + bx^2y + cxy^2 + dy^4/x$ partially with respect to x and y, we have

$$p = 3ax^2 + 2bxy + cy^2 - dy^4/x^2 \quad \text{and} \quad q = bx^2 + 2cxy + 4dy^3/x.$$

Thus, $px + qy = 3(ax^3 + bx^2y + cxy^2 + dy^4/x) = 3z$ as before.

The fact that these two equations, one with two arbitrary constants and the other with four, give rise to the same differential equation will indicate the subordinate role which the arbitrary constant will play here. In its place we will have arbitrary functions. Since (a) may be written as

$$z = ax^3 + by^3 = x^3[a + b(y/x)^3] = x^3 \cdot g(y/x),$$

while (b) may be written as

$$z = x^3[a + b(y/x) + c(y/x)^2 + d(y/x)^4] = x^3 \cdot h(y/x),$$

each is a particular case of $z = x^3 \cdot f(y/x)$ considered in Example 4.

SUPPLEMENTARY PROBLEMS

Eliminate the arbitrary constants a, b, c from each of the following equations.

13. $z = (x-a)^2 + (y-b)^2$ *Ans.* $4z = p^2 + q^2$

14. $z = axy + b$ $xp - yq = 0$

15. $ax + by + cz = 1$ $r = 0,\ s = 0,$ or $t = 0$

16. $z = axe^y + \frac{1}{2}a^2e^{2y} + b$ $q = xp + p^2$

17. $z = xy + y\sqrt{x^2 - a^2} + b$ $pq = xp + yq$

18. $x^2/a^2 + y^2/b^2 + z^2/c^2 = 1$ $xzr + xp^2 - zp = 0,\ yzt + yq^2 - zq = 0,$ or $zs + pq = 0$

Eliminate the arbtirary constants a, b and the arbitrary functions ϕ, f, g.

19. $z = x^2\phi(x-y)$ or $\psi(z/x^2, x-y) = 0$ *Ans.* $2z = xp + xq$

20. $xyz = \phi(x+y+z)$ $x(y-z)p + y(z-x)q = z(x-y)$

21. $z = (x+y)\,\phi(x^2 - y^2)$ $yp + xq = z$

22. $z = f(x) + e^y g(x)$ $t - q = 0$

23. $x = f(z) + g(y)$ $ps - qr = 0$

24. $z = f(xy) + g(x+y)$ *Ans.* $x(y-x)r - (y^2 - x^2)s + y(y-x)t + (p-q)(x+y) = 0$

25. $z = f(x+z) + g(x+y)$ *Ans.* $qr - (1+p+q)s + (1+p)t = 0$

26. $z = ax^2 + g(y)$ $p - xr = 0$ or $s = 0$

27. $z = \frac{1}{2}(a^2+2)x^2 + axy + bx + \phi(y+ax)$ $r - 2t + rt - s^2 = 2$

28. Find the differential equation of all spheres of radius 2 having their centers in the xOy plane. Hint: Eliminate a and b from $(x-a)^2 + (y-b)^2 + z^2 = 4$. *Ans.* $z^2(p^2 + q^2 + 1) = 4$

29. Find the differential equation of planes having equal x- and y-intercepts. *Ans.* $p - q = 0$

30. Find the differential equation of all surfaces of revolution having the z-axis as axis of rotation. Hint: Eliminate ϕ from $z = \phi(\sqrt{x^2+y^2}) = \psi(x^2+y^2)$. *Ans.* $yp - xq = 0$

CHAPTER 29

Linear Partial Differential Equations of Order One

THE PARTIAL DIFFERENTIAL EQUATIONS of order one

$1_1)\quad px + qy = 3z \qquad$ and $\qquad 1_2)\quad px^2 + qy = z^3$

are called *linear* to indicate that they are of the first degree in p and q. Note that, unlike linear ordinary differential equations, there is no restriction on the degree of the dependent variable z.

All partial differential equations of order one which are not linear, as

$2_1)\quad p^2 + q^2 = 1 \qquad$ and $\qquad 2_2)\quad p + \ln q = 2z^3,$

are called *non-linear*.

LINEAR PARTIAL DIFFERENTIAL EQUATIONS OF ORDER ONE. Equation 1_1) was obtained in Chapter 28, Example 4, from the arbitrary functional relation

3) $$\phi(z/x^3,\ y/x) = 0$$

or its equivalent $z/x^3 = f(y/x)$. This solution, involving an arbitrary function, is called the *general solution* of 1_1).

The differential equation was also obtained (Chapter 28, Problem 12) by eliminating the arbitrary constants from

4_1) $$z = ax^3 + by^3$$

and from

4_2) $$z = ax^3 + bx^2y + cxy^2 + dy^4/x.$$

A study of the problems of that chapter indicates that relations involving two arbitrary constants usually yield non-linear partial differential equations of order one, while those involving more than two arbitrary constants yield equations of order higher than one. However, as was pointed out in Chapter 28, Problem 12, both of these relations are particular cases of the arbitrary functional relation 3). It is clear then that the general solution of 1) yields a much greater variety of solutions than that obtained (in the case of ordinary differential equations) through the appearance of arbitrary constants; for example,

$$z/x^3 = A\sin(y/x)^2 + B\cos(y/x) + C\ln(y/x) + De^{y/x} + E(y/x)^{12}$$

is included in the general solution 3).

THE GENERAL SOLUTION. A linear partial differential equation of order one, involving a dependent variable z and two independent variables x and y, is of the form

5) $$Pp + Qq = R$$

where P, Q, R are functions of x, y, z.

If $P = 0$ or $Q = 0$, 5) may be solved easily. Thus, the equation $\dfrac{\partial z}{\partial x} = 2x + 3y$ has as solution $z = x^2 + 3xy + \phi(y)$, where ϕ is an arbitrary function.

Lagrange reduced the problem of finding the general solution of 5) to that of solving an auxiliary system (called the Lagrange system) of *ordinary* differential equations

6) $$\frac{dx}{P} = \frac{dy}{Q} = \frac{dz}{R}$$

by showing (see Problem 7) that

7) $$\phi(u,v) = 0, \qquad (\phi, \text{ arbitrary})$$

is the general solution of 5) provided $u = u(x,y,z) = a$ and $v = v(x,y,z) = b$ are two independent solutions of 6). Here, a and b are arbitrary constants and at least one of u, v must contain z.

EXAMPLE 1. Find the general solution of

1) $$px + qy = 3z.$$

The auxiliary system is $\dfrac{dx}{x} = \dfrac{dy}{y} = \dfrac{dz}{3z}$.

From $\dfrac{dx}{x} = \dfrac{dz}{3z}$, we obtain $u = z/x^3 = a$; and from $\dfrac{dx}{x} = \dfrac{dy}{y}$, we obtain $v = y/x = b$.

Thus, the general solution is $\phi(z/x^3, y/x) = 0$, where ϕ is arbitrary.

Of course, from $\dfrac{dy}{y} = \dfrac{dz}{3z}$, we obtain $z/y^3 = c$, and we may write

$$\psi(z/x^3, z/y^3) = 0 \qquad \text{or} \qquad \lambda(z/y^3, y/x) = 0,$$

where ψ and λ are arbitrary. However, these are all equivalent and we shall call any one of them *the* general solution.

The above procedure may be extended readily to solve linear first order differential equations involving more than two independent variables.

EXAMPLE 2. Find the general solution of

$$x\frac{\partial z}{\partial x} + y\frac{\partial z}{\partial y} + t\frac{\partial z}{\partial t} = xyt,$$

z being the dependent variable.

The auxiliary system is $\dfrac{dx}{x} = \dfrac{dy}{y} = \dfrac{dt}{t} = \dfrac{dz}{xyt}$.

We obtain readily $u = x/y = a, \quad v = t/y = b$.

A third independent solution may be found by using the multipliers $yt, xt, xy, -3$. Since

$$x(yt) + y(xt) + t(xy) + (xyt)(-3) = 0,$$

$$yt\,dx + xt\,dy + xy\,dt - 3\,dz = 0$$

and $$xyt - 3z = c.$$

Thus, the general solution is $\phi(x/y, t/y, xyt - 3z) = 0$.

COMPLETE SOLUTIONS. If $u = a$ and $v = b$ are two independent solutions of 6) and if α, β are arbitrary constants,

8) $$u = \alpha v + \beta$$

is called *a complete solution* of 5). Thus, for the equation of Example 1,

$$z/x^3 = \alpha(y/x) + \beta$$

is a complete solution.

A complete solution 8) represents a two-parameter family of surfaces which does not have an envelope, since the arbitrary constants enter linearly. It is possible, however, to select one-parameter families of surfaces from among 8) which have envelopes. As shown in Problem 8, these envelopes (surfaces) are merely particular surfaces of the general solution.

SOLVED PROBLEMS

1. Find the general solution of $2p + 3q = 1$.

The auxiliary system is $\dfrac{dx}{2} = \dfrac{dy}{3} = \dfrac{dz}{1}$.

From $\dfrac{dx}{2} = \dfrac{dz}{1}$, we have $x - 2z = a$; and from $\dfrac{dx}{2} = \dfrac{dy}{3}$, we have $3x - 2y = b$. Thus, the general solution is $$\phi(x - 2z,\ 3x - 2y) = 0.$$

The complete solution $x - 2z = \alpha(3x - 2y) + \beta$ is a two-parameter family of planes. The one-parameter family determined by taking $\beta = \alpha^2$ has equation

$$A)\qquad x - 2z = \alpha(3x - 2y) + \alpha^2.$$

Differentiating $A)$ with respect to α yields $0 = 3x - 2y + 2\alpha$ or $\alpha = -\frac{1}{2}(3x - 2y)$.

Substituting for α in $A)$, we obtain the envelope, a parabolic cylinder, $x - 2z = -\frac{1}{4}(3x - 2y)^2$. This cylinder is clearly a part of the general solution.

2. Find the general solution of $y^2zp - x^2zq = x^2y$.

The auxiliary equations are $\dfrac{dx}{y^2z} = \dfrac{dy}{-x^2z} = \dfrac{dz}{x^2y}$.

From $\dfrac{dz}{x^2y} = \dfrac{dy}{-x^2z}$ or $z\,dz + y\,dy = 0$, we have $y^2 + z^2 = a$; from $\dfrac{dx}{y^2z} = \dfrac{dy}{-x^2z}$, we have $x^3 + y^3 = b$.

Thus, the general solution is $\phi(y^2 + z^2,\ x^3 + y^3) = 0$.

3. Find the general solution of $(y - z)p + (x - y)q = z - x$.

The auxiliary system is $\dfrac{dx}{y - z} = \dfrac{dy}{x - y} = \dfrac{dz}{z - x}$.

Since $(y - z) + (x - y) + (z - x) = 0$, $dx + dy + dz = 0$ and $x + y + z = a$.
Since $x(y - z) + z(x - y) + y(z - x) = 0$, $x\,dx + z\,dy + y\,dz = 0$ and $x^2 + 2yz = b$.

Thus, the general solution is $\phi(x^2 + 2yz,\ x+y+z) = 0$.

The complete solution $x^2 + 2yz = \alpha(x + y + z) + \beta$ represents a family of hyperboloids.

4. Find the general solution of $(x^2 - y^2 - z^2)p + 2xyq = 2xz$.

The auxiliary system is $\dfrac{dx}{x^2 - y^2 - z^2} = \dfrac{dy}{2xy} = \dfrac{dz}{2xz}$.

From $\dfrac{dy}{2xy} = \dfrac{dz}{2xz}$, we obtain $y/z = a$.

From $\dfrac{dz}{2xz} = \dfrac{x\,dx + y\,dy + z\,dz}{x(x^2 - y^2 - z^2) + y(2xy) + z(2xz)} = \dfrac{x\,dx + y\,dy + z\,dz}{x(x^2 + y^2 + z^2)}$ or $\dfrac{dz}{z} = \dfrac{2(x\,dx + y\,dy + z\,dz)}{x^2 + y^2 + z^2}$,

we obtain $\dfrac{x^2 + y^2 + z^2}{z} = b$.

Thus, the general solution is $\phi(\dfrac{y}{z}, \dfrac{x^2 + y^2 + z^2}{z}) = 0$.

The complete solution $x^2 + y^2 + z^2 = \alpha y + \beta z$ consists of the spheres through the origin with centers on the plane yOz.

5. Solve $ap + bq + cz = 0$.

The auxiliary system is $\dfrac{dx}{a} = \dfrac{dy}{b} = \dfrac{dz}{-cz}$. From $\dfrac{dx}{a} = \dfrac{dy}{b}$, we obtain $ay - bx = A$.

If $a \neq 0$, $\dfrac{dz}{-cz} = \dfrac{dx}{a}$ yields $\ln z = -\dfrac{c}{a}x + \ln B$ or $z = Be^{-cx/a}$, and the general solution may be written as $z = e^{-cx/a}\phi(ay - bx)$. If $b \neq 0$, $\dfrac{dz}{-cz} = \dfrac{dy}{b}$ yields $z = Ce^{-cy/b}$, and the general solution may be written as $z = e^{-cy/b}\psi(ay - bx)$.

6. Solve 1) $2p + q + z = 0$, 2) $p - 3q + 2z = 0$, 3) $2p + 3q + 5z = 0$, 4) $q + 2z = 0$.

1) Comparing with Problem 5 above, $a = 2$, $b = 1$, $c = 1$.
The general solution is $z = e^{-x/2}\phi(2y - x)$ or $z = e^{-y}\psi(2y - x)$.

2) Here, $a = 1$, $b = -3$, $c = 2$. The general solution is $z = e^{-2x}\phi(y + 3x)$ or $z = e^{2y/3}\psi(y + 3x)$.

3) The general solution is $z = e^{-5x/2}\phi(2y - 3x)$ or $z = e^{-5y/3}\psi(2y - 3x)$.

4) The general solution is $z = e^{-2y}\phi(-x) = e^{-2y}\psi(x)$.

7. Show that if $u = u(x,y,z) = a$ and $v = v(x,y,z) = b$ are two independent solutions of $\dfrac{dx}{P} = \dfrac{dy}{Q} = \dfrac{dz}{R}$, where P,Q,R are functions of x,y,z, then $\phi(u,v) = 0$, with ϕ arbitrary, is the general solution of $Pp + Qq = R$.

Taking the differentials of $u = a$ and $v = b$, we have

$$\frac{\partial u}{\partial x}dx + \frac{\partial u}{\partial y}dy + \frac{\partial u}{\partial z}dz = 0, \qquad \frac{\partial v}{\partial x}dx + \frac{\partial v}{\partial y}dy + \frac{\partial v}{\partial z}dz = 0.$$

Since u and v are independent functions, we may solve for the ratios

$$dx : dy : dz = \left(\frac{\partial u}{\partial y}\frac{\partial v}{\partial z} - \frac{\partial u}{\partial z}\frac{\partial v}{\partial y}\right) : \left(\frac{\partial u}{\partial z}\frac{\partial v}{\partial x} - \frac{\partial u}{\partial x}\frac{\partial v}{\partial z}\right) : \left(\frac{\partial u}{\partial x}\frac{\partial v}{\partial y} - \frac{\partial u}{\partial y}\frac{\partial v}{\partial x}\right) = P : Q : R.$$

But these are the relations (see Chapter 28) defining P,Q,R in the equation $Pp + Qq = R$ whose general solution is $\phi(u,v) = 0$.

8. Let $u = av + \beta$ be a complete solution of $Pp + Qq = R$. From this two-parameter family of surfaces, select a one-parameter family by setting $\beta = h(\alpha)$, where h is a given function of α, and obtain the envelope.

The envelope of the family

$$1)\quad u = av + h(\alpha)$$

is obtained by eliminating α between 1) and

$$2)\quad 0 = v + h'(\alpha).$$

Solving 2) for $\alpha = \mu(v)$ and substituting in 1), we have

$$3)\quad u = v\cdot\mu(v) + h[\mu(v)] = \lambda(v).$$

Now 3) is a part of the general solution $\phi(u,v) = 0$. Thus, unlike the case of ordinary differential equations, the envelope is not a new locus.

If $h(\alpha)$ is taken as an arbitrary function of α, $\lambda(v)$ is an arbitrary function of v, and 3) *is* the general solution. Thus, the general solution of a linear partial differential equation of order one is the totality of envelopes of all one-parameter families 1) obtained from a complete solution. It is to be noted that when $h(\alpha)$ is arbitrary, the elimination of α between 1) and 2) is not possible; thus, the general solution cannot be obtained from the complete solution.

9. Show that the conditions for exactness of the ordinary differential equation

$$\mu(x,y)\,M(x,y)dx + \mu(x,y)\,N(x,y)dy = 0$$

is a linear partial differential equation of order one. Thus, show how to find an integrating factor of $M\,dx + N\,dy = 0$. (See Chapter 4.)

If $$\mu M\,dx + \mu N\,dy = 0$$

is exact, then $\dfrac{\partial}{\partial y}(\mu M) = \dfrac{\partial}{\partial x}(\mu N)$ or $M\dfrac{\partial\mu}{\partial y} - N\dfrac{\partial\mu}{\partial x} = \mu\left(\dfrac{\partial N}{\partial x} - \dfrac{\partial M}{\partial y}\right)$.

This is a linear partial differential equation of order one for which the auxiliary system is

$$1)\quad \frac{dx}{-N} = \frac{dy}{M} = \frac{d\mu}{\mu\left(\dfrac{\partial N}{\partial x} - \dfrac{\partial M}{\partial y}\right)}.$$

Any solution, involving μ, of this system is an integrating factor of $M\,dx + N\,dy = 0$.

Writing 1) in the form

$$2)\quad \frac{\dfrac{\partial N}{\partial x} - \dfrac{\partial M}{\partial y}}{-N}\,dx = \frac{\dfrac{\partial N}{\partial x} - \dfrac{\partial M}{\partial y}}{M}\,dy = \frac{d\mu}{\mu},$$ it is evident that if

$\dfrac{\dfrac{\partial N}{\partial x} - \dfrac{\partial M}{\partial y}}{-N} = f(x)$, then $\mu = e^{\int f(x)dx}$ is an integrating factor; or if $\dfrac{\dfrac{\partial N}{\partial x} - \dfrac{\partial M}{\partial y}}{M} = g(y)$, $\mu = e^{\int g(y)dy}$ is an integrating factor. Moreover, if the equation is linear (that is, $y' + Py = Q$), then $M = Py - Q$, $N = 1$ and 2) becomes $P\,dx = \dfrac{-P}{Py - Q}\,dy = \dfrac{d\mu}{\mu}$ and $\mu = e^{\int P\,dx}$ is an integrating factor.

10. Find an integrating factor for $(2x^3y - y^2)dx - (2x^4 + xy)dy = 0$. (See Problem 9 above.)

Here $M = 2x^3y - y^2$, $N = -(2x^4 + xy)$, $\dfrac{\partial M}{\partial y} = 2x^3 - 2y$, $\dfrac{\partial N}{\partial x} = -(8x^3 + y)$.

We seek a solution involving μ of $\dfrac{dx}{2x^4 + xy} = \dfrac{dy}{2x^3y - y^2} = \dfrac{d\mu}{\mu(y - 10x^3)}$.

From $\dfrac{d\mu}{\mu(y-10x^3)} = \dfrac{-2y\,dx - 3x\,dy}{-2y(2x^4 + xy) - 3x(2x^3y - y^2)} = \dfrac{-2y\,dx - 3x\,dy}{xy(y - 10x^3)}$ or $\dfrac{d\mu}{\mu} = \dfrac{-2y\,dx - 3x\,dy}{xy}$

we obtain $\ln\mu = -2\ln x - 3\ln y$. Thus, $\mu = x^{-2}y^{-3}$ is an integrating factor.

11. Find the integral surface of $x^2p + y^2q + z^2 = 0$ which passes through the hyperbola $xy = x + y,\ z = 1$.

The auxiliary system is $\dfrac{dx}{x^2} = \dfrac{dy}{y^2} = \dfrac{dz}{-z^2}$.

From $\dfrac{dx}{x^2} = \dfrac{dz}{-z^2}$ we obtain $u = \dfrac{x+z}{xz} = a$, and from $\dfrac{dy}{y^2} = \dfrac{dz}{-z^2}$ we obtain $v = \dfrac{y+z}{yz} = b$.

We first eliminate x_0, y_0, z_0 between $x_0y_0 = x_0 + y_0$, $z_0 = 1$ and $u = \dfrac{x_0+z_0}{x_0z_0} = \dfrac{x_0+1}{x_0} = a$ and $v = \dfrac{y_0+z_0}{y_0z_0} = \dfrac{y_0+1}{y_0} = b$. Solving the latter for $x_0 = \dfrac{1}{a-1}$, $y_0 = \dfrac{1}{b-1}$ and substituting in $x_0y_0 = x_0 + y_0$, we obtain $\dfrac{1}{(a-1)(b-1)} = \dfrac{1}{a-1} + \dfrac{1}{b-1}$ or $a + b = 3$ as the relation which must exist between a and b. Then the equation of the required surface is

$$a + b = u + v = \frac{x+z}{xz} + \frac{y+z}{yz} = 3 \qquad \text{or} \qquad 2xy + z(x+y) = 3xyz.$$

SUPPLEMENTARY PROBLEMS

Find the general solution of each of the following equations.

12. $p + q = z$ *Ans.* $z = e^y\phi(x-y)$
13. $3p + 4q = 2$ $3y - 4x = f(3z - 2x)$ or $\phi(3y-4x,\ 3z-2x) = 0$
14. $yq - xp = z$ $\phi(xy, xz) = 0$
15. $xzp + yzq = xy$ $y = x\phi(xy - z^2)$
16. $x^2p + y^2q = z^2$ $x - y = xy\phi(1/x - 1/z)$
17. $yp - xq + x^2 - y^2 = 0$ $\phi(x^2+y^2,\ xy - z) = 0$
18. $yzp - xzq = xy$ $\phi(x^2+y^2,\ y^2+z^2) = 0$
19. $zp + yq = x$ $x + z = y\phi(x^2 - z^2)$
20. $x(y-z)p + y(z-x)q = z(x-y)$ $\phi(xyz,\ x+y+z) = 0$
21. $x(y^2-z^2)p + y(z^2-x^2)q = z(x^2-y^2)$ $\phi(xyz,\ x^2+y^2+z^2) = 0$

22. Find the equation of all the surfaces whose tangent planes pass through the point (0,0,1).
Hint: Solve $xp + yq = z - 1$. *Ans.* $z = 1 + x\phi(y/x)$

23. Find the equation of the surface satisfying $4yzp + q + 2y = 0$ and passing through $y^2 + z^2 = 1$, $x + z = 2$. *Ans.* $y^2 + z^2 + x + z = 3$

CHAPTER 30

Non-linear Partial Differential Equations of Order One

COMPLETE AND SINGULAR SOLUTIONS. Let the non-linear partial differential equation of order one

1) $$f(x,y,z,p,q) = 0$$

be derived from

2) $$g(x,y,z,a,b) = 0$$

by eliminating the arbitrary constants a and b. Then 2) is called a (or the) *complete solution* of 1).

This complete solution represents a two-parameter family of surfaces which may or may not have an envelope. To find the envelope (if one exists) we eliminate a and b from

$$g = 0, \quad \frac{\partial g}{\partial a} = 0, \quad \frac{\partial g}{\partial b} = 0.$$

If the eliminant

3) $$\lambda(x,y,z) = 0$$

satisfies 1), it is called the *singular solution* of 1); if

$$\lambda(x,y,z) = \xi(x\ y\ z)\cdot\eta(x,y,z)$$

and if $\xi = 0$ satisfies 1) while $\eta = 0$ does not, $\xi = 0$ is the singular solution. As in the case of ordinary differential equations (Chapter 10), the singular solution may be obtained from the partial differential equation by eliminating p and q from

$$f = 0, \quad \frac{\partial f}{\partial p} = 0, \quad \frac{\partial f}{\partial q} = 0.$$

EXAMPLE 1. It is readily verified that $z = ax + by - (a^2 + b^2)$ is a complete solution of $z = px + qy - (p^2 + q^2)$. Eliminating a and b from

$$g = z - ax - by + a^2 + b^2 = 0, \quad \frac{\partial g}{\partial a} = -x + 2a = 0, \quad \frac{\partial g}{\partial b} = -y + 2b = 0,$$

we have $z = \frac{1}{2}x^2 + \frac{1}{2}y^2 - \frac{1}{4}(x^2 + y^2) = \frac{1}{4}(x^2 + y^2)$. This satisfies the differential equation and is the singular solution. The complete solution represents a two-parameter family of planes which envelope the paraboloid $x^2 + y^2 = 4z$.

GENERAL SOLUTION. If, in the complete solution 2), one of the constants, say b, is replaced by a known function of the other, say $b = \phi(a)$, then

$$g(x,y,z,a,\phi(a)) = 0$$

is a one-parameter family of the surfaces of 1). If this family has an envelope, its equation may be found as usual by eliminating a from

$$g(x,y,z,a,\phi(a)) = 0 \quad \text{and} \quad \frac{\partial}{\partial a}\, g(x,y,z,a,\phi(a)) = 0$$

and determining that part of the result which satisfies 1).

EXAMPLE 2. Set $b = \phi(a) = a$ in the complete solution of Example 1. The result of eliminating a from $g = z - a(x+y) + 2a^2 = 0$ and $\dfrac{\partial g}{\partial a} = -(x+y) + 4a = 0$ is $z = \dfrac{1}{8}(x+y)^2$ which can be readily shown to satisfy the differential equation of Example 1. This is a parabolic cylinder with its elements parallel to the xOy plane.

The totality of solutions obtained by varying $\phi(a)$ is called the *general solution* of the differential equation. Thus, from Example 2, $8z = (x+y)^2$ is included in the general solution of the differential equation of Example 1.

When $b = \phi(a)$, ϕ arbitrary, is used, the elimination of a between

$$g = 0 \quad \text{and} \quad \frac{\partial g}{\partial a} = 0$$

is not possible; hence, we are unable to express the general solution as a single equation, involving an arbitrary function, as we were in the case of the linear equation.

SOLUTIONS. Before considering a general method for obtaining a complete solution of 1), we give special procedures for handling four types of equations.

TYPE I: $f(p,q) = 0$. Example: $p^2 - q^2 = 1$.

From Problem 3, Chapter 28, it follows that a complete solution is

4) $$z = ax + h(a)\,y + c,$$

where $f(a,h(a)) = 0$, and a and c are arbitrary constants.

The equations for determining the singular solution are

$$z = ax + h(a)\,y + c, \qquad 0 = x + h'(a)\,y, \qquad 0 = 1.$$

Thus, there is no singular solution.

The general solution is obtained by putting $c = \phi(a)$, ϕ arbitrary, and eliminating a between

5) $$z = ax + h(a)\,y + \phi(a) \quad \text{and} \quad 0 = x + h'(a)\,y + \phi'(a).$$

The first equation of 5) for a stipulated function $\phi(a)$ represents a one-parameter family of planes and its envelope (a part of the general solution) is a developable surface. (See Problem 10, Chapter 28.)

EXAMPLE 3. Solve $p^2 - q^2 = 1$.

Here $f(p,q) = p^2 - q^2 - 1 = 0$, $f(a,h(a)) = a^2 - [h(a)]^2 - 1 = 0$ and $h(a) = (a^2-1)^{1/2}$. A complete solution is $z = ax + (a^2-1)^{1/2}y + c$.

A neater form is obtained by putting $a = \sec\alpha$; then $h(a) = \tan\alpha$ and we have

$$z = x\sec\alpha + y\tan\alpha + c.$$

If we set $c = \phi(\alpha) = 0$, the result of eliminating a from

$$z = x\sec\alpha + y\tan\alpha, \quad 0 = x\tan\alpha + y\sec\alpha \quad \text{or} \quad 0 = x\sin\alpha + y$$

is $$z^2 = x^2 - y^2.$$

This developable surface (cone) is a part of the general solution of the given differential equation.

Note that we might have taken $h(a) = -(a^2-1)^{1/2}$ and obtained as a complete solution

$$z = ax - (a^2-1)^{1/2}y + c.$$

See also Problems 1-2.

<u>*TYPE II:*</u> $z = px + qy + f(p,q)$. Example: $z = px + qy + 3p^{1/3}q^{1/3}$.

From Problem 4, Chapter 28, it follows that a complete solution is

6) $$z = ax + by + f(a,b).$$

This is known as the extended Clairaut type, for obvious reasons. This complete solution consists of a two-parameter family of planes. The singular solution (if one exists) is a surface having the complete solution as its tangent planes.

EXAMPLE 4. Solve $z = px + qy + 3p^{1/3}q^{1/3}$.

A complete solution is $z = ax + by + 3a^{1/3}b^{1/3}$.

The derivatives with respect to a and b are $x + a^{-2/3}b^{1/3} = 0$ and $y + a^{1/3}b^{-2/3} = 0$.

Then $ax + by = -2a^{1/3}b^{1/3}$, $xy = a^{-1/3}b^{-1/3}$,

and, substituting in the complete solution, we obtain the singular solution

$$z = a^{1/3}b^{1/3} = 1/xy \quad \text{or} \quad xyz = 1.$$

See also Problems 3-4.

<u>*TYPE III:*</u> $f(z,p,q) = 0$. Example: $z = p^2 + q^2$.

Assume $z = F(x+ay) = F(u)$, where a is the arbitrary constant. Then

$$p = \frac{\partial z}{\partial x} = \frac{dz}{du}\frac{\partial u}{\partial x} = \frac{dz}{du} \quad \text{and} \quad q = \frac{dz}{du}\frac{\partial u}{\partial y} = a\frac{dz}{du}.$$

When these are substituted in the given differential equation, we obtain an ordinary differential equation of order one

$$f(z, \frac{dz}{du}, a\frac{dz}{du}) = 0$$

whose solution is the required complete solution.

EXAMPLE 5. Solve $z = p^2 + q^2$.

Put $z = F(x+ay) = F(u)$. Then $p = dz/du$, $q = a\,dz/du$, and the given equation may be reduced to $z = (\frac{dz}{du})^2 + a^2(\frac{dz}{du})^2$.

Solving $\frac{dz}{du} = \frac{\sqrt{z}}{\sqrt{1+a^2}}$ or $\frac{dz}{\sqrt{z}} = \frac{1}{\sqrt{1+a^2}}du$, we obtain $2\sqrt{z} = \frac{1}{\sqrt{1+a^2}}u + k = \frac{1}{\sqrt{1+a^2}}(u+b)$.

Thus, a complete solution is $4(1+a^2)z = (x+ay+b)^2$, a family of parabolic cylinders.

Taking the derivatives with respect to a and b, we have

$$8az - 2(x+ay+b)y = 0, \qquad x + ay + b = 0.$$

The singular solution is $z = 0$.

See also Problems 5-7.

<u>*TYPE IV:*</u> $f_1(x,p) = f_2(y,q)$. Example: $p - x^2 = q + y^2$.

Set $f_1(x,p) = a$, $f_2(y,q) = a$, where a is an arbitrary constant, and solve to obtain

$$p = F_1(x,a) \quad \text{and} \quad q = F_2(y,a).$$

Since z is a function of x and y, $dz = p\,dx + q\,dy = F_1(x,a)\,dx + F_2(y,a)\,dy$. Thus,

7) $$z = \int F_1(x,a)\,dx + \int F_2(y,a)\,dy + b,$$

containing two arbitrary constants, is the required complete solution.

EXAMPLE 6. Solve $p - q = x^2 + y^2$ or $p - x^2 = q + y^2$.

Setting $p - x^2 = a$, $q + y^2 = a$, we obtain $p = a + x^2$, $q = a - y^2$.

Integrating $dz = p\,dx + q\,dy = (a + x^2)dx + (a - y^2)dy$, the required complete solution is $z = ax + x^3/3 + ay - y^3/3 + b$. There is no singular solution.

See also Problems 8-9.

TRANSFORMATIONS. As in the case of ordinary differential equations, it is possible at times to find a transformation of the variables which will reduce a given equation to one of the above four types.

The combination px, for example, suggests the transformation $X = \ln x$, since then

$$p = \frac{\partial z}{\partial x} = \frac{\partial z}{\partial X}\frac{dX}{dx} = \frac{1}{x}\frac{\partial z}{\partial X} \quad \text{and} \quad px = \frac{\partial z}{\partial X}.$$

Thus, $q = px + p^2x^2$ becomes $\dfrac{\partial z}{\partial y} = \dfrac{\partial z}{\partial X} + \left(\dfrac{\partial z}{\partial X}\right)^2$, of Type I.

Similarly, the combination qy suggests the transformation $Y = \ln y$.

The appearance of $\dfrac{p}{z}, \dfrac{q}{z}$ in an equation suggests the transformation $Z = \ln z$,

since then $p = \dfrac{\partial z}{\partial x} = \dfrac{dz}{dZ}\dfrac{\partial Z}{\partial x} = z\dfrac{\partial Z}{\partial x}$ and $\dfrac{p}{z} = \dfrac{\partial Z}{\partial x}$; similarly, $\dfrac{q}{z} = \dfrac{\partial Z}{\partial y}$.

Thus, $\dfrac{q}{z} = \left(\dfrac{p}{z}\right)^2$ becomes $\dfrac{\partial Z}{\partial y} = \left(\dfrac{\partial Z}{\partial x}\right)^2$, of Type I.

See also Problems 10-14.

COMPLETE SOLUTION. CHARPIT'S METHOD. Consider the non-linear partial differential equation

1) $$f(x,y,z,p,q) = 0.$$

Since z is a function of x and y, it follows that

8) $$dz = p\,dx + q\,dy.$$

Let us assume $p = u(x,y,z,a)$, where a is an arbitrary constant, substitute in 1) and solve to obtain $q = v(x,y,z,a)$. For these values of p and q, 8) becomes

8_1) $$dz = u\,dx + v\,dy.$$

Now if 8_1) can be integrated, yielding

9) $$g(x,y,z,a,b) = 0,$$

this is a complete solution of 1).

EXAMPLE 7. Solve $pq + qx = y$.

Take $p = a - x$, substitute in $pq + qx = y$, and solve for $q = y/a$.

Substituting in $dz = p\,dx + q\,dy$, we have $dz = (a - x)dx + (y/a)dy$, an integrable equation, with solution

$$z = ax - \tfrac{1}{2}x^2 + \tfrac{1}{2}y^2/a + k \quad \text{or} \quad 2az = 2a^2x - ax^2 + y^2 + b.$$

Since the success of the above procedure depends upon our making a fortunate choice for p, it cannot be suggested as a standard procedure. We turn now to a general method for solving 1). This consists in finding an equation

10) $$F(x,y,z,p,q) = 0$$

such that 1) and 10) may be solved for $p = P(x,y,z)$ and $q = Q(x,y,z)$, (that is, such that

11) $$\Delta = \begin{vmatrix} \frac{\partial f}{\partial p} & \frac{\partial f}{\partial q} \\ \frac{\partial F}{\partial p} & \frac{\partial F}{\partial q} \end{vmatrix} \neq 0, \text{ identically}),$$

and such that for these value of p and q the total differential equation

8) $$dz = p\,dx + q\,dy = P(x,y,z)dx + Q(x,y,z)dy$$

is integrable, that is, $P\frac{\partial Q}{\partial z} - Q\frac{\partial P}{\partial z} - \frac{\partial P}{\partial y} + \frac{\partial Q}{\partial x} = \frac{\partial q}{\partial x} - \frac{\partial p}{\partial y} = 0.$

Differentiating 1) and 10) partially with respect to x and y, we find

12) $$\frac{\partial f}{\partial x} + p\frac{\partial f}{\partial z} + \frac{\partial f}{\partial p}\frac{\partial p}{\partial x} + \frac{\partial f}{\partial q}\frac{\partial q}{\partial x} = 0,$$

13) $$\frac{\partial f}{\partial y} + q\frac{\partial f}{\partial z} + \frac{\partial f}{\partial p}\frac{\partial p}{\partial y} + \frac{\partial f}{\partial q}\frac{\partial q}{\partial y} = 0,$$

14) $$\frac{\partial F}{\partial x} + p\frac{\partial F}{\partial z} + \frac{\partial F}{\partial p}\frac{\partial p}{\partial x} + \frac{\partial F}{\partial q}\frac{\partial q}{\partial x} = 0,$$

15) $$\frac{\partial F}{\partial y} + q\frac{\partial F}{\partial z} + \frac{\partial F}{\partial p}\frac{\partial p}{\partial y} + \frac{\partial F}{\partial q}\frac{\partial q}{\partial y} = 0.$$

Multiplying 12) by $\frac{\partial F}{\partial p}$, 13) by $\frac{\partial F}{\partial q}$, 14) by $-\frac{\partial f}{\partial p}$, 15) by $-\frac{\partial f}{\partial q}$, and adding, we obtain (noting that $\frac{\partial p}{\partial y} = \frac{\partial q}{\partial x}$)

$$\left(\frac{\partial f}{\partial x} + p\frac{\partial f}{\partial z}\right)\frac{\partial F}{\partial p} + \left(\frac{\partial f}{\partial y} + q\frac{\partial f}{\partial z}\right)\frac{\partial F}{\partial q} - \frac{\partial f}{\partial p}\frac{\partial F}{\partial x} - \frac{\partial f}{\partial q}\frac{\partial F}{\partial y} - \left(p\frac{\partial f}{\partial p} + q\frac{\partial f}{\partial q}\right)\frac{\partial F}{\partial z} = 0.$$

This is a linear partial differential equation in F, considered as a function of the independent variables x, y, z, p, q. The auxiliary system is

16) $$\frac{dp}{\frac{\partial f}{\partial x} + p\frac{\partial f}{\partial z}} = \frac{dq}{\frac{\partial f}{\partial y} + q\frac{\partial f}{\partial z}} = \frac{dx}{-\frac{\partial f}{\partial p}} = \frac{dy}{-\frac{\partial f}{\partial q}} = \frac{dz}{-\left(p\frac{\partial f}{\partial p} + q\frac{\partial f}{\partial q}\right)} = \frac{dF}{0}.$$

Thus, we may take for 10) any solution of this system which involves p or q, or both, which contains an arbitrary constant, and for which 11) holds.

EXAMPLE 8. Solve $q = -xp + p^2$.

Here $f = p^2 - xp - q$ so that $\frac{\partial f}{\partial x} = -p$, $\frac{\partial f}{\partial y} = 0$, $\frac{\partial f}{\partial z} = 0$, $\frac{\partial f}{\partial p} = 2p - x$, $\frac{\partial f}{\partial q} = -1$, and

$$\frac{\partial f}{\partial x} + p\frac{\partial f}{\partial z} = -p, \qquad \frac{\partial f}{\partial y} + q\frac{\partial f}{\partial z} = 0, \qquad -(p\frac{\partial f}{\partial p} + q\frac{\partial f}{\partial q}) = -2p^2 + xp + q.$$

The auxiliary system (16) is $\dfrac{dp}{-p} = \dfrac{dq}{0} = \dfrac{dx}{-2p+x} = \dfrac{dy}{1} = \dfrac{dz}{-2p^2+xp+q}$.

From $\dfrac{dp}{-p} = \dfrac{dy}{1}$, we have $\ln p = -y + \ln a$ or $p = ae^{-y}$.

Using the given differential equation, $q = -xp + p^2 = -axe^{-y} + a^2e^{-2y}$.

Then $dz = p\,dx + q\,dy$ becomes $dz = ae^{-y}\,dx + (-axe^{-y} + a^2e^{-2y})dy$. Integrating,

$$z = axe^{-y} - \tfrac{1}{2}a^2e^{-2y} + b.$$

There is no singular solution.

See also Problem 15.

SOLVED PROBLEMS

(In these solutions, the equations leading to the general solution will not be given.)

TYPE I: $f(p,q) = 0$.

1. Solve $p^2 + q^2 = 9$.

A complete solution is $z = ax + by + c$, where $a^2 + b^2 = 9$.

The equations for determining the singular solution are

$z = ax + \sqrt{9-a^2}\,y + c$, $\quad 0 = x - \dfrac{a}{\sqrt{9-a^2}}\,y$, $\quad 0 = 1$. Thus, there is no singular solution.

2. Solve $pq + p + q = 0$.

A complete solution is $z = ax + by + c$, where $ab + a + b = 0$, or $z = ax - \dfrac{a}{a+1}\,y + c$.

There is no singular solution.

TYPE II: $z = px + qy + f(p,q)$.

3. Solve $z = px + qy + p^2 + pq + q^2$.

A complete solution is $z = ax + by + a^2 + ab + b^2$.

Differentiating the complete solution with respect to a and b, we have

$$0 = x + 2a + b, \qquad 0 = y + a + 2b.$$

Solving to obtain $a = (y-2x)/3$, $b = (x-2y)/3$ and substituting in the complete solution, the singular solution is $3z = xy - x^2 - y^2$.

4. Solve $z = px + qy + p^2q^2$.

A complete solution is $z = ax + by + a^2b^2$. The equations obtained by differentiating with respect to a and b are $0 = x + 2ab^2$ and $0 = y + 2a^2b$. Then $a = -\sqrt[3]{\dfrac{y^2}{2x}}$, $b = -\sqrt[3]{\dfrac{x^2}{2y}}$

and the singular solution is $z = -x\sqrt[3]{\dfrac{y^2}{2x}} - y\sqrt[3]{\dfrac{x^2}{2y}} + \sqrt[3]{\dfrac{x^2y^2}{16}} = -\dfrac{3}{4}\sqrt[3]{4}\,x^{2/3}y^{2/3}$.

TYPE III: $f(z,p,q) = 0$.

5. Solve $4(1+z^3) = 9z^4pq$.

Assume $z = F(x+ay) = F(u)$. Then $p = \dfrac{dz}{du}$, $q = a\dfrac{dz}{du}$, and the given equation becomes

$$4(1+z^3) = 9az^4(\frac{dz}{du})^2 \quad \text{or} \quad \frac{3\sqrt{a}\,z^2}{\sqrt{1+z^3}}\,dz = 2\,du.$$

Integrating, $\sqrt{a(1+z^3)} = u+b$, and a complete solution is $a(1+z^3) = (x+ay+b)^2$.

Using the results of differentiating this with respect to a and b,

$$1 + z^3 = 2(x+ay+b)y \quad \text{and} \quad 0 = 2(x+ay+b),$$

the singular solution is $z^3 + 1 = 0$.

6. Solve $p(1-q^2) = q(1-z)$.

Assume $z = F(x+ay) = F(u)$. Then $p = \dfrac{dz}{du}$, $q = a\dfrac{dz}{du}$, and the given equation becomes

$$(\frac{dz}{du})[1-a^2(\frac{dz}{du})^2] = a\frac{dz}{du}(1-z) \quad \text{or} \quad (\frac{dz}{du})[1-a+az-a^2(\frac{dz}{du})^2] = 0.$$

Then $\dfrac{dz}{du} = 0$ and $z = c$; or $1-a+az-a^2(\dfrac{dz}{du})^2 = 0$, $\dfrac{a\,dz}{\sqrt{1-a+az}} = du$ and

$$2\sqrt{1-a+az} = u+b = x+ay+b \quad \text{or} \quad 4(1-a+az) = (x+ay+b)^2.$$

Each of $z = c$ and $4(1-a+az) = (x+ay+b)^2$ is a solution; the latter is a complete solution. Using it, the equations for obtaining the singular solution are

$g = 4(1-a+az) - (x+ay+b)^2 = 0$, $\dfrac{\partial g}{\partial a} = 4(-1+z) - 2y(x+ay+b) = 0$, $\dfrac{\partial g}{\partial b} = -2(x+ay+b) = 0$;

there is no singular solution.

7. Solve $1 + p^2 = qz$.

Assume $z = F(x+ay) = F(u)$. Then $p = \dfrac{dz}{du}$, $q = a\dfrac{dz}{du}$, and the given equation becomes

$$(\frac{dz}{du})^2 - az\frac{dz}{du} + 1 = 0 \quad \text{or} \quad \frac{dz}{az - \sqrt{a^2z^2-4}} = \tfrac{1}{2}\,du.$$

Rationalizing the left member of the latter equation, we obtain $(az + \sqrt{a^2z^2-4})dz = 2\,du$.

whose solution is $\frac{1}{2}az^2 + \dfrac{1}{a}[\dfrac{az}{2}\sqrt{a^2z^2-4} - 2\ln(az + \sqrt{a^2z^2-4})] = 2(u+b)$.

A complete solution is then $a^2z^2 + az\sqrt{a^2z^2-4} - 4\ln(az+\sqrt{a^2z^2-4}) = 4a(x+ay+b)$.

Note that $a^2z^2 - az\sqrt{a^2z^2-4} + 4\ln(az+\sqrt{a^2z^2-4}) = 4a(x+ay+b)$, obtained from

$\dfrac{dz}{az+\sqrt{a^2z^2-4}} = \frac{1}{2}\,du$, is also a complete solution.

TYPE IV: $f_1(x,p) = f_2(y,q)$.

8. Solve $\sqrt{p} - \sqrt{q} + 3x = 0$ or $\sqrt{p} + 3x = \sqrt{q}$.

Set $\sqrt{p} + 3x = a$ and $\sqrt{q} = a$. Then $p = (a-3x)^2$ and $q = a^2$. A complete solution is $z = \int p\,dx + \int q\,dy + b = \int (a-3x)^2\,dx + a^2\int dy + b$ or $z = -\frac{1}{9}(a-3x)^3 + a^2y + b$.

There is no singular solution.

9. Solve $q = -px + p^2$.

Set $p^2 - px = a$ and $q = a$. Then $p = \frac{1}{2}(x + \sqrt{x^2+4a}\,)$.

A complete solution is $z = \frac{1}{2}\int (x + \sqrt{x^2+4a}\,)dx + a\int dy + b$

or $z = \frac{1}{4}(x^2 + x\sqrt{x^2+4a}\,) + a\ln(x + \sqrt{x^2+4a}\,) + ay + b$.

Another complete solution is obtained by the method of Charpit in Example 8.
There is no singular solution.

USE OF TRANSFORMATIONS.

10. Solve $pq = x^m y^n z^{2l}$ or $\dfrac{pz^{-l}}{x^m}\cdot\dfrac{qz^{-l}}{y^n} = 1$.

The transformation

$$Z = \frac{z^{1-l}}{1-l},\quad X = \frac{x^{m+1}}{m+1},\quad Y = \frac{y^{n+1}}{n+1},\quad \frac{\partial Z}{\partial X} = \frac{\partial Z}{\partial x}\frac{dx}{dX} = z^{-l}p\frac{1}{x^m},\quad \frac{\partial Z}{\partial Y} = \frac{\partial Z}{\partial y}\frac{dy}{dY} = z^{-l}q\frac{1}{y^n}$$

reduces the given differential equation to $\dfrac{\partial Z}{\partial X}\cdot\dfrac{\partial Z}{\partial Y} = 1$.

This equation is of Type I and its solution is $Z = aX + \frac{1}{a}Y + c$.

A complete solution of the given equation is $\dfrac{z^{1-l}}{1-l} = a\dfrac{x^{m+1}}{m+1} + \dfrac{y^{n+1}}{a(n+1)} + c$.

There is no singular solution.

11. Solve $x^2p^2 + y^2q^2 = z$.

1) The transformation

$$X = \ln x,\quad Y = \ln y,\quad Z = 2z^{\frac{1}{2}},\quad \frac{\partial Z}{\partial X} = \frac{\partial Z}{\partial x}\frac{dx}{dX} = pxz^{-\frac{1}{2}},\quad \frac{\partial Z}{\partial Y} = \frac{\partial Z}{\partial y}\frac{dy}{dY} = qyz^{-\frac{1}{2}}$$

reduces the given equation to $z(\frac{\partial Z}{\partial X})^2 + z(\frac{\partial Z}{\partial Y})^2 = z$ or $(\frac{\partial Z}{\partial X})^2 + (\frac{\partial Z}{\partial Y})^2 = 1$, of Type I.

A complete solution is $Z = aX + bY + c$ or $4z = (a\ln x + b\ln y + c)^2$, where $a^2 + b^2 = 1$. The singular solution is $z = 0$.

2) The transformation $X = \ln x$, $Y = \ln y$, $p = \dfrac{\partial z}{\partial x} = \dfrac{\partial z}{\partial X}\dfrac{dX}{dx} = \dfrac{1}{x}\dfrac{\partial z}{\partial X}$, $q = \dfrac{1}{y}\dfrac{\partial z}{\partial Y}$

reduces the given differential equation to $(\frac{\partial z}{\partial X})^2 + (\frac{\partial z}{\partial Y})^2 = z$, of Type III.

We set $z = F(X+aY) = F(u)$. Then $\dfrac{\partial z}{\partial X} = \dfrac{dz}{du}\dfrac{\partial u}{\partial X} = \dfrac{dz}{du}$, $\dfrac{dz}{dY} = a\dfrac{dz}{du}$, and

$$(\frac{dz}{du})^2 + a^2(\frac{dz}{du})^2 = z \qquad \text{or} \qquad \sqrt{1+a^2}\,\frac{dz}{\sqrt{z}} = du.$$

Integrating, $2\sqrt{1+a^2}\,z^{\frac{1}{2}} = u + b = X + aY + b = \ln x + a \ln y + b.$

A complete solution is $4(1+a^2)z = (\ln x + a \ln y + b)^2$. The singular solution is $z = 0$.

12. Solve $4xyz = pq + 2px^2y + 2qxy^2$.

Let $x = X^{\frac{1}{2}}$, $y = Y^{\frac{1}{2}}$. Then $p = \dfrac{\partial z}{\partial x} = \dfrac{\partial z}{\partial X}\dfrac{dX}{dx} = 2X^{\frac{1}{2}}\dfrac{\partial z}{\partial X}$ and $q = \dfrac{\partial z}{\partial Y}\dfrac{dY}{dy} = 2Y^{\frac{1}{2}}\dfrac{\partial z}{\partial Y}$.

Substituting in the given equation, we have $z = X\dfrac{\partial z}{\partial X} + Y\dfrac{\partial z}{\partial Y} + \dfrac{\partial z}{\partial X}\dfrac{\partial z}{\partial Y}$ of Type II.

A complete solution is $z = aX + bY + ab$ or $z = ax^2 + by^2 + ab$.

Eliminating a and b from this and $0 = x^2 + b$, $0 = y^2 + a$, obtained by differentiating it with respect to a and b, the singular solution is found to be $z + x^2y^2 = 0$.

13. Solve $p^2x^2 = z(z - qy)$.

The transformation $Y = \ln y$, $X = \ln x$, $p = \dfrac{\partial z}{\partial x} = \dfrac{\partial z}{\partial X}\dfrac{dX}{dx} = \dfrac{1}{x}\dfrac{\partial z}{\partial X}$, $q = \dfrac{1}{y}\dfrac{\partial z}{\partial Y}$

reduces the given equation to A) $(\dfrac{\partial z}{\partial X})^2 = z(z - \dfrac{\partial z}{\partial Y})$, of Type III.

We set $z = F(X+aY) = F(u)$. Then $\dfrac{\partial z}{\partial X} = \dfrac{dz}{du}$, $\dfrac{\partial z}{\partial Y} = a\dfrac{dz}{du}$, and A) becomes $(\dfrac{dz}{du})^2 = z^2 - az\dfrac{dz}{du}$.

Then $\dfrac{dz}{du} = \frac{1}{2}z(\sqrt{a^2+4} - a)$, $2\dfrac{dz}{z} = (\sqrt{a^2+4} - a)du$, and $\ln z^2 = (\sqrt{a^2+4} - a)(u+b)$.

A complete solution is $\ln z^2 = (\sqrt{a^2+4} - a)(\ln x + a \ln y + b)$.
There is no singular solution.

14. Solve $p^2 + q^2 = z^2(x+y)$ or $(\frac{p}{z})^2 + (\frac{q}{z})^2 = x + y$.

The transformation $Z = \ln z$, $p = z\dfrac{\partial Z}{\partial x}$, $q = z\dfrac{\partial Z}{\partial y}$ reduces the given equation to

$$(\frac{\partial Z}{\partial x})^2 + (\frac{\partial Z}{\partial y})^2 = x + y \qquad \text{or} \qquad (\frac{\partial Z}{\partial x})^2 - x = y - (\frac{\partial Z}{\partial y})^2, \quad \text{of Type IV.}$$

Set $(\dfrac{\partial Z}{\partial x})^2 - x = a = y - (\dfrac{\partial Z}{\partial y})^2$. Then $\dfrac{\partial Z}{\partial x} = (a+x)^{\frac{1}{2}}$ and $\dfrac{\partial Z}{\partial y} = (y-a)^{\frac{1}{2}}$.

A complete solution is $Z = \int (a+x)^{\frac{1}{2}}dx + \int (y-a)^{\frac{1}{2}}dy + b$

or $\ln z = \frac{2}{3}(a+x)^{3/2} + \frac{2}{3}(y-a)^{3/2} + b.$

CHARPIT'S METHOD.

15. Solve $16p^2z^2 + 9q^2z^2 + 4z^2 - 4 = 0$.

Let $f(x,y,z,p,q) = 16p^2z^2 + 9q^2z^2 + 4z^2 - 4$.

Then $\dfrac{\partial f}{\partial x} = 0 = \dfrac{\partial f}{\partial y}$, $\dfrac{\partial f}{\partial z} = 32p^2z + 18q^2z + 8z$, $\dfrac{\partial f}{\partial p} = 32pz^2$, $\dfrac{\partial f}{\partial q} = 18qz^2$, and the auxiliary system

$$\frac{dp}{\dfrac{\partial f}{\partial x} + p\dfrac{\partial f}{\partial z}} = \frac{dq}{\dfrac{\partial f}{\partial y} + q\dfrac{\partial f}{\partial z}} = \frac{dx}{-\dfrac{\partial f}{\partial p}} = \frac{dy}{-\dfrac{\partial f}{\partial q}} = \frac{dz}{-(p\dfrac{\partial f}{\partial p} + q\dfrac{\partial f}{\partial q})}$$ is

$$\frac{dp}{32p^3z + 18pq^2z + 8pz} = \frac{dq}{32p^2qz + 18q^3z + 8qz} = \frac{dx}{-32pz^2} = \frac{dy}{-18qz^2} = \frac{dz}{-32p^2z^2 - 18q^2z^2}.$$

Using the multipliers $4z, 0, 1, 0, 4p$, we find

$$4z(32p^3z + 18pq^2z + 8pz) + 1(-32pz^2) + 4p(-32p^2z^2 - 18q^2z^2) = 0$$

and so
$$dx + 4p\,dz + 4z\,dp = 0.$$

Then $x + 4pz = a$ and $p = -\dfrac{x-a}{4z}$. Substituting for p in the given differential equation, we find $(x-a)^2 + 9q^2z^2 + 4z^2 - 4 = 0$. Using the root $q = \dfrac{2}{3z}\sqrt{1-z^2-\frac{1}{4}(x-a)^2}$,

$$dz = p\,dx + q\,dy = -\frac{x-a}{4z}dx + \frac{2}{3z}\sqrt{1-z^2-\tfrac{1}{4}(x-a)^2}\,dy \quad \text{or} \quad dy = \frac{3[z\,dz + \frac{1}{4}(x-a)dx]}{2\sqrt{1-z^2-\frac{1}{4}(x-a)^2}}.$$

Then $y - b = -\dfrac{3}{2}\sqrt{1-z^2-\frac{1}{4}(x-a)^2}$ or $\dfrac{(x-a)^2}{4} + \dfrac{(y-b)^2}{9/4} + z^2 = 1$ is a complete solution. This is a family of ellipsoids with centers on the xOy plane. The semi-axes of the ellipsoids are 2 units parallel to the x-axis, 3/2 units parallel to the y-axis, and 1 unit parallel to the z-axis. The singular solution consists of the parallel planes $z = \pm 1$.

Another complete solution may be found by noting that the equation is of Type III. Using $F(x+ay) = F(u)$ and setting $p = \dfrac{dz}{du}$ and $q = a\dfrac{dz}{du}$, the given equation becomes

$$16z^2(\frac{dz}{du})^2 + 9a^2z^2(\frac{dz}{du})^2 + 4z^2 - 4 = 0 \quad \text{or} \quad \frac{z\,dz}{\sqrt{1-z^2}} = \frac{2}{\sqrt{16+9a^2}}du. \quad \text{Then}$$

$$-\sqrt{1-z^2} = \frac{2}{\sqrt{16+9a^2}}(u+b) = \frac{2}{\sqrt{16+9a^2}}(x+ay+b).$$

This complete solution $(16+9a^2)(1-z^2) = 4(x+ay+b)^2$ represents a family of elliptic cylinders with elements parallel to the xOy plane. The major axis of a cross section lies in the xOy plane and the minor axis is 2 units parallel to the z-axis.

SUPPLEMENTARY PROBLEMS

Find a complete solution and the singular solution (if any).

16. $p = q^2$ *Ans.* $z = b^2x + by + c$

17. $p^2 + p = q^2$ $z = ax + by + c$ where $b^2 = a^2 + a$

18. $pq = p + q$ $(b-1)z = bx + b(b-1)y + c$

19. $z = px + qy + pq$ $z = ax + by + ab$; s.s., $z = -xy$

20. $p^2 + q^2 = 4z$ $z(1+a^2) = (x+ay+b)^2$; s.s., $z = 0$

21. $pz = 1 + q^2$ $z^2 - z\sqrt{z^2-4a^2} + 4a^2 \ln(z + \sqrt{z^2-4a^2}) = 4(x+ay+b)$

22. $z^2(p^2+q^2+1) = 1$ $(1+a^2)(1-z^2) = (x+ay+b)^2$; s.s., $z^2 - 1 = 0$

23. $p^2 + pq = 4z$ $(1+a)z = (x+ay+b)^2$; s.s., $z = 0$

24. $p^2 - x = q^2 - y$ $3(z-b) = 2(x+a)^{3/2} + 2(y+a)^{3/2}$

25. $yp - x^2q^2 = x^2y$ $4(a-1)y^3 = (3z - ax^3 - b)^2$

26. $(1-y^2)xq^2 + y^2p = 0$ $(2z - ax^2 + b)^2 = 4a(y^2-1)$

27. $x^4p^2 - yzq - z^2 = 0$ $x \ln z = a + (a^2-1)x \ln y + bx$
Hint: Use $X = 1/x$, $Y = \ln y$, $Z = \ln z$.

28. $x^4p^2 + y^2zq - 2z^2 = 0$ $xy \ln z = ay + (a^2-2)x + bxy$
Hint: Use $X = 1/x$, $Y = 1/y$, $Z = \ln z$.

29. $x^4p^2 + y^2q = 0$ $x^2(zy + a + by)^2 + ay^2 = 0$

30. $2py^2 - q^2z = 0$ $z^2 = a^2x + ay^2 + b$

31. $q = xp + p^2$ $z = 2axe^y + 2a^2e^{2y} + b$

32. $zp^2 - y^2p + y^2q = 0$ $yz^2 = 2(axy + ay^2 + a^2 + by)$

Hint: $\dfrac{dp}{p^3} = \dfrac{dz}{-p^2z}$; $pz = a$ and $q = \dfrac{a}{z}(1 - \dfrac{a}{y^2})$.

33. $pq + 2x(y+1)p + y(y+2)q - 2(y+1)z = 0$

Ans. $z = ax + b(y^2 + 2y + a)$; s.s., $z + x(y^2 + 2y) = 0$

CHAPTER 31

Homogeneous Partial Differential Equations of Higher Order with Constant Coefficients

AN EQUATION SUCH AS

1) $$(x^2+y^2)\frac{\partial^3 z}{\partial x^3} + 2x\frac{\partial^3 z}{\partial x\,\partial y^2} + \frac{\partial^3 z}{\partial y^3} - \frac{\partial^2 z}{\partial x^2} + 5xy\frac{\partial^2 z}{\partial x\,\partial y} + x^3\frac{\partial z}{\partial x} + x\frac{\partial z}{\partial y} + yz = e^{x+y}$$

which is linear in the dependent variable z and its partial derivatives is called a *linear partial differential equation*. The order of 1) is three, being the order of the highest ordered derivative.

A linear partial differential equation such as

2) $$x^2\frac{\partial^3 z}{\partial x^3} + xy\frac{\partial^3 z}{\partial x^2\,\partial y} + 2\frac{\partial^3 z}{\partial x\,\partial y^2} + \frac{\partial^3 z}{\partial y^3} = x^2 + y^3,$$

in which the derivatives involved are all of the same order, will be called homogeneous, although there is no agreement among authors in the use of this term.

HOMOGENEOUS LINEAR DIFFERENTIAL EQUATIONS WITH CONSTANT COEFFICIENTS. Consider

3) $$A\frac{\partial z}{\partial x} + B\frac{\partial z}{\partial y} = 0,$$

4) $$A\frac{\partial^2 z}{\partial x^2} + B\frac{\partial^2 z}{\partial x\,\partial y} + C\frac{\partial^2 z}{\partial y^2} = 0,$$

5) $$A\frac{\partial^2 z}{\partial x^2} + B\frac{\partial^2 z}{\partial x\,\partial y} + C\frac{\partial^2 z}{\partial y^2} = x + 2y,$$

in which the numbers A, B, C are real constants.

It will be seen as we proceed that the methods for solving equations 3)-5) parallel those used in solving the linear ordinary differential equation

$$f(D)y = Q(x) \quad \text{where} \quad D = \frac{d}{dx}.$$

We shall employ two operators, $D_x = \dfrac{\partial}{\partial x}$ and $D_y = \dfrac{\partial}{\partial y}$, so that equations 3)-5) may be written as

3') $$f(D_x, D_y)z = (AD_x + BD_y)z = 0,$$

4') $$f(D_x, D_y)z = (AD_x^2 + BD_xD_y + CD_y^2)z = 0,$$

5') $$f(D_x, D_y)z = (AD_x^2 + BD_xD_y + CD_y^2)z = x + 2y.$$

Equation 3') is of order one and the general solution (Chapter 29) is

$$z = \phi(y - \frac{B}{A}x), \quad \phi \text{ arbitrary.}$$

Suppose now that $z = \phi(y + mx) = \phi(u)$, ϕ arbitrary, is a solution of 4'); then substituting

$$D_x z = \frac{\partial z}{\partial x} = \frac{d\phi}{du}\frac{\partial u}{\partial x} = m\frac{d\phi}{du}, \qquad D_y z = \frac{\partial z}{\partial y} = \frac{d\phi}{du}\frac{\partial u}{\partial y} = \frac{d\phi}{du}$$

in 4') we obtain

$$\frac{d^2\phi}{du^2}(Am^2 + Bm + C) = 0.$$

Since ϕ is arbitrary, $d^2\phi/du^2$ is not zero identically; hence, m is one of the roots $m = m_1, m_2$ of $Am^2 + Bm + C = 0$. If $m_1 \neq m_2$, $z = \phi_1(y + m_1x)$ and $z = \phi_2(y + m_2x)$ are distinct solution of 4'). Clearly,

$$z = \phi_1(y + m_1x) + \phi_2(y + m_2x)$$

is also a solution; it contains two arbitrary functions and is the *general solution*.

More generally, if

6) $$f(D_x, D_y)z = (D_x - m_1D_y)(D_x - m_2D_y)\cdots\cdots\cdot(D_x - m_nD_y)z = 0$$

and if $m_1 \neq m_2 \neq \cdots\cdots\cdot \neq m_n$, then

7) $$z = \phi_1(y + m_1x) + \phi_2(y + m_2x) + \cdots\cdots\cdot + \phi_n(y + m_nx)$$

is the general solution of $f(D_x, D_y)z = 0$.

EXAMPLE 1. Solve $(D_x^2 - D_xD_y - 6D_y^2)z = (D_x + 2D_y)(D_x - 3D_y)z = 0$.
Here, $m_1 = -2$, $m_2 = 3$, and the general solution is $y = \phi_1(y - 2x) + \phi_2(y + 3x)$.

See also Problems 1-2.

If $m_1 = m_2 = \cdots\cdot = m_k \neq m_{k+1} \neq \cdots\cdot \neq m_n$, so that 6) becomes

6') $$f(D_x, D_y)z = (D_x - m_1D_y)^k(D_x - m_{k+1}D_y)\cdots\cdot(D_x - m_nD_y)z = 0,$$

the part of the general solution given by the corresponding k equal factors is

$$\phi_1(y + m_1x) + x\phi_2(y + m_1x) + x^2\phi_3(y + m_1x) + \cdots\cdot + x^{k-1}\phi_k(y + m_1x),$$

and the general solution of 6') is

$$z = \phi_1(y + m_1x) + x\phi_2(y + m_1x) + \cdots\cdot + x^{k-1}\phi_k(y + m_1x) + \phi_{k+1}(y + m_{k+1}x) + \cdots\cdot + \phi_n(y + m_nx),$$

where $\phi_1, \phi_2, \cdots\cdot\phi_n$ are arbitrary functions.

EXAMPLE 2. Solve $(D_x^3 - D_x^2D_y - 8D_xD_y^2 + 12D_y^3)z = (D_x - 2D_y)^2(D_x + 3D_y)z = 0$.
Here, $m_1 = m_2 = 2$, $m_3 = -3$ and the general solution is $z = \phi_1(y + 2x) + x\phi_2(y + 2x) + \phi_3(y - 3x)$.

See also Problems 3-4.

If one of the numbers, say m_1, of 6) is imaginary then another, say m_2, is the conjugate of m_1. Let $m_1 = a + bi$ and $m_2 = a - bi$ so that 6) becomes

6'') $$f(D_x, D_y)z = [D_x - (a+bi)D_y][D_x - (a-bi)D_y](D_x - m_3D_y)\cdots\cdots(D_x - m_nD_y)z = 0.$$

The part of the general solution given by the first two factors is

$$\phi_1(y + ax + ibx) + \phi_1(y + ax - ibx) + i[\phi_2(y + ax + ibx) - \phi_2(y + ax - ibx)],$$

(ϕ_1, ϕ_2 arbitrary, real functions), and the general solution of 6'') is

$$z = \phi_1(y + ax + ibx) + \phi_1(y + ax - ibx) + i[\phi_2(y + ax + ibx) - \phi_2(y + ax - ibx)] + \phi_3(y + m_3x) + \cdots\cdots\cdots + \phi_n(y + m_nx).$$

EXAMPLE 3. Solve $$(D_x^4 - D_x^3D_y + 2D_x^2D_y^2 - 5D_xD_y^3 + 3D_y^4)z = (D_x - D_y)^2[D_x + \tfrac{1}{2}(1 + i\sqrt{11})D_y][D_x + \tfrac{1}{2}(1 - i\sqrt{11})D_y]z = 0.$$

Here, $m_1 = m_2 = 1$, $m_3 = -\frac{1}{2}(1 + i\sqrt{11})$, $m_4 = -\frac{1}{2}(1 - i\sqrt{11})$, and the general solution is

$$z = \phi_1(y+x) + x\phi_2(y+x) + \phi_3[y - \tfrac{1}{2}(1 + i\sqrt{11})x] + \phi_3[y - \tfrac{1}{2}(1 - i\sqrt{11})x] + i[\phi_4\{y - \tfrac{1}{2}(1 + i\sqrt{11})x\} - \phi_4\{y - \tfrac{1}{2}(1 - i\sqrt{11})x\}].$$

See also Problem 5.

The general solution of

5') $$f(D_x, D_y)z = (AD_x^2 + BD_xD_y + CD_y^2)z = x + 2y$$

consists of the general solution of the reduced equation

4') $$f(D_x, D_y)z = (AD_x^2 + BD_xD_y + CD_y^2)z = 0$$

plus any particular integral of 5'). We shall speak of the general solution of 4') as the *complementary function* of 5').

In setting up procedures for obtaining a particular integral of

8) $$f(D_x, D_y)z = (D_x - m_1D_y)(D_x - m_2D_y)\cdots\cdots(D_x - m_nD_y)z = F(x, y),$$

we define the operator $\dfrac{1}{f(D_x, D_y)}$ by the identity

$$f(D_x, D_y)\frac{1}{f(D_x, D_y)}F(x, y) = F(x, y).$$

The particular integral, denoted by

$$z = \frac{1}{f(D_x, D_y)}F(x, y) = \frac{1}{(D_x - m_1D_y)(D_x - m_2D_y)\cdots\cdots(D_x - m_nD_y)}F(x, y),$$

may be found, as in Chapter 13, by solving n equations of the first order

9) $$u_1 = \frac{1}{D_x - m_nD_y}F(x, y), \quad u_2 = \frac{1}{D_x - m_{n-1}D_y}u_1, \quad \cdots\cdot, \quad z = u_n = \frac{1}{D_x - m_1D_y}u_{n-1}.$$

Note that each of the equations of 9) is of the form

10) $$p - mq = g(x,y)$$

and that we need only a solution, the simpler the better. In Problem 6 below, the following rule is established for obtaining one such solution of 10): Evaluate $z = \int g(x, a-mx)\,dx$, omitting the arbitrary constant of integration, and then replace a by $y+mx$.

EXAMPLE 4. Solve $(D_x^2 - D_xD_y - 6D_y^2)z = (D_x + 2D_y)(D_x - 3D_y)z = x + y.$

From Example 1, the complementary function is $z = \phi_1(y-2x) + \phi_2(y+3x)$.

To obtain the particular integral denoted by $z = \dfrac{1}{D_x + 2D_y}\left[\dfrac{1}{D_x - 3D_y}(x+y)\right]$:

a) Set $u = \dfrac{1}{D_x - 3D_y}(x+y)$ and obtain a particular integral of $(D_x - 3D_y)u = x+y$.

Using the procedure of Problem 6, we have $u = \int (x + a - 3x)\,dx = ax - x^2$ and, replacing a by $y+3x$, $u = xy + 2x^2$.

b) Set $z = \dfrac{1}{D_x + 2D_y}u = \dfrac{1}{D_x + 2D_y}(xy + 2x^2)$ and obtain a particular integral of

$$(D_x + 2D_y)z = xy + 2x^2.$$

Then $z = \int [x(a+2x) + 2x^2]\,dx = \frac{1}{2}ax^2 + \frac{4}{3}x^3$ and, replacing a by $y-2x$, $z = \frac{1}{2}x^2y + \frac{1}{3}x^3$.

Thus the general solution is $z = \phi_1(y-2x) + \phi_2(y+3x) + \frac{1}{2}x^2y + \frac{1}{3}x^3$.

See also Problems 8-9.

The method of undetermined coefficients may be used if $F(x,y)$ involves $\sin(ax+by)$ or $\cos(ax+by)$.

EXAMPLE 5. Solve

$$(D_x^2 + 5D_xD_y + 5D_y^2)z = [D_x - \tfrac{1}{2}(-5+\sqrt{5})D_y][D_x - \tfrac{1}{2}(-5-\sqrt{5})D_y]z = x\sin(3x-2y).$$

The complementary function is $z = \phi_1[y + \frac{1}{2}(-5+\sqrt{5})x] + \phi_2[y + \frac{1}{2}(-5-\sqrt{5})x]$.

Take as a particular integral

$$z = Ax\sin(3x-2y) + Bx\cos(3x-2y) + C\sin(3x-2y) + D\cos(3x-2y).$$ Then

$$D_x^2z = (6A-9D)\cos(3x-2y) - (6B+9C)\sin(3x-2y) - 9Ax\sin(3x-2y) - 9Bx\cos(3x-2y),$$

$$D_xD_yz = (-2A+6D)\cos(3x-2y) + (2B+6C)\sin(3x-2y) + 6Ax\sin(3x-2y) + 6Bx\cos(3x-2y),$$

$$D_y^2z = -4D\cos(3x-2y) - 4C\sin(3x-2y) - 4Ax\sin(3x-2y) - 4Bx\cos(3x-2y),$$

and $$(D_x^2 + 5D_xD_y + 5D_y^2)z = Ax\sin(3x-2y) + Bx\cos(3x-2y) + (C+4B)\sin(3x-2y) + (D-4A)\cos(3x-2y) = x\sin(3x-2y).$$

Then $A = 1$, $B = C = 0$, $D = 4$ and the particular integral is

$$z = x\sin(3x-2y) + 4\cos(3x-2y).$$ The general solution is

$$z = \phi_1[y + \tfrac{1}{2}(-5+\sqrt{5})x] + \phi_2[y + \tfrac{1}{2}(-5-\sqrt{5})x] + x\sin(3x-2y) + 4\cos(3x-2y).$$

See also Problem 10.

Short methods for obtaining particular integrals, analogous to those of Chapter 16, may be used.

a) $$\frac{1}{f(D_x, D_y)} e^{ax+by} = \frac{1}{f(a,b)} e^{ax+by}, \quad \text{provided } f(a,b) \neq 0.$$

If $f(a,b) = 0$, write $f(D_x, D_y) = (D_x - \frac{a}{b}D_y)^r g(D_x, D_y)$, where $g(a,b) \neq 0$; then

$$\frac{1}{(D_x - \frac{a}{b}D_y)^r} \frac{1}{g(D_x, D_y)} e^{ax+by} = \frac{1}{g(a,b)} \frac{1}{(D_x - \frac{a}{b}D_y)^r} e^{ax+by} = \frac{1}{g(a,b)} \frac{x^r}{r!} e^{ax+by}.$$

b) $$\frac{1}{f(D_x^2, D_x D_y, D_y^2)} \sin(ax+by) = \frac{1}{f(-a^2, -ab, -b^2)} \sin(ax+by) \quad \text{and}$$

$$\frac{1}{f(D_x^2, D_x D_y, D_y^2)} \cos(ax+by) = \frac{1}{f(-a^2, -ab, -b^2)} \cos(ax+by),$$

provided $f(-a^2, -ab, -b^2) \neq 0$.

EXAMPLE 6. Solve $(D_x^2 - 3D_xD_y + 2D_y^2)z = (D_x - D_y)(D_x - 2D_y)z = e^{2x+3y} + e^{x+y} + \sin(x-2y)$.

The complementary function is $z = \phi_1(y+x) + \phi_2(y+2x)$.

Now $\dfrac{1}{D_x^2 - 3D_xD_y + 2D_y^2} e^{2x+3y} = \dfrac{1}{2^2 - 3\cdot 2\cdot 3 + 2\cdot 3^2} e^{2x+3y} = \dfrac{1}{4} e^{2x+3y}$ is one term of the particular integral. Since $\phi_1(y+x)$ includes e^{x+y}, we write

$$\frac{1}{D_x^2 - 3D_xD_y + 2D_y^2} e^{x+y} = \frac{1}{D_x - D_y}\left(\frac{1}{D_x - 2D_y} e^{x+y}\right) = \frac{1}{D_x - D_y}\left(\frac{1}{1 - 2\cdot 1} e^{x+y}\right) = -\frac{1}{D_x - D_y} e^{x+y} = -xe^{x+y}.$$

Also, $$\frac{1}{D_x^2 - 3D_xD_y + 2D_y^2} \sin(x-2y) = \frac{1}{-1 - 3(2) + 2(-1)(-2)^2} \sin(x-2y) = -\frac{1}{15} \sin(x-2y).$$

Thus, the general solution is $z = \phi_1(y+x) + \phi_2(y+2x) + \frac{1}{4}e^{2x+3y} - xe^{x+y} - \frac{1}{15}\sin(x-2y)$.

c) If $F(x,y)$ is a polynomial, that is, $F(x,y) = \Sigma p_{ij} x^i y^j$, where i, j are positive integers or zero and p_{ij} are constants, the procedure illustrated below may be used.

EXAMPLE 7. Solve $(D_x^2 - D_xD_y - 6D_y^2)z = x+y$. (Example 4.)

For a particular integral, write

$$\frac{1}{D_x^2 - D_xD_y - 6D_y^2}(x+y) = \frac{1}{D_x^2} \frac{1}{1 - \frac{D_y}{D_x} - 6\frac{D_y^2}{D_x^2}}(x+y) = \frac{1}{D_x^2}\{[1 + \frac{D_y}{D_x} + \cdots](x+y)\} = \frac{1}{D_x^2}(x + y + \frac{1}{D_x})$$

$$= \frac{1}{D_x^2}(x+y+x) = \frac{1}{D_x^2}(2x+y) = \frac{1}{3}x^3 + \frac{1}{2}x^2 y.$$ Note that $D_y(x+y) = 1$ and $\frac{1}{D_x} = \int dx$.

See also Problems 11-13.

SOLVED PROBLEMS

1. Solve $(D_x^3 + 2D_x^2D_y - D_xD_y^2 - 2D_y^3)z = (D_x - D_y)(D_x + D_y)(D_x + 2D_y)z = 0.$

Here $m_1 = 1,\ m_2 = -1,\ m_3 = -2$ and the general solution is

$$z = \phi_1(y+x) + \phi_2(y-x) + \phi_3(y-2x).$$

2. Solve $(D_x^3 - 5D_x^2D_y + 5D_xD_y^2 + 3D_y^3)z = (D_x - 3D_y)[D_x - (1+\sqrt{2})D_y][D_x - (1-\sqrt{2})D_y]z = 0.$

Here $m_1 = 3,\ m_2 = 1+\sqrt{2},\ m_3 = 1-\sqrt{2}$ and the general solution is

$$z = \phi_1(y+3x) + \phi_2[y + (1+\sqrt{2})x] + \phi_3[y + (1-\sqrt{2})x].$$

3. Solve $(D_x^3 + 3D_x^2D_y - 4D_y^3)z = (D_x - D_y)(D_x + 2D_y)^2 z = 0.$

Since $m_1 = 1,\ m_2 = m_3 = -2$, the general solution is

$$z = \phi_1(y+x) + \phi_2(y-2x) + x\,\phi_3(y-2x).$$

Another form of the general solution is $z = \phi_1(y+x) + \phi_2(y-2x) + y\,\phi_3(y-2x).$

4. Solve $(D_x^4 - 2D_x^2D_y^2 + D_y^4)z = (D_x - D_y)^2(D_x + D_y)^2 z = 0.$

Here $m_1 = m_2 = 1,\ m_3 = m_4 = -1$ and the general solution is

$$z = \phi_1(y+x) + x\,\phi_2(y+x) + \phi_3(y-x) + x\,\phi_4(y-x).$$

5. Solve $(D_x^2 - 2D_xD_y + 5D_y^2)z = [D_x - (1+2i)D_y][D_x - (1-2i)D_y]z = 0.$

Since $m_1 = 1+2i,\ m_2 = 1-2i$, the general solution is

$$z = \phi_1(y+x+2ix) + \phi_1(y+x-2ix) + i[\phi_2(y+x+2ix) - \phi_2(y+x-2ix)],$$

where ϕ_1, ϕ_2 are real functions.

If we take $\phi_1(u) = \cos u$ and $\phi_2(u) = e^u$, then since

$$e^{ibx} = \cos bx + i\sin bx, \qquad \sin bx = \frac{1}{2i}(e^{ibx} - e^{-ibx}),$$

$$e^{-ibx} = \cos bx - i\sin bx, \qquad \cos bx = \frac{1}{2}(e^{ibx} + e^{-ibx}),$$

$$\begin{aligned}\phi_1(y+x+2ix) &= \cos(y+x)\cos(2ix) - \sin(y+x)\sin(2ix)\\ &= \cos(y+x)\cosh 2x - i\sin(y+x)\sinh 2x,\end{aligned}$$

$$\begin{aligned}\phi_1(y+x-2ix) &= \cos(y+x)\cos(2ix) + \sin(y+x)\sin(2ix)\\ &= \cos(y+x)\cosh 2x + i\sin(y+x)\sinh 2x,\end{aligned}$$

$$\phi_2(y+x+2ix) - \phi_2(y+x-2ix) = e^{y+x+2ix} - e^{y+x-2ix} = e^{y+x}(e^{2ix} - e^{-2ix}) = 2ie^{y+x}\sin 2x.$$

Thus, we obtain as a particular integral

$$\begin{aligned}z &= [\cos(y+x)\cosh 2x - i\sin(y+x)\sinh 2x] + [\cos(y+x)\cosh 2x + i\sin(y+x)\sinh 2x]\\ &\quad + i(2ie^{y+x}\sin 2x) = 2\cos(y+x)\cosh 2x - 2e^{y+x}\sin 2x.\end{aligned}$$

Note that z is a real function of x and y.

6. Show that a particular integral of $p - mq = g(x,y)$ may be found by integrating $dz = g(x,a-mx)dx$, omitting the arbitrary constant of integration, and then replacing a by $y + mx$.

The auxiliary system is $\dfrac{dx}{1} = \dfrac{dy}{-m} = \dfrac{dz}{g(x,y)}$. Integrating the equation formed with the first two terms, we have $y + mx = a$. Using this relation, the equation

$$\frac{dx}{1} = \frac{dz}{g(x,y)} \quad \text{becomes} \quad \frac{dx}{1} = \frac{dz}{g(x,a-mx)}.$$

Then $z = \int g(x,a-mx)dx$ and, in order that no arbitrary constants be involved, we replace a by $y + mx$ in the solution.

7. Using the procedure of Problem 6, find particular integrals of

$$a)\ p + 3q = \cos(2x+y), \qquad b)\ p - 2q = (y+1)e^{3x}.$$

$a)$ Here $m = -3$ and $g(x,y) = \cos(2x+y)$.

Then $z = \int g(x,a-mx)dx = \int \cos(2x+a+3x)dx = \frac{1}{5}\sin(5x+a)$ and, replacing a by $y-3x$, the required particular integral is $z = \frac{1}{5}\sin(2x+y)$.

$b)$ $z = \int g(x,a-mx)dx = \int (a-2x+1)e^{3x}\,dx = \frac{1}{3}(a+1)e^{3x} - \frac{2}{3}xe^{3x} + \frac{2}{9}e^{3x}$.

Replacing a by $y+2x$, we have $z = \frac{1}{3}(y+2x+1)e^{3x} - \frac{2}{3}xe^{3x} + \frac{2}{9}e^{3x} = \frac{1}{3}(y+\frac{5}{3})e^{3x}$.

8. Solve $(D_x^2 + 2D_xD_y - 8D_y^2)z = (D_x - 2D_y)(D_x + 4D_y)z = \sqrt{2x+3y}$.

The complementary function is $z = \phi_1(y+2x) + \phi_2(y-4x)$.

To obtain the particular integral denoted by $\dfrac{1}{(D_x-2D_y)(D_x+4D_y)}\sqrt{2x+3y}$, we obtain from $(D_x+4D_y)u = \sqrt{2x+3y}$ the solution

$$u = \int [2x+3(a-mx)]^{1/2}\,dx = \int [2x+3(a+4x)]^{1/2}\,dx$$
$$= \int (14x+3a)^{1/2}\,dx = \frac{1}{21}(14x+3a)^{3/2} = \frac{1}{21}(2x+3y)^{3/2}$$

and from $(D_x-2D_y)z = u = \frac{1}{21}(2x+3y)^{3/2}$, the solution

$$z = \frac{1}{21}\int [(2x+3(a-2x)]^{3/2}dx = -\frac{1}{210}(3a-4x)^{5/2} = -\frac{1}{210}(2x+3y)^{5/2}.$$

The general solution is $z = \phi_1(y+2x) + \phi_2(y-4x) - \frac{1}{210}(2x+3y)^{5/2}$.

9. Solve $(D_x - 2D_y)^2(D_x + 3D_y)z = e^{2x+y}$.

The complementary function is $z = \phi_1(y+2x) + x\phi_2(y+2x) + \phi_3(y-3x)$.

To obtain the particular integral denoted by $\dfrac{1}{(D_x-2D_y)(D_x-2D_y)(D_x+3D_y)}e^{2x+y}$, we obtain from $(D_x+3D_y)u = e^{2x+y}$ the solution $u = \int e^{2x+(a+3x)}dx = \int e^{5x+a}\,dx = \frac{1}{5}e^{5x+a} = \frac{1}{5}e^{2x+y}$;

from $(D_x - 2D_y)v = u = \frac{1}{5}e^{2x+y}$ the solution $v = \frac{1}{5}\int e^{2x+(a-2x)}\,dx = \frac{1}{5}x e^{a} = \frac{1}{5}x e^{2x+y}$;

and from $(D_x - 2D_y)z = v = \frac{1}{5}x e^{2x+y}$ the solution $z = \frac{1}{5}\int x e^{a}\,dx = \frac{1}{10}x^2 e^{a} = \frac{1}{10}x^2 e^{2x+y}$.

The general solution is $z = \phi_1(y+2x) + x\,\phi_2(y+2x) + \phi_3(y-3x) + \frac{1}{10}x^2 e^{2x+y}$.

10. Solve $(D_x^3 + D_x^2D_y - D_xD_y^2 - D_y^3)z = (D_x + D_y)^2(D_x - D_y)z = e^x \cos 2y$.

The complementary function is $z = \phi_1(y-x) + x\phi_2(y-x) + \phi_3(y+x)$.

Take as a particular integral $z = Ae^x \cos 2y + Be^x \sin 2y$. Then

$D_x^3 z = Ae^x \cos 2y + Be^x \sin 2y$, $\quad D_xD_y^2 z = -4Ae^x \cos 2y - 4Be^x \sin 2y$,

$D_x^2D_y z = -2Ae^x \sin 2y + 2Be^x \cos 2y$, $\quad D_y^3 z = 8Ae^x \sin 2y - 8Be^x \cos 2y$.

Substituting in the given differential equation, we have

$(5A + 10B)e^x \cos 2y + (5B - 10A)e^x \sin 2y = e^x \cos 2y$, so that $A = 1/25$ and $B = 2/25$.

The particular integral is $z = \frac{1}{25}e^x \cos 2y + \frac{2}{25}e^x \sin 2y$, and the general solution is $z = \phi_1(y-x) + x\phi_2(y-x) + \phi_3(y+x) + \frac{1}{25}e^x \cos 2y + \frac{2}{25}e^x \sin 2y$.

11. Solve $(D_x^2 - 2D_xD_y)z = D_x(D_x - 2D_y)z = e^{2x} + x^3y$.

The complementary function is $z = \phi_1(y) + \phi_2(y+2x)$.

A particular integral is given by $\frac{1}{D_x^2 - 2D_xD_y}e^{2x} + \frac{1}{D_x^2 - 2D_xD_y}x^3y$. The first term yields $\frac{1}{(2)^2 - 2(2)(0)}e^{2x} = \frac{1}{4}e^{2x}$. Writing the second term

$$\frac{1}{D_x^2}\,\frac{1}{1-2\frac{D_y}{D_x}}\,x^3y = \frac{1}{D_x^2}\left(1 + 2\frac{D_y}{D_x} + \cdots\cdot\right)x^3y = \frac{1}{D_x^2}\left(x^3y + \frac{2}{D_x}x^3\right) = \frac{1}{D_x^2}\left(x^3y + \frac{1}{2}x^4\right),$$

we obtain $\frac{x^5y}{20} + \frac{x^6}{60}$. The general solution is $z = \phi_1(y) + \phi_2(y+2x) + \frac{1}{4}e^{2x} + \frac{x^5y}{20} + \frac{x^6}{60}$.

12. Solve $(D_x^3 - 7D_xD_y^2 - 6D_y^3)z = (D_x + D_y)(D_x + 2D_y)(D_x - 3D_y)z = \sin(x+2y) + e^{3x+y}$.

The complementary function is $z = \phi_1(y-x) + \phi_2(y-2x) + \phi_3(y+3x)$. A particular integral is given by $\frac{1}{(D_x + D_y)(D_x^2 - D_xD_y - 6D_y^2)}\sin(x+2y) + \frac{1}{(D_x - 3D_y)(D_x^2 + 3D_xD_y + 2D_y^2)}e^{3x+y}$.

(Note. The separation in the first term is one of convenience, i.e., we could have written $\frac{1}{(D_x + 2D_y)(D_x^2 - 2D_xD_y - 3D_y^2)}\sin(x+2y)$. The separation in the second term is necessary, however, since e^{3x+y} is part of the term $\phi_3(y+3x)$ of the complementary function.)

For the first term: $\dfrac{1}{(D_x+D_y)(D_x^2-D_xD_y-6D_y^2)}\sin(x+2y) = \dfrac{1}{D_x+D_y}\,\dfrac{1}{-1+2+24}\sin(x+2y)$

$$= \frac{1}{25}\frac{D_x-D_y}{D_x^2-D_y^2}\sin(x+2y) = \frac{1}{25(3)}(D_x-D_y)\sin(x+2y) = -\frac{1}{75}\cos(x+2y).$$

For the second term: $\dfrac{1}{(D_x-3D_y)(D_x^2+3D_xD_y+2D_y^2)}e^{3x+y} = \dfrac{1}{D_x-3D_y}\,\dfrac{e^{3x+y}}{9+9+2}$

$$= \frac{1}{20}\frac{1}{D_x-3D_y}e^{3x+y} = \frac{1}{20}xe^{3x+y}.$$

The general solution is $z = \phi_1(y-x) + \phi_2(y-2x) + \phi_3(y+3x) - \dfrac{1}{75}\cos(x+2y) + \dfrac{1}{20}xe^{3x+y}$.

13. Solve $(D_x^3 - 7D_xD_y^2 - 6D_y^3)z = \cos(x-y) + x^2 + xy^2 + y^3$.

The reduced equation is that of Problem 12. A particular integral is given by

$$\frac{1}{(D_x+D_y)(D_x^2-D_xD_y-6D_y^2)}\cos(x-y) + \frac{1}{D_x^3-7D_xD_y^2-6D_y^3}(x^2+xy^2+y^3).$$

(Note that $\cos(x-y)$ is part of the complementary function; hence, the corresponding factor (D_x+D_y) must be treated separately.)

For the first term: $\dfrac{1}{(D_x+D_y)(D_x^2-D_xD_y-6D_y^2)}\cos(x-y) = \dfrac{1}{4}\,\dfrac{1}{D_x+D_y}\cos(x-y)$. We must solve $(D_x+D_y)u = \frac{1}{4}\cos(x-y)$, obtaining $u = \frac{1}{4}\int\cos[x-(a+x)]\,dx = \frac{1}{4}\int\cos(-a)\,dx$

$$= \frac{1}{4}x\cos(-a) = \frac{1}{4}x\cos(x-y).$$

For the second term: $\dfrac{1}{D_x^3-7D_xD_y^2-6D_y^3}(x^2+xy^2+y^3) = \dfrac{1}{D_x^3\left(1-7\dfrac{D_y^2}{D_x^2}-6\dfrac{D_y^3}{D_x^3}\right)}(x^2+xy^2+y^3)$

$$= \frac{1}{D_x^3}\left(1+7\frac{D_y^2}{D_x^2}+6\frac{D_y^3}{D_x^3}\right)(x^2+xy^2+y^3) = \frac{1}{D_x^3}\left[x^2+xy^2+y^3+\frac{7}{D_x^2}(2x+6y)+\frac{6}{D_x^3}(6)\right]$$

$$= \frac{1}{D_x^3}(x^2+xy^2+y^3) + \frac{7}{D_x^5}(2x+6y) + \frac{36}{D_x^6} = \frac{5}{72}x^6 + \frac{1}{60}x^5(1+21y) + \frac{1}{24}x^4y^2 + \frac{1}{6}x^3y^3.$$

The general solution is

$$z = \phi_1(y-x) + \phi_2(y-2x) + \phi_3(y+3x) + \frac{1}{4}x\cos(x-y) + \frac{5}{72}x^6 + \frac{1}{60}x^5(1+21y) + \frac{1}{24}x^4y^2 + \frac{1}{6}x^3y^3.$$

SUPPLEMENTARY PROBLEMS

Solve each of the following equations.

14. $(D_x^2 - 8D_xD_y + 15D_y^2)z = 0.$ *Ans.* $z = \phi_1(y+3x) + \phi_2(y+5x)$

15. $(D_x^2 - 2D_xD_y - D_y^2)z = 0.$ *Ans.* $z = \phi_1[y + x(1+\sqrt{2})] + \phi_2[y + x(1-\sqrt{2})]$

16. $(D_x^2 - 4D_xD_y + 4D_y^2)z = 0.$ *Ans.* $z = \phi_1(y+2x) + x\phi_2(y+2x)$

17. $(D_x^3 + 2D_x^2D_y - D_xD_y^2 - 2D_y^3)z = 0.$ *Ans.* $z = \phi_1(y+x) + \phi_2(y-x) + \phi_3(y-2x)$

18. $(D_x^3D_y^2 + D_x^2D_y^3)z = 0.$ *Ans.* $z = \phi_1(y) + x\phi_2(y) + \phi_3(x) + y\phi_4(x) + \phi_5(y-x)$

19. $(D_x^2 + 5D_xD_y + 6D_y^2)z = e^{x-y}.$ *Ans.* $z = \phi_1(y-2x) + \phi_2(y-3x) + \dfrac{1}{2}e^{x-y}$

20. $(D_x^2 + D_y^2)z = x^2y^2.$

    Ans. $z = \phi_1(y+ix) + \phi_1(y-ix) + i[\phi_2(y+ix) - \phi_2(y-ix)] + \dfrac{1}{180}(15x^4y - x^6)$

21. $(D_x^3 - 3D_x^2D_y + 4D_y^3)z = e^{y+2x}.$ *Ans.* $z = \phi_1(y-x) + \phi_2(y+2x) + x\phi_3(y+2x) + \dfrac{1}{6}x^2e^{y+2x}$

22. $(D_x^3 + 2D_x^2D_y - D_xD_y^2 - 2D_y^3)z = (y+2)e^x.$ *Ans.* $z = \phi_1(y+x) + \phi_2(y-x) + \phi_3(y-2x) + ye^x$

23. $(D_x^3 - 3D_x^2D_y - 4D_xD_y^2 + 12D_y^3)z = \sin(y+2x).$

    Ans. $z = \phi_1(y-2x) + \phi_2(y+2x) + \phi_3(y+3x) + \dfrac{1}{4}x\sin(y+2x)$

24. $(D_x^3 - 3D_xD_y^2 + 2D_y^3)z = \sqrt{x+2y}.$ *Ans.* $z = \phi_1(y+x) + x\phi_2(y+x) + \phi_3(y-2x) + \dfrac{8}{525}(x+2y)^{7/2}$

25. $(D_x^3 + D_x^2D_y - 6D_xD_y^2)z = x^2 + y^2.$

    Ans. $z = \phi_1(y) + \phi_2(y+2x) + \phi_3(y-3x) + \dfrac{2}{15}x^5 - \dfrac{1}{12}x^4y + \dfrac{1}{6}x^3y^2$

26. $(D_x^3 - 4D_x^2D_y + 5D_xD_y^2 - 2D_y^3)z = e^{y+x} + e^{y-2x} + e^{y+2x}.$

    Ans. $z = \phi_1(y+x) + x\phi_2(y+x) + \phi_3(y+2x) - \dfrac{1}{2}x^2e^{y+x} - \dfrac{1}{36}e^{y-2x} + xe^{y+2x}$

27. $(D_x^3 - 2D_x^2D_y)z = 2e^{2x} + 3x^2y.$

    Ans. $z = \phi_1(y) + x\phi_2(y) + \phi_3(y+2x) + \dfrac{1}{4}e^{2x} + \dfrac{1}{20}x^5y + \dfrac{1}{60}x^6$

28. $(D_x^3 - 3D_xD_y^2 - 2D_y^3)z = \cos(x+2y) - e^y(3+2x).$

    Ans. $z = \phi_1(y-x) + x\phi_2(y-x) + \phi_3(y+2x) + \dfrac{1}{27}\sin(x+2y) + xe^y$

CHAPTER 32

Non-homogenous Linear Equations with Constant Coefficients

A NON-HOMOGENEOUS LINEAR partial differential equation with constant coefficients such as

$$f(D_x, D_y)z = (D_x^2 - D_y^2 + 3D_x + D_y + 2)z = (D_x + D_y + 1)(D_x - D_y + 2)z = x^2 + xy$$

is called *reducible*, since the left member can be resolved into factors each of which is of the first degree in D_x, D_y, while

$$f(D_x, D_y)z = (D_x D_y + 2D_y^3)z = D_y(D_x + 2D_y^2)z = \cos(x-2y),$$

which cannot be so resolved, is called *irreducible*.

REDUCIBLE NON-HOMOGENEOUS EQUATIONS. Consider the reducible non-homogeneous equation

1) $f(D_x, D_y)z = (a_1D_x + b_1D_y + c_1)(a_2D_x + b_2D_y + c_2)\cdots\cdot(a_nD_x + b_nD_y + c_n)z = 0,$

where the a_i, b_i, c_i are constants. Any solution of

2) $$(a_iD_x + b_iD_y + c_i)z = 0$$

is a solution of 1). From Problem 5, Chapter 29, the general solution of 2) is

3) $$z = e^{-c_ix/a_i}\,\phi(a_iy - b_ix), \qquad a_i \neq 0,$$

or

3') $$z = e^{-c_iy/b_i}\,\psi(a_iy - b_ix), \qquad b_i \neq 0,$$

with ϕ and ψ arbitrary functions of their argument. Thus, if no two factors of 1) are linearly dependent (that is, if no factor is a mere multiple of another), the general solution of 1) consists of the sum of n arbitrary functions of the types 3) and 3').

EXAMPLE 1. Solve $(2D_x + D_y + 1)(D_x - 3D_y + 2)z = 0.$

The general solution is $z = e^{-y}\phi_1(2y-x) + e^{-2x}\phi_2(y+3x)$. Note that the first term on the right may be replaced by $e^{-x/2}\psi_1(2y-x)$ and the second by $e^{2y/3}\psi_2(y+3x)$.

EXAMPLE 2. Solve $(2D_x + 3D_y - 5)(D_x + 2D_y)(D_x - 2)(D_y + 2)z = 0.$

The general solution is $z = e^{5x/2}\phi_1(2y-3x) + \phi_2(y-2x) + e^{2x}\phi_3(y) + e^{-2y}\phi_4(x).$

See also Problems 1-2.

If

4) $f(D_x, D_y)z = (a_1D_x + b_1D_y + c_1)^k(a_{k+1}D_x + b_{k+1}D_y + c_{k+1})\cdots(a_nD_x + b_nD_y + c_n)z = 0,$

where no two of the n factors are linearly dependent except as indicated, the part of the general solution corresponding to the k repeated factors is

$$e^{-c_1x/a_1}[\phi_1(a_1y-b_1x) + x\phi_2(a_1y-b_1x) + \cdots + x^{k-1}\phi_k(a_1y-b_1x)].$$

EXAMPLE 3. Solve $(2D_x+D_y+5)(D_x-2D_y+1)^2z = 0$.

The general solution is $z = e^{-5y}\phi_1(2y-x) + e^{-x}[\phi_2(y+2x) + x\phi_3(y+2x)]$.

See also Problem 3.

THE GENERAL SOLUTION OF

5) $f(D_x,D_y)z = (a_1D_x+b_1D_y+c_1)(a_2D_x+b_2D_y+c_2)\cdots(a_nD_x+b_nD_y+c_n)z = F(x,y)$

is the sum of the general solution of 1), (now called the complementary function of 5)), and a particular integral of 5),

6) $$z = \frac{1}{f(D_x,D_y)}F(x,y).$$

The general procedure for evaluating 6) as well as short methods applicable to particular forms of $F(x,y)$ are those of the previous chapter.

EXAMPLE 4. Solve $f(D_x,D_y)z = (D_x^2 - D_xD_y - 2D_y^2 + 2D_x - 4D_y)z$
$= (D_x-2D_y)(D_x+D_y+2)z = ye^x + 3xe^{-y}$.

The complementary function is $z = \phi_1(y+2x) + e^{-2x}\phi_2(y-x)$.

To evaluate $\dfrac{1}{f(D_x,D_y)}ye^x = \dfrac{1}{(D_x-2D_y)(D_x+D_y+2)}ye^x$, we first solve $(D_x+D_y+2)u = ye^x$

whose auxiliary system is $\dfrac{dx}{1} = \dfrac{dy}{1} = \dfrac{du}{ye^x-2u}$. We obtain $y = x+a$ readily and the equation $\dfrac{du}{ye^x-2u} = \dfrac{dx}{1}$ or $\dfrac{du}{dx} + 2u = ye^x = (x+a)e^x$. This linear equation has e^{2x} as integrating factor; hence, $ue^{2x} = \int(x+a)e^{3x}dx = \frac{1}{3}xe^{3x} - \frac{1}{9}e^{3x} + \frac{1}{3}ae^{3x} = \frac{1}{3}xe^{3x} - \frac{1}{9}e^{3x} + \frac{1}{3}(y-x)e^{3x}$ and $u = \frac{1}{3}ye^x - \frac{1}{9}e^x$.

We then solve $(D_x-2D_y)z = u = \frac{1}{3}ye^x - \frac{1}{9}e^x$ obtaining the required particular integral (see Problem 6, Chapter 31)

$$z = \int[\tfrac{1}{3}(a-2x)e^x - \tfrac{1}{9}e^x]dx = \tfrac{1}{3}ae^x - \tfrac{2}{3}xe^x + \tfrac{2}{3}e^x - \tfrac{1}{9}e^x$$
$$= \tfrac{1}{3}(y+2x)e^x - \tfrac{2}{3}xe^x + \tfrac{5}{9}e^x = \tfrac{1}{3}(y+\tfrac{5}{3})e^x.$$

To evaluate $\dfrac{1}{(D_x-2D_y)(D_x+D_y+2)}(3xe^{-y})$, we solve $(D_x+D_y+2)u = 3xe^{-y}$ whose auxiliary system is $\dfrac{dx}{1} = \dfrac{dy}{1} = \dfrac{du}{3xe^{-y}-2u}$. Then $y = x+a$, and from $\dfrac{du}{3xe^{-y}-2u} = \dfrac{dy}{1}$ or

$\dfrac{du}{dy} + 2u = 3xe^{-y} = 3(y-a)e^{-y}$, $ue^{2y} = 3\int(y-a)e^{y}\,dy = 3(y-1-a)e^{y} = 3(x-1)e^{y}$ and $u = 3(x-1)e^{-y}$. Solving in turn $(D_x - 2D_y)z = u = 3(x-1)e^{-y}$, the required particular integral is $z = 3\int(x-1)e^{-a+2x}\,dx = \frac{3}{2}(xe^{-a+2x} - \frac{3}{2}e^{-a+2x}) = \frac{3}{2}(x-\frac{3}{2})e^{-y}$.

The general solution is $z = \phi_1(y+2x) + e^{-2x}\phi_2(y-x) + \frac{1}{3}(y+\frac{5}{3})e^{x} + \frac{3}{2}(x-\frac{3}{2})e^{-y}$.

EXAMPLE 5. Solve

$$f(D_x,D_y)z = (D_x^2 - D_xD_y - 2D_y^2 + 6D_x - 9D_y + 5)z = (D_x + D_y + 5)(D_x - 2D_y + 1)z = e^{2x+y} + e^{x+y}.$$

The complementary function is $z = e^{-5x}\phi_1(y-x) + e^{-x}\phi_2(y+2x)$.

For the particular integral corresponding to the first term of $F(x,y)$, we use

$$\frac{1}{f(D_x,D_y)}e^{ax+by} = \frac{1}{f(a,b)}e^{ax+by}, \quad f(a,b) \neq 0,$$

and obtain $\dfrac{1}{D_x^2 - D_xD_y - 2D_y^2 + 6D_x - 9D_y + 5}e^{2x+y} = \dfrac{1}{4-2-2+12-9+5}e^{2x+y} = \dfrac{1}{8}e^{2x+y}$.

In evaluating $\dfrac{1}{f(D_x,D_y)}e^{x+y}$, we note that $f(1,1) = 0$. This means that e^{x+y} is a part of the complementary function. (To see this, take $\phi_2(y+2x) = e^{y+2x} + \psi_2(y+2x)$; then $e^{-x}\phi_2(y+2x) = e^{-x}[e^{y+2x} + \psi_2(y+2x)] = e^{y+x} + e^{-x}\psi_2(y+2x)$.) We write

$$\frac{1}{f(D_x,D_y)}e^{x+y} = \frac{1}{D_x - 2D_y + 1}\,\frac{1}{D_x + D_y + 5}e^{x+y} = \frac{1}{7}\,\frac{1}{D_x - 2D_y + 1}e^{x+y} = \frac{1}{7}xe^{x+y}.$$

The general solution is $z = e^{-5x}\phi_1(y-x) + e^{-x}\phi_2(y+2x) + \frac{1}{8}e^{2x+y} + \frac{1}{7}xe^{x+y}$.

See also Problems 4-5.

The use of the formula

$$7)\qquad \frac{1}{f(D_x,D_y)}Ve^{ax+by} = e^{ax+by}\frac{1}{f(D_x+a, D_y+b)}V, \quad V = V(x,y),$$

is illustrated below.

EXAMPLE 6. Solve $(D_x^3 + 3D_x^2D_y - 2D_x^2)z = D_x^2(D_x + 3D_y - 2)z = (x^2 + 2y)e^{2x+y}$.

The complementary function is $z = \phi_1(y) + x\phi_2(y) + e^{2x}\phi_3(y-3x)$. A particular integral is $z = \dfrac{1}{D_x^2(D_x + 3D_y - 2)}(x^2 + 2y)e^{2x+y} = e^{2x+y}\dfrac{1}{(D_x+2)^2(D_x + 3D_y + 3)}(x^2 + 2y)$.

Setting $(D_x + 3D_y + 3)u = x^2 + 2y$, the auxiliary system is $\dfrac{dx}{1} = \dfrac{dy}{3} = \dfrac{du}{x^2 + 2y - 3u}$.

Then $y = 3x + a$, and from $\dfrac{du}{x^2 + 2y - 3u} = \dfrac{dx}{1}$ or $\dfrac{du}{dx} + 3u = x^2 + 2y$, we have

$$ue^{3x} = \int (x^2 + 6x + 2a)e^{3x}\,dx = e^{3x}(\tfrac{1}{3}x^2 + \tfrac{16}{9}x - \tfrac{16}{27} + \tfrac{2}{3}a) \quad \text{and} \quad u = \tfrac{1}{3}x^2 - \tfrac{2}{9}x - \tfrac{16}{27} + \tfrac{2}{3}y.$$

Next, setting $(D_x + 2)v = u$ and making use of the integrating factor e^{2x}, y being regarded as a constant, $ve^{2x} = \int e^{2x}(\tfrac{1}{3}x^2 - \tfrac{2}{9}x - \tfrac{16}{27} + \tfrac{2}{3}y)\,dx = (\tfrac{1}{6}x^2 - \tfrac{5}{18}x - \tfrac{17}{108} + \tfrac{1}{3}y)e^{2x}$

and $v = \tfrac{1}{6}x^2 - \tfrac{5}{18}x - \tfrac{17}{108} + \tfrac{1}{3}y$.

Finally, setting $(D_x + 2)w = v$, we have

$$we^{2x} = \int e^{2x}(\tfrac{1}{6}x^2 - \tfrac{5}{18}x - \tfrac{17}{108} + \tfrac{1}{3}y)\,dx = (\tfrac{1}{12}x^2 - \tfrac{2}{9}x + \tfrac{7}{216} + \tfrac{1}{6}y)e^{2x}$$

and $w = \tfrac{1}{12}x^2 - \tfrac{2}{9}x + \tfrac{7}{216} + \tfrac{1}{6}y$.

Then $z = we^{2x+y}$ and the general solution is

$$z = \phi_1(y) + x\phi_2(y) + e^{2x}\phi_3(3y - x) + (\tfrac{1}{12}x^2 - \tfrac{2}{9}x + \tfrac{7}{216} + \tfrac{1}{6}y)e^{2x+y}.$$

See also Problems 6-7.

IRREDUCIBLE EQUATIONS WITH CONSTANT COEFFICIENTS. Consider the linear equation with constant coefficients

8) $$f(D_x, D_y)z = 0.$$

Since $D_x^r D_y^s(ce^{ax+by}) = ca^r b^s e^{ax+by}$, where a, b, c are constants, the result of substituting

9) $$z = ce^{ax+by}$$

in 8) is $cf(a,b)e^{ax+by} = 0$. Thus, 9) is a solution of 8) provided

10) $$f(a,b) = 0,$$

with c arbitrary. Now for any chosen value of a (or b) one or more values of b (or a) are obtained by means of 10). Thus, there exist infinitely many pairs of numbers (a_i, b_i) satisfying 10). Moreover,

11) $$z = \sum_{i=1}^{\infty} c_i e^{a_i x + b_i y}, \quad \text{where } f(a_i, b_i) = 0,$$

is a solution of 8).

If $$f(D_x, D_y)z = (D_x + hD_y + k)\,g(D_x, D_y)z,$$

then any pair (a,b) for which $a + hb + k = 0$ satisfies 10). Consider all such pairs $(a_i, b_i) = (-hb_i - k, b_i)$. By 11),

$$z = \sum_{i=1}^{\infty} c_i e^{-(hb_i+k)x + b_i y} = e^{-kx}\sum_{i=1}^{\infty} c_i e^{b_i(y-hx)}$$

is a solution of 8) corresponding to the linear factor $(D_x + hD_y + k)$ of $f(D_x, D_y)$. This is, of course, $e^{-kx}\phi(y-hx)$, ϕ arbitrary, used above. Thus, if $f(D_x, D_y)$ has no linear factor, 11) will be called the solution of 8); however, if $f(D_x, D_y)$ has $m < n$ linear factors, we shall write part of the solution involving arbitrary functions (corresponding to the linear factors) and the remainder involving arbitrary constants.

EXAMPLE 7. Solve $f(D_x, D_y)z = (D_x^2 + D_x + D_y)z = 0$.

The equation is irreducible. Here $f(a,b) = a^2 + a + b = 0$ so that for any $a = a_i$, $b_i = -a_i(a_i + 1)$. Thus the solution is

$$z = \sum_{i=1}^{\infty} c_i e^{a_i x + b_i y} = \sum_{i=1}^{\infty} c_i e^{a_i x - a_i(a_i+1)y}, \quad \text{with } c_i \text{ and } a_i \text{ arbitrary constants.}$$

EXAMPLE 8. Solve $(D_x + 2D_y)(D_x - 2D_y + 1)(D_x - D_y^2)z = 0$.

Corresponding to the linear factors we have $\phi_1(y-2x)$ and $e^{-x}\phi_2(y+2x)$ respectively.

For the irreducible factor $D_x - D_y^2$ we have $a - b^2 = 0$ or $a = b^2$.

The required solution is

$$z = \phi_1(y-2x) + e^{-x}\phi_2(y+2x) + \sum_{i=1}^{\infty} c_i e^{b_i^2 x + b_i y}, \quad \text{with } c_i \text{ and } b_i \text{ arbitrary constants.}$$

In obtaining a particular integral of $f(D_x, D_y)z = F(x,y)$, all procedures used heretofore are available.

EXAMPLE 9. Solve $f(D_x, D_y)z = (D_x - D_y^2)z = e^{2x+3y}$.

From Example 8, the complementary function is $z = \sum_{i=1}^{\infty} c_i e^{b_i^2 x + b_i y}$.

For the particular integral: $\dfrac{1}{D_x - D_y^2} e^{2x+3y} = \dfrac{1}{2-(3)^2} e^{2x+3y} = -\dfrac{1}{7} e^{2x+3y}$.

The required solution is $z = \sum_{i=1}^{\infty} c_i e^{b_i^2 x + b_i y} - \dfrac{1}{7} e^{2x+3y}$

See also Problems 8-11.

THE CAUCHY (ORDINARY) DIFFERENTIAL EQUATION $f(xD)y = F(x)$ is transformed into a linear equation with constant coefficients by means of the substitution $x = e^z$ (see Chapter 17). The analogue in partial differential equations is an equation of the form

$$f(xD_x, yD_y)z = \sum_{r,s} c_{rs} x^r y^s D_x^r D_y^s z = F(x,y), \qquad c_{rs} = \text{constant},$$

which is reduced to a linear partial differential equation with constant coefficients by the substitution

$$x = e^u, \quad y = e^v.$$

EXAMPLE 10. Solve $(x^2D_x^2 + 2xyD_xD_y - xD_x)z = x^3/y^2$.

The substitution $x = e^u$, $y = e^v$, $xD_x z = D_u z$, $yD_y z = D_v z$, $x^2D_x^2 z = D_u(D_u - 1)z$, $xyD_xD_y = D_uD_v z$, $y^2D_y^2 z = D_v(D_v - 1)z$ transforms the given equation into

$$[D_u(D_u - 1) + 2D_uD_v - D_u]z = D_u(D_u + 2D_v - 2)z = e^{3u-2v}$$

whose solution is $z = \phi_1(v) + e^{2u}\phi_2(v - 2u) - \frac{1}{9}e^{3u-2v}$.

Thus, the general solution (expressed in the original variables) is

$$z = \phi_1(\ln y) + x^2\phi_2(\ln \frac{y}{x^2}) - \frac{1}{9}\frac{x^3}{y^2} \quad \text{or} \quad z = \psi_1(y) + x^2\psi_2(\frac{y}{x^2}) - \frac{1}{9}\frac{x^3}{y^2}.$$

See also Problems 12-13.

SOLVED PROBLEMS

REDUCIBLE EQUATIONS.

1. Solve $(D_x^2 - D_y^2 + 3D_x - 3D_y)z = (D_x - D_y)(D_x + D_y + 3)z = 0$.

The general solution is $z = \phi_1(y+x) + e^{-3x}\phi_2(y-x)$.

2. Solve $D_x(2D_x - D_y + 1)(D_x + 2D_y - 1)z = 0$.

The general solution is $z = \phi_1(y) + e^y\phi_2(2y+x) + e^x\phi_3(y-2x)$.

3. Solve $(2D_x + 3D_y - 1)^2(D_x - 3D_y + 3)^3 z = 0$. The general solution is

$$z = e^{\frac{1}{2}x}[\phi_1(2y-3x) + x\phi_2(2y-3x)] + e^y[\phi_3(y+3x) + y\phi_4(y+3x) + y^2\phi_5(y+3x)].$$

4. Solve $(2D_xD_y + D_y^2 - 3D_y)z = D_y(2D_x + D_y - 3)z = 3\cos(3x - 2y)$.

The complementary function is $z = \phi_1(x) + e^{3y}\phi_2(2y-x)$. A particular integral is

$$\frac{1}{2D_xD_y + D_y^2 - 3D_y}3\cos(3x-2y) = \frac{3}{2(6) - 4 - 3D_y}\cos(3x-2y) = \frac{3}{8-3D_y}\cos(3x-2y)$$

$$= \frac{3(8+3D_y)}{64 - 9D_y^2}\cos(3x-2y) = \frac{3}{100}(8+3D_y)\cos(3x-2y) = \frac{3}{50}[4\cos(3x-2y) + 3\sin(3x-2y)].$$

The general solution is $z = \phi_1(x) + e^{3y}\phi_2(2y-x) + \frac{3}{50}[4\cos(3x-2y) + 3\sin(3x-2y)]$.

5. Solve $D_x(D_x + D_y - 1)(D_x + 3D_y - 2)z = x^2 - 4xy + 2y^2$.

The complementary function is $z = \phi_1(y) + e^x\phi_2(y-x) + e^{2x}\phi_3(y-3x)$.

A particular integral is denoted by $z = \dfrac{1}{D_x(D_x + D_y - 1)(D_x + 3D_y - 2)}(x^2 - 4xy + 2y^2)$.

To evaluate it, consider $\frac{1}{D_x + 3D_y - 2}(x^2 - 4xy + 2y^2) = \frac{1}{2}\frac{1}{-1 + \frac{1}{2}(D_x + 3D_y)}(x^2 - 4xy + 2y^2)$

$= \frac{1}{2}[-1 - \frac{1}{2}(D_x + 3D_y) - \frac{1}{4}(D_x + 3D_y)^2 - \cdots](x^2 - 4xy + 2y^2)$

$= \frac{1}{2}[-(x^2 - 4xy + 2y^2) - (-5x + 4y) - 7/2] = -\frac{1}{2}(x^2 - 4xy + 2y^2 - 5x + 4y + 7/2).$

Consider next $\frac{-\frac{1}{2}}{D_x + D_y - 1}(x^2 - 4xy + 2y^2 - 5x + 4y + 7/2) = \frac{1}{2}\frac{1}{1 - (D_x + D_y)}(x^2 - 4xy + 2y^2 - 5x + 4y + 7/2)$

$= \frac{1}{2}[1 + (D_x + D_y) + (D_x + D_y)^2 + \cdots](x^2 - 4xy + 2y^2 - 5x + 4y + 7/2) = \frac{1}{2}(x^2 - 4xy + 2y^2 - 7x + 4y + \frac{1}{2}).$

Finally, $z = \frac{\frac{1}{2}}{D_x}(x^2 - 4xy + 2y^2 - 7x + 4y + \frac{1}{2}) = \frac{1}{2}(x^3/3 - 2x^2y + 2xy^2 - 7x^2/2 + 4xy + x/2).$

The general solution is

$$z = \phi_1(y) + e^x\phi_2(y - x) + e^{2x}\phi_3(y - 3x) + \frac{1}{12}(2x^3 - 12x^2y + 12xy^2 - 21x^2 + 24xy + 3x).$$

TYPE: $\frac{1}{f(D_x, D_y)} e^{ax+by} V(x,y).$

6. Solve $(D_x + D_y - 1)(D_x + D_y - 3)(D_x + D_y)z = e^{x+y+2}\cos(2x - y).$

The complementary function is $z = e^x\phi_1(y - x) + e^{3x}\phi_2(y - x) + \phi_3(y - x).$

For the particular integral, $\frac{1}{(D_x + D_y - 1)(D_x + D_y - 3)(D_x + D_y)} e^{x+y+2}\cos(2x - y)$

$= e^{x+y}\frac{1}{(D_x + D_y + 1)(D_x + D_y - 1)(D_x + D_y + 2)} e^2 \cos(2x - y)$

$= e^{x+y+2}\frac{1}{(D_x^2 + 2D_xD_y + D_y^2 - 1)(D_x + D_y + 2)}\cos(2x - y) = -\frac{1}{2}e^{x+y+2}\frac{1}{D_x + D_y + 2}\cos(2x - y)$

$= -\frac{1}{2}e^{x+y+2}\frac{D_x + D_y - 2}{D_x^2 + 2D_xD_y + D_y^2 - 4}\cos(2x - y) = \frac{1}{10}e^{x+y+2}(D_x + D_y - 2)\cos(2x - y)$

$= -\frac{1}{10}e^{x+y+2}[\sin(2x - y) + 2\cos(2x - y)].$ The general solution is

$$z = e^x\phi_1(y - x) + e^{3x}\phi_2(y - x) + \phi_3(y - x) - \frac{1}{10}e^{x+y+2}[\sin(2x - y) + 2\cos(2x - y)].$$

7. Solve $D_x(D_x - 2D_y)(D_x + D_y)z = e^{x+2y}(x^2 + 4y^2).$

The complementary function is $z = \phi_1(y) + \phi_2(y + 2x) + \phi_3(y - x).$

For the particular integral

$\frac{1}{D_x(D_x - 2D_y)(D_x + D_y)} e^{x+2y}(x^2 + 4y^2) = e^{x+2y}\frac{1}{(D_x + 1)(D_x - 2D_y - 3)(D_x + D_y + 3)}(x^2 + 4y^2),$ we first find $u = \dfrac{1}{D_x + D_y + 3}(x^2 + 4y^2) = \dfrac{1}{3}\dfrac{1}{1 + \frac{1}{3}(D_x + D_y)}(x^2 + 4y^2)$

$$= \frac{1}{3}[1 - \frac{1}{3}(D_x + D_y) + \frac{1}{9}(D_x + D_y)^2 + \cdots\cdots](x^2 + 4y^2)$$

$$= \frac{1}{3}[x^2 + 4y^2 - \frac{2}{3}(x + 4y) + \frac{10}{9}] = \frac{1}{27}(9x^2 + 36y^2 - 6x - 24y + 10),$$

then $v = \dfrac{1}{D_x - 2D_y - 3}u = -\dfrac{1}{3}\dfrac{1}{1 + \frac{1}{3}(2D_y - D_x)}u = -\dfrac{1}{3}[1 - \frac{1}{3}(2D_y - D_x) + \frac{1}{9}(2D_y - D_x)^2 - \cdots]u$

$$= -\frac{1}{81}(9x^2 + 36y^2 - 72y + 58),$$

and finally, $z = \dfrac{1}{D_x + 1}v = (1 - D_x + D_x^2 + \cdots)v = -\dfrac{1}{81}(9x^2 + 36y^2 - 18x - 72y + 76).$

The general solution is

$$z = \phi_1(y) + \phi_2(y + 2x) + \phi_3(y - x) - \frac{1}{81}(9x^2 + 36y^2 - 18x - 72y + 76)e^{x+2y}.$$

TYPE: IRREDUCIBLE EQUATIONS.

8. Solve $f(D_x, D_y)z = (D_x - D_y^2)z = e^{x+y}$.

The complementary function is $z = \sum_{i=1}^{\infty} c_i e^{b_i^2 x + b_i y}$ from Example 9.

The short method for evaluating the particular integral $\dfrac{1}{f(D_x, D_y)}e^{x+y}$ cannot be used, since $f(a,b) = f(1,1) = 0$. We shall use the method of undetermined coefficients, assuming the particular integral to be of the form $z = Axe^{x+y} + Bye^{x+y}$.

Now $D_x z = (A + Ax + By)e^{x+y}$, $D_y^2 z = (Ax + 2B + By)e^{x+y}$ and $(D_x - D_y^2)z = (A - 2B)e^{x+y} = e^{x+y}$; hence $A - 2B = 1$. Taking $A = 1$, $B = 0$, we have as particular integral $z = xe^{x+y}$; taking $A = 0$, $B = -\frac{1}{2}$, we have $z = -\frac{1}{2}ye^{x+y}$; and so on. Choosing the first, the required solution is

$$z = \sum_{i=1}^{\infty} c_i e^{b_i^2 x + b_i y} + xe^{x+y}.$$

9. Solve $(2D_x^2 - D_y^2 + D_x)z = x^2 - y$.

The complementary function is $z = \sum_{i=1}^{\infty} c_i e^{a_i x + b_i y}$, $2a_i^2 - b_i^2 + a_i = 0$.

The particular integral $\dfrac{1}{2D_x^2 - D_y^2 + D_x}(x^2 - y) = -\dfrac{1}{D_y^2}\dfrac{1}{1 - \dfrac{D_x + 2D_x^2}{D_y^2}}(x^2 - y)$

$$= -\frac{1}{D_y^2}[1 + \frac{D_x + 2D_x^2}{D_y^2} + \frac{(D_x + 2D_x^2)^2}{D_y^4} + \cdots](x^2 - y) = -\frac{1}{D_y^2}[x^2 - y + \frac{2x + 4}{D_y^2} + \frac{2}{D_y^4}]$$

$$= -\frac{1}{D_y^2}(x^2 - y + xy^2 + 2y^2 + y^4/12) = -\frac{1}{2}x^2y^2 + \frac{1}{6}y^3 - \frac{1}{12}xy^4 - \frac{1}{6}y^4 - \frac{1}{360}y^6.$$

The required solution is $z = \sum_{i=1}^{\infty} c_i e^{a_i x \pm \sqrt{2a_i^2 + a_i}\, y} - \frac{1}{2}x^2y^2 + \frac{1}{6}y^3 - \frac{1}{12}xy^4 - \frac{1}{6}y^4 - \frac{1}{360}y^6.$

10. Find a particular integral of $(D_x^2 + D_y)(D_x - D_y - D_y^2)z = \sin(2x + y)$.

A particular integral is given by

$$\frac{1}{(D_x^2 + D_y)(D_x - D_y - D_y^2)}\sin(2x+y) = \frac{1}{(-4 + D_y)(D_x - D_y + 1)}\sin(2x+y)$$

$$= \frac{1}{D_xD_y - D_y^2 - 4D_x + 5D_y - 4}\sin(2x+y) = \frac{1}{5D_y - 4D_x - 5}\sin(2x+y)$$

$$= \frac{5D_y - 4D_x + 5}{25D_y^2 - 40D_xD_y + 16D_x^2 - 25}\sin(2x+y) = -\frac{1}{34}[5\sin(2x+y) - 3\cos(2x+y)].$$

The method of undetermined coefficients with $z = A\sin(2x+y) + B\cos(2x+y)$ may also be used here.

11. Find a particular integral of $(D_x - 2D_y + 5)(D_x^2 + D_y + 3)z = e^{3x+4y}\sin(x - 2y)$.

A particular integral is $\frac{1}{(D_x - 2D_y + 5)(D_x^2 + D_y + 3)}e^{3x+4y}\sin(x-2y)$

$$= e^{3x+4y}\frac{1}{(D_x - 2D_y)(D_x^2 + 6D_x + D_y + 16)}\sin(x-2y) = e^{3x+4y}\frac{1}{(D_x - 2D_y)(6D_x + D_y + 15)}\sin(x-2y)$$

$$= e^{3x+4y}\frac{1}{6D_x^2 - 11D_xD_y - 2D_y^2 + 15D_x - 30D_y}\sin(x-2y) = \frac{1}{5}e^{3x+4y}\frac{1}{3D_x - 6D_y - 4}\sin(x-2y)$$

$$= \frac{1}{5}e^{3x+4y}\frac{3D_x - 6D_y + 4}{9D_x^2 - 36D_xD_y + 36D_y^2 - 16}\sin(x-2y) = -\frac{1}{1205}e^{3x+4y}(3D_x - 6D_y + 4)\sin(x-2y)$$

$$= -\frac{1}{1205}e^{3x+4y}[15\cos(x-2y) + 4\sin(x-2y)].$$

TYPE: $f(xD_x, yD_y)z = 0.$

12. Solve $(xD_x^3D_y^2 - yD_x^2D_y^3)z = 0$ or $(x^3y^2D_x^3D_y^2 - x^2y^3D_x^2D_y^3)z = 0$.

The substitution $x = e^u$, $y = e^v$, $x^3y^2D_x^3D_y^2z = D_u(D_u - 1)(D_u - 2)D_v(D_v - 1)z$, $x^2y^3D_x^2D_y^3z = D_u(D_u - 1)D_v(D_v - 1)(D_v - 2)z$ transforms the given equation into

$D_uD_v(D_u - 1)(D_v - 1)(D_u - D_v)z = 0.$ The required solution is

$z = \phi_1(v) + \phi_2(u) + e^u\phi_3(v) + e^v\phi_4(u) + \phi_5(v + u)$ or, in the original variables,

$$z = \phi_1(\ln y) + \phi_2(\ln x) + x\phi_3(\ln y) + y\phi_4(\ln x) + \phi_5(\ln xy)$$
$$= \psi_1(y) + \psi_2(x) + x\psi_3(y) + y\psi_4(x) + \psi_5(xy).$$

13. Solve $(x^2D_x^2 - 4y^2D_y^2 - 4yD_y - 1)z = x^2y^3 \ln y$.

The substitution $x = e^u$, $y = e^v$ transforms the given equation into

$$[D_u(D_u - 1) - 4D_v(D_v - 1) - 4D_v - 1]z = (D_u^2 - 4D_v^2 - D_u - 1)z = ve^{2u+3v}.$$

A particular integral of this equation is given by $\dfrac{1}{D_u^2 - 4D_v^2 - D_u - 1}\, ve^{2u+3v}$

$$= e^{2u+3v}\frac{1}{(D_u+2)^2 - 4(D_v+3)^2 - (D_u+2) - 1}\, v = e^{2u+3v}\frac{1}{D_u^2 - 4D_v^2 + 3D_u - 24D_v - 35}\, v.$$

By inspection, a solution of $(D_u^2 - 4D_v^2 + 3D_u - 24D_v - 35)w = v$ is found to be $w = -\dfrac{1}{35}v + \dfrac{24}{(35)^2}$.

Hence, the particular integral is $z = -\dfrac{1}{(35)^2}e^{2u+3v}(35v - 24)$.

The required solution of the given differential equation is

$$z = \sum_{i=1}^{\infty} c_i e^{a_i u + b_i v} - \frac{1}{1225}e^{2u+3v}(35v - 24)$$ or, in the original variables,

$$z = \sum_{i=1}^{\infty} c_i x^{a_i} y^{b_i} - \frac{1}{1225}x^2y^3(35 \ln y - 24), \quad a_i^2 - 4b_i^2 - a_i - 1 = 0.$$

SUPPLEMENTARY PROBLEMS

Solve each of the following equations.

14. $(D_x + D_y + 1)(D_x - 2D_y - 1)z = 0$. *Ans.* $z = e^{-x}\phi_1(y - x) + e^x\phi_2(y + 2x)$

15. $(D_x + 2D_y - 3)(D_x + D_y - 1)z = 0$. *Ans.* $z = e^{3x}\phi_1(y - 2x) + e^x\phi_2(y - x)$

16. $(2D_x + D_y + 1)(D_x^2 + 3D_xD_y - 3D_x)z = 0$. *Ans.* $z = \phi_1(y) + e^{-y}\phi_2(2y - x) + e^y\phi_3(y - 3x)$

17. $(D_xD_y + D_y^2)(D_x - D_y - 2)z = 0$. *Ans.* $z = \phi_1(x) + \phi_2(y - x) + e^{2x}\phi_3(y + x)$

18. $(D_x + 2D_y)(D_x + 2D_y + 1)(D_x + 2D_y + 2)^2 z = 0$.

Ans. $z = \phi_1(y - 2x) + e^{-x}\phi_2(y - 2x) + e^{-y}[\phi_3(y - 2x) + y\phi_4(y - 2x)]$

19. $(D_x + D_y)(D_x + D_y - 2)z = \sin(x + 2y)$.

Ans. $z = \phi_1(y - x) + e^{2x}\phi_2(y - x) + \dfrac{1}{117}[6\cos(x + 2y) - 9\sin(x + 2y)]$

20. $(D_x + D_y - 1)(D_x + 2D_y + 2)z = e^{3x+4y} + y(1 - 2x)$.

Ans. $z = e^x\phi_1(y - x) + e^{-y}\phi_2(y - 2x) + xy + \dfrac{3}{2} + \dfrac{1}{78}e^{3x+4y}$

21. $(D_x^2 + D_xD_y + D_y - 1)z = e^x + e^{-x}$. *Ans.* $z = e^{-x}\phi_1(y) + e^x\phi_2(y - x) + \dfrac{1}{2}xe^x - \dfrac{1}{2}xe^{-x}$

22. $(D_x^3 - D_xD_y^2 - D_x^2 + D_xD_y)z = (x+2)/x^3$. Ans. $z = \phi_1(y) + \phi_2(y+x) + e^x\phi_3(y-x) + \ln x$

23. $(3D_xD_y - 2D_y^2 - D_y)z = \cos(3y+2x)$. Ans. $z = \phi_1(x) + e^{\frac{1}{2}y}\phi_2(3y+2x) - \frac{1}{3}\sin(3y+2x)$

24. $(D_x^2 + D_xD_y - D_y^2 + D_x - D_y)z = e^{2x-3y}$. Ans. $z = \sum c_i e^{a_ix+b_iy} - \frac{1}{6}e^{2x-3y}$, $a_i^2 + a_ib_i - b_i^2 + a_i - b_i = 0$

25. $(3D_x^2 - 2D_y^2 + D_x - 1)z = 3e^{x+y}\sin(x+y)$.

Ans. $z = \sum c_i e^{a_ix+b_iy} - e^{x+y}\cos(x+y)$, $3a_i^2 - 2b_i^2 + a_i - 1 = 0$

26. $(D_x^2 + 2D_xD_y^2 - 2D_y + 3)z = e^{x+y}\cos(x+2y)$.

Ans. $z = \sum c_i e^{a_ix+b_iy} - \frac{1}{13}e^{x+y}\cos(x+2y)$, $a_i^2 + 2a_ib_i - 2b_i + 3 = 0$

27. $(D_x^2 + D_xD_y + D_x + D_y + 1)z = e^{-2x}(x^2+2y^2)$.

Ans. $z = \sum c_i e^{a_ix+b_iy} + \frac{1}{27}e^{-2x}(9x^2 + 18y^2 + 18x + 12y + 16)$, $a_i^2 + a_ib_i + a_i + b_i + 1 = 0$

28. $(D_x^2D_y + D_y^2 - 2)z = e^{2y}\cos 3x + e^x\sin 2y$.

Ans. $z = \sum c_i e^{a_ix+b_iy} - \frac{1}{16}e^{2y}\cos 3x - \frac{1}{20}e^x(\cos 2y + 3\sin 2y)$, $a_i^2b_i + b_i^2 - 2 = 0$

29. $(xyD_xD_y - y^2D_y^2 - 3xD_x + 2yD_y)z = 0$. Ans. $z = \phi_1(\ln xy) + y^3\phi_2(\ln x) = \psi_1(xy) + y^3\psi_2(x)$

30. $(x^2D_x^2 - 2xyD_xD_y - 3y^2D_y^2 + xD_x - 3yD_y) = x^2y\sin(\ln x^2)$.

Ans. $z = \phi_1(x^3y) + \phi_2(y/x) - \frac{1}{65}x^2y[4\cos(\ln x^2) + 7\sin(\ln x^2)]$

31. $(x^2D_x^2 + xyD_xD_y - 2y^2D_y^2 - xD_x - 6yD_y)z = 0$. Ans. $z = \phi_1(y/x^2) + x^2\phi_2(xy)$

32. $(x^2D_x^2 - xyD_xD_y - 2y^2D_y^2 + xD_x - 2yD_y)z = \ln(y/x) - 1/2$.

Ans. $z = \phi_1(x^2y) + \phi_2(y/x) + \frac{1}{2}(\ln x)^2\ln y + \frac{1}{2}\ln x\ln y$

33. $(x^2yD_x^2D_y - xy^2D_xD_y^2 - x^2D_x^2 + y^2D_y^2)z = \dfrac{x^3+y^3}{xy}$.

Ans. $z = x\phi_1(y) + y\phi_2(x) + \phi_3(xy) - \frac{1}{6}\left(\dfrac{x^3-y^3}{xy}\right)$

CHAPTER 33

Partial Differential Equations of Order Two with Variable Coefficients

THE MOST GENERAL LINEAR PARTIAL DIFFERENTIAL EQUATION of order two in two independent variables has the form

1) $$Rr + Ss + Tt + Pp + Qq + Zz = F$$

where R, S, T, P, Q, Z, F are functions of x and y only and not all R, S, T are zero.

Before considering the general equation, a number of special types will be treated.

TYPE I.

2a) $$r = \frac{\partial^2 z}{\partial x^2} = F/R = F_1(x,y)$$

2b) $$s = \frac{\partial^2 z}{\partial x\, \partial y} = F/S = F_2(x,y)$$

2c) $$t = \frac{\partial^2 z}{\partial y^2} = F/T = F_3(x,y).$$

These are reducible equations with constant coefficients (Chapter 32), but a more direct method of solving will be used here.

EXAMPLE 1. Solve $s = x - y$.

Integrating $s = \dfrac{\partial^2 z}{\partial x\, \partial y} = x - y$ with respect to y, $p = \dfrac{\partial z}{\partial x} = xy - \frac{1}{2}y^2 + \psi(x)$, ψ arbitrary.

Integrating this relation with respect to x, $z = \frac{1}{2}x^2 y - \frac{1}{2}xy^2 + \phi_1(x) + \phi_2(y)$,

where $\dfrac{d}{dx}\phi_1(x) = \psi(x)$ and $\phi_2(y)$ are arbitrary functions.

TYPE II.

3a) $$Rr + Pp = R\frac{\partial p}{\partial x} + Pp = F$$

3b) $$Ss + Pp = S\frac{\partial p}{\partial y} + Pp = F$$

3c) $$Ss + Qq = S\frac{\partial q}{\partial x} + Qq = F$$

3d) $$Tt + Qq = T\frac{\partial q}{\partial y} + Qq = F.$$

These are essentially linear *ordinary* differential equations of order one in which p (or q) is the dependent variable.

EXAMPLE 2. Solve $xr + 2p = (9x+6)e^{3x+2y}$.

Considering p as the dependent variable, x as the independent variable, and y as constant, the equation is $x\dfrac{\partial p}{\partial x} + 2p = (9x+6)e^{3x+2y}$ for which x is an integrating factor.

Integrating $x^2\dfrac{\partial p}{\partial x} + 2xp = (9x^2+6x)e^{3x+2y}$, we have

$$x^2 p = \frac{1}{D_x}(9x^2+6x)e^{3x+2y} = \frac{1}{3}e^{3x+2y}\left(1 - \frac{D_x}{3} + \frac{D_x^2}{9} - \cdots\right)(9x^2+6x)$$

$$= 3x^2 e^{3x+2y} + \phi_1(y) \quad \text{or} \quad p = \frac{\partial z}{\partial x} = 3e^{3x+2y} + \frac{1}{x^2}\phi_1(y).$$

Then $z = e^{3x+2y} - \dfrac{1}{x}\phi_1(y) + \phi_2(y)$ is the required solution.

TYPE III.

4*a*) $Rr + Ss + Pp = F$ or $R\dfrac{\partial p}{\partial x} + S\dfrac{\partial p}{\partial y} = F - Pp$

4*b*) $Ss + Tt + Qq = F$ or $S\dfrac{\partial q}{\partial x} + T\dfrac{\partial q}{\partial y} = F - Qq.$

These are linear partial differential equations of order one with p (or q) as dependent variable and x, y as independent variables.

EXAMPLE 3. Solve $2xr - ys + 2p = 4xy^2$ or $2x\dfrac{\partial p}{\partial x} - y\dfrac{\partial p}{\partial y} = 4xy^2 - 2p.$

Using the method of Lagrange (Chapter 29), the auxiliary system is $\dfrac{dx}{2x} = \dfrac{dy}{-y} = \dfrac{dp}{4xy^2 - 2p}$.

From the first two ratios, we obtain readily $xy^2 = a$.

By inspection, $2y^4(2x) + 2py(-y) - y^2(4xy^2 - 2p) = 0$. Thus,

$$2y^4\,dx + 2py\,dy - y^2\,dp = 0 \quad \text{or} \quad 2\,dx - \frac{y^2\,dp - 2py\,dy}{y^4} = 0, \quad \text{and} \quad \frac{p}{y^2} - 2x = b.$$

The general solution is $p/y^2 - 2x = \psi(xy^2)$. Then

$$p = \frac{\partial z}{\partial x} = 2xy^2 + y^2\psi(xy^2) \quad \text{and} \quad z = x^2y^2 + \phi_1(xy^2) + \phi_2(y), \quad \text{where} \quad \frac{\partial}{\partial x}\phi_1(xy^2) = y^2\psi(xy^2).$$

TYPE IV.

5*a*) $Rr + Pp + Zz = F$ or $R\dfrac{\partial^2 z}{\partial x^2} + P\dfrac{\partial z}{\partial x} + Zz = F$

5*b*) $Tt + Qq + Zz = F$ or $T\dfrac{\partial^2 z}{\partial y^2} + Q\dfrac{\partial z}{\partial y} + Zz = F.$

These are essentially linear ordinary differential equations of order two with x as independent variable in 5*a*) and y as independent variable in 5*b*).

EXAMPLE 4. Solve $t - 2xq + x^2 z = (x-2)e^{3x+2y}$.

The equation may be written as $(D_y^2 - 2xD_y + x^2)z = (D_y - x)^2 z = (x-2)e^{3x+2y}$.

The complementary function is $z = e^{xy}\phi_1(x) + xe^{xy}\phi_2(x)$ and a particular integral is

$$\frac{1}{(D_y - x)^2}(x-2)e^{3x+2y} = \frac{x-2}{(2-x)^2}e^{3x+2y} = \frac{e^{3x+2y}}{x-2}.$$

The required solution is $z = e^{xy}\phi_1(x) + xe^{xy}\phi_2(x) + \dfrac{e^{3x+2y}}{x-2}$.

See also Problems 1-8.

LAPLACE'S TRANSFORMATION. This transformation on

1) $$Rr + Ss + Tt + Pp + Qq + Zz = G(u,v)$$

consists of changing from the independent variables x,y to a new set u,v, where

6) $$u = u(x,y), \quad v = v(x,y)$$

are to be chosen so that the resulting equation is simpler than 1). By means of 6), we obtain

$$p = \frac{\partial z}{\partial x} = \frac{\partial z}{\partial u}\frac{\partial u}{\partial x} + \frac{\partial z}{\partial v}\frac{\partial v}{\partial x} = z_u u_x + z_v v_x, \qquad q = \frac{\partial z}{\partial y} = z_u u_y + z_v v_y,$$

$$r = \frac{\partial p}{\partial x} = z_u u_{xx} + (z_{uu}u_x + z_{uv}v_x)u_x + z_v v_{xx} + (z_{uv}u_x + z_{vv}v_x)v_x$$

$$= z_{uu}(u_x)^2 + 2z_{uv}u_x v_x + z_{vv}(v_x)^2 + z_u u_{xx} + z_v v_{xx},$$

$$s = \frac{\partial p}{\partial y} = z_u u_{xy} + (z_{uu}u_y + z_{uv}v_y)u_x + z_v v_{xy} + (z_{uv}u_y + z_{vv}v_y)v_x$$

$$= z_{uu}u_x u_y + z_{uv}(u_x v_y + u_y v_x) + z_{vv}v_x v_y + z_u u_{xy} + z_v v_{xy},$$

$$t = \frac{\partial q}{\partial y} = z_{uu}(u_y)^2 + 2z_{uv}u_y v_y + z_{vv}(v_y)^2 + z_u u_{yy} + z_v v_{yy}.$$

Let

1') $$R'z_{uu} + S'z_{uv} + T'z_{vv} + P'z_u + Q'z_v + Zz = F$$

be obtained by making the above replacements in 1) and rearranging. We shall need only the coefficients

$$R' = R(u_x)^2 + Su_x u_y + T(u_y)^2 \quad \text{and} \quad T' = R(v_x)^2 + Sv_x v_y + T(v_y)^2.$$

We note that both are of the form

7) $$R(\xi_x)^2 + S\xi_x\xi_y + T(\xi_y)^2 = (a\xi_x + b\xi_y)(e\xi_x + f\xi_y).$$

i) Suppose $b/a \neq f/e$; then, if for u we take any solution of $a\xi_x + b\xi_y = 0$ and for v any solution of $e\xi_x + f\xi_y = 0$, 1) is transformed into 1') with $R' = T' = 0$.

EXAMPLE 5. Solve a) $x^2(y-1)r - x(y^2-1)s + y(y-1)t + xyp - q = 0$,

b) $y(x+y)(r-s) - xp - yq - z = 0$.

a) Here 7) is $x^2(y-1)(\xi_x)^2 - x(y^2-1)\xi_x\xi_y + y(y-1)(\xi_y)^2 = 0$

or $x^2(\xi_x)^2 - x(y+1)\xi_x\xi_y + y(\xi_y)^2 = (x\xi_x - y\xi_y)(x\xi_x - \xi_y) = 0$.

Now $x\xi_x - y\xi_y = 0$ is satisfied by $\xi = u = xy$ and $x\xi_x - \xi_y = 0$ is satisfied by $\xi = v = xe^y$. Moreover, it is easily shown that these solutions also satisfy the given differential equation. Hence, the required solution is

$$z = \phi_1(xy) + \phi_2(xe^y).$$

b) Here 7) is $y(x+y)[(\xi_x)^2 - \xi_x\xi_y] = 0$ or $(\xi_x - \xi_y)\xi_x = 0$.

Now $\xi_x - \xi_y = 0$ is satisfied by $\xi = x+y$ and $\xi_x = 0$ by $\xi = y$. However, neither of these solutions will satisfy the given differential equation.

We take $u = x+y$ and $v = y$. Then $p = z_u$, $q = z_u + z_v$, $r = z_{uu}$, $s = z_{uu} + z_{uv}$, and the given differential equation becomes

$$-y(x+y)z_{uv} - xz_u - yz_u - yz_v - z = 0 \quad \text{or} \quad uvz_{uv} + uz_u + vz_v + z = 0.$$

This may be written as

$$z_{uv} + \frac{1}{v}z_u + \frac{1}{u}z_v + \frac{1}{uv}z = \frac{\partial}{\partial u}(\frac{\partial z}{\partial v} + \frac{1}{v}z) + \frac{1}{u}(\frac{\partial z}{\partial v} + \frac{1}{v}z) = (\frac{\partial}{\partial u} + \frac{1}{u})(\frac{\partial z}{\partial v} + \frac{1}{v}z) = 0.$$

Let $\frac{\partial z}{\partial v} + \frac{1}{v}z = w$; then $\frac{\partial w}{\partial u} + \frac{1}{u}w = 0$ and $wu = \psi(v)$. Now

$$\frac{\partial z}{\partial v} + \frac{1}{v}z = w = \frac{1}{u}\psi(v), \quad zv = \frac{1}{u}\lambda(v) + \phi_2(u), \quad \text{and} \quad z = \frac{1}{u}\phi_1(v) + \frac{1}{v}\phi_2(u),$$

where $\frac{d}{dv}\lambda(v) = v\cdot\psi(v)$ and $\phi_1(v) = \frac{1}{v}\lambda(v)$. The required solution is $z = \frac{\phi_1(y)}{x+y} + \frac{\phi_2(x+y)}{y}$.

EXAMPLE 6. Solve $x^2r - y^2t + px - qy = x^2$.

Here 7) is $x^2(\xi_x)^2 - y^2(\xi_y)^2 = (x\xi_x - y\xi_y)(x\xi_x + y\xi_y) = 0$.

Now $x\xi_x - y\xi_y = 0$ is satisfied by $\xi = xy$ and $x\xi_x + y\xi_y = 0$ by $\xi = x/y$. It is found readily that these solutions satisfy the reduced equation $x^2r - y^2t + px - qy = 0$; hence, the complementary function is $z = \phi_1(x/y) + \phi_2(xy)$. However, this complementary function may be obtained along with the particular integral as follows. Take $u = xy$ and $v = x/y$; then

$$p = yz_u + \frac{1}{y}z_v, \quad q = xz_u - \frac{x}{y^2}z_v, \quad r = y^2z_{uu} + 2z_{uv} + \frac{1}{y^2}z_{vv}, \quad t = x^2z_{uu} - 2\frac{x^2}{y^2}z_{uv} + \frac{x^2}{y^4}z_{vv} + \frac{2x}{y^3}z_v,$$

and the given equation becomes $4x^2z_{uv} = x^2$ or $z_{uv} = \frac{1}{4}$.

Integrating first with respect to u, $z_v = \psi(v) + \frac{1}{4}u$,

and then with respect to v, $z = \phi_1(v) + \phi_2(u) + \frac{1}{4}uv = \phi_1(x/y) + \phi_2(xy) + \frac{1}{4}x^2$,

where $\frac{d}{dv}\phi_1(v) = \psi(v)$.

See Problems 9-10.

ii) Suppose $b/a = f/e$; then $R(\xi_x)^2 + S\xi_x\xi_y + T(\xi_y)^2 = m(a\xi_x + b\xi_y)^2$. This case is treated in Problem 11.

NON-LINEAR PARTIAL DIFFERENTIAL EQUATIONS OF ORDER TWO. One possible method for solving a given non-linear partial differential equation of order two

8) $$F(x, y, z, p, q, r, s, t) = 0$$

is suggested by several of the examples of linear equations above. In each of Examples 1-3, the first step consisted in finding a relation of the form

9) $$u = \psi(v), \quad \psi \text{ arbitrary},$$

where $u = u(x,y,z,p,q)$ and $v = v(x,y,z,p,q)$, from which the given differential equation could be derived by eliminating the arbitrary function. Such a relation 9) is called an *intermediate integral* of 8). For example, $p - xy + \frac{1}{2}y^2 = \psi(x)$ is an intermediate integral of $s = x - y$, (Example 1).

It can be shown that the most general partial differential equation having

$$u = \psi(v), \quad \psi \text{ arbitrary},$$

where $u = u(x,y,z,p,q)$ and $v = v(x,y,z,p,q)$, as intermediate integral has the form

10) $$Rr + Ss + Tt + U(rt - s^2) = V,$$

where R, S, T, U, V are functions of x, y, z, p, q. However, it is evident from the definitions of $R, S, \cdots, V$ that not every equation of the form 10) has an intermediate integral. The discussion below concerns Monge's method for determining an intermediate integral of 10), assuming that one exists.

TYPE: $Rr + Ss + Tt = V$. Consider the equation

11) $$Rr + Ss + Tt = V,$$

that is, 10) with U identically zero. Since we seek z as a function of x and y, we have always

12_1) $$dz = \frac{\partial z}{\partial x}dx + \frac{\partial z}{\partial y}dy = p\,dx + q\,dy,$$

12_2) $$dp = \frac{\partial p}{\partial x}dx + \frac{\partial p}{\partial y}dy = r\,dx + s\,dy,$$

12_3) $$dq = \frac{\partial q}{\partial x}dx + \frac{\partial q}{\partial y}dy = s\,dx + t\,dy.$$

Solving the latter two for $r = \dfrac{dp - s\,dy}{dx}$, $t = \dfrac{dq - s\,dx}{dy}$ and substituting in 11), we obtain $R\dfrac{dp - s\,dy}{dx} + Ss + T\dfrac{dq - s\,dx}{dy} = V$ or

13) $$s[R(dy)^2 - S\,dx\,dy + T(dx)^2] = R\,dy\,dp + T\,dx\,dq - V\,dx\,dy.$$

The equations

14_1) $$R(dy)^2 - S\,dx\,dy + T(dx)^2 = 0$$

14_2) $$R\,dy\,dp + T\,dx\,dq - V\,dx\,dy = 0$$

are called *Monge's equations*.

Suppose $R(dy)^2 - S\,dx\,dy + T(dx)^2 = (A\,dy + B\,dx)^2 = 0$. If now $u = u(x,y,z,p,q) = a$, $v = v(x,y,z,p,q) = b$ satisfy the system

$$\left[\begin{array}{l} A\,dy + B\,dx = 0 \\ R\,dy\,dp + T\,dx\,dq - V\,dx\,dy = 0, \end{array}\right.$$

then $$u = \psi(v)$$
is an intermediate integral of 11) since $u=a$, $v=b$ satisfy 13) and, hence, 11).

Suppose $R(dy)^2 - S\,dx\,dy + T(dx)^2 = (A_1\,dy + B_1\,dx)(A_2\,dy + B_2\,dx) = 0$, where $A_1B_2 - A_2B_1 \neq 0$ identically. We now have two systems

$$\left[\begin{array}{l} A_1\,dy + B_1\,dx = 0 \\ R\,dy\,dp + T\,dx\,dq - V\,dx\,dy = 0 \end{array}\right. \quad \text{and} \quad \left[\begin{array}{l} A_2\,dy + B_2\,dx = 0 \\ R\,dy\,dp + T\,dx\,dq - V\,dx\,dy = 0. \end{array}\right.$$

If either system is integrable, we are led to an intermediate integral of 11); if both are integrable, we have two intermediate integrals at our disposal. Procedures for finding a solution of a given equation for which intermediate integrals have been obtained will be discussed in the examples and solved problems.

EXAMPLE 7. Solve $q(yq+z)r - p(2yq+z)s + yp^2t + p^2q = 0$.

Here $R = q(yq+z)$, $S = -p(2yq+z)$, $T = yp^2$, $V = -p^2q$; Monge's equations are

$$\begin{aligned} R(dy)^2 - S\,dx\,dy + T(dx)^2 &= q(yq+z)(dy)^2 + p(2yq+z)dx\,dy + yp^2(dx)^2 \\ &= (q\,dy + p\,dx)[(yq+z)dy + yp\,dx] = 0 \end{aligned}$$

and $$R\,dy\,dp + T\,dx\,dq - V\,dx\,dy = q(yq+z)dy\,dp + yp^2dx\,dq + p^2q\,dx\,dy = 0.$$

We seek first a solution of the system $\left[\begin{array}{l} q\,dy + p\,dx = 0 \\ q(yq+z)dy\,dp + yp^2dx\,dq + p^2q\,dx\,dy = 0. \end{array}\right.$

Combining the first equation and 12_1), we have $dz = 0$ and $z = a$. Substituting in the second equation $dy = -p\,dx/q$, obtained from the first, we obtain

$$(yq+z)dp - p(y\,dq + q\,dy) = 0.$$

We add $-p\,dz = 0$ to this, obtaining

$$(yq+z)dp - p(y\,dq + q\,dy + dz) = 0 \quad \text{or} \quad \frac{dp}{p} = \frac{y\,dq + q\,dy + dz}{yq+z}$$

with solution $\dfrac{yq+z}{p} = b$. Then $yq + z = p\cdot f(z)$ is an intermediate integral. The Lagrange system for this first order equation is $\dfrac{dx}{f(z)} = \dfrac{dy}{-y} = \dfrac{dz}{z}$. From $\dfrac{dy}{-y} = \dfrac{dz}{z}$ we obtain $yz = a$, and from $\frac{dx}{f(z)} = \frac{dz}{z}$ we obtain $x = \int f(z)\frac{dz}{z} = \phi_1(z) + b$. Thus, the required solution is

$$x = \phi_1(z) + \phi_2(yz).$$

Consider next the second system $\left[\begin{aligned} (yq+z)dy + yp\,dx &= 0 \\ q(yq+z)dy\,dp + yp^2 dx\,dq + p^2 q\,dx\,dy &= 0. \end{aligned}\right.$

From the first equation, $p\,dx + q\,dy = -z\,dy/y$; then $dz = -z\,dy/y$ and $yz = a$. Substituting from the first equation, the second becomes

$$qy\,dp - py\,dq - pq\,dy = 0 \qquad \text{or} \qquad \frac{dp}{p} - \frac{dq}{q} - \frac{dy}{y} = 0$$

with solution $qy/p = b$. Then $qy = p\cdot g(yz)$ is an intermediate integral. The Lagrange system is $\frac{dx}{g(yz)} = \frac{dy}{-y}$, $dz = 0$. Then $z = a$ and the first equation $\frac{dx}{g(ya)} = \frac{dy}{-y}$ has solution $x = -\int g(ya)\frac{dy}{y} = \phi_2(ya) + b$. We thus obtain $x = \phi_1(z) + \phi_2(yz)$ as before.

The solution may also be obtained by using the two intermediate integrals simultaneously. Upon solving them for $p = \frac{z}{f(z) - g(yz)}$, $q = \frac{z\cdot g(yz)}{y[f(z) - g(yz)]}$

and substituting in $p\,dx + q\,dy = dz$, we have $yz\,dx + zg(yz)dy = yf(z)dz - yg(yz)dz$. Writing $f(z) = zf_1(z)$ and $g(yz) = -yz\,g_1(yz)$, this equation becomes

$$dx = f_1(z)dz + g_1(yz)[z\,dy + y\,dz]$$

and, integrating, $$x = \phi_1(z) + \phi_2(yz).$$

See also Problems 12-16.

TYPE: $Rr + Ss + Tt + U(rt - s^2) = V$. Consider equation 10) with $U \neq 0$. By substituting $r = \frac{dp - s\,dy}{dx}$, $t = \frac{dq - s\,dx}{dy}$ as in the preceding type, we obtain

$$s[R(dy)^2 - S\,dx\,dy + T(dx)^2 + U(dx\,dp + dy\,dq)] = R\,dy\,dp + T\,dx\,dq + U\,dp\,dq - V\,dx\,dy.$$

The equations

15_1) $$R(dy)^2 - S\,dx\,dy + T(dx)^2 + U(dx\,dp + dy\,dq) = 0$$

15_2) $$R\,dy\,dp + T\,dx\,dq + U\,dp\,dq - V\,dx\,dy = 0$$

are called *Monge's equations*. Note that when $U = 0$, these equations are 14_1) and 14_2). However, unlike 14_1) and 14_2), neither can be factored.

We shall attempt to choose $\lambda = \lambda(x,y,z,p,q)$ so as to obtain a factorable combination

16) $$\begin{aligned} \lambda[R(dy)^2 - S\,dx\,dy + T(dx)^2 + U(dx\,dp + dy\,dq)] + R\,dy\,dp + T\,dx\,dq + U\,dp\,dq - V\,dx\,dy \\ = (a\,dy + b\,dx + c\,dp)(\alpha\,dy + \beta\,dx + \gamma\,dq) \\ = a\alpha(dy)^2 + (a\beta + b\alpha)dx\,dy + b\beta(dx)^2 + c\beta\,dx\,dp + a\gamma\,dy\,dq + c\alpha\,dy\,dp \\ + b\gamma\,dx\,dq + c\gamma\,dp\,dq = 0. \end{aligned}$$

Comparing coefficients, we have

$$a\alpha = R\lambda,\quad a\beta + b\alpha = -S\lambda - V,\quad b\beta = T\lambda,\quad c\beta = U\lambda = a\gamma,\quad c\alpha = R,\quad b\gamma = T,\quad c\gamma = U.$$

The first relation will be satisfied by taking $a = \lambda$ and $\alpha = R$; this choice determines $b = T/U$, $\beta = \lambda U$, $c = 1$, $\gamma = U$. The remaining relation $a\beta + b\alpha = -S\lambda - V$ takes the form

$$U\lambda^2 + \frac{TR}{U} = -S\lambda - V \qquad \text{or}$$

17) $$U^2\lambda^2 + SU\lambda + TR + UV = 0.$$

In general 17) will have two distinct roots $\lambda = \lambda_1$, $\lambda = \lambda_2$; thus, 16) can be factored as

18_1) $$(\lambda_1 U\,dy + T\,dx + U\,dp)(R\,dy + \lambda_1 U\,dx + U\,dq) = 0 \qquad \text{and}$$

18_2) $$(\lambda_2 U\,dy + T\,dx + U\,dp)(R\,dy + \lambda_2 U\,dx + U\,dq) = 0.$$

There are four systems to be considered. The system $\lambda_1 U\,dy + T\,dx + U\,dp = 0$, $\lambda_2 U\,dy + T\,dx + U\,dp = 0$ implies $(\lambda_1 - \lambda_2)U\,dy = 0$ and, hence, unless $\lambda_1 = \lambda_2$, $U\,dy = 0$ identically. Similarly, the system $R\,dy + \lambda_1 U\,dx + U\,dq = 0$, $R\,dy + \lambda_2 U\,dx + U\,dq = 0$ implies $U\,dx = 0$ identically. We therefore shall use only the systems

19) $$\left[\begin{array}{l}\lambda_1 U\,dy + T\,dx + U\,dp = 0\\ R\,dy + \lambda_2 U\,dx + U\,dq = 0\end{array}\right. \qquad \text{and} \qquad \left[\begin{array}{l}\lambda_2 U\,dy + T\,dx + U\,dp = 0\\ R\,dy + \lambda_1 U\,dx + U\,dq = 0.\end{array}\right.$$

Each system, if integrable, yields an intermediate integral of 10).

EXAMPLE 8. Solve $3s - 2(rt - s^2) = 2$.

Here, $R = 0$, $S = 3$, $T = 0$, $U = -2$, $V = 2$. Then $U^2\lambda^2 + SU\lambda + TR + UV = 4\lambda^2 - 6\lambda - 4 = 0$, $\lambda_1 = -\frac{1}{2}$ and $\lambda_2 = 2$. We seek solutions of the systems

$$\left[\begin{array}{l}\lambda_1 U\,dy + T\,dx + U\,dp = dy - 2\,dp = 0\\ R\,dy + \lambda_2 U\,dx + U\,dq = -4\,dx - 2\,dq = 0\end{array}\right. \qquad \text{and} \qquad \left[\begin{array}{l}\lambda_2 U\,dy + T\,dx + U\,dp = -4\,dy - 2\,dp = 0\\ R\,dy + \lambda_1 U\,dx + U\,dq = dx - 2\,dq = 0.\end{array}\right.$$

From the first system, $y - 2p = a$ and $2x + q = b$; then (i) $y - 2p = f(2x + q)$ is an intermediate integral. From the second system, $2y + p = a$ and $x - 2q = b$; then (ii) $2y + p = g(x - 2q)$ is an intermediate integral. Since q appears in the argument of both f and g, it is no longer possible to obtain a solution of the given equation involving two arbitrary functions by solving for p and q and substituting in $dz = p\,dx + q\,dy$.

We shall attempt to find a solution involving arbitrary constants from the intermediate integral $y - 2p = f(2x + q)$. To obtain an integrable equation, take $f(2x + q) = \alpha(2x + q) + \beta$, where α and β are arbitrary constants. The Lagrange system for

$$y - 2p = \alpha(2x + q) + \beta \qquad \text{or} \qquad 2p + \alpha q = y - 2\alpha x - \beta$$

is

$$\frac{dx}{2} = \frac{dy}{\alpha} = \frac{dz}{y - 2\alpha x - \beta}.$$

From the first two members, $\alpha x = 2y + \xi$. Substituting for αx, the last two members become

$$\frac{dy}{\alpha} = \frac{dz}{-3y - 2\xi - \beta}$$

or $\alpha\,dz = (-3y - 2\xi - \beta)dy$ and $\alpha z = -\frac{3}{2}y^2 - 2\xi y - \beta y + \eta$.

Thus, $\alpha z = \frac{5}{2}y^2 - (2\alpha x + \beta)y + \phi_1(\alpha x - 2y)$ is a solution of the given equation involving one arbitrary function and two arbitrary constants.

Treating the second intermediate integral similarly, we take $2y + p = \gamma(x - 2q) + \delta$ or $p + 2\gamma q = \gamma x - 2y + \delta$, where γ and δ are arbitrary constants. The corresponding Lagrange system is $\dfrac{dx}{1} = \dfrac{dy}{2\gamma} = \dfrac{dz}{\gamma x - 2y + \delta}$. From the first two members, $y = 2\gamma x + \xi$. Now the first and third members become $\dfrac{dx}{1} = \dfrac{dz}{-3\gamma x - 2\xi + \delta}$ and $z = -\dfrac{3}{2}\gamma x^2 - 2\xi x + \delta x + \eta$. Thus, $z = \dfrac{5}{2}\gamma x^2 - (2y - \delta)x + \phi_2(y - 2\gamma x)$ is also a solution involving one arbitrary function and two arbitrary constants.

A solution involving two arbitrary functions of parameters λ and μ will next be found. Set $2x + q = \lambda$ and $x - 2q = \mu$ so that $x = (2\lambda + \mu)/5$. Then (i) and (ii) become $y - 2p = f(\lambda)$ and $2y + p = g(\mu)$, and $y = [f(\lambda) + 2g(\mu)]/5$. Now

(iii) $\qquad p = \tfrac{1}{2}[y - f(\lambda)] = -2y + g(\mu)$ and

(iv) $\qquad q = \lambda - 2x = \tfrac{1}{2}(x - \mu)$.

Substituting the second value of p and the first value of q in $dz = p\,dx + q\,dy$, we have

$$\begin{aligned} dz &= [-2y + g(\mu)]dx + (\lambda - 2x)dy \\ &= -2(y\,dx + x\,dy) + \frac{1}{5}g(\mu)[2\,d\lambda + d\mu] + \frac{1}{5}\lambda[f'(\lambda)d\lambda + 2g'(\mu)d\mu] \\ &= -2(y\,dx + x\,dy) + \frac{2}{5}[\lambda g'(\mu)d\mu + g(\mu)d\lambda] + \frac{1}{5}[\lambda f'(\lambda) + f(\lambda)]d\lambda - \frac{1}{5}f(\lambda)d\lambda + \frac{1}{5}g(\mu)d\mu \end{aligned}$$

and

$$\begin{aligned} z &= -2xy + \frac{2}{5}\lambda g(\mu) + \frac{1}{5}\lambda f(\lambda) - \phi_1(\lambda) + \phi_2(\mu) \\ &= -2xy + \lambda y - \phi_1(\lambda) + \phi_2(\mu). \end{aligned}$$

This solution may have been obtained by using the first value of p in (iii) and the second value of q in (iv).

See also Problems 17-18.

SOLVED PROBLEMS

1. Solve $r = x^2 e^{-y}$ or $\dfrac{\partial^2 z}{\partial x^2} = x^2 e^{-y}$.

One integration with respect to x yields $p = \dfrac{\partial z}{\partial x} = \dfrac{x^3}{3}e^{-y} + \phi_1(y)$, and the second integration with respect to x yields $z = \dfrac{x^4}{12}e^{-y} + x\phi_1(y) + \phi_2(y)$.

2. Solve $xy^2 s = 1 - 4x^2 y$.

Integrating $\dfrac{\partial^2 z}{\partial x\,\partial y} = x^{-1}y^{-2} - 4xy^{-1}$ with respect to y, $\dfrac{\partial z}{\partial x} = -x^{-1}y^{-1} - 4x\ln y + \psi(x)$.

Integrating this with respect to x, $z = -\dfrac{1}{y}\ln x - 2x^2\ln y + \phi_1(x) + \phi_2(y)$,

where $\dfrac{d}{dx}\phi_1(x) = \psi(x)$.

3. Solve $xys - px = y^2$.

Integrating $\dfrac{y\dfrac{\partial p}{\partial y} - p}{y^2} = \dfrac{1}{x}$ with respect to y, we get $\dfrac{p}{y} = \dfrac{y}{x} + \psi(x)$ or $\dfrac{\partial z}{\partial x} = \dfrac{y^2}{x} + y\psi(x)$.

Integrating with respect to x, we get $z = y^2 \ln x + y\phi_1(x) + \phi_2(y)$, where $\dfrac{d}{dx}\phi_1(x) = \psi(x)$.

4. Solve $t - xq = -\sin y - x\cos y$.

Integrating $\dfrac{\partial q}{\partial y} - xq = -(\sin y + x\cos y)$, using the integrating factor e^{-xy}, we obtain

$$e^{-xy}q = -\int e^{-xy}(\sin y + x\cos y)dy = e^{-xy}\cos y + \psi(x) \quad \text{or} \quad q = \frac{\partial z}{\partial y} = \cos y + e^{xy}\psi(x).$$

A second integration, with respect to y, yields $z = \sin y + e^{xy}\phi_1(x) + \phi_2(x)$, where $\phi_1(x) = \psi(x)/x$.

5. Solve $sy - 2xr - 2p = 6xy$.

The auxiliary system for the equation $2x\dfrac{\partial p}{\partial x} - y\dfrac{\partial p}{\partial y} = -6xy - 2p$ is $\dfrac{dx}{2x} = \dfrac{dy}{-y} = \dfrac{dp}{-6xy-2p}$.

From the first and second ratios, we find $xy^2 = a$. By inspection,

$$2y^3(2x) - (2yp + 2xy^2)(-y) + y^2(-6xy - 2p) = 0$$

so that
$$2y^3dx - (2yp + 2xy^2)dy + y^2dp = 0,$$

or
$$\frac{y^2(dp + 2x\,dy + 2y\,dx) - 2y(p + 2xy)dy}{y^4} = 0, \quad \text{and} \quad \frac{p + 2xy}{y^2} = b.$$

Thus, we obtain as solution $p + 2xy = y^2\psi(xy^2)$. Then

$$\frac{\partial z}{\partial x} = -2xy + y^2\psi(xy^2) \quad \text{and} \quad z = -x^2y + \phi_1(xy^2) + \phi_2(y), \quad \text{where} \quad \frac{\partial}{\partial x}\phi_1(xy^2) = y^2\psi(xy^2).$$

6. Solve $xs + yt + q = 10x^3y$.

The auxiliary system for the equation $x\dfrac{\partial q}{\partial x} + y\dfrac{\partial q}{\partial y} = 10x^3y - q$ is $\dfrac{dx}{x} = \dfrac{dy}{y} = \dfrac{dq}{10x^3y - q}$.

From the first two ratios, $x/y = a$. By inspection,

$$(q - 8x^3y)x - 2x^4(y) + x(10x^3y - q) = 0$$

so that $(q - 8x^3y)dx - 2x^4dy + x\,dq = 0$, or $x\,dq + q\,dx = 8x^3y\,dx + 2x^4dy$,

and
$$qx = 2x^4y + b.$$

The general solution is $qx = 2x^4y + \psi(y/x)$. Thus,

$$\frac{\partial z}{\partial y} = 2x^3y + \frac{1}{x}\psi(\frac{y}{x}) \quad \text{and} \quad z = x^3y^2 + \phi_1(\frac{y}{x}) + \phi_2(x), \quad \text{where} \quad \frac{\partial}{\partial y}\phi_1(\frac{y}{x}) = \frac{1}{x}\psi(\frac{y}{x}).$$

7. Solve $t - q - \frac{1}{x}(\frac{1}{x} - 1)z = xy^2 - x^2y^2 + 2x^3y - 2x^3$.

The equation may be written as $[D_y^2 - D_y - \frac{1}{x}(\frac{1}{x} - 1)]z = xy^2 - x^2y^2 + 2x^3y - 2x^3$.

The complementary function is $z = e^{y/x}\phi_1(x) + e^{y-y/x}\phi_2(x)$.

For a particular integral we try $z = Ay^2 + By + C$, where A, B, C are functions of x or constants. Then $[D_y^2 - D_y - \frac{1}{x}(\frac{1}{x} - 1)]z = 2A - 2Ay - B - (\frac{1}{x^2} - \frac{1}{x})(Ay^2 + By + C) = xy^2 - x^2y^2 + 2x^3y - 2x^3$, identically. Equating coefficients of the several powers of y, we have

$$-(\frac{1}{x^2} - \frac{1}{x})A = x(1 - x), \quad -2A - (\frac{1}{x^2} - \frac{1}{x})B = 2x^3, \quad 2A - B - (\frac{1}{x^2} - \frac{1}{x})C = -2x^3.$$

Then $A = -x^3$, $B = C = 0$ and the required solution is $z = e^{y/x}\phi_1(x) + e^{y-y/x}\phi_2(x) - x^3y^2$.

8. Solve $ys + p - yq - z = (1 - x)(1 + \ln y)$.

This equation is solved readily by noting that it may be put in the form

$$\frac{\partial^2 z}{\partial x\, \partial y} + \frac{1}{y}\frac{\partial z}{\partial x} - \frac{\partial z}{\partial y} - \frac{z}{y} = \frac{\partial}{\partial x}(\frac{\partial z}{\partial y} + \frac{1}{y}z) - (\frac{\partial z}{\partial y} + \frac{1}{y}z) = \frac{1-x}{y}(1 + \ln y).$$

Setting $w = \frac{\partial z}{\partial y} + \frac{1}{y}z$, the equation becomes $\frac{\partial w}{\partial x} - w = \frac{1-x}{y}(1 + \ln y)$ for which e^{-x} is an integrating factor. Then

$$e^{-x}w = \frac{1 + \ln y}{y}\int^x (e^{-x} - xe^{-x})dx = \frac{1 + \ln y}{y}(xe^{-x}) + \psi(y) \quad \text{and} \quad w = x\frac{1 + \ln y}{y} + e^x\psi(y).$$

In turn, integrating $\frac{\partial z}{\partial y} + \frac{1}{y}z = x\frac{1 + \ln y}{y} + e^x\psi(y)$, using the integrating factor y, we find

$$yz = x\int^y (1 + \ln y)dy + e^x\int^y y\psi(y)\,dy = xy\ln y + e^x\phi_1(y) + \phi_2(x).$$

LAPLACE'S TRANSFORMATION.

9. Solve $t - s + p - q(1 + 1/x) + z/x = 0$.

Setting $(\xi_y)^2 - \xi_x\xi_y = 0$ and solving, we have $\xi = x$ and $\xi = x + y$.

For the choice $u = x$ and $v = x + y$, $p = z_u + z_v$, $q = z_v$, $s = z_{uv} + z_{vv}$, and $t = z_{vv}$. Substituting in the given equation, we have $z_{uv} - z_u + \frac{1}{x}(z_v - z) = \frac{\partial}{\partial u}(\frac{\partial z}{\partial v} - z) + \frac{1}{x}(\frac{\partial z}{\partial v} - z) = 0$.

Let $\frac{\partial z}{\partial v} - z = w$; then $\frac{\partial w}{\partial u} + \frac{w}{u} = 0$ and $uw = u(\frac{\partial z}{\partial v} - z) = \psi(v)$.

Integrating $\frac{\partial z}{\partial v} - z = \frac{1}{u}\psi(v)$, we have $e^{-v}z = \frac{1}{u}\phi_1(v) + \phi(u)$ or $z = \frac{e^v}{u}\phi_1(v) + e^v\phi(u)$.

In the original variables, $z = \dfrac{e^{x+y}}{x}\phi_1(x+y) + e^{x+y}\phi(x) = \dfrac{1}{x}f(x+y) + e^y g(x)$,

where $f(x+y) = e^{x+y}\phi_1(x+y)$ and $g(x) = e^x\phi(x)$.

10. Solve $xys - x^2r - px - qy + z = -2x^2y$.

From $xy\xi_x\xi_y - x^2(\xi_x)^2 = x\xi_x(y\xi_y - x\xi_x) = 0$, we obtain $\xi = y$ and $\xi = xy$.

Using $u = xy$, $v = y$, $p = yz_u$, $q = xz_u + z_v$, $r = y^2z_{uu}$, $s = z_u + xyz_{uu} + yz_{uv}$, the given differential equation becomes

$$z_{uv} - \frac{1}{v}z_u - \frac{1}{u}z_v + \frac{1}{uv}z = -\frac{2u}{v^2} \quad \text{or} \quad \frac{\partial}{\partial u}\left(\frac{\partial z}{\partial v} - \frac{1}{v}z\right) - \frac{1}{u}\left(\frac{\partial z}{\partial v} - \frac{1}{v}z\right) = -\frac{2u}{v^2}.$$

Let $\dfrac{\partial z}{\partial v} - \dfrac{1}{v}z = w$; then $\dfrac{\partial w}{\partial u} - \dfrac{w}{u} = -\dfrac{2u}{v^2}$, and $\dfrac{w}{u} = -\dfrac{2u}{v^2} + \psi(v)$ or $w = -\dfrac{2u^2}{v^2} + u\,\psi(v)$.

Integrating $w = \dfrac{\partial z}{\partial v} - \dfrac{1}{v}z = -\dfrac{2u^2}{v^2} + u\,\psi(v)$, we have $\dfrac{z}{v} = \dfrac{u^2}{v^2} + u\,\psi_1(v) + \phi_2(u)$ or

$$z = \frac{u^2}{v} + uv\,\psi_1(v) + v\,\phi_2(u) = \frac{u^2}{v} + u\,\lambda_1(v) + v\,\phi_2(u).$$

In the original variables, $z = xy\,\lambda_1(y) + y\,\phi_2(xy) + x^2y = x\phi_1(y) + y\phi_2(xy) + x^2y$.

11. Solve $x^2r - 2xys + y^2t - xp + 3yq = 8y/x$.

Here $x^2(\xi_x)^2 - 2xy\xi_x\xi_y + y^2(\xi_y)^2 = (x\xi_x - y\xi_y)^2 = 0$, and since the factors are not distinct we obtain only $\xi = xy$.

We set $u = xy$ and take $v = y$; then $p = yz_u$, $q = xz_u + z_v$, $r = y^2z_{uu}$, $s = z_u + xyz_{uu} + yz_{uv}$, $t = x^2z_{uu} + 2xz_{uv} + z_{vv}$ and the given differential equation becomes

$$y^2z_{vv} + 3yz_v = 8y/x \quad \text{or} \quad v^2z_{vv} + 3vz_v = 8v^2/u,$$

an equation of the Cauchy type. However, it is seen that v is an integrating factor; hence

$$v^3z_{vv} + 3v^2z_v = 8v^3/u \quad \text{and} \quad v^3z_v = 2v^4/u + \phi(u).$$

Then $z_v = \dfrac{2v}{u} + \dfrac{1}{v^3}\phi(u)$ and

$$z = \frac{v^2}{u} - \frac{1}{2v^2}\phi(u) + \phi_1(u)$$

$$= \frac{v^2}{u} + \frac{1}{v^2}\psi(u) + \phi_1(u)$$

$$= \phi_1(xy) + \frac{1}{y^2}\psi(xy) + \frac{y}{x}$$

or $z = \phi_1(xy) + x^2\,\phi_2(xy) + \dfrac{y}{x}$, where $\psi(xy) = x^2y^2\,\phi_2(xy)$.

MONGE'S METHOD.

12. Solve $qs - pt = q^3$.

The Monge equations are $q\,dx\,dy + p(dx)^2 = 0$ and $p\,dx\,dq + q^3\,dx\,dy = 0$.

From the first equation, $q\,dy + p\,dx = 0$; then $dz = p\,dx + q\,dy = 0$ and $z = a$.

Substituting $q\,dy = -p\,dx$ in the second equation yields $dq - q^2dx = 0$; thus $1/q + x = b$ and $1/q + x = f(z)$ or $[x - f(z)]q = -1$ is an intermediate integral.

The required solution is obtained by solving this first order equation; thus

$xz - \int f(z)dz = -y + \phi_2(x)$ or $y + xz = \phi_1(z) + \phi_2(x)$, where $\phi_1'(z) = f(z)$.

13. Solve $q^2r - 2pqs + p^2t = pq^2$.

The Monge equations are $(q\,dy + p\,dx)^2 = 0$ and $q^2\,dy\,dp + p^2\,dx\,dq - pq^2\,dx\,dy = 0$.

From the first equation, $q\,dy + p\,dx = 0$; then $dz = p\,dx + q\,dy = 0$ and $z = a$.

Substituting $q\,dy = -p\,dx$ in the second yields $-q\,dp + p\,dq + pq\,dx = 0$ or $-\dfrac{dp}{p} + \dfrac{dq}{q} + dx = 0$ and $e^xq/p = b$. Thus $e^xq - p\,f(z) = 0$ is an intermediate integral. The Lagrange system for this equation is $\dfrac{dx}{f(z)} = \dfrac{dy}{-e^x}$, $dz = 0$.

From the second equation, $z = c$. Then the first becomes $\dfrac{dx}{f(c)} = \dfrac{dy}{-e^x}$ with solution $e^x/f(c) + y = d$. As required solution, we find

$$y = -e^x/f(z) + \phi_2(z) = e^x\,\phi_1(z) + \phi_2(z), \quad \text{where } \phi_1(z) = -1/f(z).$$

14. Solve $x(r + 2xs + x^2t) = p + 2x^3$.

The Monge equations are $(dy)^2 - 2x\,dx\,dy + x^2(dx)^2 = (dy - x\,dx)^2 = 0$

and $x\,dy\,dp + x^3\,dx\,dq - (p + 2x^3)dx\,dy = 0$.

We seek a solution of the system $dy - x\,dx = 0$, $x\,dy\,dp + x^3\,dx\,dq - (p + 2x^3)dx\,dy = 0$.

From the first equation, $x^2 - 2y = a$. Substituting $dy = x\,dx$ in the second, we get

$$x\,dp + x^2dq - (p + 2x^3)dx = 0.$$

Using the integrating factor $1/x^2$, we obtain the intermediate integral $p + xq = x^3 + x\,f(x^2 - 2y)$.

The Lagrange system is $\dfrac{dx}{1} = \dfrac{dy}{x} = \dfrac{dz}{x^3 + x\,f(x^2 - 2y)}$. The first two members yield $x^2 - 2y = c$

and then the first and third become $\dfrac{dx}{1} = \dfrac{dz}{x^3 + x\,f(c)}$. Solving,

$$z = \tfrac{1}{4}x^4 + \tfrac{1}{2}x^2f(c) + \phi(c) \quad \text{or} \quad z = \tfrac{1}{4}x^4 + \tfrac{1}{2}x^2f(x^2 - 2y) + \phi(x^2 - 2y).$$

15. Solve $q(1+q)r - (1+2q)(1+p)s + (1+p)^2t = 0$.

The Monge equations are

$q(1+q)(dy)^2 + (1+2q)(1+p)dx\,dy + (1+p)^2(dx)^2 = [q\,dy + (1+p)dx][(1+q)dy + (1+p)dx] = 0$

and $q(1+q)dy\,dp + (1+p)^2\,dx\,dq = 0$.

Consider first the system

$$q\,dy + (1+p)dx = 0$$
$$q(1+q)dy\,dp + (1+p)^2\,dx\,dq = 0.$$

From the first equation, $p\,dx + q\,dy = -dx$; then $dz = -dx$ and $x+z=a$. The substitution of $q\,dy = -(1+p)dx$ in the second yields

$$-(1+q)dp + (1+p)dq = 0$$

from which we obtain $\dfrac{1+p}{1+q} = b$. Thus, $\dfrac{1+p}{1+q} = f(x+z)$ is an intermediate integral.

Consider next the system

$$(1+q)dy + (1+p)dx = 0$$
$$q(1+q)dy\,dp + (1+p)^2\,dx\,dq = 0.$$

From the first, $p\,dx + q\,dy = -(dx+dy)$ so that $dz = -(dx+dy)$ and $x+y+z=a$. The substitution of $(1+q)dy = -(1+p)dx$ in the second gives $-q\,dp + (1+p)dq = 0$ which is satisfied by $\dfrac{1+p}{q} = b$. Thus, $\dfrac{1+p}{q} = g(x+y+z)$ is an intermediate integral.

Solving the two intermediate integrals for $p = \dfrac{fg+f-g}{g-f}$, $q = \dfrac{f}{g-f}$ and substituting in the relation $p\,dx + q\,dy = dz$, we have

$$(fg+f-g)dx + f\,dy = (g-f)dz, \qquad fg\,dx = -f(dx+dy+dz) + g(dx+dz),$$

$$dx = -\frac{dx+dy+dz}{g(x+y+z)} + \frac{dx+dz}{f(x+z)}, \qquad \text{and} \qquad x = \phi_1(x+y+z) + \phi_2(x+z).$$

16. Solve $(x-z)[xq^2r - q(x+z+2px)s + (z+px+pz+p^2x)t] = (1+p)q^2(x+z)$.

Monge's equations are

$$xq^2(dy)^2 + q(x+z+2px)dx\,dy + (1+p)(z+px)(dx)^2 = [q\,dy + (1+p)dx][xq\,dy + (z+px)dx] = 0$$

and $$(x-z)[xq^2dy\,dp + (1+p)(z+px)dx\,dq] - (1+p)q^2(x+z)dx\,dy = 0.$$

Consider first the system

$$q\,dy + (1+p)dx = 0$$
$$(x-z)xq^2dy\,dp + (1+p)(z+px)(x-z)dx\,dq - (1+p)q^2(x+z)dx\,dy = 0.$$

From the first equation, $p\,dx + q\,dy = -dx$; then $dz = -dx$ and $x+z=a$. Substituting $q\,dy = -(1+p)dx$, $z = a-x$ in the second, we have

i) $$-(2x-a)xq\,dp + (2x-a)(a-x+px)dq + (1+p)qa\,dx = 0.$$

To solve this equation, consider x as a constant so that $dx = 0$. Then i) becomes

$$-(2x-a)xq\,dp + (2x-a)(a-x+px)dq = 0 \qquad \text{or} \qquad x(q\,dp - p\,dq) - (a-x)dq = 0$$

and $\dfrac{xp+a-x}{q} = \psi(x)$. To determine $\psi(x)$, we take the differential of this relation,

$$q(x\,dp + p\,dx - dx) - (xp+a-x)dq = q^2d\psi$$

and obtain $$xq\,dp - xp\,dq = q^2d\psi - pq\,dx + q\,dx + a\,dq - x\,dq.$$

From i), $$xq\,dp - xp\,dq = \frac{(2x-a)(a-x)dq + (1+p)qa\,dx}{2x-a} = (a-x)dq + \frac{(1+p)qa\,dx}{2x-a};$$

then $$q^2 d\psi - pq\,dx + q\,dx + a\,dq - x\,dq = (a-x)dq + \frac{(1+p)qa\,dx}{2x-a},$$

$$d\psi = \frac{2(px+a-x)}{q(2x-a)}dx = \frac{2\psi}{2x-a}dx \quad \text{and} \quad \frac{\psi}{2x-a} = b = f(x+z).$$

Thus, $\dfrac{xp+a-x}{q(2x-a)} = \dfrac{xp+z}{q(x-z)} = f(x+z)$ is an intermediate integral.

Consider next the system

$$xq\,dy + (z+px)dx = 0$$

$$(x-z)xq^2 dy\,dp + (1+p)(z+px)(x-z)dx\,dq - (1+p)q^2(x+z)dx\,dy = 0.$$

From the first equation, $p\,dx + q\,dy = -z\,dx/x$; then. $dz = -z\,dx/x$ and $xz = a$. Substituting $xq\,dy = -(z+px)dx$, $z = a/x$ in the second, we have

ii) $$-xq(x^2-a)dp + x(1+p)(x^2-a)dq + (1+p)q(x^2+a)dx = 0.$$

Considering x as a constant, this becomes $q\,dp - (1+p)dq = 0$ and we have $\dfrac{1+p}{q} = \psi(x)$. From this relation we find $q\,dp - (1+p)dq = q^2 d\psi$, while from ii) $q\,dp-(1+p)dq = \dfrac{(1+p)q(x^2+a)}{x(x^2-a)}dx$.

Then $d\psi = \dfrac{(1+p)q(x^2+a)}{q^2x(x^2-a)}dx = \dfrac{\psi(x^2+a)}{x(x^2-a)}dx = (-\dfrac{dx}{x} + \dfrac{2x\,dx}{x^2-a})\psi$, $\ln\psi = -\ln x + \ln(x^2-a) + \ln b$,

and $\psi = \dfrac{b(x^2-a)}{x} = \dfrac{1+p}{q}$. Thus, $\dfrac{1+p}{q(x-z)} = g(xz)$ is an intermediate integral.

Solving the two intermediate integrals, we find $p = \dfrac{f-zg}{xg-f}$ and $q = \dfrac{1}{xg-f}$; then

$$dz = p\,dx + q\,dy = \frac{f-zg}{xg-f}dx + \frac{1}{xg-f}dy \quad \text{or} \quad f(x+z)(dx+dz) + dy = zg(xz)dx + xg(xz)dz.$$

Thus, $y + \phi_1(x+z) = \phi_2(xz)$ is the required solution.

17. Solve $3r + s + t + (rt - s^2) = -9$.

Here, $R = 3$, $S = T = U = 1$, $V = -9$; then

$$U^2\lambda^2 + SU\lambda + TR + UV = \lambda^2 + \lambda - 6 = 0 \quad \text{and} \quad \lambda_1 = 2,\ \lambda_2 = -3.$$

We seek solutions of the systems (see equations 19))

$$\lambda_1 U\,dy + T\,dx + U\,dp = 2\,dy + dx + dp = 0, \quad R\,dy + \lambda_2 U\,dx + U\,dq = 3\,dy - 3\,dx + dq = 0$$

and $$\lambda_2 U\,dy + T\,dx + U\,dp = -3\,dy + dx + dp = 0, \quad R\,dy + \lambda_1 U\,dx + U\,dq = 3\,dy + 2\,dx + dq = 0.$$

From the first system, we have $2y + x + p = a$, $3y - 3x + q = b$; thus, $p + 2y + x = f(q + 3y - 3x)$ is an intermediate integral. From the second system, we have $-3y + x + p = c$, $3y + 2x + q = d$; thus, $p - 3y + x = g(q + 3y + 2x)$ is an intermediate integral. Since q appears in the argument of both f and g, it will not be possible to solve for p and q as before, and it will not be possible to find a solution involving two arbitrary functions. We give two solutions involving arbitrary constants.

Replacing the arbitrary function f of the first intermediate integral by $\alpha(q + 3y - 3x) + \beta$, we obtain

$$p + 2y + x = \alpha(q + 3y - 3x) + \beta \quad \text{or} \quad p - \alpha q = (3\alpha - 2)y - (3\alpha + 1)x + \beta$$

for which the Lagrange system is $\dfrac{dx}{1} = \dfrac{dy}{-\alpha} = \dfrac{dz}{(3\alpha-2)y-(3\alpha+1)x+\beta}$. From $\dfrac{dx}{1} = \dfrac{dy}{-\alpha}$, we find

$y+\alpha x = \xi$; then $\dfrac{dx}{1} = \dfrac{dz}{(3\alpha-2)y-(3\alpha+1)x+\beta} = \dfrac{dz}{-(3\alpha^2+\alpha+1)x+3\alpha\xi-2\xi+\beta}$ and

$$z = -\tfrac{1}{2}(3\alpha^2+\alpha+1)x^2 + (3\alpha\xi-2\xi+\beta)x + \eta = -\tfrac{1}{2}(3\alpha^2+\alpha+1)x^2 + (3\alpha y+3\alpha^2x-2y-2\alpha x+\beta)x+\eta.$$

Thus, $z = \tfrac{1}{2}(3\alpha^2-5\alpha-1)x^2 + (3\alpha-2)xy + \beta x + \phi_1(y+\alpha x)$ is a solution involving one arbitrary function and two arbitrary constants.

Replacing the arbitrary function $g(q+3y+2x)$ of the second intermediate integral by the linear function $\gamma(q+3y+2x)+\delta$, we obtain

$$p-3y+x = \gamma(q+3y+2x)+\delta \qquad \text{or} \qquad p-\gamma q = 3(\gamma+1)y+(2\gamma-1)x+\delta$$

for which the Lagrange system is $\dfrac{dx}{1} = \dfrac{dy}{-\gamma} = \dfrac{dz}{3(\gamma+1)y+(2\gamma-1)x+\delta}$. From $\dfrac{dx}{1} = \dfrac{dy}{-\gamma}$, we get

$y+\gamma x = \xi$; then $\dfrac{dx}{1} = \dfrac{dz}{3(\gamma+1)y+(2\gamma-1)x+\delta} = \dfrac{dz}{-(3\gamma^2+\gamma+1)x+3\gamma\xi+3\xi+\delta}$ and

$$z = -\tfrac{1}{2}(3\gamma^2+\gamma+1)x^2 + (3\gamma\xi+3\xi+\delta)x+\eta.$$

Thus, $z = \tfrac{1}{2}(3\gamma^2+5\gamma-1)x^2 + 3(\gamma+1)xy + \delta x + \phi_2(y+\gamma x)$ is also a solution.

18. Solve $xqr + (p+q)s + ypt + (xy-1)(rt-s^2) + pq = 0$.

Here, $R = xq$, $S = p+q$, $T = yp$, $U = xy-1$, $V = -pq$; then

$U^2\lambda^2 + SU\lambda + TR + UV = (xy-1)^2\lambda^2 + (p+q)(xy-1)\lambda + pq = 0$ and $\lambda_1 = \dfrac{-p}{xy-1}$, $\lambda_2 = \dfrac{-q}{xy-1}$.

Consider first the system $\left[\begin{array}{l} -p\,dy + yp\,dx + (xy-1)dp = 0 \\ xq\,dy - q\,dx + (xy-1)dq = 0 \end{array}\right.$. The system is not integrable since neither equation is integrable.

Consider next the system $-q\,dy + yp\,dx + (xy-1)dp = 0$, $xq\,dy - p\,dx + (xy-1)dq = 0$. We multiply the second equation by y, add the first, and divide by $xy-1$ to obtain $q\,dy + dp + y\,dq = 0$ and thus $p+yq = a$. Again, we multiply the first equation by x, add the second, and divide by $xy-1$ to obtain $p\,dx + x\,dp + dq = 0$ and thus $xp+q = b$. However, the form of the resulting intermediate integral $xp+q = f(yq+p)$ or $yq+p = g(xp+q)$ does not permit a solution involving two arbitrary functions.

To obtain a solution, involving one arbitrary function and two arbitrary constants, we replace $f(yq+p)$ by the linear function $\alpha(yq+p)+\beta$ in the first form of the intermediate integral above and have

$$(x-\alpha)p + (1-\alpha y)q = \beta.$$

The corresponding Lagrange system is $\dfrac{dx}{x-\alpha} = \dfrac{dy}{1-\alpha y} = \dfrac{dz}{\beta}$. From the first two members we obtain $\alpha\ln(x-\alpha) + \ln(1-\alpha y) = \ln\xi$ or $(x-\alpha)^{\alpha}(1-\alpha y) = \xi$, and from the first and third members we get $z = \beta\ln(x-\alpha)+\eta$. Thus, the solution is

$$z = \beta\ln(x-\alpha) + \phi[(x-\alpha)^{\alpha}(1-\alpha y)].$$

SUPPLEMENTARY PROBLEMS

Solve.

19. $r = xy$ Ans. $z = x\,\phi_1(y) + \phi_2(y) + \frac{1}{6}x^3y$

20. $s = x^2 + y^2$ $z = \phi_1(x) + \phi_2(y) + \frac{1}{3}(x^3y + xy^3)$

21. $t = -x^2\sin(xy)$ $z = y\,\phi_1(x) + \phi_2(x) + \sin(xy)$

22. $xr - p = 0$ $z = x^2\,\phi_1(y) + \phi_2(y)$

23. $xr + p = 1/x^2$ $z = \phi_1(y)\ln x + \phi_2(y) + 1/x$

24. $yt - q = 2x^2y$ $z = y^2\,\phi_1(x) + \phi_2(x) + x^2y^2\ln y$

25. $ys - p = xy^2\sin(xy)$ $z = y\,\phi_1(x) + \phi_2(y) - \sin(xy)$

26. $t + q = xe^{-y}$ $z = e^{-y}\phi_1(x) + \phi_2(x) - xye^{-y}$

27. $r + s = 3y^2$ $z = \phi_1(x-y) + \phi_2(y) + xy^3$

28. $xyr + x^2s - yp = x^3e^y$ $z = \phi_1(x^2-y^2) + \phi_2(y) + \frac{1}{2}x^2e^y$

29. $2yt - xs + 2q = x^2y$ $z = \phi_1(x^2y) + \phi_2(x) + \frac{1}{4}x^2y^2$

30. $xr + ys + p = 8xy^2 + 9x^2$ $z = \phi_1(x/y) + \phi_2(y) + x^2y^2 + x^3$

LAPLACE'S TRANSFORMATION.

31. $6r - s - t = 18y - 4x$ Ans. $z = \phi_1(x-3y) + \phi_2(x+2y) + y(2x^2+y^2)$

32. $x(xy-1)r - (x^2y^2-1)s + y(xy-1)t + (x-1)p + (y-1)q = 0$ Ans. $z = \phi_1(xe^y) + \phi_2(ye^x)$

33. $x(y-x)r - (y^2-x^2)s + y(y-x)t + (y+x)(p-q) = 2(x+y+1)$
Hint: Let $x+y=u$, $xy=v$. Ans. $z = \phi_1(x+y) + \phi_2(xy) + x - y + \ln x$

34. $(y-1)r - (y^2-1)s + y(y-1)t + p - q = 2ye^{2x}(1-y)^3$
Ans. $z = \phi_1(x+y) + \phi_2(ye^x) + (x+y)y^2e^{2x}$

35. $xyr - (x^2-y^2)s - xyt + py - qx = 2(x^2-y^2)$ Ans. $z = \phi_1(x^2+y^2) + \phi_2(y/x) - xy$

36. $r - 2s + t + p - q = e^x(2y-3) - e^y$ Ans. $z = \phi_1(x+y) + e^y\,\phi_2(x+y) + xe^y + ye^x$
Hint: Let $x+y=u$, $y=v$.

37. $y^2(r-2s+t) - y(p-q) - z = y^2$ Ans. $z = y\,\phi_1(x+y) + \frac{1}{y}\,\phi_2(x+y) + \frac{1}{3}y^2$

MONGE'S METHOD.

38. $(e^x-1)(qr-ps) = pqe^x$ I.I.: $p = \psi(z)$. G.S.: $x = \phi_1(z) + \phi_2(y) + e^x$

39. $r - 3s - 10t = -3$ I.I.: $p + 2q = \psi_1(y + 5x), \quad p - 5q = \psi_2(y - 2x)$
G.S.: $z = \phi_1(y + 5x) + \phi_2(y - 2x) + xy$

40. $q^2 r - 2pqs + p^2 t = 0$ I.I.: $p = q\,\psi(z)$. G.S.: $x\phi_1(z) + y = \phi_2(z)$.

41. $qr - (1 + p + q)s + (1 + p)t = 0$ I.I.: $p - q = \psi_1(x + z), \quad p + 1 = q\,\psi_2(x + y)$
G.S.: $z = f(x + z) + g(x + y)$

42. $(1 - q)^2 r - 2(2 - p - 2q + pq)s + (2 - p)^2 t = 0$ I.I.: $\dfrac{1 - q}{2 - p} = \psi(y + 2x - z)$
G.S.: $x + y\,\phi_1(y + 2x - z) = \phi_2(y + 2x - z)$

43. $5r - 10s + 4t - (rt - s^2) = -1$
I.I.: $3y + 4x - p = f(5y + 7x - q), \quad 7y + 4x - p = g(5y + 3x - q)$
Sol.: $z = 2x^2 + 3xy + \frac{5}{2}y^2 - 2\alpha x^2 - \beta x + \phi_1(y + \alpha x)$ or $z = 2x^2 + 7xy + \frac{5}{2}y^2 + 2\gamma x^2 - \delta x + \phi_2(y + \gamma x)$

44. $2r - 6s + 2t + (rt - s^2) = 4$
I.I.: $2y + 2x + p = f(2y + 4x + q), \quad 4y + 2x + p = g(2y + 2x + q)$
Sol.: $z = \alpha x^2 + \beta x - (x + y)^2 + \phi_1(y + \alpha x)$ or $z = -\gamma x^2 + \delta x - x^2 - 4xy - y^2 + \phi_2(y + \gamma x)$

45. $3r - 6s + 4t - (rt - s^2) = 3$
I.I.: $3y + 4x - p = f(3y + 3x - q)$. Sol.: $z = 2x^2 + 3xy + \frac{3}{2}y^2 + \beta x + \phi(y + \alpha x)$.

46. $yr - ps + t + y(rt - s^2) = -1$
I.I.: $yp + x = f(q + y)$. Sol.: $6\alpha^2 z = 2y^3 - 3\alpha^2 y^2 + 6\alpha xy + 6\beta y + \phi(\alpha x + \frac{1}{2}y^2)$.

47. $xqr - (x + y)s + ypt + xy(rt - s^2) = 1 - pq$
I.I.: $xp + y = f(yq + x)$. Sol.: $z = \alpha x + y/\alpha + \beta \ln x + \phi(x^\alpha y)$.

Index

SCHAUM'S OUTLINE SERIES

COLLEGE PHYSICS
including 625 SOLVED PROBLEMS
Edited by CAREL W. van der MERWE, Ph.D.,
Professor of Physics, New York University

COLLEGE CHEMISTRY
including 385 SOLVED PROBLEMS
Edited by JEROME L. ROSENBERG, Ph.D.,
Professor of Chemistry, University of Pittsburgh

GENETICS
including 500 SOLVED PROBLEMS
By WILLIAM D. STANSFIELD, Ph.D.,
Dept. of Biological Sciences, Calif. State Polytech.

MATHEMATICAL HANDBOOK
including 2400 FORMULAS and 60 TABLES
By MURRAY R. SPIEGEL, Ph.D.,
Professor of Math., Rensselaer Polytech. Inst.

First Yr. COLLEGE MATHEMATICS
including 1850 SOLVED PROBLEMS
By FRANK AYRES, Jr., Ph.D.,
Professor of Mathematics, Dickinson College

COLLEGE ALGEBRA
including 1940 SOLVED PROBLEMS
By MURRAY R. SPIEGEL, Ph.D.,
Professor of Math., Rensselaer Polytech. Inst.

TRIGONOMETRY
including 680 SOLVED PROBLEMS
By FRANK AYRES, Jr., Ph.D.,
Professor of Mathematics, Dickinson College

MATHEMATICS OF FINANCE
including 500 SOLVED PROBLEMS
By FRANK AYRES, Jr., Ph.D.,
Professor of Mathematics, Dickinson College

PROBABILITY
including 500 SOLVED PROBLEMS
By SEYMOUR LIPSCHUTZ, Ph.D.,
Assoc. Prof. of Math., Temple University

STATISTICS
including 875 SOLVED PROBLEMS
By MURRAY R. SPIEGEL, Ph.D.,
Professor of Math., Rensselaer Polytech. Inst.

ANALYTIC GEOMETRY
including 345 SOLVED PROBLEMS
By JOSEPH H. KINDLE, Ph.D.,
Professor of Mathematics, University of Cincinnati

DIFFERENTIAL GEOMETRY
including 500 SOLVED PROBLEMS
By MARTIN LIPSCHUTZ, Ph.D.,
Professor of Mathematics, University of Bridgeport

CALCULUS
including 1175 SOLVED PROBLEMS
By FRANK AYRES, Jr., Ph.D.,
Professor of Mathematics, Dickinson College

DIFFERENTIAL EQUATIONS
including 560 SOLVED PROBLEMS
By FRANK AYRES, Jr., Ph.D.,
Professor of Mathematics, Dickinson College

SET THEORY and Related Topics
including 530 SOLVED PROBLEMS
By SEYMOUR LIPSCHUTZ, Ph.D.,
Assoc. Prof. of Math., Temple University

FINITE MATHEMATICS
including 750 SOLVED PROBLEMS
By SEYMOUR LIPSCHUTZ, Ph.D.,
Assoc. Prof. of Math., Temple University

MODERN ALGEBRA
including 425 SOLVED PROBLEMS
By FRANK AYRES, Jr., Ph.D.,
Professor of Mathematics, Dickinson College

LINEAR ALGEBRA
including 600 SOLVED PROBLEMS
By SEYMOUR LIPSCHUTZ, Ph.D.,
Assoc. Prof. of Math., Temple University

MATRICES
including 340 SOLVED PROBLEMS
By FRANK AYRES, Jr., Ph.D.,
Professor of Mathematics, Dickinson College

PROJECTIVE GEOMETRY
including 200 SOLVED PROBLEMS
By FRANK AYRES, Jr., Ph.D.,
Professor of Mathematics, Dickinson College

GENERAL TOPOLOGY
including 650 SOLVED PROBLEMS
By SEYMOUR LIPSCHUTZ, Ph.D.,
Assoc. Prof. of Math., Temple University

GROUP THEORY
including 600 SOLVED PROBLEMS
By B. BAUMSLAG, B. CHANDLER, Ph.D.,
Mathematics Dept., New York University

VECTOR ANALYSIS
including 480 SOLVED PROBLEMS
By MURRAY R. SPIEGEL, Ph.D.,
Professor of Math., Rensselaer Polytech. Inst.

ADVANCED CALCULUS
including 925 SOLVED PROBLEMS
By MURRAY R. SPIEGEL, Ph.D.,
Professor of Math., Rensselaer Polytech. Inst.

COMPLEX VARIABLES
including 640 SOLVED PROBLEMS
By MURRAY R. SPIEGEL, Ph.D.,
Professor of Math., Rensselaer Polytech. Inst.

LAPLACE TRANSFORMS
including 450 SOLVED PROBLEMS
By MURRAY R. SPIEGEL, Ph.D.,
Professor of Math., Rensselaer Polytech. Inst.

NUMERICAL ANALYSIS
including 775 SOLVED PROBLEMS
By FRANCIS SCHEID, Ph.D.,
Professor of Mathematics, Boston University

DESCRIPTIVE GEOMETRY
including 175 SOLVED PROBLEMS
By MINOR C. HAWK, *Head of Engineering Graphics Dept., Carnegie Inst. of Tech.*

ENGINEERING MECHANICS
including 460 SOLVED PROBLEMS
By W. G. McLEAN, B.S. in E.E., M.S.,
Professor of Mechanics, Lafayette College
and E. W. NELSON, B.S. in M.E., M. Adm. E.,
Engineering Supervisor, Western Electric Co.

THEORETICAL MECHANICS
including 720 SOLVED PROBLEMS
By MURRAY R. SPIEGEL, Ph.D.,
Professor of Math., Rensselaer Polytech. Inst.

LAGRANGIAN DYNAMICS
including 275 SOLVED PROBLEMS
By D. A. WELLS, Ph.D.,
Professor of Physics, University of Cincinnati

STRENGTH OF MATERIALS
including 430 SOLVED PROBLEMS
By WILLIAM A. NASH, Ph.D.,
Professor of Eng. Mechanics, University of Florida

FLUID MECHANICS and HYDRAULICS
including 475 SOLVED PROBLEMS
By RANALD V. GILES, B.S., M.S. in C.E.,
Prof. of Civil Engineering, Drexel Inst. of Tech.

FLUID DYNAMICS
including 100 SOLVED PROBLEMS
By WILLIAM F. HUGHES, Ph.D.,
Professor of Mech. Eng., Carnegie Inst. of Tech.
and JOHN A. BRIGHTON, Ph.D.,
Asst. Prof. of Mech. Eng., Pennsylvania State U.

ELECTRIC CIRCUITS
including 350 SOLVED PROBLEMS
By JOSEPH A. EDMINISTER, M.S.E.E.,
Assoc. Prof. of Elec. Eng., University of Akron

ELECTRONIC CIRCUITS
including 160 SOLVED PROBLEMS
By EDWIN C. LOWENBERG, Ph.D.,
Professor of Elec. Eng., University of Nebraska

FEEDBACK & CONTROL SYSTEMS
including 680 SOLVED PROBLEMS
By J. J. DiSTEFANO III, A. R. STUBBERUD,
and I. J. WILLIAMS, Ph.D.,
Engineering Dept., University of Calif., at L.A.

TRANSMISSION LINES
including 165 SOLVED PROBLEMS
By R. A. CHIPMAN, Ph.D.,
Professor of Electrical Eng., University of Toledo

REINFORCED CONCRETE DESIGN
including 200 SOLVED PROBLEMS
By N. J. EVERARD, MSCE, Ph.D.,
Prof. of Eng. Mech. & Struc., Arlington State Col.
and J. L. TANNER III, MSCE,
Technical Consultant, Texas Industries Inc.

MECHANICAL VIBRATIONS
including 225 SOLVED PROBLEMS
By WILLIAM W. SETO, B.S. in M.E., M.S.,
Assoc. Prof. of Mech. Eng., San Jose State College

MACHINE DESIGN
including 320 SOLVED PROBLEMS
By HALL, HOLOWENKO, LAUGHLIN
Professors of Mechanical Eng., Purdue University

BASIC ENGINEERING EQUATIONS
including 1400 BASIC EQUATIONS
By W. F. HUGHES, E. W. GAYLORD, Ph.D.,
Professors of Mech. Eng., Carnegie Inst. of Tech.

ELEMENTARY ALGEBRA
including 2700 SOLVED PROBLEMS
By BARNETT RICH, Ph.D.,
Head of Math. Dept., Brooklyn Tech. H.S.

PLANE GEOMETRY
including 850 SOLVED PROBLEMS
By BARNETT RICH, Ph.D.,
Head of Math. Dept., Brooklyn Tech. H.S.

TEST ITEMS IN EDUCATION
including 3100 TEST ITEMS
By G. J. MOULY, Ph.D., L. E. WALTON, Ph.D.,
Professors of Education, University of Miami